Climate and Weather in the Tropics

Climate and Weather in the Tropics

HERBERT RIEHL

1979

ACADEMIC PRESS
London   New York   San Francisco
A Subsidiary of Harcourt Brace Jovanovich, Publishers

ACADEMIC PRESS INC. (LONDON) LTD.
24/28 Oval Road,
London NW1

United States Edition published by
ACADEMIC PRESS INC.
111 Fifth Avenue
New York, New York 10003

Copyright © 1979 by
ACADEMIC PRESS INC. (LONDON) LTD.

All rights Reserved
No part of this book may be reproduced in any form by photostat, microfilm, or any other means, without written permission from the publishers

Library of Congress Catalog Card Number 78-73890
Riehl, Herbert
   Climate and weather in the tropics.
   1. Tropics—Climate
   I. Title
   551.6'9'13   QC993.5   78-73890
   ISBN 0-12-588180-0

Text set in 11/12pt. Compugraphic English, and printed in Great Britain by
The Lavenham Press Limited, Lavenham, Suffolk.

Preface

The years since 1950 have seen the propulsion of tropical countries around the globe into positions of prime importance in relation to world politics, economics and the future development of countries outside the tropics. Air communication and space observations have closed vast distances on earth, so that an increasingly coherent understanding of the world's climate and weather has been achieved, and it has become more and more clear that tropical atmospheric regimes, with their internal convolutions, affect not only areas within the tropical boundaries but eventually the entire earth. Our expanded understanding of the tropics along with their enhanced political importance, then, are the chief reasons for the appearance of this volume. Furthermore, my earlier book "Tropical Meteorology" (1954) needed to be updated and actually replaced.

The atmospheric scientist is now asked to stop the encroachment of deserts, create more water in arid areas, move hurricanes away from populated shores and, in general, make life easier for everyone. Progress within the quarter century since my previous publication means that some of these demands can now be achieved. This volume draws together new and old information from many sources to present a coherent picture of our current comprehension and to suggest, overtly as well as by implication, some of the needed directions for further advance, so that the science of weather can achieve the exalted goals set for it. It will be of particular use to atmospheric science students, provided that they are already familiar with the basics of the subject, including elementary theory, and it will also serve as a handbook for professionals in atmospheric science and in related fields such as physics and engineering. This by no means excludes those working in the social sciences who may also find it necessary to improve and update their knowledge of the tropics.

The information explosion in this, as in other fields, has created an urgency for a comprehensive new volume on the subject. Even the usually quiescent tropics have proved to be not at all sleepy in this area of research during the last quarter century. Advances have been most evident in techniques of weather analysis, now largely objective, and in the analysis of convective processes. Emphasis in the book is on the presentation of facts and their interpretation; a solid body of factual information underlies all twelve chapters. However, controversial interpretation and uncertainties of observations are by no means glossed over. Certain topics are caught in a state of rapid transition, such as weather modification through cloud seeding. Subjects such as these should be pursued through texts and articles dedicated to the problem alone; as nothing would be gained for tomorrow by attempting to present the whole of today's highly conflicting state of speculation within the confines of a single book. On other topics such as the high atmosphere and air-sea interaction, speciality textbooks exist to which the reader is referred.

New technology—satellite, radar, computer—has greatly furthered the advance of atmospheric science. I have been fortunate in obtaining the collaboration of Professor Ferdinand Baer who contributed Chapter 11, a thorough discourse on numerical weather prediction. His specific subject is tropical hurricanes, but the methods he has outlined are relevant to all attempts to apply advanced numerical methods to the problems of the tropics and their relation to other latitude belts. Satellite and radar information has been woven into the text in many places; this greatly enhances the pictorial presentation.

A decision had to be made regarding the inclusion of references literature. Although the number of references does not fall far short of one thousand, it was impossible to include a fully comprehensive literature survey without altering the main purpose of the book. The reader will find that most references are to widely disseminated journals which are accessible in most parts of the world. Much useful material is contained in monographs and articles which even I have had difficulty in locating and, while a few such references were unavoidable—largely for historical reasons—I am aware that they are practically useless to the general student. Regional and local investigations likewise have been mostly omitted except in Chapter 6 which is dedicated to the local scene. An apology is due to all friends and colleagues whose works were not included, some, no doubt, inadvertently.

Professor Hermann Flohn, Bonn, was especially active in encouraging the writing of this volume and its publication through Academic Press, London. Permission by several publishing houses to reproduce copyrighted illustrations is acknowledged, as is the contribution of numerous illustrations by friends and professional acquaintances. Editorial and manuscript assistance was generously provided by J. M. Van Valin.

30 March, 1979   HERBERT RIEHL
Professor (emer.)
Department of Atmospheric Science,
Colorado State University,
Fort Collins, Colorado 80303

Contents

Preface v

Symbols x

Note on Experiments xii

1. Wind Systems of the Tropics   1
    The low-level general circulation. An overview . . . 2
    Seasonal variations of the surface circulation . . . 8
    World distribution of surface winds and pressures . . 11
    Surface parameters following the equatorial trough . . 16
    Mean winds of the troposphere 21
    Stratospheric circulation 29
    References 36

2. Radiation, Temperature and Humidity   39
    Radiation 40
    Surface climate 48
    Upper air climate 60
    References 78
    Appendix 80

3. Precipitation and Evaporation   81
    Annual rainfall in the tropics 82
    Annual precipitation and evaporation 83
    Seasonal march of rainfall 89
    References 120

4. Vertical Energy Transfer   123
    The subcloud layer 125
    The cloud layer 144
    Turbulent exchange in the cloud layer 169
    References 198

5. The Trade Wind Inversion   202
    Vertical structure 205
    Variations of trade wind inversion structure . . . 211

Dissolution of the trade wind inversion 218
The northeast trade of the Pacific Ocean 225
Trade wind energy budgets 234
References 248

6. Diurnal and Local Controls 250
Ground-air interaction through heating 251
Diurnal wind variation 263
Diurnal variation of rainfall 271
The control of climate by orography 276
Possible increases of precipitation through reservoir construction 280
Air pollution 284
Fog 285
Semi-diurnal pressure variation at the surface . . . 286
References 287

7. Weather Observations and Analyses 289
Observation 289
Methods of analysis 303
References 313

8. Synoptic Scale Weather Systems 315
Introduction 315
Scales of turbulence 316
Uncoupling in the troposphere 317
Influence of radiation 319
The tropical rainstorm as a synoptic system 321
Synoptic envelopes 340
References 389

9. Tropical Cyclones—Structure and Mechanics 394
Summary of life cycle 396
Surface structure 398
Upper air structure 418
Budgets of momentum and energy 440
References 452
Appendix: discussion of local heat source 455

10. Tropical Cyclones—Formation and Movement 459
Some observations during hurricane formation . . . 462
Variability of hurricane intensity 463
Formation of hurricanes 464
Late hurricane stages 483
Motion of hurricanes 486
References 493

CONTENTS

11. Numerical Hurricane Prediction by Ferdinand Baer — 497
    Introduction — 497
    Dynamical method — 500
    Internal stress — 505
    Heating — 508
    Boundary conditions — 511
    Numerics — 514
    Initial conditions — 520
    Steady state models — 521
    Statistical method — 532
    Results and potential — 534
    References — 542

12. The General Circulation — 545
    The classical model — 546
    Critique of the classical model: radiation — 547
    Critique of the classical model: motions — 549
    Numerical modelling — 556
    General circulation budgets — 557
    Variations of the general circulation — 583
    References — 596

Appendix — 601
    Text books, monographs and articles containing extensive data sources or comprehensive literature surveys.

Index — 603

Symbols

All mathematical symbols are explained the first time they appear. Frequently used symbols are listed here, rarely used ones are defined throughout the text.

A	Area
a	Radius of earth
C	Unit in Celsius temperature scale
C	1) Propagation speed (or vector) or
	2) Circulation about closed curve
C_d	Drag coefficient
c_p	Specific heat at constant pressure
D	Thickness between isobaric surfaces
E	Evaporation
e	Vapour pressure
F	1) Latent heat of fusion
	2) Frictional force
F_l	Flux of latent heat
F_s	Flux of sensible heat
F_z	Vertical flux
f	Coriolis parameter
g	Gram
g	Acceleration of gravity
h_p	Height of constant pressure surface
h_s	Sensible heat content per unit mass
J	Joule, kJ kilojoule
K	Unit in absolute temperature scale
K	1) Kinetic energy ($V^2/2$)
	2) Exchange coefficient
K_h	Eddy transfer coefficient for heat
K_m	Eddy transfer coefficient for momentum
K_v	Eddy transfer coefficient for water vapour
L	1) Latent heat of vaporization;
	2) Wave length
M	Mass
M_r, M_ϕ	Horizontal mass flow
M_z	Vertical mass flow
m_d	Molecular weight of dry air

SYMBOLS

m_v	Molecular weight of water vapour
P	1) Precipitation
	2) Potential energy
p	Pressure
Q	Thermodynamic heat content of air
Q_e	Latent heat exchange between ground and air
Q_s	Sensible heat exchange between ground and air
q	Specific humidity
q_{as}	Specific humidity of surface air
q_s	Saturation specific humidity
q_w	Specific humidity over water surface
R	Gas constant for air
R_a	Atmospheric radiation
r	1) Radius in cylindrical coordinates
	2) Distance from earth's axis
s, n	Natural horizontal coordinates; s points along wind
T	Temperature
T_{as}	Surface air temperature
T_v	Virtual temperature
T_w	Water temperature
t	Time
U	Basic zonal current
u, v	Horizontal velocity components in Cartesian coordinates
V	Wind speed
$\mathbf{V}$	Wind vector
v_θ, v_r	Horizontal velocity components in cylindrical coordinates
W	Watts, kW kilowatts
W	Mechanical work
w	Vertical motion
x, y	Horizontal Cartesian coordinates; x points east, y north
z	Vertical coordinate

β	Variation of Coriolis parameter with latitude
δ	General difference symbol
Δ	Finite difference
γ	Lapse rate of temperature
ζ	Relative vorticity about vertical axis
ζ_a	Absolute vorticity
η	Vorticity about horizontal axis
θ	Angular measure in cylindrical coordinates
θ	Potential temperature
θ_e	Equivalent potential temperature
λ	Longitude
μ	Coefficient of eddy viscosity

ϱ	Density of air
ϱ_v	Density of water vapour
τ	Horizontal shearing stress
τ_s	Surface stress
ϕ	Latitude
ω	1) Earth's angular rotation rate
	2) Individual pressure change dp/dt
Ω	Angular momentum
Ω_r, Ω_ϕ	Lateral momentum transport
Ω_z	Vertical momentum transport
div ( )	Divergence of
$\nabla \cdot \mathbf{V}$ or $\text{div}_2 \mathbf{V}$	Horizontal velocity divergence
⎯⎯	Time or line average; primes denote deviations
∼	Area average; asterisk denotes deviations
∧	Vertical average (caret)

Note on Experiments

Several large tropical experiments, which have been conducted are mentioned frequently in the text. The ship array for three expeditions employing research fleets is shown in Fig. 5.2. The acronyms, periods and locations of the major experiments are as follows:

LIE	Line Islands Experiment, north central equatorial Pacific, March-April 1967.
ATEX	Atlantic Tropical Experiment, February 1969 in central equatorial Atlantic.
BOMEX	Barbados Oceanographic and Meteorological Experiment, May-July 1969 in the western tropical Atlantic east of the Lesser Antilles.
VIMHEX	Venezuela International Meteorological and Hydrologic Experiment—Northern Venezuela, June-October 1969, May-September 1972.
GATE	GARP (Global Atmospheric Research Program) Atlantic Tropical Experiment, off West Africa, June-September 1974.

1
Wind Systems of the Tropics

The tropics straddle the earth's equator, but where are their limits toward the middle latitudes in both hemispheres? With the meaning of the Greek-derived word "Tropics" (i.e. turning) in mind, the encyclopedia strictly positions the Tropics of Cancer and Capricorn near 23·5° north and south. At these latitudes the zenithal position of the sun in the sky turns or reverses in the course of its annual "march". Another definition would be the latitudes 30° north and south; these then divide the global surface into two halves, tropics and extra-tropics. So bounded, the tropics are the source of all momentum and most heat for the atmosphere.

From the viewpoint of atmospheric science, one might define the tropics as that part of the world where atmospheric processes differ decidedly and sufficiently from those in higher latitudes, so that one is justified in writing a separate book on tropical weather and climates alone (see 28a for the geographer's viewpoint). If this is accepted, a dividing line between easterly and westerly winds in the middle troposphere (700 mb) may serve as a useful guide for the boundaries. This fluctuating line allows for seasonal variations and for differences between one part of the world and another, within the same season.

The "tropics" so outlined will be the major concern in this book. One knows, of course, that weather does not stop at any arbitrary boundary, neither that between two countries nor an imagined one between sources and sinks, or a change in predominant physical mechanisms. Indeed, we are very interested in the connections between tropical and extratropical zones. No part of the atmosphere exists alone or can be understood without examining wider regions. Therefore, "excursions" from any narrow definition of the tropics into the higher latitudes will be made throughout this volume.

The low-level general circulations: an overview

Mean zonal winds

A remarkable feature catches the eye at a cursory glance over the average surface wind systems of the globe. By and large, easterly winds (from east) prevail equatorward of latitudes 30°, and westerly winds (from west) over the middle and higher latitudes. Why this remarkable division of tropical and extratropical wind belts? The roots for understanding date from antiquity.

From measurements in the Nile Valley, the Ancients were able to prove that the earth was round as early as 500 B.C.; they suspected that it moved and rotated around the sun. When these concepts won general acceptance, from the sixteenth and seventeenth century onward, dynamical theory was not slow in developing. In the minds of early explorers, the quest for observation coupled with understanding was always uppermost. One of the valuable concepts to emerge was that of conservation of *angular momentum*.

Consider a basin with radius r rotating around the centre with angular velocity ω. The velocity of the rim, then, is $r\omega$ (Fig. 1.1 left).

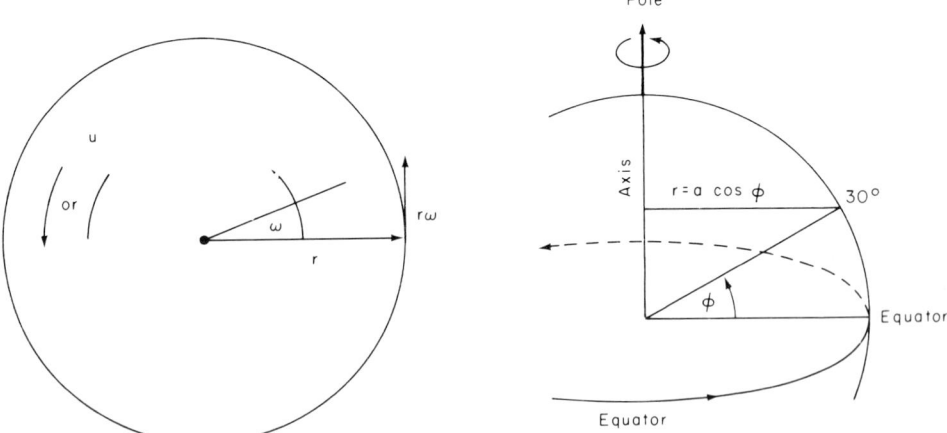

Fig. 1.1. Illustrating angular momentum for rotating basin (left) and for globe (right).

Multiplying again with r the length of the arm, the angular momentum of the basin per unit mass Ω is ωr^2. If fluid, such as water, moves at a different velocity u (the wind in the atmosphere) around the pan, the additional or *relative angular momentum* of the water is ur so that the total or *absolute angular momentum* of the water per unit mass $\Omega = ur + \omega r^2$. The farther toward the centre of the pan a measurement is made, the smaller ωr^2 is, which decreases with the radius.

Passing now to the earth and its atmosphere, the angular momentum of the surface about the rotating earth's axis will also be denoted by ωr^2. Now $\omega = 2\pi/86\,400$ rad s^{-1} or $7 \cdot 29 \times 10^{-5}$ s^{-1}. The arm $r = \cos \phi$ (Fig. 1.1 right); by convention, the symbol a is used for the earth's radius and ϕ is latitude. The value of ω, of course, is constant everywhere, but r varies from $r = a$ on the equator to $r = 0$ at the pole, so that the earth's angular momentum decreases poleward from the equator as $\cos^2\phi$. The relative angular momentum is given by the east-west or zonal component of the wind and will again be denoted by ur, always per unit mass.

The total or absolute angular momentum of earth plus atmosphere must take the masses into account. The mass of the earth is about 6×10^{21} tons and that of the atmosphere is 5×10^{15} tons (about one-millionth of the earth's mass). The momentum $\Omega = ur + \omega r^2$ integrated over the masses of earth plus atmosphere will be conserved in the absence of outside forces acting on the system; these need not be considered.

The conservation theorem may be applied not only to the total mass of earth and atmosphere but also to an individual column or parcel of air if it is not subject to any forces. Then

$$\frac{d}{dt}(ur + \omega r^2) = 0 \qquad (1.1)$$

along the trajectory where d/dt is the time change following an individual mass. Air moving toward higher latitudes should acquire westerly relative velocity as ωr^2 decreases; for equatorward-moving air the winds should become easterly. Surface wind charts show easterly winds in the tropics and westerly winds at higher latitudes; therefore, momentum is less than that of the earth's surface in the tropics and higher beyond latitudes 30°. The suggestion is that the air in the tropics has moved equatorward and the air of the middle and high latitudes, poleward.

As the reader may compute from Eq. (1.1), the mean zonal wind distribution given by average seasonal profiles (Fig. 1.2) does not show nearly so large an equatorward increase of east winds as is demanded by conservation of momentum. The air, in contrast to the combined system earth-atmosphere, is subject to forces acting on it. Moving

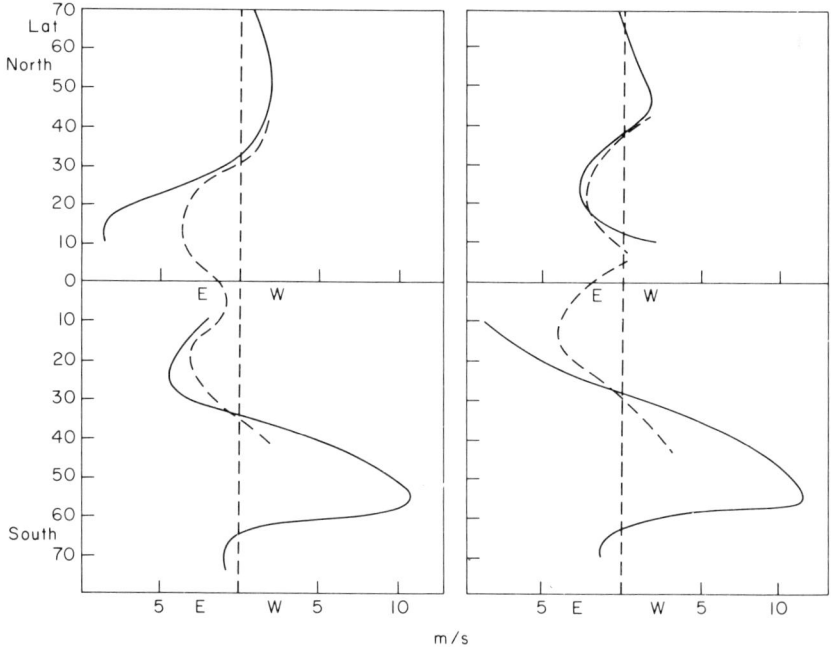

Fig. 1.2. Mean zonal winds (m/s, broken) and geostrophic wind (solid) for January (left) and July (right) over the oceans (23).

toward lower latitudes, air is slowed through ground friction so that very strong east winds in a belt around the globe will not develop; also, air moving toward higher latitudes is retarded so that the westerlies increase more slowly toward the pole than would be demanded by Eq. (1.1). In spite of the retardation, however, a basic tendency toward conservation of momentum is apparent; this tendency is unmistakeable and dominates the mean global wind field at the surface.

Mean meridional circulation

The mean zonal wind profiles imply north-south or meridional motions. These motions indeed exist (Fig. 1.3); they are directed predominantly toward the equator within latitudes 30° north and south (7, 43, 49). Such motions can exist only if some force accelerates the air in that direction, and the only force in the atmosphere known to produce accelerations of large-scale windfields is the pressure force.

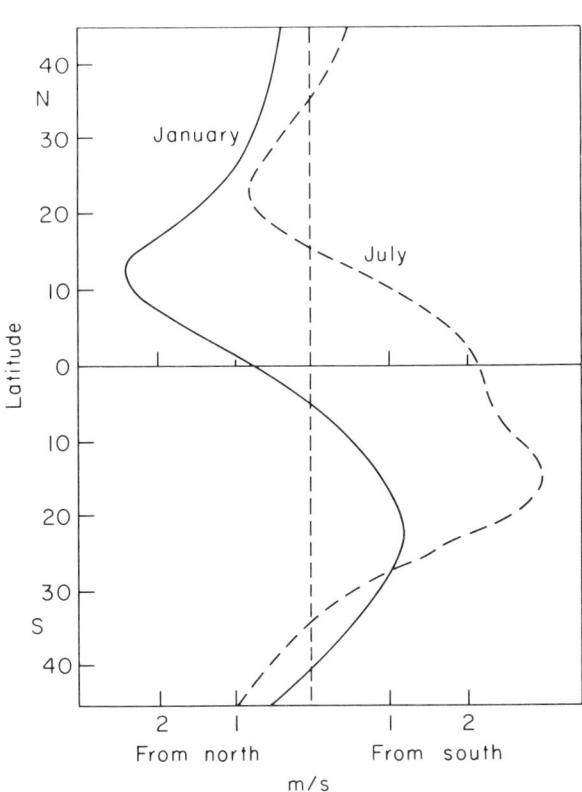

Fig. 1.3. Mean meridional winds at the surface for January and July over the oceans (43).

Therefore, one should expect to find high pressure in the subtropics and low pressure in the equatorial zone associated with the equatorward air motion. Such a pressure distribution does prevail (Fig. 1.4); a broad high-pressure belt overlies the subtropics and another, with pressure about 5-10 mb lower, often called the *equatorial trough* of low pressure, is found in the heart of the tropics. In the history of meteorology the low pressure trough, which migrates seasonally with the sun, has been related to maximum solar heating of the earth in equatorial latitudes.

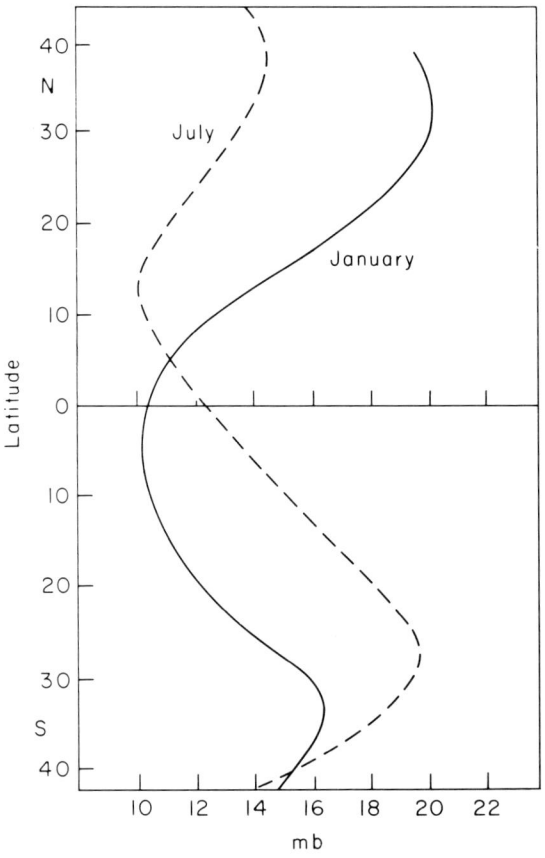

Fig. 1.4. Mean sea-level pressure profiles (mb).

Vertical motion and rain belts

Another step will relate the meridional circulation to weather distribution in the tropics in broadest outline. From Fig. 1.3 the horizontal divergence $\nabla \cdot \mathbf{V}$ (\\mathbf{V} wind vector) can be computed for latitude belts:

$$\nabla \cdot \overline{\mathbf{V}} = \frac{1}{r} \frac{\partial \overline{v}r}{a\partial\phi} = \frac{\partial \overline{v}}{a\partial\phi} - \frac{\overline{v}\tan\phi}{a}. \tag{1.2}$$

Here the overbar indicates averages around the globe; the symbol ∂ denotes partial differentiation. The second term arises from the spreading out of the meridians with decreasing latitude on the globe. It is exactly zero at the equator; inside the tropics it is much smaller than the first term which determines the divergence profile (Fig. 1.5).

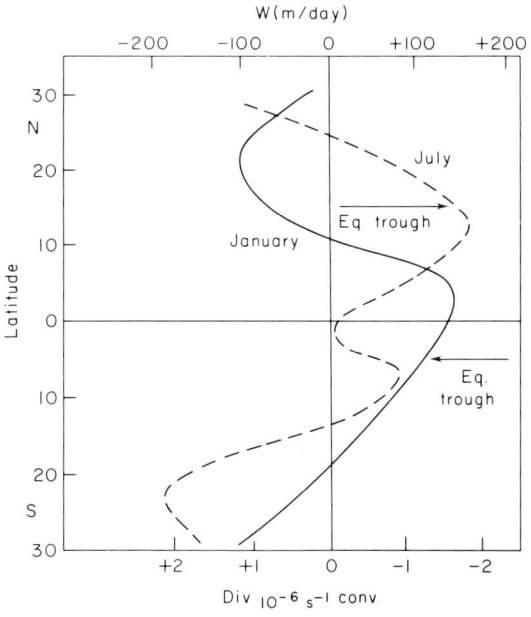

Fig. 1.5. Mean velocity divergence ($10^{-6}s^{-1}$) over the oceans and mean vertical motion at 900 mb (m/day) if divergence profiles represent average for the layer between the surface and 900 mb (43).

In the outer tropics divergence is positive, but in the inner equatorial zone one finds negative divergence, usually termed "convergence".

Applying now the concept of conservation of mass in the atmosphere,

$$\frac{\partial(\overline{\varrho w})}{\partial z} + \nabla \cdot (\varrho \overline{\mathbf{V}}) = 0 \tag{1.3}$$

for steady state as may seasonally be assumed to exist. In Eq. (1.3), ϱ is air density, z the vertical coordinate and w vertical motion. If we now consider a small depth, say one kilometre, over which the profiles of Figs 1.3 and 1.5 may hold, density varies little and may be dropped from Eq. (1.3). Integrating with respect to height,

$$\overline{w_h} - \overline{w_0} = -\nabla \cdot \hat{\overline{\mathbf{V}}} h = \overline{w_h} \ . \tag{1.4}$$

since vertical motion at the ground itself is zero. The caret (∧) indicates the vertical mean value over the layer from $h = 0$ to $h = 1$ km as specified above. It follows that in the regions of divergence, w_h must be downward near the earth's surface, and in regions of convergence, w_h must be upward. Now sinking motion brings drying and clear skies, rising motion clouds and precipitation. Thus, Fig. 1.5 outlines a belt with high rainfall in the equatorial zone and clear skies in the outer tropics and subtropics, exactly as observed in gross terms.

It is quite amazing how much of the broad structure of the tropics can be deduced correctly from simple, classical reasoning. Based on these successes whole theories of the tropical atmosphere have been evolved. However, modern observations prove that the story of the tropics is far more complex than was believed, even in the 1930s. For the present we shall elicit more facts about the tropical atmosphere.

Seasonal variations of the surface circulation

On a globe with uniform surface and perfectly circular orbit around the sun, the seasonal variations could be compressed into a small table. The great complexity that actually faces the climatologist owes its existence—apart from the inclination of the earth's axis—in part to the irregular distribution of continents and oceans with very different surface responses to the incident solar radiation, also to the role of the water and water vapour cycles in the atmosphere and to the distribution of mountain chains. Firstly, the variations of the mean circulation with latitude, seen in the preceding illustrations, will be examined further. Then the world-wide view will be opened. Though the longitudinal patterns are pronounced, average conditions still change much more with latitude than with longitude. From ancient times, men have so viewed the climate and coined names such as "doldrums" and "trades".

Asymmetry of the Hemispheres

The average latitude of the equatorial low pressure trough is about 5°S in the northern winter and 15°N in the southern winter (Figs 1.4 and 1.10). In the annual mean its latitude is 5°N, called the *meteorological equator*, so that the meteorological Southern Hemisphere is larger than the northern one. As is generally known, the tropics export heat to higher latitudes. It follows that the Southern Hemisphere draws more on the tropical heat source than the northern one, but why this is so remains a matter for speculation. The broad expanse of ocean around the Antarctic continent, undisturbed by any other continent, provides a convenient mechanism for "enlarging" the Southern Hemisphere. Westerly winds develop unhindered in the southern ocean and remain equally strong both summer and winter, much stronger than the northern westerlies (Fig. 1.2). Thus, southern weather systems can sweep to the equatorial zone with superior strength and extend their influence well beyond the geographical equator.

The trade wind belts

Between the equatorial low pressure trough and the subtropical high pressure belt lies the belt of tropical easterlies (Fig. 1.2) coupled with the equatorward meridional flow of Fig. 1.3. One speaks of northeast trades in the Northern Hemisphere and of southeast trades in the Southern Hemisphere (Fig. 1.6). These wind systems are the steadiest,

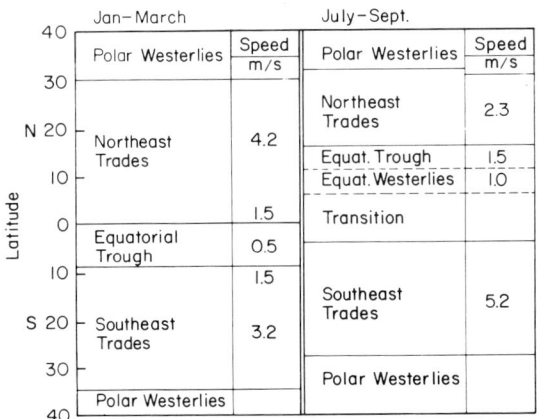

Fig. 1.6. Latitudinal distribution of surface winds in summer and winter.

or most invariant ones of the globe at the surface and can be relied upon to be encountered 80-90% of the time. Hence, they received their name "trade" winds because they provided reliable passage for the merchant vessels of sailing days.

Life has adjusted to the uniform and often monotonous trade wind flow in numerous ways. On many islands the towns lie on the leeward side, which provides protection to shipping from wind and ocean. Many busy airports have only one runway in the direction of the "prevailing" or most frequent wind direction, though this at times has proved to be a mistake on account of local wind circulations. The trade winds are weaker in summer than in winter. In the Northern Hemisphere the highest speed is 2·4 m/s in summer and 4·3 m/s in winter, in the Southern Hemisphere 3·3 m/s in summer and 5·3 m/s in winter. The increase in both cases is 2 m/s, but the southern trades exceed those of the north by 1 m/s.

Seasonal changes are by far the greatest in the Northern Hemisphere. The northern trades extend from the equator to about 28°N in winter and from 18°N to 31°N in summer. Not only is their area coverage cut in half from winter to summer, but their central latitude also moves from 14°N to 25°N. The situation is complicated by the vast Asiatic monsoon regime; over the Pacific and Atlantic Oceans seasonal changes are much smaller and more comparable to the Southern Hemisphere. There the trades extend over about 20° latitude in both seasons and the poleward shift in the centre of the belt is 5° latitude. The seasonal change in angular momentum transport is produced by greater strength of the trades, together with a shift toward the equator in winter, in comparison with summer.

Geostrophic and ageostrophic winds

Outside the tropics and above the friction layer, wind direction usually coincides with the direction of contours of pressure surfaces within 10° and speeds agree with geostrophic speeds within 10%, except in strongly curving flow. Computations of the surface wind must always involve frictional drag. In the tropics, geostrophic control near the surface decreases with approach to the equator in the presence of deep layers with convective clouds (Fig. 1.2). Nevertheless, we have noted some relation between the pressure profiles of Fig. 1.4 and the mean zonal winds of Fig. 1.2. The subtropical boundary between easterly and westerly winds agrees quite well with the high pressure mark in the subtropics. Seasonal variations coincide closely. The subtropical ridges shift poleward in summer, about 5° latitude in the Northern Hemisphere and a little less in the Southern Hemisphere. Since the equatorial trough moves 20° latitude, the areas of actual and of geostrophic east wind expand and contract together. The summer subtropical high pressure is 5-6 mb lower than that of winter in both hemispheres. The pressure gradient across the trades is 10 mb in winter and 5 mb in summer. Altogether, the climatic seasonal east wind and north-south pressure profile variations agree remarkably well by sign. Even a detailed comparison of actual and geostrophic east wind reveals surprising agreement in wind speed from about latitude 20° to 25° poleward (Fig. 1.2). Only equatorward from there does the averaged geostrophic zonal wind in winter greatly exceed observed zonal wind.

The meridional wind profiles (Fig. 1.3) portray a completely ageostrophic wind directed along the pressure gradient. The transition from north to south winds in Fig. 1.3 agrees very well with the equatorial low pressure positions in Fig. 1.4; the meridional circulation is always directed from high to low pressure. Nevertheless, the reality of the net meridional circulation has been doubted. North and south components of the mean wind could alternate around the globe on any latitude circle, so that the result could be a statistically insignificant difference between large numbers. A test can be made by counting the number of 5° latitude-longitude squares with north and south components on each latitude circle (Table 1.1). In winter, equatorward components dominate so strongly that the validity of Fig. 1.3 is assured in that season. In summer, especially in the Northern Hemisphere, north and south components alternate much more, in relation to the meanderings of the equatorial low pressure trough. Figure 1.3, as much as any illustration, brings out the fact that in the tropics one must not add all seasons together for a "yearly" picture which is obviously meaningless.

World distribution of surface winds and pressures

Anyone who sets out to prepare a set of charts such as Figs 1.7 and 1.8 will soon realize that this is not a small or wholly satisfactory task. Although every atlas or textbook carries a picture of the mean wind field, there are some differences not easy to reconcile. This book has drawn mainly on the "Climatological Atlas" (53) for the oceans and various special publications for the land areas.

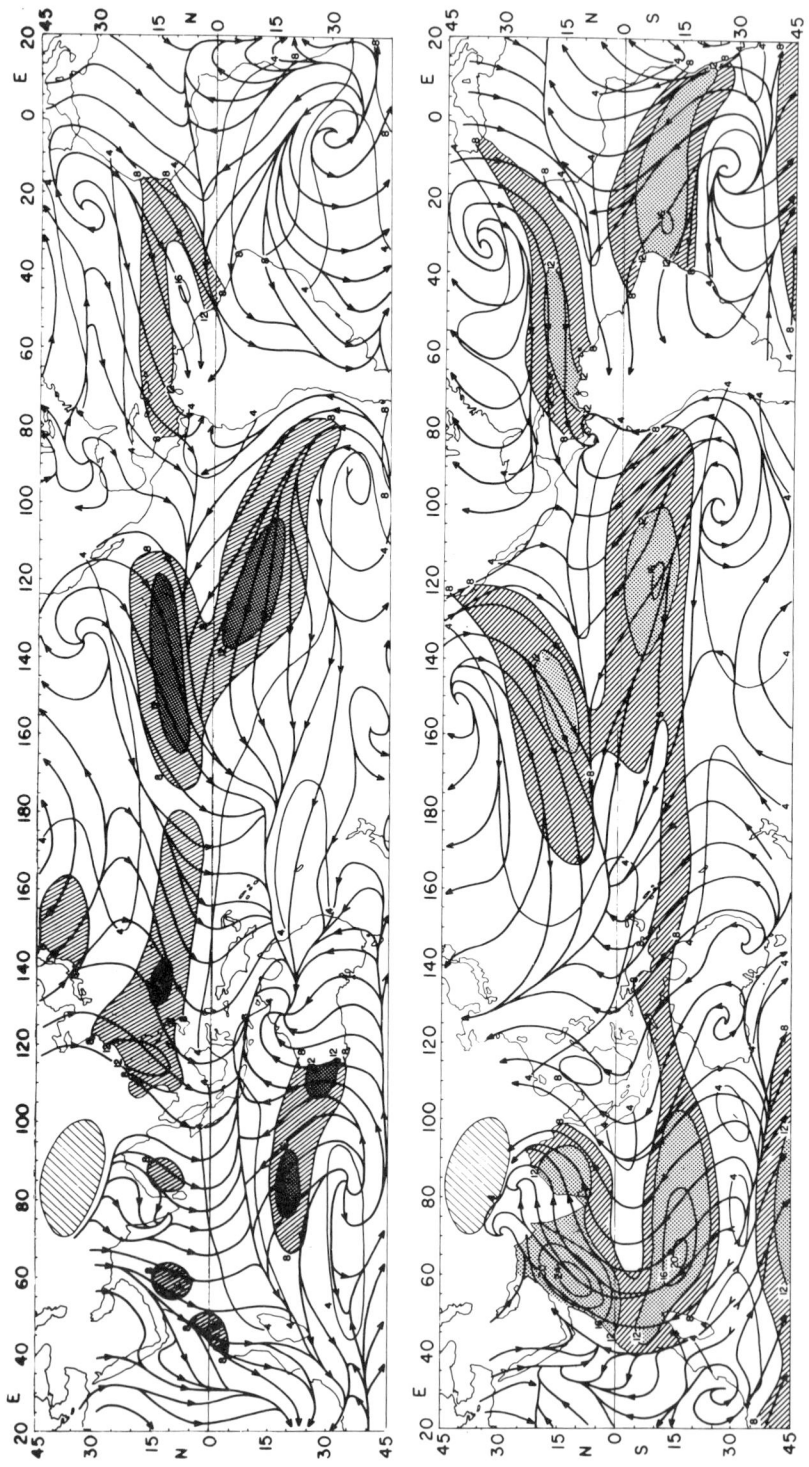

Fig. 1.7. Resultant streamlines and isotachs (knots). Top: January. Bottom: July. Light shading denotes areas with wind speed greater than 8 knots; heavy shading, greater than 12 knots.

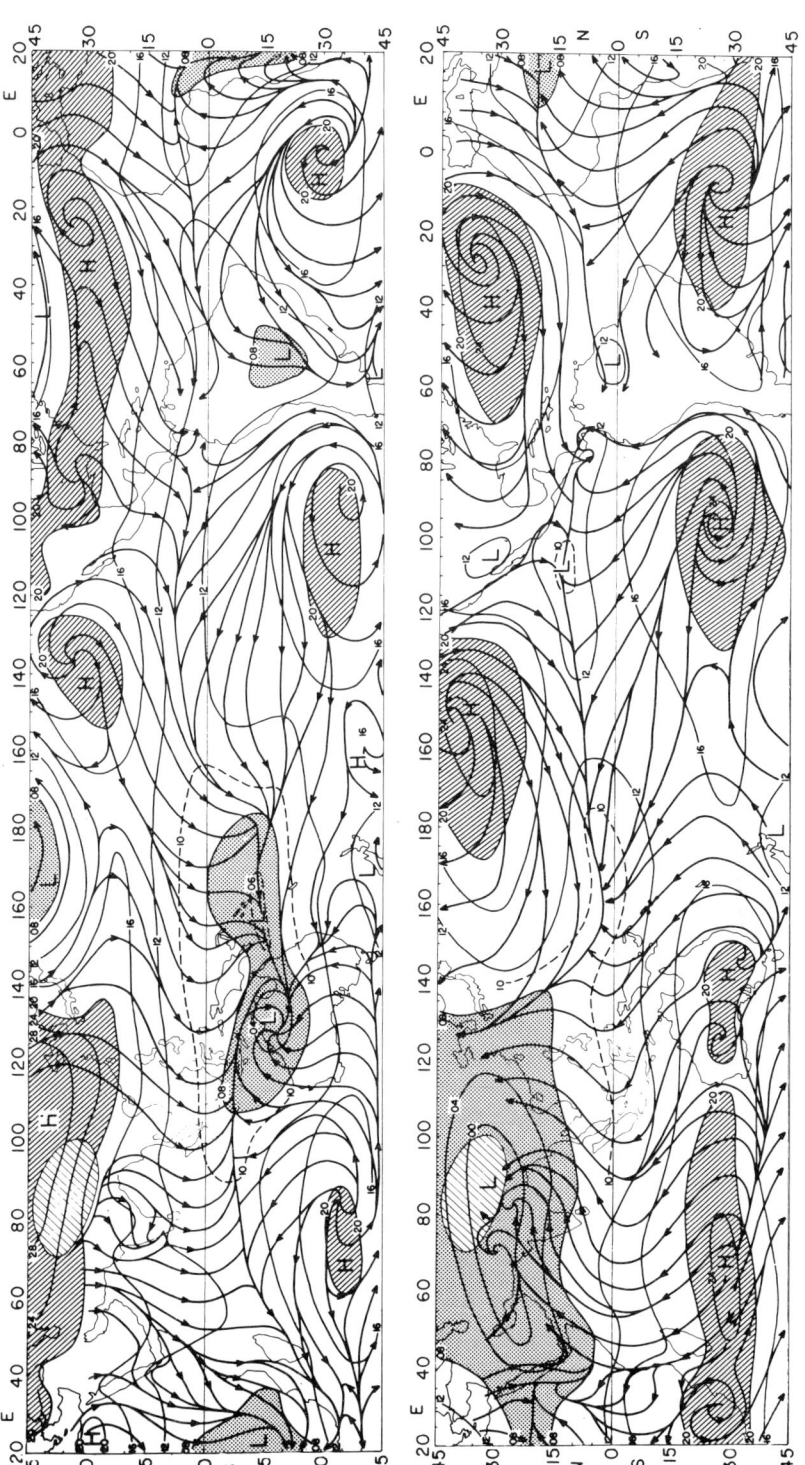

Fig. 1.8. Surface streamlines and sea-level isobars (mb, first two digits omitted). Top: January. Bottom: July. Areas with pressure above 1020 mb are shaded; below 1008 mb, hatched.

Table 1.1. Number of 5° latitude-longitude squares with (A) mean poleward and (B) equatorward wind component in the trades, in winter.

	A	B
July		
5-10°S	4	54
10-15°S	2	58
15-20°S	1	54
20-25°S	7	50
January		
5-10°N	3	53
10-15°N	0	59
15-20°N	0	54
20-25°N	0	52

Analysis of Fig. 1.7 was carried out using the streamline technique developed by Bjerknes (3) and extensively applied to tropical work by Palmer (33). In view of high wind constancy, the streamlines in the trades represent mean air trajectories; those of the subtropics merely indicate net mass displacements. Nevertheless, the anticyclones in the eastern parts of the subtropical oceans are one of the most striking features of the maps. Such centres occur in all oceans in both seasons except the South Indian Ocean in July. There the data suggest an anticyclonic line source.

The streamlines which emanate from the subtropics diverge widely as they pass through the trades, especially in the Southern Hemisphere in July. Approaching the equator from south, they curve clockwise over the eastern parts of the oceans. Eventually the southeast trades move from southwest, west and even northwest in some areas. This turning once again is related to a basic tendency for the air to conserve its angular momentum as it crosses the equator toward the equatorial trough zone which, over the monsoon belt from Africa to eastern Asia, is far removed from the equator in that season. However, the term "monsoonally deflected trades" is not fully appropriate. The clockwise turning in July occurs also in both the eastern and western Pacific away from any heated continent. During the southern summer corresponding counterclockwise turning is observed, for instance, across the entire Indian Ocean. The phenomenon, therefore, is a part of the general circulation, though enhanced when coupled with a strong monsoon regime. After crossing the equator, the air moves toward lower earth's angular momentum. It should, therefore, lose its east component and acquire relative momentum from west, which in fact happens irrespective of whether the crossing is northward or southward.

In general, the equatorial trough is situated where the streamlines from both hemispheres converge. Its position is clear-cut in January, apart from the central South Pacific, where a convergence line oriented SE-NW emanates from the subtropics and becomes part of the equatorial trough between 160°W and 180°. In July the trough can be located with ease over Africa, the Atlantic and the Pacific to about 150°E. In the eastern hemisphere it lies over the mainland of Asia and appears in the streamlines only over India and Pakistan, where detailed data exist. Analysis is most difficult in the western Pacific. The zone of streamline convergence that extends southeastward from Korea does not represent the mean positions found on daily charts. These extend, on average, across the Philippines and the South China Sea.

Each trade wind region contains definite centres of high resultant speed that reach 6-8 m/s. In contrast, resultant speeds are low in the subtropics and in the equatorial trough zone. There, wind direction is variable and this is the reason for the low resultant speeds. When wind speed is averaged without regard to direction, an almost uniform value of 7 m/s appears everywhere except within the equatorial trough. Strongest mass transport at the surface and also in the lower troposphere are found along the western edge of the Indian Ocean in July, over 12 m/s in the Persian Gulf. About half of the low-tropospheric mass transport across the equator is due to this current which has been termed a "low-level jet stream" (14, 15).

Figure 1.8 shows the relations between world streamlines and isobars. This set of charts is difficult to draw, especially in the equatorial zone where pressure gradients are weak, and many published maps leave this area blank. The oceanic isobars here are based on analyses by Schott (46, 47) and on additional pressures recorded largely at island stations and collected in "World Weather Records" (5). After completion of preliminary isobars, these were superimposed on the finished streamlines; definite relations between the two sets of lines appeared which resemble those found on daily charts. The subtropical ridges coincide fairly well with the anticyclonic singular points. In the trades, air blows across the isobars toward lower pressure; the equatorial troughs, as determined from streamline and isobaric analysis, coincide fairly well. Further, the streamlines cut equatorward across the subtropical high-pressure belt, especially in winter. This occurs in all parts of the globe, indicating that most longitudes take part in the equatorward flow of air across the subtropics.

Correspondence between streamlines and isobars was deficient in some areas. This may not be open to objection as it is not obvious what relations between isobars and streamlines should be found on mean maps, especially in regions of unsteady flow. The areas of disagreement, however, were also those with sparse data. Agreement was excellent where data were plentiful, as in the North Atlantic. For this reason, the pressure distribution was adjusted to the wind field wherever the discrepancies were not too great and where isobaric gradients were so weak that the lines could have been drawn one way as well as another.

Surface parameters following the equatorial trough

In Figs 1.2 to 1.6 some of the most cogent features of the tropical surface circulation are presented as a function of latitude. However, the equatorial trough meanders substantially with longitude, and its relation to the zenithal position of the sun is also complex. Its annual march lags the sun by two months (Fig. 1.9) which is readily explained: the heating of the atmosphere does not stop at the solstice; highest temperatures are reached a month later over large, continental areas, but two months later over the oceans and in the upper air. The trough motion reflects this lag. In addition, the latitudinal excursion is

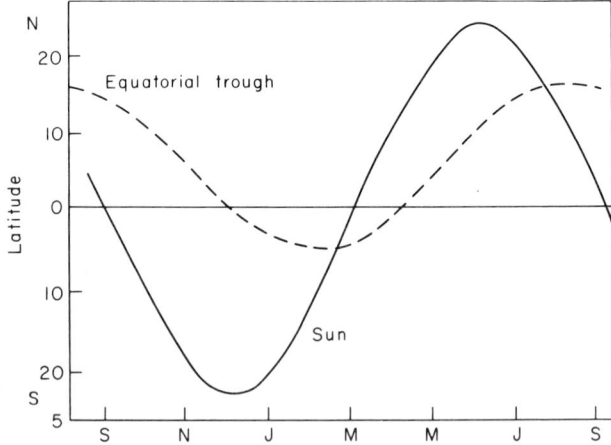

Fig. 1.9. Annual march of sun's zenithal position and of equatorial trough.

only half that of the sun, a feature which will be examined in Chapter 2.

The meandering with longitude—in January from 17°S to 8°N, in July from 2°N to 27°N (Fig. 1.10)— is derived largely from wide oscillations in the primary monsoon regions, that is, the whole southern belt of Asia plus North Africa in the northern summer, and southern Africa and Australia in the southern summer. The excursions into the Southern Hemisphere are smaller than those into the Northern Hemisphere. It appears that the more constant westerly circulation of the Southern Hemisphere's middle latitudes constrains the equatorial trough zone largely to what is often called the "near-equatorial" belt. Deep intrusions into central Australia, however, do occur in rare years. The great northward extension over Asia is undoubtedly aided by the elevated heat source of the great Tibetan Plateau. Mean seasonal positions are 15°N and 5°S, a difference of 20° latitude, probably 22-23° latitude in February and August at the time of extreme trough positions. Note that the trough is quasi-stationary over the oceanic half of the Pacific and Atlantic where seasonal displacement is restricted to 5° latitude or less.

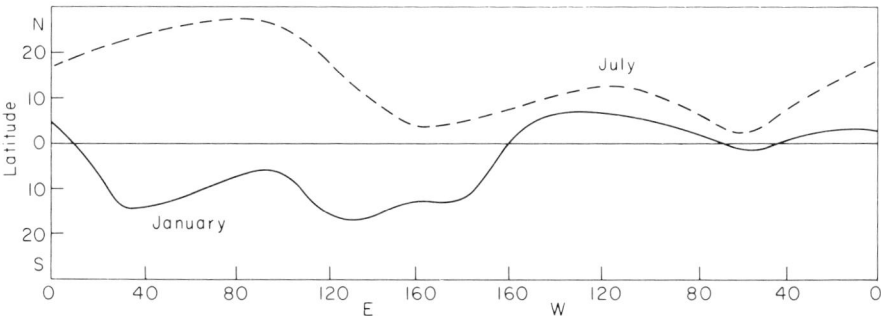

Fig. 1.10. Mean positions of equatorial low pressure trough in January and July.

Because of the meandering, Fig. 1.11 presents pressure, meridional wind, velocity divergence and wind constancy† in coordinates relative to the trough. These computations, which are often very rewarding and clarifying, are made as follows: at intervals of 10° to 20° longitude, values of pressure and other variables are tabulated on the trough line and on lines parallel to the trough situated 5, 10, ..., degrees northward and southward from the centre. Then, all values on the trough are averaged, and the same is done on each of the parallel lines.

†The percentage frequency of all observations at a given point with wind direction within 45° of the most frequent direction, or mode.

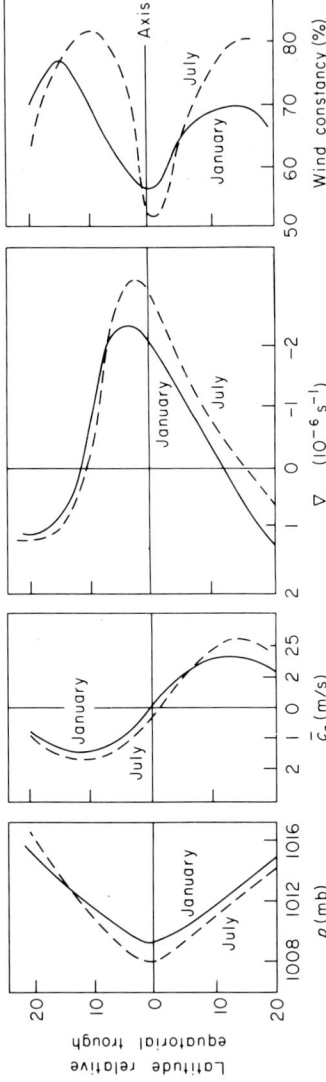

Fig. 1.11. Profiles relative to equatorial trough. Left: pressure. Centre: meridional wind over the oceans (m/s) and velocity divergence (10^{-6} s^{-1}). Right: wind constancy (per cent).

Very simple distributions result. All quantities are nearly symmetrical on both sides of the trough and, most interestingly, seasonal variations disappear, indicating that the functions of the equatorial trough zone are constant throughout the year. Pressure, by definition, has its minimum at the trough line; pressure gradients on both sides are quite uniform. The inward-bound velocity component, closely approximated by the meridional wind component, shows the same picture and, hence, also its derivative, the horizontal divergence. The last diagram, wind constancy, indicates marked unsteadiness of flow within the trough zone itself, which must be related to the circumstances producing heavy rain (Chapter 8).

One feature not evident in Fig. 1.11 is brought out by the regional distribution of velocity divergence in July (Fig. 1.12). Both in the Indian and Pacific Oceans there are two very distinct belts of convergence, but less so in the Atlantic; the secondary zone is centred near 5-10°S. Satellite cloud observations confirm the existence of the secondary convergence in these large areas and, on a transient basis, also in other parts of the equatorial belt. Between the convergence zones, marked divergence is computed on the equator itself, associated with dry zones—deserts if they occur over land. Hantel (21) has computed the wind stress over the Indian Ocean.

We have so far referred to the boundary between the meteorological hemispheres in terms of pressure, a quantity related to dynamics through the change in pressure gradient, and to heating through its mean position and seasonal migration. Others call it the equatorial or intertropical convergence zone. That convergence occurs in or near the low pressure trough is confirmed by Figs 1.11 and 1.12, except that the secondary convergence zone is omitted. Also convergence is only intermittent, so there is a degree of inaccuracy attending this definition. The term "intertropical" is more acceptable, especially when a sharp discontinuity of dust content is observed north and south of the line, but this term is really derived from old times when it was thought that the meeting of northern and southern air masses occurred right on the equator, when there would be a discontinuity of temperature between winter and summer hemispheres. The seasonal trough displacement negates such a discontinuity (Chapter 2). The argument for considering "intertropical" as important must thus fall back on the tendency toward conservation of momentum. Such an argument may have validity, but then a better term than intertropical convergence zone (ITCZ) should be proposed. Here the definition in terms of pressure will be retained; the low pressure between hemispheres can be found on all synoptic charts.

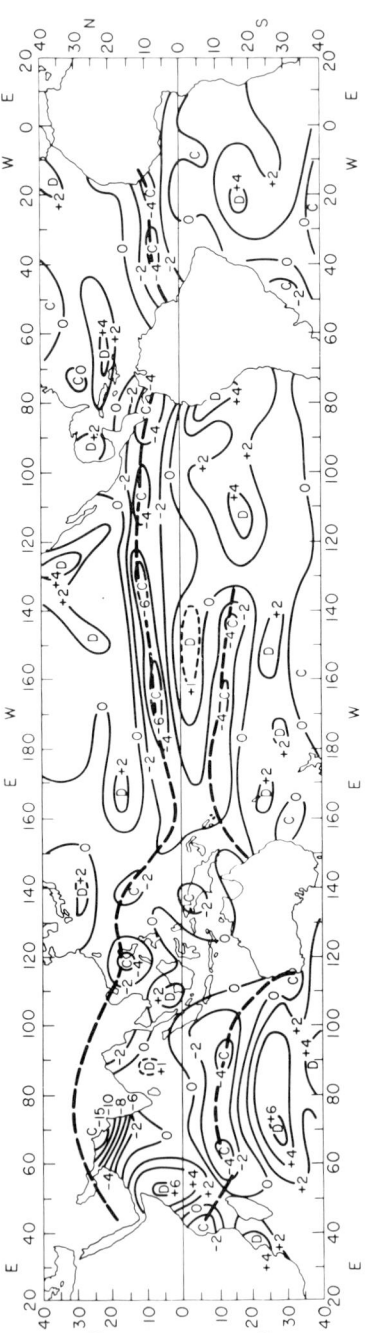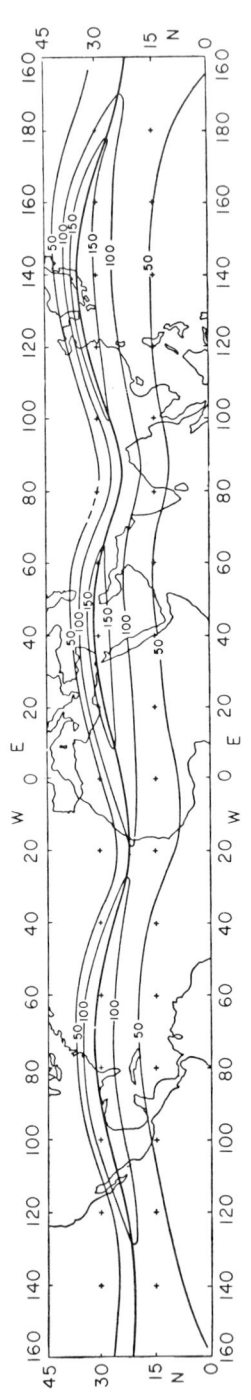

Fig. 1.12. Upper: chart of mean horizontal velocity divergence ($10^{-6} s^{-1}$) for July over the oceans.
Fig. 1.15. Lower: mean subtropical jet stream for winter 1955-1956. Isotach analysis at 200 mb surface. Isotachs are drawn for every 50 knots. The mean latitude of the jet axis is 27.5°N (28).

Mean winds in the troposphere

One of the great achievements of meteorology since about 1940 has been the vast increase in the number of places where rawinsonde stations have been erected around the globe, including on many islands and climatically unfavourable continental locations in the tropics and subtropics. One of the main motivators of this expansion has been the World Weather Watch programme of the World Meteorological Organization, which deserves universal acknowledgement. In addition, vast quantities of upper "winds" are now computed from geostationary satellites by following the motion of individual clouds over time steps of about half an hour (Chapter 7). In former days there was only a selection of high-level balloon ascents so that only good weather ascents reached the high troposphere. In the author's earlier book, "Tropical Meteorology", a warning against the acceptance of these good weather balloons as being representative of the general circulation was given. Now it is necessary to warn against upper-air climatologies biased by an overwhelming mass of measurements made only in cloudy areas, a rather dramatic and also somewhat ludicrous change.

Winds at 850 and 200 mb

A good example of the efficiency of the rawin station network is provided by Figs 1.13 and 1.14 showing mean winds at 850 and 200 mb in July. In contrast to earlier days, the number of observations decreases only very slightly with height. The compilation of data is due to van de Boogaard (A-23); some stations with very short records, such as the Galapagos Islands, are not included. Over continental areas (India, China, Australia) the density of observations is so great that only a representative sample can be shown. Over the oceans, gaps remain in some important places, but the number of islands where a station can be erected is dwindling quickly.

On the whole, the distribution of winds appears to be adequate to outline the important circulation features, and thus they are presented without any "analysis" except for the demarcation curves between easterly and westerly winds and principal oceanic troughs at 200 mb. The reader who wishes to see streamlines, will find them in (45); alternatively, he can draw his own lines on Figs 1.13 and 1.14.

At 850 mb the boundary between polar westerlies and tropical easterlies, the subtropical ridge, lies near 30-35°N and near 20°S in the southern winter. Thus, the subtropical ridge slopes equatorward

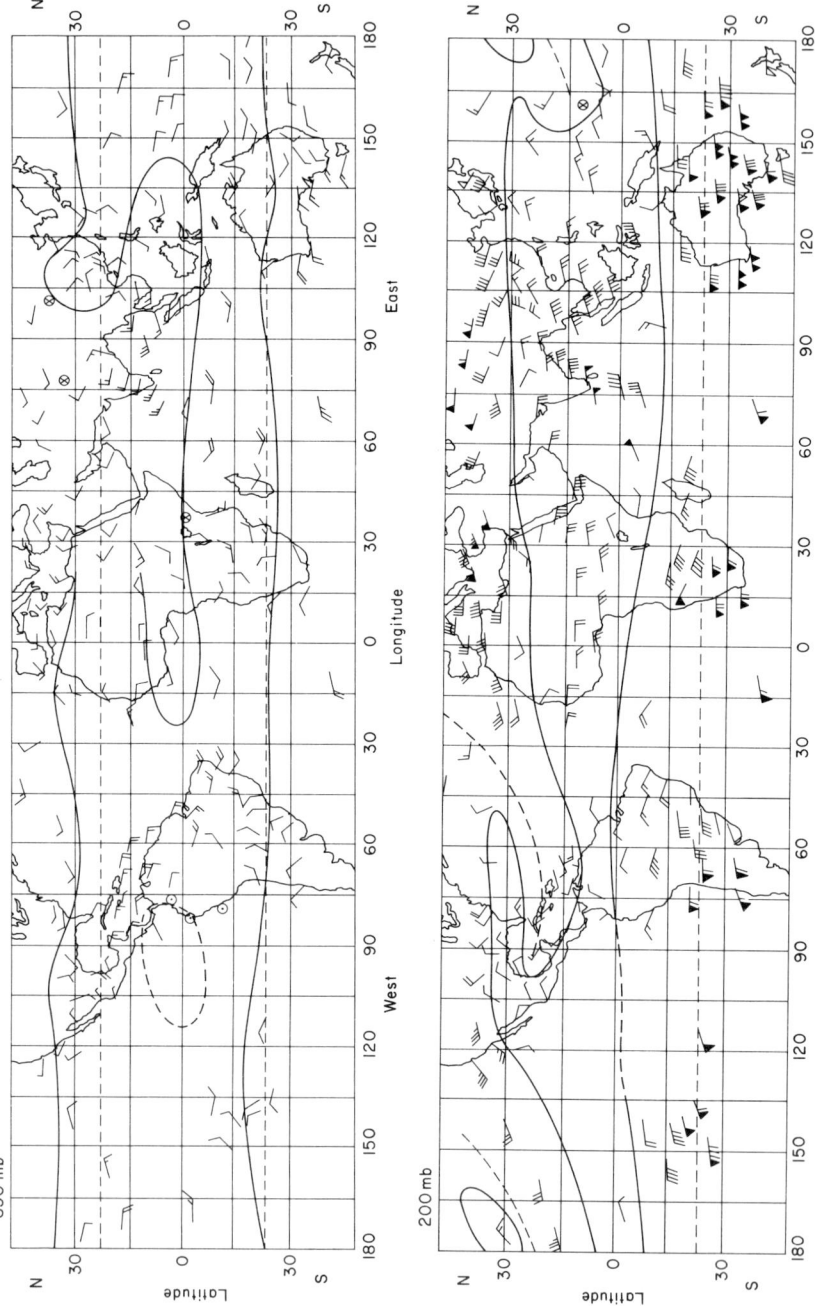

Fig. 1.13. Top: Chart of average 850 mb winds (knots) for July. Lines separate areas with east and west wind components. Courtesy of Henry van de Boogard (A-23).
Bottom: Chart of average 200 mb winds (knots) for July. Lines separate areas with east and west components. Dashed: positions of semi-permanent troughs in oceans (A-23).

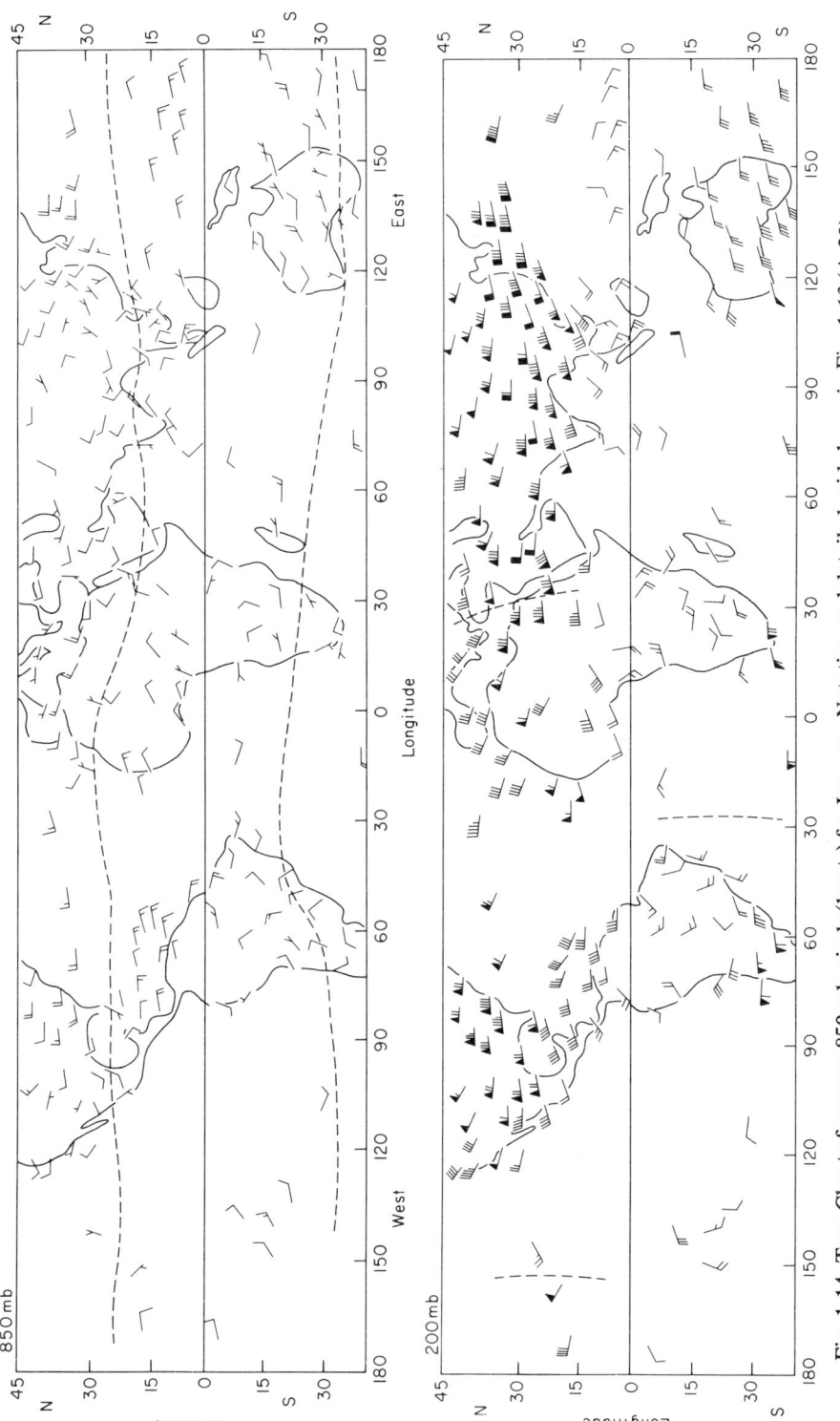

Fig. 1.14. Top: Chart of average 850 mb winds (knots) for January. Notation and detailed grid shown in Fig. 1.13 (A-23). Bottom: Chart of average 200 mb winds (knots) for January (A-23).

on the winter side from the surface to 850 mb, much more so than on the summer side. From 850 to 200 mb, the southern subtropical ridge approaches and even reaches the equator over South America. At 200 mb, the subtropical jet stream of winter is best seen over Australia in Fig. 1.14, with lower speeds both to north and south (54). Fig. 1.15 (p. 20) shows the Northern Hemisphere counterpart in the northern winter with a stationary set of three long waves (28, see also 26). Both northern and southern subtropical jet streams have a core velocity of over 50 m/s; their mean latitude is 26° in the respective seasons. The northern westerlies also approach the equator at that time of year, whereas they have wide and complicated north-south excursions in July. Between the subtropical ridges lies the belt of easterlies at 200 mb, very well pronounced in the half-hemisphere 30°W to 150°E, which includes all of Africa and Asia. Over South America, in contrast, the easterlies almost vanish and they may be non-existent over the eastern Pacific.

The high-tropospheric easterlies, with strongest wind often at 150 mb, at times attain 35-40 m/s and have been labelled the "easterly jet stream of summer" (27); the analogy with a current as depicted in Fig. 1.15, however, remains feeble. The outstanding fact about these easterlies is that they lie for the most part above surface equatorial westerlies, and thereby denote what may be termed the "monsoon" profile—lower southwesterlies and upper northeasterlies in the Northern Hemisphere with mass compensation, very evident at Singapore (29) during May-October when the NE winds attain 20-25 m/s in the layer 200-100 mb. In contrast, the trades usually have the reverse structure: lower easterlies, upper westerlies.

Basic current structures

The complications of the upper mean flow (not to speak of daily charts!) are very great. In the higher latitudes of both hemispheres there is no diversity of basic current structures; rather, we find a single baroclinic westerly current with undulations. Therefore, the daily weather systems of northern and southern westerlies are all rather similar. In the tropics, in contrast, a large diversity of weather systems is encountered. Variable basic current structure appears to be partly responsible. San Juan, Puerto Rico, for example, has typical trade wind profiles (Fig. 1.16); strongest easterlies at 2-3 km height, decreasing to the high troposphere with weak westerlies in summer and strong westerlies in winter. At Bangkok the "monsoon" profile prevails in summer, lower westerlies to as high as 6 km, easterlies increasing strongly to the high troposphere. At Singapore, on the equator, seasonal changes are small; the east wind remains constant in the high troposphere and local mass compensation is evident throughout the year (29). When the low-level winds change from SW to NE in the northern winter, the high-tropospheric winds change from NE to SE. Fernando Noronha provides an Atlantic counterpart to Singapore on the equator. Low and mid-tropospheric easterlies persist throughout the year, while the upper circulation remains part of the southern westerlies.

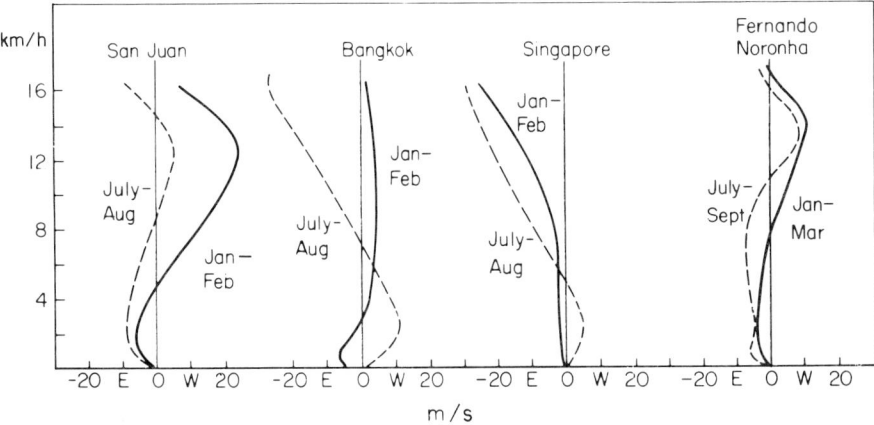

Fig. 1.16. Selected vertical profiles of mean zonal wind: San Juan, Puerto Rico (18·5°N, 66°W); Bangkok (14°N, 100·5°E); Singapore (1°N, 104°E); Fernando Noronha (4°S, 32·5°W).

North-south cross-sections of wind components

For the mean zonal winds, cross-sections are available from many sources (for instance 6, 7, 13, 22, 48, 52, A-12). They show the seasonal variation in the strength of the subtropical westerly current with centre near 200 mb at about 40 m/s in winter and half of that in summer (Fig. 1.17). A deep belt of easterlies, broadest in the northern summer, remains at all altitudes.

The meridional wind component cannot be treated so easily, especially in the zone 5°S to 25°N in the northern summer, where the oscillation of the equatorial trough vitiates the value of averaging around latitude circles. During December-February the situation, while not perfect, does permit construction of a cross-section of the mean meridional component (Fig. 1.18, compare Fig. 1.3). Great care must be exercised in drawing the section since, apart from the very small residual arising from net mass transport from summer to winter hemisphere, the integral over the pressure (*p*) interval

$$\int_{p=1000}^{p=100} \bar{v}\, \delta p = 0 \tag{1.5}$$

by necessity; otherwise large holes would soon begin to appear in the atmosphere. Most critical is the low troposphere where, in the

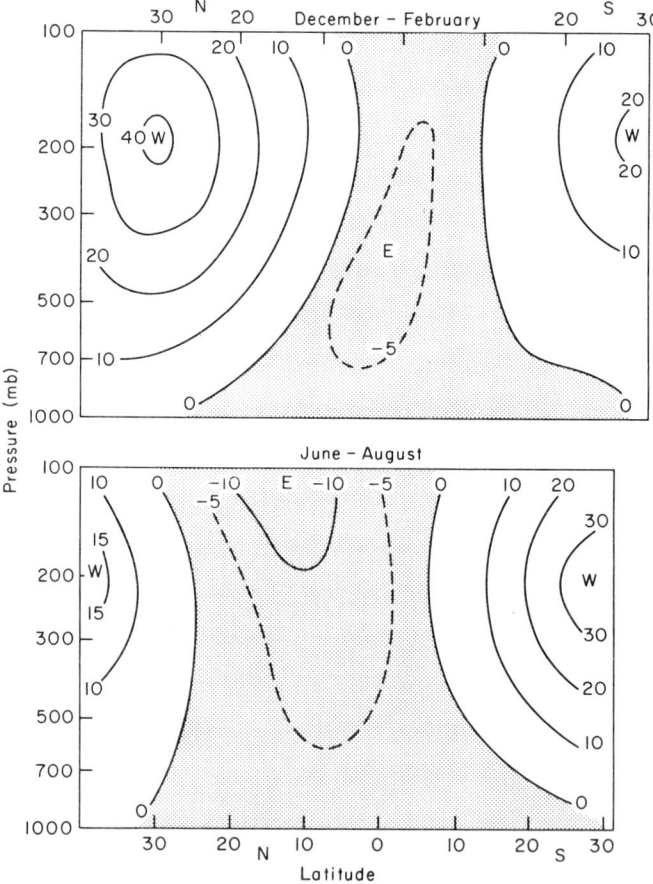

Fig. 1.17. Vertical cross-sections of mean zonal wind averaged around the globe (m/s). "W" and "E" denote centres of westerly and easterly currents. Shaded: regions with east wind.

computer age, most stations report 1000, 850 and 700 mb as standard levels in their climatologies, which is completely unsatisfactory for defining $\bar{v}$ in the low troposphere. That this is so will readily be seen from Fig. 1.18, carefully constructed by Palmén (32); the mean meridional wind is strongest just a small distance above the surface and decreases upward from there, partly a function of the boundary layer effect and the tendency toward a vertical wind structure, with some approach to the Ekman spiral in many parts of the tropics not too close to the equator. The upper return current and its centre have been well localized near 200 mb with rawin observations; at the latitudes with the largest northward current the zonal component is weak enough, so that problems of accuracy of measurement arising from very small elevation angles of balloons above the surface do not occur. For budget calculation of momentum and energy in equatorial regions (Chapter 12) it becomes very critical exactly how the meridional component $\bar{v}$ is distributed along the vertical. Even with present knowledge about the winds, computations are barely stable and any shift of the centres of meridional transport up and down in the atmosphere will affect balance calculations drastically. Publication of other mean meridional circulations may be consulted to show

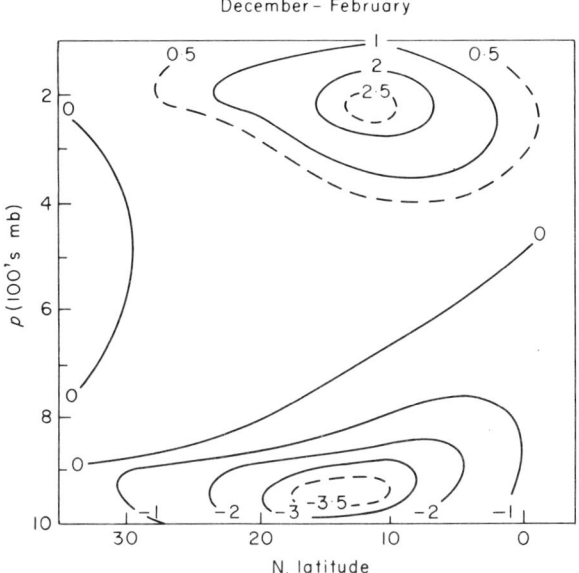

Fig. 1.18. Vertical cross-section of the mean meridional wind (m/s) in the northern winter (32).

the range of possible disagreement (7, 22, 50). In the Southern Hemisphere, the winter meridional circulation inferred from balance calculations (18) appears to be somewhat weaker than that in the Northern Hemisphere.

For the northern summer, the inward component to the equatorial trough centre is a quantity superior to the mean meridional mass flow.

Transport of tracers

It is fortunate that the mean wind charts can be supplemented by observations which follow particular masses of air around large parts of the globe. One of the world's major sources of dust in the atmosphere is the Sahara desert. During the northern winter, mean streamlines generally point toward northern South America from the eastern Atlantic and North Africa, while in the northern summer they point toward the Caribbean (Fig. 1.7). At that time of year, dust from Africa is carried westward across the Atlantic in many individual situations. Reports of haze in the air are found in early travel reports and its origin was already assigned to Africa by Frolov, who headed the French Meteorological Service on Martinique during the earlier part of the twentieth century. Continuing measurements at Barbados since the 1960s (34) have confirmed dust loading in the air from May to September. Surface concentrations have been as high as 20 $\mu g/m^3$, which equals the average background count of particulate matter in open country in the western United States. Aircraft measurements have found dust concentrations at various altitudes up to 700 mb and more (4, 35, 36).

Since the transit time of the Atlantic averages about one week, not only does it follow that the particles can remain suspended in substantial concentration for that much time, but we also obtain confirmation for the very low precipitation regime in the centre of the trades across the ocean (Chapter 3). The change in underlying surface does not matter. It is also of interest that the Barbados dust loading has increased with drought over West Africa and decreased with higher rainfall there (37). To the extent that dust transport imports a continental condensation nuclei distribution right across the ocean to North America, there should be considerable effects on cumulus growth. Indeed, when the dust is heavy, oceanic cumuli are virtually unable to form at all, as has been observed on various aircraft missions along the Greater Antilles.

The Sahara exports dust not only westwards but also eastwards across the Persian Gulf toward India in the northern summer. Very low visibilities have been reported there too by Andrew F. Bunker, Woods Hole Oceanog. Inst., as a result of research aircraft traverses down to latitude 10°. There the dust stops abruptly as the aircraft enters air trajectories coming from the Southern Hemisphere. In the northern winter, continuous heavy dust has been observed by aircraft in the upper westerlies over southeast Asia up to about 650 mb where the potential temperature corresponds to surface maximum temperatures in the Sahara at that time of year. Here also, then, is a chance for African dust to spread widely over the globe. Numerous other examples of such mobility can be given. Fission products from French atomic tests in Polynesia (22°S, 138°W) were picked up at the west coast (!) of India during the summer monsoon season there two to three weeks after the explosion (19). The material must have travelled in the easterlies across Australia and the South Indian Ocean and then have been swept northward with the intense low-level southerly flow near the African east coast (Fig. 1.7). Concentration at Bombay was comparable to that measured in Australia, a much smaller distance from the explosion site. Again there is evidence that material can travel compactly, without much dilution of synoptic-scale processes and rainfall, over long distances in the steady tropical currents, affecting very distant locations.

Stratospheric circulation

Coincident with the annual cycle of heating and cooling, stratospheric winds from the pole to tropical latitudes are easterly in summer and westerly in winter. Somewhere there must be a transition in the tropics which can take many diffuse forms and even lead to occasional, unexpected synoptic systems, like those encountered during forecasting for a high stratospheric experiment in the Caribbean in January 1960 (44). A special high-level balloon network was created for this experiment, a considerable achievement in 1960. Wind direction at 10 mb was predominantly easterly, even during the northern winter and steady conditions were expected during the experiment. Suddenly, however, the wind turned to southwest, first at Curaçao and then progressively farther north, down to levels well below 25 km height. Over an interval of four days a sharp shearline could be followed advancing northward at an average rate of 4° latitude per day. On 29 January (Fig. 1.19), the westerlies at 10mb clearly extended across the whole Caribbean; a large skyhook balloon could also be followed moving from west (in the Atlantic in Fig. 1.19). On 31 January,

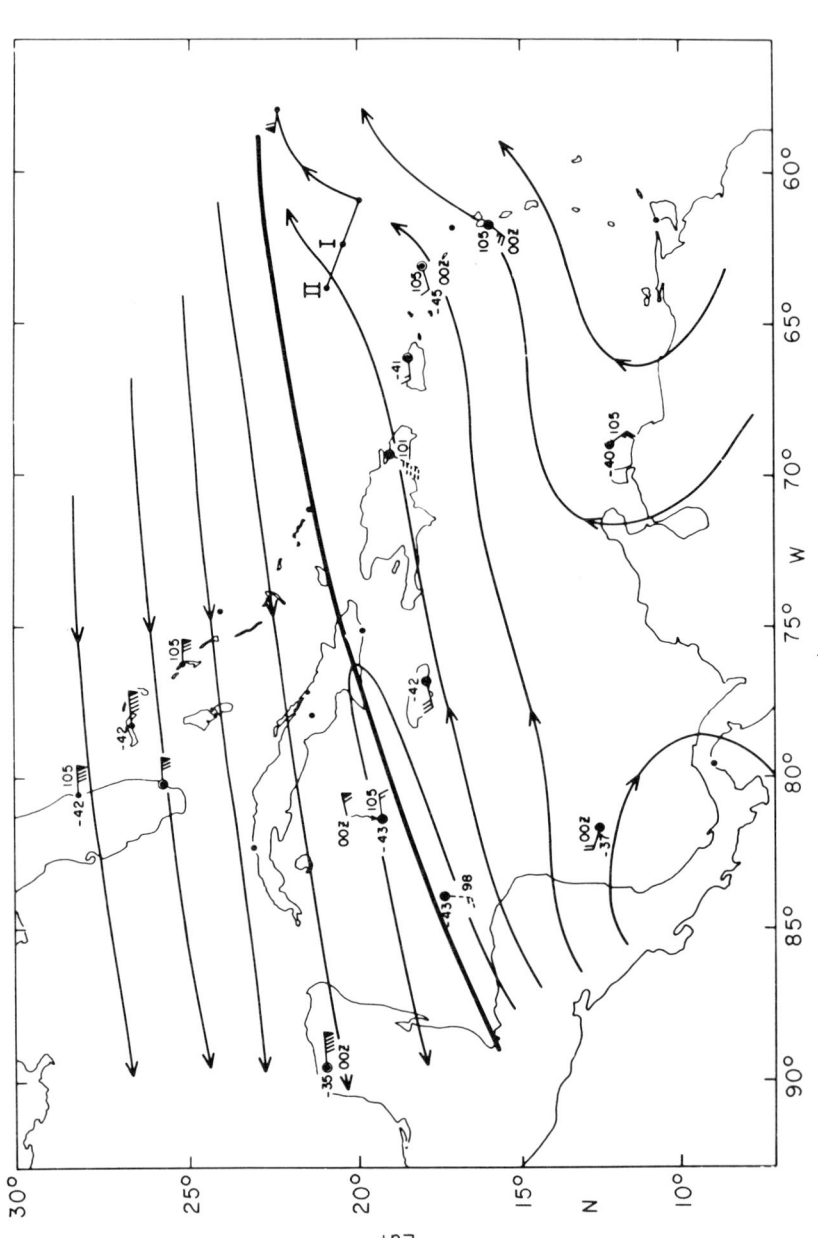

Fig. 1.19. Shearline between easterly and westerly winds (knots) at 10 mb in the Caribbean-West Atlantic area, 29 January, 1960 (1200 Z). 10 mb temperatures (°C). Track in Atlantic shows movement of skyhook tracer balloon (44).

easterlies had returned to the Caribbean, but westerly winds were reported from the Bahamas and the shearline was approaching Florida from the south. Occurrence of such synoptic systems, reminiscent of the low troposphere, is not the only surprise which stratospheric balloon observations have served on the atmospheric science community.

Quasi-biennial oscillation

For many years Canton Island (3°S, 172°W) served as a vital stop for aircraft on the route between Hawaii and Australia and New Zealand. Nowadays jet planes fly over the small island and the weather station has had a mixed history, but Canton Island will never be forgotten, for it is here that a strange phenomenon was first recognized (9). In one month of a given year the 50 mb winds tended to be all from east, in the next year from west. Analysts from the British Meteorological Office, drawing charts at 50 mb and other levels with a view to constructing monthly climatic charts, quickly became aware of these year-to-year changes and a large amount of research developed on what came to be called the equatorial, quasi-biennial, stratospheric wind variation (10, 11, 12, 39, 40, 51, A-1, A-16).

Historically, the course of stratospheric monthly winds at Canton Island has served as illustration, but, because of interruptions of the observations there, our example is Ascension Island (Fig. 1.20)

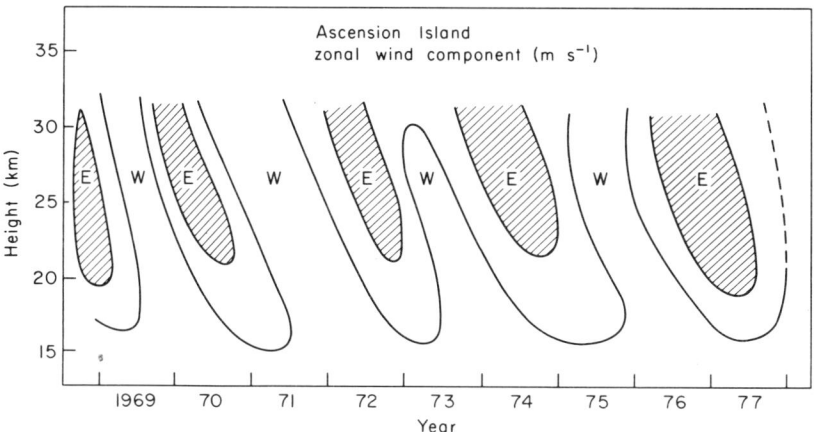

Fig. 1.20. The quasi-biennial stratospheric wind variation at Ascension Island (8°S, 14°W) for 1969 to 1977. Heavy lines denote crossovers between regimes, periods with wind speed greater than 20 m/s shaded (courtesy Henry van de Boogaard).

demonstrating that the same type of time history can be picked up in any part of the equatorial zone. Periods of about one year of easterlies and westerlies alternate in the middle stratosphere. The change in zonal speed by as much as 40 m/s from one year to the next leaves no doubt about the reality of the phenomenon. Easterly maxima are stronger than westerly ones (20-30 m/s, compared to 10-15 m/s). Downward propagation of the maxima, somewhat variable, occurs at an average rate of about 1 km per month, so that the tropopause region and the middle stratosphere are out of phase. The amplitude is largest at 25-30 km and dies out toward the tropopause and also the high stratosphere at 50-60 km.

The two big problems are: why does the phenomenon occur at all with the strange period averaging 26 months, and how do we get westerlies in a closed ring around the equator possessing higher absolute angular momentum than the earth itself at its maximum? The slightly larger radius than that of the earth is insignificant. Doubts persisted regarding the reality of the circumpolar belt until enough stations sent high-level balloons in sufficient quantity for global charts to be drawn and profiled. A four-year period is shown in Fig. 1.21 for the whole belt 0-5°N. A very regular succession of westerly and easterly maxima was obtained, with stronger easterlies and the typical downward trend from high to low stratosphere in the course of a year. In contrast, the belt 15-20°N exhibited only the normal seasonal alternation between an eastwind summer and a westwind winter hemisphere. In the Southern Hemisphere, traces of the oscillation have been found as far as 40°S.

A remarkable observational step was achieved when high-level balloons from the National Center for Atmospheric Research circumnavigated the earth eastbound close to the equator at 30 mb (Fig. 1.22). The circumnavigation took 40 days, i.e. average balloon speed was 1000 km per day. At the mean track latitude of 3°N, circumnavigation at constant angular momentum of the earth at the equator would have taken nearly ten times as long. Only countergradient momentum transport, from higher latitudes to the equator can produce absolute angular momentum higher than that of the earth for a closed ring of air. Reed and German (42) consider the problem with respect to sensible heat and ozone transports which also appear to be countergradient.

Variations in ozone and temperature accompany the zonal wind cycle. Ozone is high when the mid-stratospheric winds are mostly westerly (38), and then there is a positive temperature anomaly compared with the subtropics, suggestive of some approach to geostrophic wind (41). From a dense observation network along 80°W in summer geostrophic equilibrium could also be confirmed up to 25 mb, the limit of the data, when computing monthly averages (20).

Correlations of surface weather and rainfall with the cycle have been investigated; even the semi-permanent surface "centres of action" in the Northern Hemisphere show some response (1, 2). It would lead much too far to take up all these connections with extratropical regions. The 26-month cycle has persisted through the 1960s, with

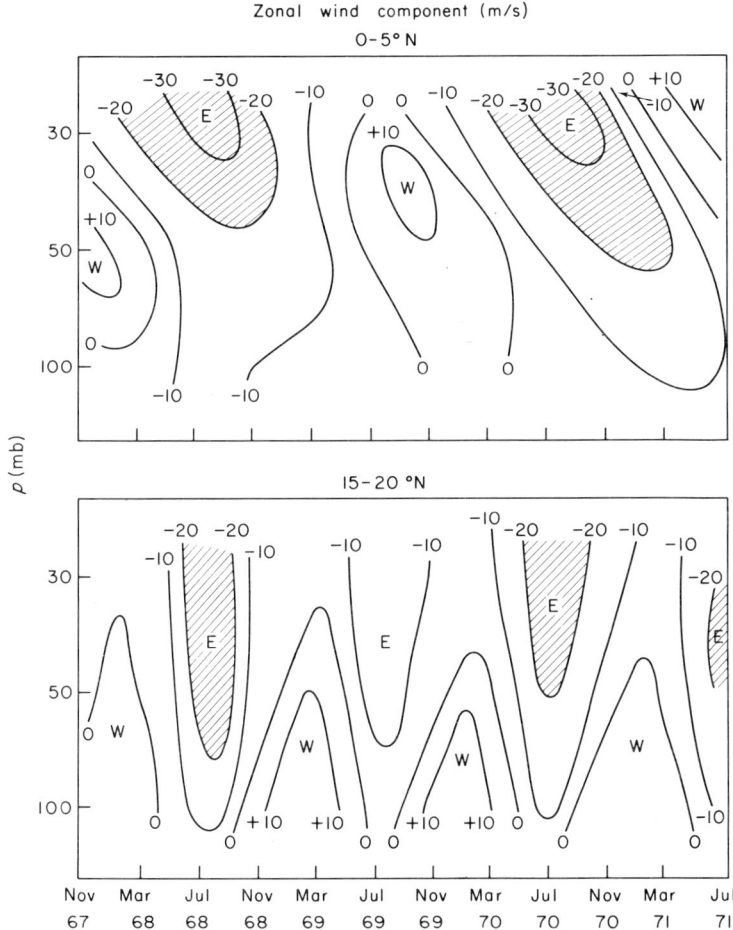

Fig. 1.21. Upper: Global average of the biennial variation (m/s) for the belt 0-5°N, 1968 to 1971. Lower: Seasonal wind variation at 15-20°N.

some disturbance after the eruption of Mt Agung in Indonesia in 1963, suggesting a connection with radiative anomalies. The cycle was soon back in place, however. It has also been traced back to the early parts of the century, based on fragmentary evidence from the early days of soundings. Thus, the "cycle" may be more permanent than some of the random periods the atmosphere exhibits from time to time and which then suddenly disappear.

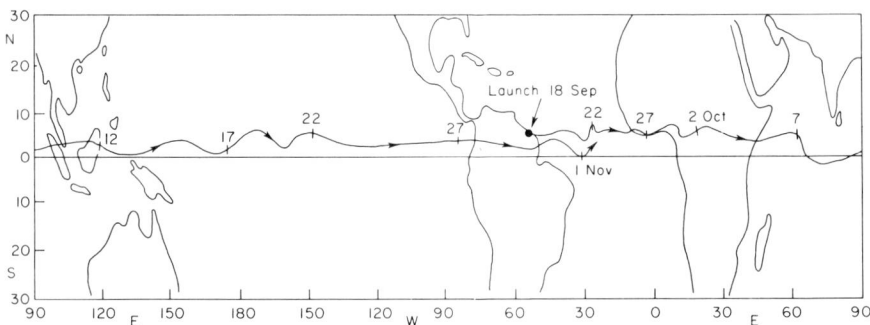

Fig. 1.22. Track of stratospheric balloon at 30 mb eastward around the equator September-October 1975 (Courtesy A. Shuman, GHOST Balloon Project, National Center for Atmospheric Research, Boulder, Colorado).

Tracers

In view of the general dearth of stratospheric soundings before the 1960s, tracer movements have served as a means of learning about stratospheric winds and circulations. In this respect, the great eruption of the Krakatao volcano in 1883 has been of outstanding value. No less than 13 cubic miles of lava, ash and mud were thrown into the air, of which one-third survived in the stratosphere up to 30 km height for several years, reducing surface incoming radiation by as much as 25% (55, 56). The coordinates of the eruption were 7°S, 105°E and the date was August 27, in the middle of the northern summer. As described by Wexler, the cloud made at least two circuits of the earth in equatorial latitudes at about 32 m/s. During that time it lengthened from 5000 to 7000 miles, presumably due to vertical wind shear in the stratosphere. The average limits of latitudinal spread were 16°N and 22°S after the first circuit, and 24°N and 40°S after the second circuit. Subsequently, the cloud spread only slowly northward until late November 1883 when it was observed over Europe and North America and soon could no longer be followed as an entity.

In the Southern Hemisphere the poleward advance was most rapid in spring. Wexler feels that the advance would have been more rapid in the Northern Hemisphere had the explosion occurred in March rather than in August.

Another explosion on a grand scale happened almost a century later in the same area, on Mt Agung, Bali (8°S, 115°E) on 17 March, 1963. Again the effects were worldwide; surface incoming radiation decreased for several years, and the biennial stratospheric oscillation was disrupted. Various methods of estimation, including very ingenious photography (16), identified the top of the volcanic cloud at no higher than 30 km in the tropics; average height dropped to 20 km when the cloud reached 30°N, and later it was found at less than 10 km in the North Polar zone. By means of its effect on solar radiation, total mass of the cloud in both hemispheres could be computed (8) and followed for several years. It turned out that most mass remained in the Southern Hemisphere, but a significant fraction escaped to northern latitudes. The mass apparently spread over both hemispheres completely; a significant reduction in dust content occurred in the Southern Hemisphere already in 1963-1964 and radiation values were normal again in 1966 (17). Temperature at 50 mb, the probable centre of the mass of material thrown up, reached a maximum up to 6°C higher than before (31) at various locations, especially over Australia which should be related at least in part to the eruption. On the other hand, longer term trends in temperature, in evidence well before 1963, complicate the picture (30). At Gan (1°S, 73°E), for instance, continuing low-stratospheric warming of 2-3°C was observed at least until 1968, in contrast to Ascension Island and other stations in the equatorial zone. Nevertheless, comparison of the consequences of the Krakatao and Agung explosions suggests rather similar histories and, above all, makes it clear that substantial injections of mass from a single volcano will have world-wide effects on stratospheric radiation and circulation for two or three years afterwards.

Much speculation about large-scale exchanges across the equator has also arisen from the transports of fission products mainly injected in the 1950s and early 1960s. The evidence is not always consistent or put into a ready circulation and exchange scheme. As Junge (24) tellingly remarks, the large number and arbitrary distribution in time (also space!) of atomic tests results in the exact opposite of a well-planned, large-scale experiment for the exploration of our atmosphere. The average residence time of some products is given as two years for the lower stratosphere; mixing within one hemisphere occurs within a few months, whereas transport across the equator is much slower (25).

Several years may be needed subsequent to a Northern Hemisphere test before slowly increasing contamination of the Southern Hemisphere's atmosphere will match slowly decreasing values in the Northern Hemisphere. These descriptions agree rather well with the volcanic evidence, including the discontinuity in mixing rate near the equator, further confirmed by numerous trajectories of constant pressure balloons released into the tropical atmosphere which, as in Fig. 1.22, kept moving in a narrow latitude belt rather than escaping across the equator or to higher latitudes. The reader will find much additional information in (A-17).

References

(1) Angell, J. K. and Korshover, J. (1962). The biennial wind and temperature oscillations of the equatorial stratosphere and their possible extension to higher latitudes. *Mon. Wea. Rev.* **90**: 127-132.

(2) Angell, J. K., Korshover, J. and Cotten, G. F. (1969). Quasi-biennial variations in the "Centers of Action". *Mon. Wea. Rev.* **97**: 867-872.

(3) Bjerknes, V. F. K. *et al.* (1911). "Dynamic Meteorology." Vol. II. Carnegie Institution of Washington, Washington D.C.

(4) Carlson, T. N. and Prospero, J. M. (1972). The large-scale movement of Saharan air outbreaks over the Northern Equatorial Atlantic. *J. Appl. Meteor.* **11**: 283-297.

(5) Clayton, H. H. "World Weather Records." *Smithson. Misc. Collns* (up to 1940). Later volumes published by the National Oceanic and Atmospheric Agency, U.S.A.

(6) Crutcher, H. L. (1961). Meridional cross-sections: upper winds over the Northern Hemisphere." US Department of Commerce, Weather Bureau Tech. Paper No. 41, Washington D.C.

(7) Defant, F. and van de Boogaard, H. (1963). The global circulation features of the troposphere between the equator and 40°N, based on a single day's data. *Tellus* **15**: 251-260.

(8) Dyer, A. J. and Hicks, B. B. (1968). Global spread of volcanic dust from the Bali eruption of 1963. *Quart. J. Roy. Meteor. Soc.* **94**: 545-554.

(9) Ebdon, R. A. (1960). Notes on the wind flow at 50 mb in tropical and subtropical regions in January 1957 and January 1958. *Quart. J. Roy. Meteor. Soc.* **86**: 540-543.

(10) Ebdon, R. A. (1961). Some notes on the stratospheric winds at Canton Island and Christmas Island. *Quart. J. Roy. Meteor. Soc.* **87**: 322-331.

(11) Ebdon, R. A. and Veryard, R. G. (1961). Fluctuations in equatorial stratospheric winds. *Nature, London,* **189**: 791-793.

(12) Farkas, E. (1962). Stratospheric wind reversals over Nandi, Fiji. *Meteor. Mag.* **91**: 66-68.

(13) Faust, Heinrich (1967). Meridionalschnitte des Zonalwindes bis 60 km Hoehe von 60°N bis 10°S. *Meteor. Rundsch.* **20**: 104-108.

(14) Findlater, J. (1969). A major low-level air current near the Indian Ocean during the northern summer. *Quart. J. Roy. Meteor. Soc.* **95:** 362-380.
(15) Findlater, J. (1969). Interhemispheric transport of air in the low troposphere over the western Indian Ocean. *Quart. J. Roy. Meteor. Soc.* **95:** 400-403.
(16) Flohn, H. and Henning, D. (1964). Stratosphärische Staubwolken vulkanischer Herkunft. *Meteor. Rundsch.* **17:** 89.
(17) Flowers, E. C. and Viebrock, H. (1965). Solar radiation: an anomalous decrease of direct solar radiation. *Science* **148:** 493-494.
(18) Gilman, P. A. (1965). The mean meridional circulation of the Southern Hemisphere inferred from momentum and mass balance. *Tellus* **17:** 277-284.
(19) Gopalakrishnan, S. and Rangarajan, C. (1972). The role of the Indian monsoon in the interhemispheric transport of radioactive debris from nuclear tests. *J. Geoph. Res.* **77:** 1012-1016.
(20) Groening, H. U. (1959). Wind and Pressure Fields in the Stratosphere over the West Indies Region in August 1958. Natl. Hurricane Proj. Rep. No. 35, US Dept. of Commerce, Washington, D.C. (9 pp.).
(21) Hantel, Michael (1970). Monthly charts of surface wind stress curl over the Indian Ocean. *Mon. Wea. Rev.* **98:** 765-773.
(22) Hastenrath, S. L. (1969). A study of the atmospheric circulation between equator and 60°N during the winter and summer seasons. *Pure Appl. Geoph.* **77:** 207-225.
(23) Jordan, C. L. (1950). North-south profiles of the average zonal wind component at the surface. *Quart. J. Roy. Meteor. Soc.* **76:** 343-344.
(24) Junge, C. E. (1960). Weltweiter radioaktiver Ausfall. *Umschau* **1:** 8-10.
(25) Junge, C. E. (1962). Note on the exchange rate between the northern and southern atmospheres. *Tellus* **14:** 242-246.
(26) Koteswaram, P. (1953). An analysis of the high tropospheric wind circulation over India in winter. *Indian J. Meteor. Geoph.* **4:** 13-21.
(27) Koteswaram, P. (1958). The easterly jet stream in the tropics. *Tellus* **10:** 43-58.
(28) Krishnamurti, T. N. (1961). The subtropical jet stream of winter. *J. Meteor.* **18:** 172-191.
(28a) Lauer, W. (1975). "Vom Wesen der Tropen." Akad. der Wissenschaften und Literatur, Mainz. 53 pp.
(29) McAllen, P. F. (1962). Mean winds over Singapore. *Meteor. Mag.* **91:** 157-162.
(30) McIntuff, R. M., Miller, A. J., Angell, J. K. and Korshover, J. (1971). Possible effects on the stratosphere of the 1963 Mt Agung volcanic eruption. *J. Atmos. Sci.* **28:** 1304-1307.
(31) Newell, R. E. (1970). Stratospheric temperature change from the Mt Agung volcanic eruption of 1963. *J. Atmos. Sci.* **27:** 977-978.
(32) Palmén, E. and Vuorela, L. A. (1963). On the mean meridional circulation in the Northern Hemisphere during the winter seasons. *Quart. J. Roy. Meteor. Soc.* **89:** 131-138.
(33) Palmer, C. E. (1951). Tropical meteorology. *In* "Compendium of Meteorology". (T. F. Malone, Ed.) pp. 859-881. American Meteorological Society, Boston.
(34) Prospero, J. M. (1968). Atmospheric dust studies on Barbados. *Bull. Amer. Meteor. Soc.* **49:** 645-652.
(35) Prospero, J. M. and Carlson, T. N. (1970). Radon-222 in the North Atlantic trade winds: its relationship to dust transport from Africa. *Science* **167:** 974-977.
(36) Prospero, J. M. and Carlson, T. N. (1972). Vertical and aerial distribution of Saharan dust over the western equatorial North Atlantic Ocean. *J. Geoph. Res.* **77:** 5255-5265.

(37) Prospero, J. M. and Nees, R. T. (1977). Dust concentrations in the atmosphere of the equatorial North Atlantic: possible relationships to the Sahelian drought. *Science* **196:** 1196-1198.
(38) Ramanathan, K. R. (1963). Bi-annual variation of atmospheric ozone over the tropics. *Quart. J. Roy. Meteor. Soc.* **89:** 540-542.
(39) Reed, R. J. et al. (1961). Evidence of a downward propagating annual wind reversal in the equatorial stratosphere. *J. Geoph. Res.* **66:** 813-818.
(40) Reed, R. J. (1962). Wind and temperature oscillations in the tropical stratosphere. *Trans. Amer. Geoph. Union* **43:** 105-109.
(41) Reed, R. J. (1962). Evidence of geostrophic motion in the equatorial stratosphere. *Quart. J. Roy. Meteor. Soc.* **88:** 324-327.
(42) Reed, R. J. and German, K. E. (1965). A contribution to the problem of stratospheric diffusion by large-scale mixing. *Mon. Wea. Rev.* **93:** 313-321.
(43) Riehl, H. and Yeh, T. C. (1950). The intensity of the net meridional circulation. *Quart. J. Roy. Meteor. Soc.* **76:** 182-188.
(44) Riehl, H. and Higgs, R. (1960). Unrest in the upper stratosphere over the Caribbean sea during January 1960. *J. Meteor.* **17:** 555-561.
(45) Sadler, J. C. (1975). The Upper Tropospheric Circulation over the Global Tropics. UHMET-75-05 Dept. Meteor., Univ. Hawaii, Honolulu.
(46) Schott, G. (1935). "Geographie des Indischen and Stillen Ozeans." C. Boysen, Hamburg.
(47) Schott, G. (1944). "Geographie des Atlantischen Ozeans." C. Boysen, Hamburg.
(48) Schwerdtfeger, W. and Martin, D. W. (1964). Zonal flow of the free atmosphere between 10°N and 80°S in the South American sector. *J. Appl. Meteor.* **3:** 726-733.
(49) Tucker, G. B. (1957). Evidence of a mean meridional circulation in the atmosphere from surface wind observations. *Quart. J. Roy. Meteor. Soc.* **83:** 290-302.
(50) Tucker, G. B. (1959). Mean meridional circulations in the atmosphere. *Quart. J. Roy. Meteor. Soc.* **85:** 209-224.
(51) Tucker, G. B. (1964). Zonal winds over the equator. *Quart. J. Roy. Meteor. Soc.* **90:** 405-423.
(52) Tucker, G. B. (1965). The equatorial tropospheric wind regime. *Quart. J. Roy. Meteor. Soc.* **91:** 140-150.
(53) US Department of Commerce, NOAA. (1938). "Atlas of Climatological Charts of the Oceans." US Government Printing Office, Washington D.C.
(54) Weinert, R. A. (1968). Statistics of the subtropical jet stream over the Australian region. *Aust. Meteor. Mag.* **16:** 137-148.
(55) Wexler, H. (1951). On the effects of volcanic dust on insolation and weather. *Bull. Amer. Meteor. Soc.* **32:** 10-14.
(56) Wexler, H. (1951). Spread of the Krakatao dust cloud as related to the high-level circulation. *Bull. Amer. Meteor. Soc.* **32:** 48-51.

2
Radiation, Temperature and Humidity

The middle latitudes go through a cycle of seasons which, seen climatically, is both definite and regular. As the noon sun rises toward the zenith, spring and summer follow winter; as it sinks, autumn and winter succeed summer. Closer to the equator this seasonal rhythm, which so strongly patterns middle-latitude life, becomes less and less sharply defined. The growing season, as seen thermally at least, lasts through the whole year; the need for heated homes and seasonal clothing disappears. Overhead, the noon sun reaches the zenith not only once, but, over most of the tropics, twice (Fig. 1.9). The middle-latitude traveller needs a long time to accustom himself to the fact that his shadow points the "wrong" way in the noon hours.

In mountainous countries the slopes offer relief from the heat of the lowlands. Many cities and summer residences lie at an altitude of 1000 metres, or more, where living is pleasant and refreshing. In Central and South America one speaks of the "temperate zone" at this altitude, as distinct from the "torrid" zone near the ground and the "frigid zone" high up. On leaving the coastal city of Trujillo in northern Peru (latitude 8°S), for instance, inland bound, one passes at first the coastal desert zone with gross, horseshoe sand dunes, followed by huge, organ pipe cactus upon turning into the valley. After a climb to around 2000 metres lovely pastures and forests appear; this is where the whole population lives and the breeding farms of some of the world's famous horses are situated here. Still higher, above 3000 metres, one reaches the upper tree line. On the grassy slopes ruins of old Inca fortifications are outlined, and in the distance lies the white, snow-covered central Cordillera with its very high peaks.

Since the sun's position changes only gradually at the solstices, the time spent by the sun poleward of a given latitude increases very rapidly equatorward from the Tropics of Cancer and Capricorn. Already at latitudes 20° this time amounts to two months, and these are months of nearly constant overhead sun. At latitudes 15° the sun passes the zenith at the end of April and in August in the Northern Hemisphere, at the end of October and in February in the Southern Hemisphere. A simple rise and fall of surface temperature as the sun advances and recedes in the sky is not to be expected in the heart of the tropics.

In place of the temperature cycle, a moisture cycle comes into the foreground, as life in the tropics is completely dependent on abundant rain. Annual rainfall varies much more widely from place to place than mean annual temperatures; the concept of "dry" and "rainy" seasons replaces the four-season cycle of middle latitudes. The forecaster is seldom asked what the temperature will be, but everyone wants to know about the expected rainfall, especially the seasonal rainfall that spells success or failure for the crops. We begin with some facts about radiation, one underlying cause of all weather and climate.

Radiation

The history of radiation observations and calculations is long indeed. Since 1960, a powerful new observing tool has been added in support of both practical use and theory—radiation measurements from satellites. These permit, foremost for our purposes in this chapter, determination of the net radiation balance of the system earth-atmosphere for any place or for any area or latitude belt on a daily, monthly, seasonal or annual basis, or longer. Fortunately, we are able to use such computations from the satellite *Nimbus III* in this chapter; with the passage of time and the development of satellite technology, further improvements and solidification of the annual radiation march are constantly being achieved.

The basic radiation statement for the system earth-atmosphere is simple enough at the high-level vantage point of the satellite outside the atmosphere:

$$\text{radiation excess or deficit} = \text{incoming solar radiation} - \left(\text{reflected} + \text{long-wave radiation} \right) \quad (2.1)$$

For the earth as a whole, and over a year, the radiation balance must be close to, but not exactly zero. For shorter time periods and smaller areas, this will by no means hold. Indeed, the atmosphere runs on differential heating! It is of the greatest importance that, for any radiation excess to be stored, a reservoir capable of absorption and retention must exist. The storage capacity of the atmosphere and of the solid ground is so small as to be negligible for most purposes. Only water can store large amounts of heat. Therefore, it is vital that over three-quarters of the earth's surface in the tropics is oceanic (36). Over deserts like the Sahara the satellite shows an immediate balance of the terms on the right side of Eq. (2.1).

Seasonal radiation balance

Many texts and articles contain tables or charts of the annual radiation balance of the earth-atmosphere system as a function of latitude. Here we are most interested in the seasonal course in order to relate the changing radiation to the migration of the equatorial trough and the expansion and contraction of the meteorological hemispheres with the seasons. The first estimates of the radiation budget from satellite observations are due to Vonder Haar and Suomi (37). Then Raschke *et al.* (26) published a chart which depicts the seasonal and latitudinal course of net radiation for four fragmentary periods from *Nimbus III* during 1969 and 1970 (Fig. 2.1). We see the typical course of strong winter heat loss and summer heat gain in higher latitudes, as well as heat gain on the equator throughout the year, all known in principle for many years, but now there are definite measurements. The processes contributing to the right side of Eq. (2.1) have all been observed individually with increased reliability, especially the reflected radiation or *albedo*, the ratio of reflected to incident solar radiation. The reflection is very dependent on cloud cover, types and altitudes. Estimates of these parameters have varied greatly in earlier calculations since the cloud observations were of necessity imperfect. Cirrus, for instance, with its radiation properties (12) cannot be seen when there is a low overcast. Low cloudiness is usually overestimated from the ground, especially in the tropics where an observer sees the cumuli "banked" against the horizon. Climatic values of cumulus cover have had to be reduced substantially over the years, from as much as five or six tenths to as little as one tenth in some areas. Albedo close to 20% is now measured in the tropics (Fig. 2.2), in some regions even less. The average global albedo is a little above 30%, down from around 43% a generation ago—a large change indeed!

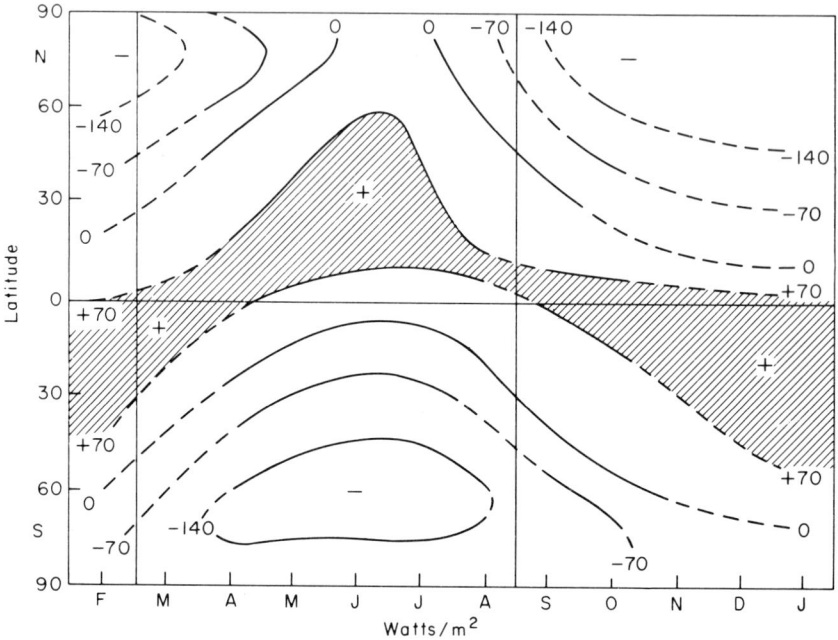

Fig. 2.1. Seasonal march of the net radiational energy gain or loss of the systems earth-atmosphere (Watts/m²). Shaded: energy gain greater than 70 Watts/m² (after 26).

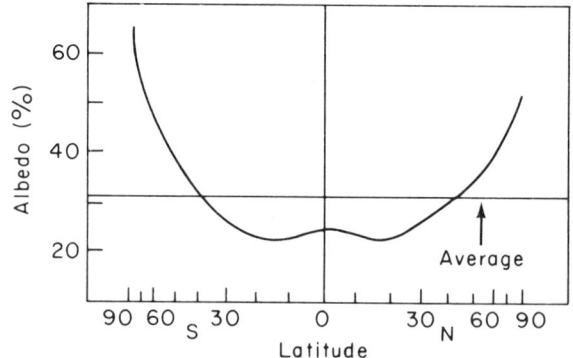

Fig. 2.2. Latitudinal profile of the satellite-measured mean annual albedo (per cent) of the system earth-atmosphere (after 37).

Meridional heat transport

One major purpose of radiation measurements and balances on a global scale is the determination of the meridional heat transport, intimately linked to the observed temperatures of the atmosphere and their seasonal course. Without such transport the tropics would evidently get continually warmer and the high latitudes colder, until some breaking point were reached. The observed heat balance probably lies close to this breaking point, considering that we witness great eruptions of heat transfer with time scale of days from the tropics into higher latitudes, especially in winter, with intermittent very quiet periods. It is easy to determine the heat transport for the whole year when the storage term on the left side of Eq. (2.1) vanishes, or nearly so. Annual transport, satellite-measured and calculated (20), is nearly symmetrical about the equator (Fig. 2.3), with a small displacement of zero transfer to 5°S by the satellite. Since the satellite radiation balance of Fig. 2.1 is interpolated over good portions of the year, the accuracy of this displacement remains uncertain.

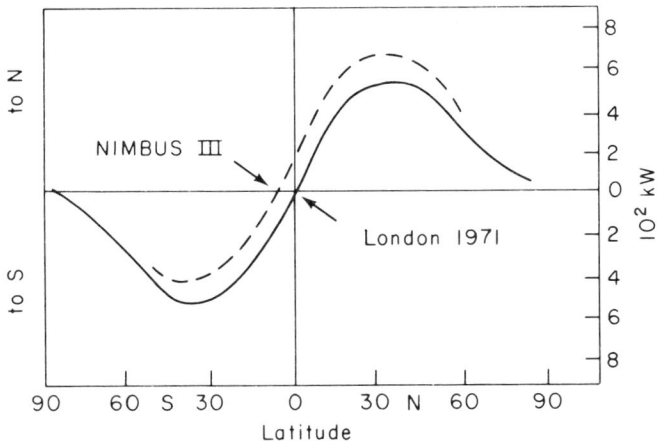

Fig. 2.3. Mean annual latitudinal heat transport (after 20, 37).

If the flux of energy across latitudes 30° took place entirely in the atmosphere, it would warm the extratropical atmospheres by 160°C a year which would balance half of the radiation cooling poleward from 30° and contribute greatly toward maintaining tolerable temperatures for growing seasons outside the tropics. In past years, this conclusion was readily drawn, but with the observation of reduced albedo and increased absorption of solar radiation in the tropical oceans, the picture has become more complicated. Increasingly large fractions of the heat transport are being assigned to the oceans, not only in the subtropics and middle latitudes but also deep within the tropics (38).

For determination of seasonal heat transport, the oceans have always been a major problem because of their large seasonal energy storage or loss. The reader will recall that a water depth of only ten metres corresponds to one atmosphere and that the calorie was defined as the quantity of heat needed to raise the temperature of one gram of water by 1°C under specified conditions. This heat capacity is large compared to that of air (0·24 cal/g) and even water vapour (0·44 cal/g). During summer, most of the energy gained from the sun warms the oceans; during winter, cooling of the oceans will counteract a significant part of the negative energy balance in that season. Therefore, the oceans strongly act to mitigate climate everywhere, as recognized from the earliest days of climatology. Now, to understand the tropics it is essential to determine seasonal energy transfer there. Unless a direct approach is attempted (23), a method must be found that circumvents the large, uncertain, positive and negative oceanic heat storages. One way that has been proposed (27) is to compute latitudinal energy flux at times when oceanic storage is zero and then to assume that the flux so determined may be regarded as valid for a whole season.

Ocean temperatures are well known to lag the solstices by about two months since heat accumulation and depletion, of course, carry far beyond the sun's highest and lowest altitudes. Two vertical lines have been drawn in Fig. 2.1 at the end of August and February when ocean temperatures may be assumed to have attained their extreme values. The profiles of net heating and cooling along these lines then (Fig. 2.4) must yield the meridional heat transfer for the system earth-atmosphere, if steady state is also assumed for the atmosphere; at most a small term is omitted, mainly in high latitudes. Both seasonal profiles are quite symmetrical and nearly identical when referred to the respective winter and summer seasons. Heat storage takes place throughout the year in the equatorial zone. At the end of summer, the poleward limit of heating is near 40° latitude in the Southern Hemisphere and even higher in the Northern Hemisphere where summer energy gain extends briefly to the pole from Fig. 2.1. At the end of winter, the crossover latitude is 30° in both hemispheres.

The transport, or flux of energy is obtained from Fig. 2.4 by integrating from one pole to the other along the two vertical lines in both seasons. If the hypothesis of no storage is correct, all transport must vanish at both poles. This limiting condition was nearly fulfilled

RADIATION, TEMPERATURE AND HUMIDITY 45

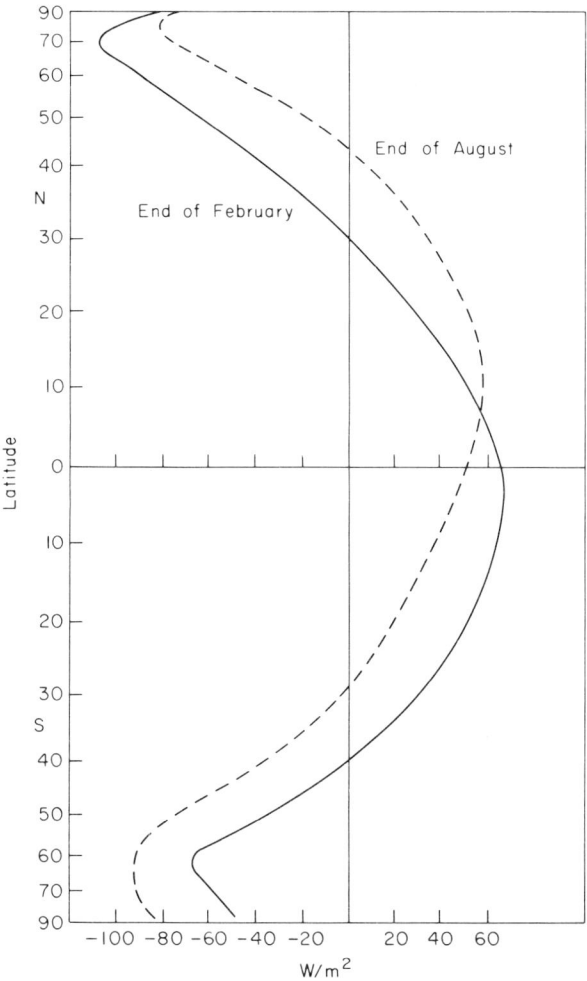

Fig. 2.4. Latitudinal profile of radiational heat storage by the system earth-atmosphere at the end of February and of August, from Fig. 2.1.

in the integration (Fig. 2.5); only a minor adjustment was required at one pole, and both flux profiles are very similar. The crossover latitudes between transport toward north and south are near 15°N in the northern summer and near 10°S in the southern summer, i.e. the crossover latitudes are quite near the equatorial trough positions attained about the time of highest ocean temperatures. The crossover latitude of the two profiles is close to 5°N, the average latitude of the meteorological equator.

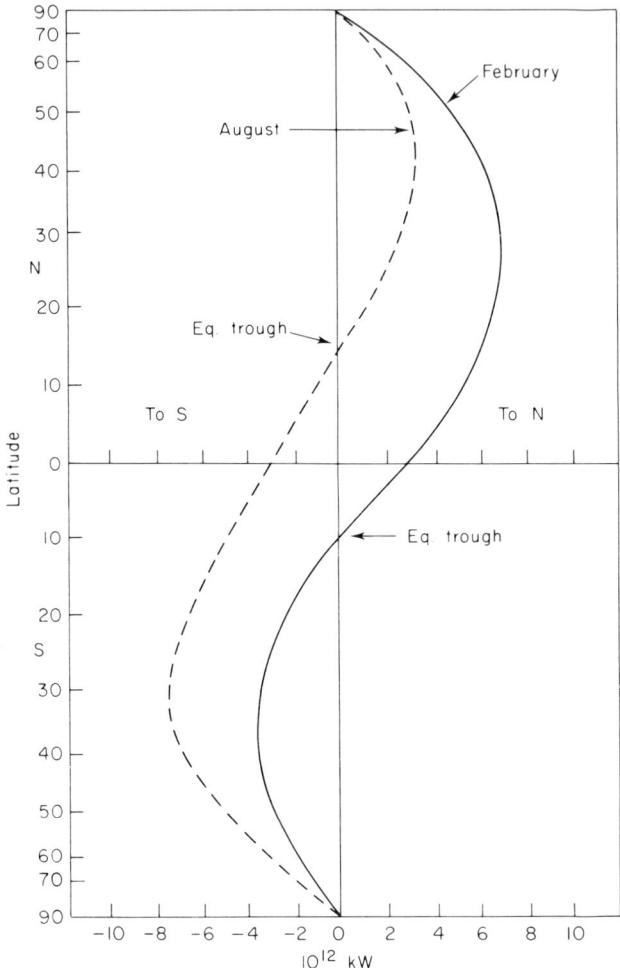

Fig. 2.5. Latitudinal heat transport by the system earth-atmosphere at the end of February and of August, from Fig. 2.4.

In view of the great similarity of the profiles, we may combine them to yield a single profile of meridional heat transport from summer to winter pole (Fig. 2.6), which may be considered representative of summer and winter seasons. In this profile, transport to the winter pole is $7\cdot5 \times 10^{12}$ kW and slightly more than half that amount to the summer pole. Average transport then is 5·8 of the above units, in agreement with the maximum transport rate of Fig. 2.3 and a large percentage increase over flux estimates in the 1950s.

RADIATION, TEMPERATURE AND HUMIDITY

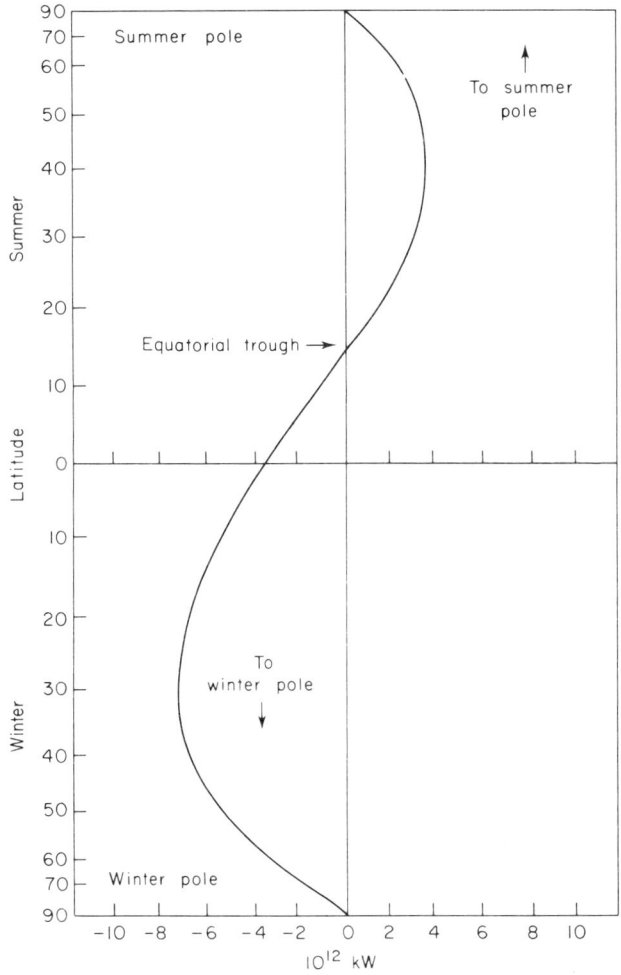

Fig. 2.6. Latitudinal heat transport by the system earth-atmosphere to summer and winter pole determined from Fig. 2.5.

If, from Fig. 2.4, we assess the mean tropical energy source as 60 W/m^2, we can calculate the latitude interval in the tropics, whose energy source must be diverted seasonally from summer to winter hemisphere in order to enable the transports of Fig. 2.6 to take place. This interval is 25° latitude from Fig. 2.5, therefore only slightly more than the estimated equatorial trough displacement. It follows that, in the first approximation, the equatorial trough zone separates energy flow to northern and southern poles; the latitude belt 15°N-10°S can be viewed as being always in the winter hemisphere, changing hands every six months. Very little energy transfer takes place across the trough, whereas the cross-equator transport is large in both seasons, from Fig. 2.5, while balancing to zero for the year. It should also be true, then, that the atmospheric general circulation arranges itself so that equatorial trough position, latitude of warmest air temperature and latitude of warmest ocean temperature all coincide closely at each longitude, without assigning any causal relations between these three quantities. We shall find these suppositions verified in this chapter. Further, the trough migration can now be understood to a large extent by "forcing", due to the meridional temperature gradient and heat flow requirement developing in both hemispheres during the winter season. The temperature gradient doubles over most of the troposphere from summer to winter at latitude 30° (Fig. 2.27) just as the heat transfer does. Thus, the assumption of a gross meridional exchange coefficient $K =$ Transport/Gradient (8) gives a satisfactory picture in principle. Returning to Fig. 1.9, if the amplitude of the seasonal equatorial trough migration equalled that of the sun, a much greater change in heat transfer from winter to summer would take place, and a very different general circulation would be required. However, one may well speculate about variations in the seasonal trough migration as related to climates of the past (18). Further, changes in meridional temperature gradient in different years, and hence different heating requirements of the polar zone, may influence trough migration significantly in one year. Kraus (17) suggests that the great droughts in the Northern Hemisphere tropics in 1972 may have been related to very warm Antarctic conditions in that year, a reduced meridional temperature gradient, reduced trough migration and, hence, low equatorial trough rains. This type of approach to climate changes on short and long time scales does appear to be rationally based and very promising. Gruber (15) has noted considerable fluctuations of the trough over the Atlantic and Pacific Ocean for the period 1967-1970, which supports the concept of short-term unsteadiness of the tropical general circulation.

Surface climate

World distribution of surface temperature

In spite of the complexity of factors governing surface temperature and its variations, a very simple, large-scale climatic pattern prevails. The mean annual temperature range, defined as the temperature

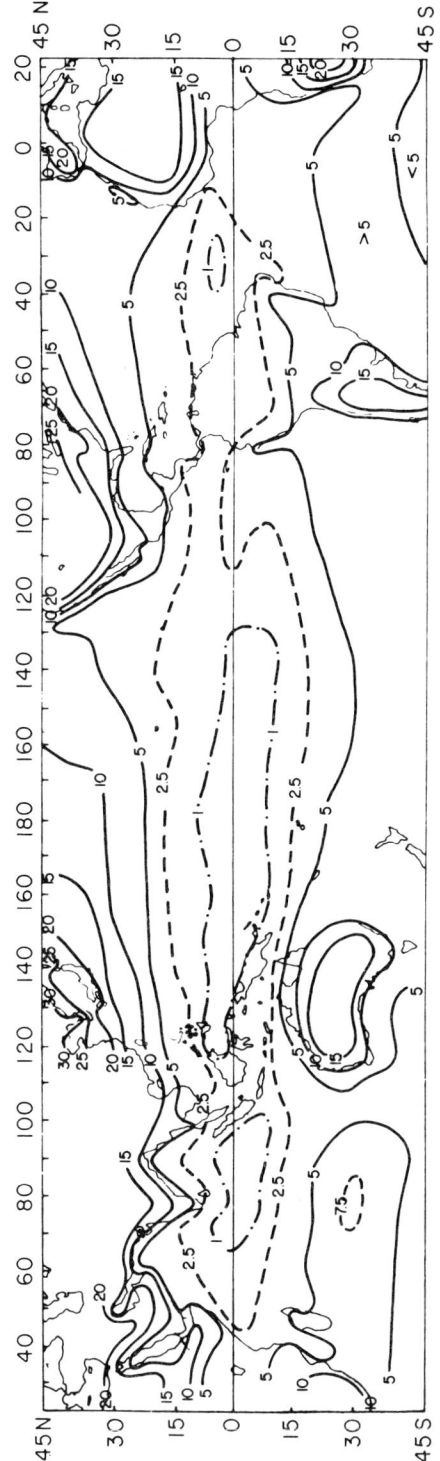

Fig. 2.7. Mean annual temperature range (°C) of the air near sea level.

difference between warmest and coldest month approaches the vanishing point everywhere near the equator (Fig. 2.7). One should not infer from this that the minute changes pass unnoticed by the residents—natives and high-latitude born alike. As the fluctuations of temperature decrease, man's sensitivity to small variations multiplies. A drop of temperature of 1°C may be felt as acutely in the tropics as the change from October to November in northern climates.

It is impossible to guarantee the shape of the 1°C line in Fig. 2.7; even the course of the 2·5°C line is dubious in places. As might be anticipated, the temperature range not only decreases equatorward from middle latitudes, but also on any latitude circle, except quite close to the equator, it is much greater over the continents than over the oceans. The continental range may be 5-10°C or more where the oceanic range is 2-3°C. The relatively small oceanic ranges have been explained as being due to the heat capacity and permeability of the water, the heat transport by the ocean currents and ready energy exchange with the atmosphere. Over land, the temperature range depends greatly on the availability of water for evaporation. Where and when rainfall is plentiful, the temperature is held to relatively low values, not only because of heavy cloudiness and attendant high cloud albedo but also, and more important, because the solar energy absorbed in the ground is returned to the atmosphere through the powerful evaporation mechanism (Fig. 2.8), also observed in the Australian desert (7). When the soil is dry, only contact heating of the air at the earth's surface provides for heat exchange; thus the temperature difference between land and air must become very large, often 20°C and more in daytime in the lowest metre.

Consider the continent of Australia during summer (Fig. 2.9). Over the surrounding oceans, large net heat storage takes place and the ocean current carries the stored heat away. Over land, the heat absorbed in the ground decreases abruptly at the coasts where the albedo rises from small values to 20% and more; but the energy reaching the ground must be returned at once since there is almost no storage capacity and no helpful current to carry any heat surplus elsewhere. In the north, evaporation is the principal mechanism for the energy transfer to the atmosphere. With increasing distance from the coast southward, rainfall and evaporation decrease; hence the sensible heat transport and the temperature difference between ground and air must increase. Quite far south, where there is a plain desert and no rainfall, all heat exchange must be carried out through this process. In the absence of any cooling winds, for instance in the central and southern Sahara, both ground and air temperature will

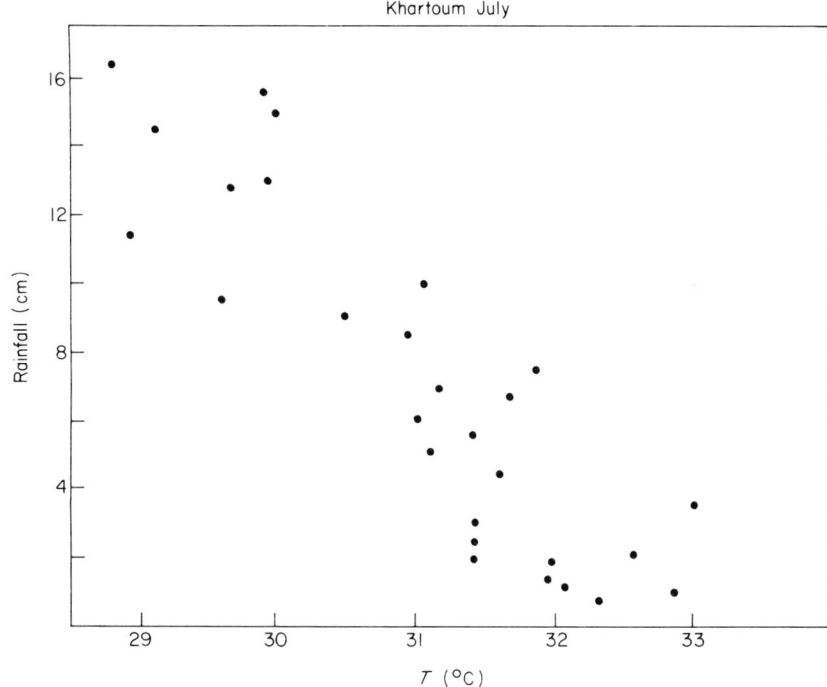

Fig. 2.8. Relation between July rainfall and mean monthly temperature at Khartoum (15°N, 32°E), 1921-1950.

rise to extreme values. In southern Australia temperatures do not rise quite so high, as a broad stream of air from the cold Southern Ocean keeps blowing inland throughout the summer (Fig. 1.7). If a portion of the Sahara was flooded with ocean water as has sometimes been proposed, the very high, daytime, summer temperatures would undoubtedly vanish from this region and its surroundings, as evaporation would become a large factor in the ground-air heat exchange.

Because of the small temperature range, a cursory comparison of the temperature field in different seasons cannot reveal changes as striking as the wind field. Therefore, only a map of mean annual surface temperature (reduced to sea level) is offered (Fig. 2.10). The isotherms so nearly follow the latitude circles that the predominant solar control of the temperature climate is placed beyond doubt. Various anomalies exist, however. Over the continents the isotherms bulge poleward for the reason just advanced: the requirement for large sensible heat exchange and attendant ground warming since the moisture supply for evaporation is not as ample as over sea.

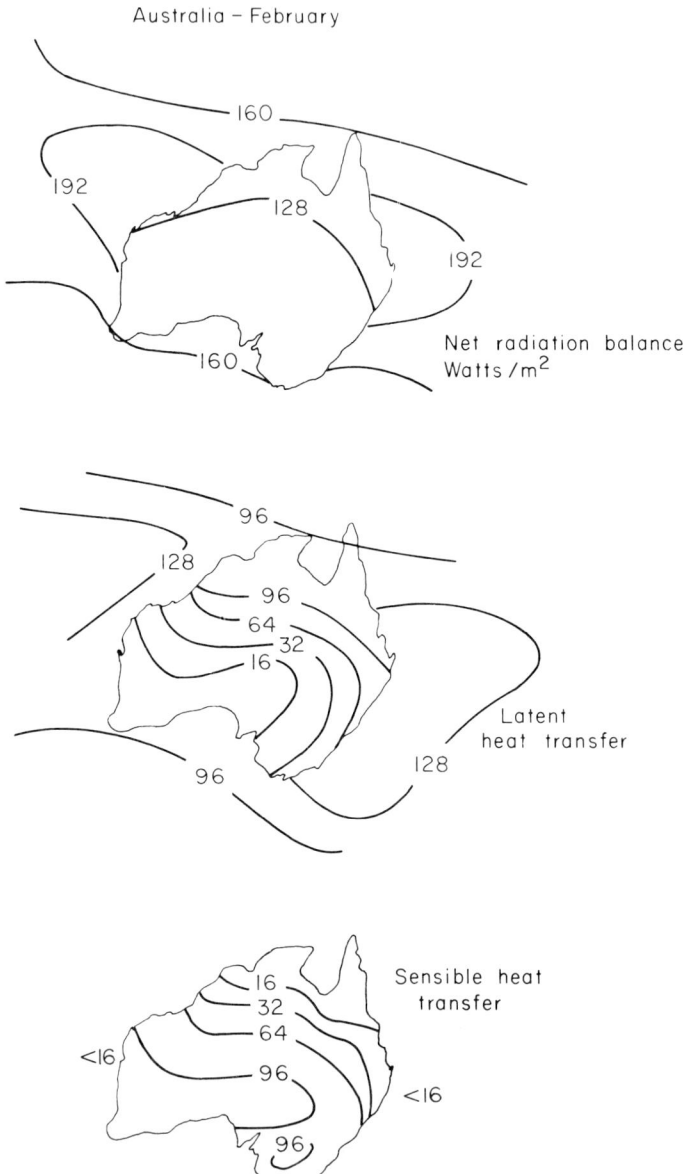

Fig. 2.9. Net radiation (Watts/m²) absorbed by Australia and surrounding oceans during February; transfer of latent and sensible heat from ground to atmosphere (4, 5).

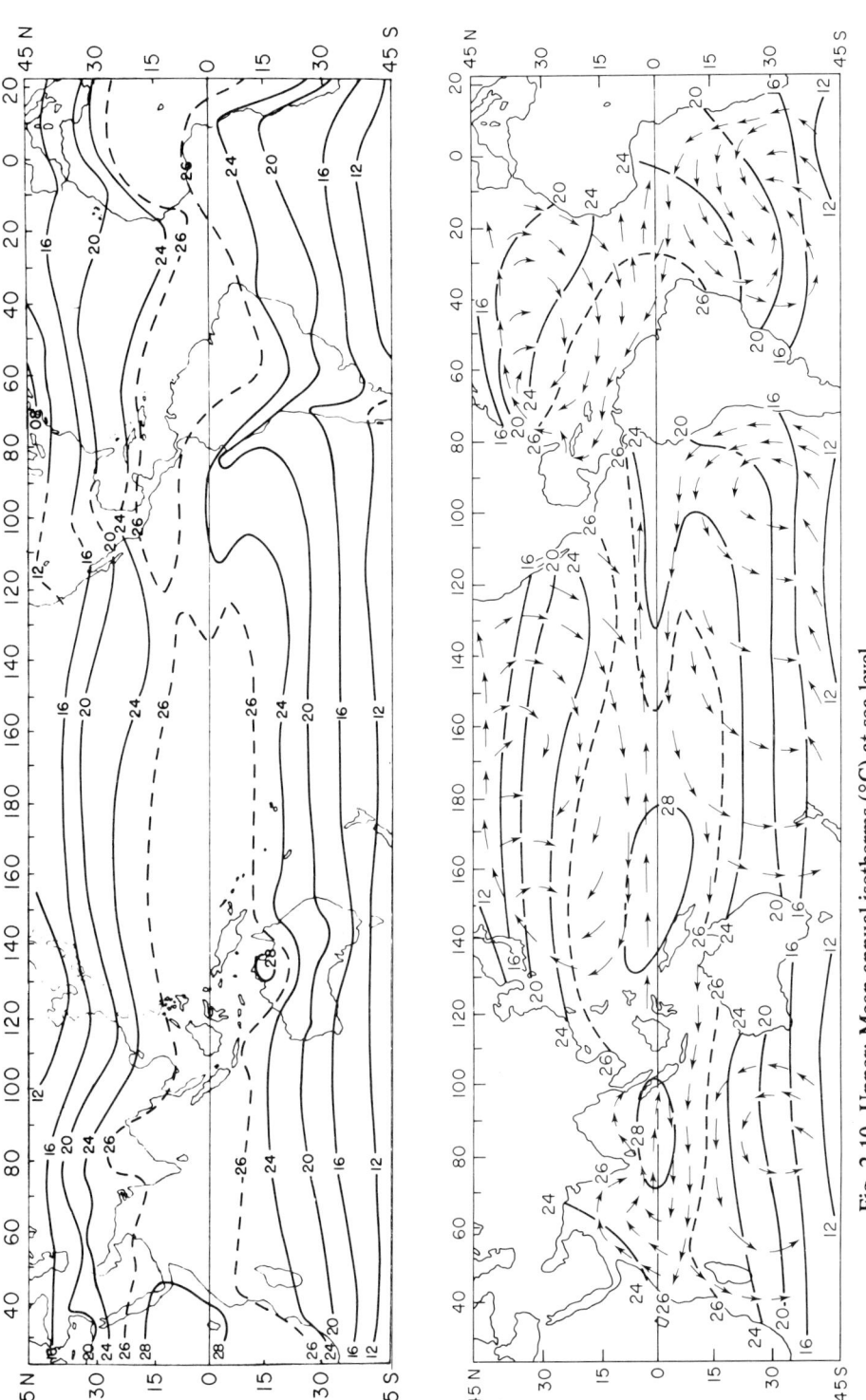

Fig. 2.10. Upper: Mean annual isotherms (°C) at sea level.
Fig. 2.11. Lower: mean annual sea surface temperatures and sketch of ocean currents, August (9, 35).

Over the oceans the outstanding fact is the equatorward trend of the isotherms from west to east (Fig. 2.11). In broad terms the isotherm orientation can be related to the wind circulation. The strongest equatorward transport of air by the trade winds takes place in the eastern parts of the oceans; this air, coming from middle latitudes, is relatively cold. In addition, air and ocean surface temperatures are highly correlated. Mean temperature differences between sea and air greater than 1°C are rare and it is impossible to draw reliable lines of this temperature difference. The largest part of the tropical oceans appears to be slightly warmer—near 0·5°C on average—than the air, both in the annual mean and in seasonal breakdowns. Thus, sea and air are very nearly in thermal equilibrium.

Approximate vectors showing the surface ocean currents have been superimposed on the isotherms of Fig. 2.11. These follow Sverdrup (35) and Defant (9). Where large seasonal changes occur, the vectors for August were chosen to depict the equatorial countercurrents plus the monsoon-produced flow from west in the North Indian Ocean and the strong winter circulation off the west coast of South America. The main objective is to observe the poleward heat transport by the ocean currents. Such transport, which continues through the seasons, is evident in all oceans except the North Indian Ocean as far as the belt 10-15° latitude. Cold water advection into the tropics extends even deeper along west coasts where the well-known "upwelling" contributes to the low temperatures. Largest seasonal change of ocean currents takes place in the North Indian Ocean where the current changes with the wind stress imposed upon the water by the reversing monsoon winds.

Seasonal march of surface temperature

Global picture

Since the isotherms of the air near the surface run nearly parallel to the latitude circles, we can average temperatures around the globe as we did for pressure (Fig. 1.4). On a graph of mean seasonal temperature against latitude (Fig. 2.12) the air is warmest near 5°S in January and at 20°N in July, the latter an effect of the heating of the large, subtropical deserts. The crossover point between the January and July curves is near 5°N. From there, the seasonal temperature range increases poleward, but at a more rapid rate in the continental Northern Hemisphere than in the maritime Southern Hemisphere. The range becomes 13°C at 30°N compared to 7°C at 30°S. Mean annual temperatures are distributed fairly symmetrically, not with respect to the geographical equator, but with respect to the meteorological or "heat" equator—5°N. Throughout the tropics the Southern Hemisphere is slightly cooler than the Northern Hemisphere: 1·4°C at latitude 10°, 2·3° at 20° and 1·9°C at 30°.

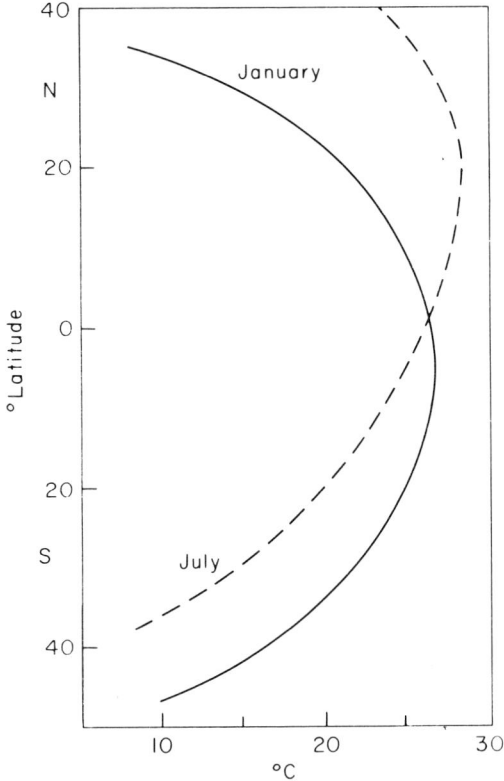

Fig. 2.12. Mean sea level temperature profiles (°C).

As anticipated, the latitudes of warmest temperature and of equatorial trough location closely coincide in both seasons. Clearly, the trough is not a boundary between cold winter and warm summer hemisphere air, so that the "intertropical" feature becomes immaterial seen from the thermal viewpoint. The trough migrates so far that air from the winter hemisphere on a long tropical trajectory draws even in temperature with summer hemisphere air on a short trajectory; the trough position is determined by this.

A revealing picture of some marked longitudinal variations is furnished by an east-west cross-section at 15°S (Fig. 2.13).

Outstanding are the tremendous depressions of temperature just west of the African and American continents, accentuated in winter. They result partly from the equatorward transport of polar maritime air over the eastern parts of the oceans and partly from upwelling of cold water. Since both South America and South Africa protrude westward with decreasing latitude, in contrast to North America and North Africa, and since the strength of the mean winds at the coasts does not decrease from summer to winter as in the Northern Hemisphere, the effect of upwelling is greatest in the Southern Hemisphere. It extends to much lower latitudes and the subtropical dryness is also carried farthest equatorward there.

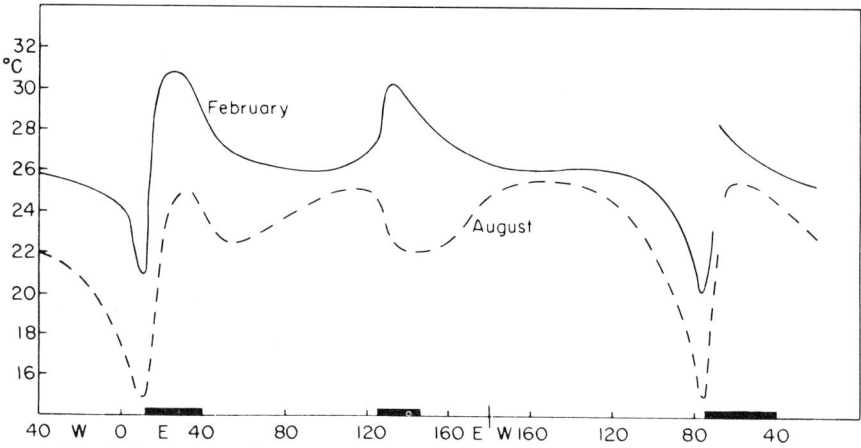

Fig. 2.13. Longitudinal profile of mean sea level temperature (°C) at 15°S.

Elsewhere Fig. 2.13 confirms the features of the temperature distribution discussed: the small temperature range over the oceans compared with the continents (especially Africa and Australia), the peak temperatures over land and the gradual lowering of temperature from west to east across the oceans, evident everywhere, except in the Indian Ocean in August where temperature in the west is strongly depressed by the rapid monsoon current from the Southern Hemisphere (Fig. 1.7).

Cross-sections of individual stations

Next we glance at the seasonal march of temperature at individual places. Our interest lies in discovering something about the typical behaviour of the temperature cycle; climatography is not the

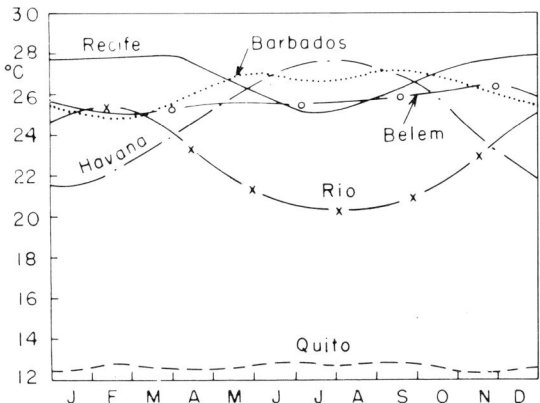

Fig. 2.14. Seasonal course of mean monthly temperature (°C) along the eastern shore of the Americas: Havana (24°N, 82°W), Barbados (13°N, 59°W), Belem (1°S, 48°W), Recife (8°S, 35°W), Rio de Janeiro (23°S, 43°W). Also Quito, Ecuador (0°, 78°W; elevation 2800 m).

objective. The first section extends along the east coast of South America to the Antilles, the second north-south in the middle of the Pacific Ocean and the third north-south across Africa.

Both on the South American coast, exposed to the easterly trades of the Atlantic (Fig. 2.14), and in the Pacific (Fig. 2.15) we should look for "oceanic"-type temperature curves in the subtropics with late summer maxima and late winter minima. Such curves are found at Rio de Janeiro and at Havana, its Caribbean counterpart, also at Tahiti and Honolulu, in the South and North Pacific. In accentuated

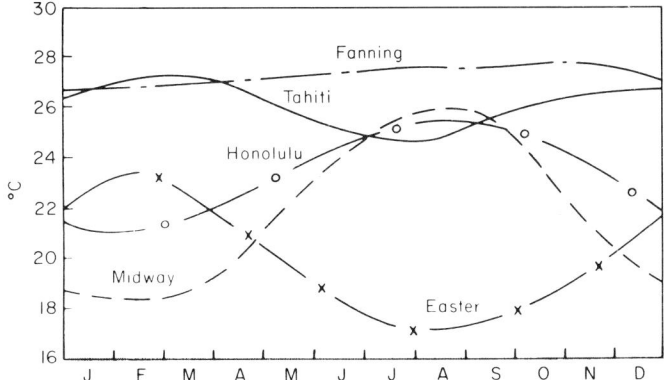

Fig. 2.15. Seasonal course of mean monthly temperature (°C) in the east central Pacific Ocean: Easter Island (29°S, 109°W), Tahiti (18°S, 148°W), Fanning Island (4°N, 159°W), Honolulu, Hawaii (21°N, 158°W), Midway Island (28°N, 177°W).

form, similar curves appear at Midway and Easter Islands, located near the equatorial margin of the belts of polar westerlies in the Pacific.

Closer to the equator the subtropical summer maxima associated with the main rainy season should flatten. This is observed at Barbados where even a slight secondary summer minimum appears. Near the equator itself, at Fanning Island and Belem, the curves are quite indifferent, as also at Quito, Ecuador, situated at about 3000 m altitude in the Altiplano of the Andes. Mean temperature there is low, yet about 2°C more than free air temperature at the same height. The temperature curve, one of the steadiest in the world, oscillates irregularly over a range of about 1°C. Combination of the agreeable mean temperature with so small an annual range, yet with warm afternoon temperatures, gives Quito one of the most desirable climates in the tropics.

If the mean temperature on the equator varied strictly with the sun, there should be two minima at the solstices and two maxima at the equinoxes. Cloudiness and rainfall, however, enter as factors of equal importance. In the rainy season, frequent daytime showers hold the maximum temperatures down; a corresponding increase of minimum temperature due to reduced nocturnal radiation is seldom found. Changes in the maximum temperature largely determine the course of the mean temperature, which then has a tendency to drop during the rainy season.

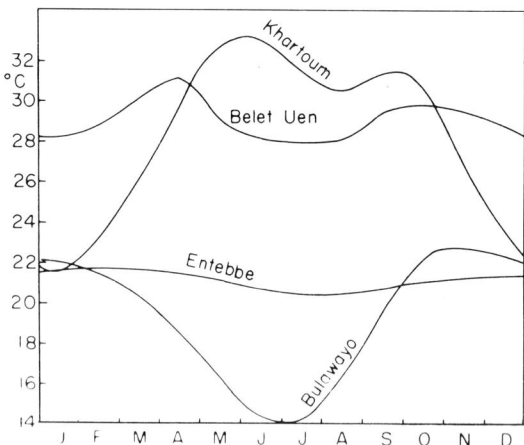

Fig. 2.16. Seasonal course of mean monthly temperature (°C) in Africa: Bulawayo (20°S, 29°E; elevation 1350 m), Entebbe (0°, 32°E; elevation 1150 m), Khartoum (15°N, 32°E; elevation 370 m). Also Belet Uen (5°N, 45°E; elevation 180 m).

RADIATION, TEMPERATURE AND HUMIDITY

The third section, oriented north-south through the mountainous interior of Africa near 30°E (Fig. 2.16) departs strikingly from the oceanic regimes. Khartoum and Bulawayo have strong seasonal temperature changes; the minima quite properly fall in the centre of the respective winters. In these dry locations, all cloudiness and rainfall are concentrated in a brief summer period. Temperatures rise sharply through the spring months, but this increase ends abruptly with the advent of the rainy season. Then, temperatures drop under the influence of the increased cloud cover, even though the precipitation may be quite small. Khartoum also has a secondary maximum after the monsoon rains end, a feature widely prominent in the monsoon regime, as evidenced by Bombay and Nagpur in India (Fig. 2.17). At the interior station, Nagpur, the temperature drop is much larger than at Bombay at the western coast with sea breeze influence. The more continental climate of Nagpur is also marked by lower winter temperatures, hence, by an annual temperature range well over double that of Bombay. Whether temperature actually declines or whether there is just that feeling of cold and dampness associated with long cloudy periods, the rainy season has come to be named "winter" in such widely different countries as Ethiopia and Venezuela.

Entebbe, situated at the northern shore of Lake Victoria, lacks a marked temperature cycle, as is typical for the other equatorial stations. On the equator the difference between continental and

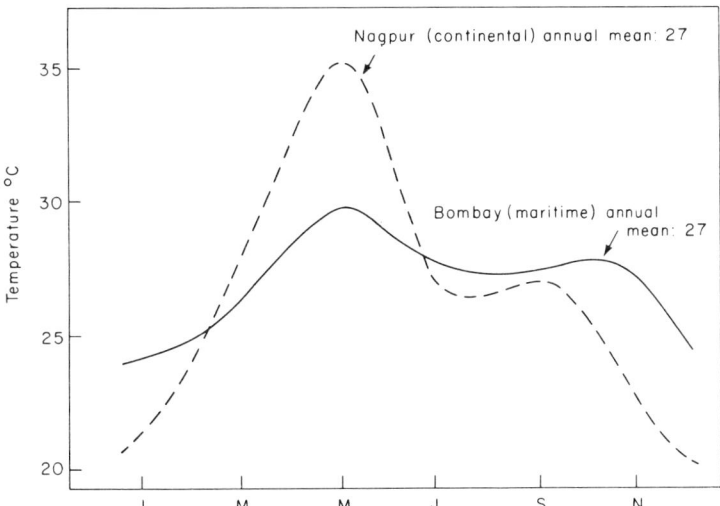

Fig. 2.17. Seasonal course of surface air temperature at Bombay (19°N, 73°E), on the west coast of India, and at Nagpur (21°N, 79°E), in the middle of the subcontinent.

oceanic regimes vanishes. However, at Entebbe the rainy season coincides with the highest temperature, while the cooler season is the dry season. At that time of year, southerly winds, which form the western edge of the Indian Ocean monsoon flow (Fig. 1.7), blow persistently from the expanse of Lake Victoria to Entebbe. The air is cooled by contact with the lake surface and reaches the city as a lake breeze.

The temperature trace from Belet Uen, located on the barren desert shore of East Africa near 5°N, emphasizes this point. Superficially, this trace looks fairly similarly to that of Khartoum. Yet the explanation of the low May to August temperatures must be different since Belet Uen is completely dry in summer. Again, the cause is the strong monsoon current across the equator which advects cool water and, in addition, forces upwelling of cold ocean water along the coast, north of the equator. The cold air advection is borne out at other places, such as Zanzibar (6°S, 39°E) and the Seychelles Islands (5°S, 55°E) which, without upwelling, also report lower temperatures from May to August. The upwelling along the shore north of the equator intensifies with increasing latitude. Near Cape Guardafui (12°N, 51°E) there exists a fog regime which is fully comparable with those of the subtropical west coasts. Cold coastal water below 20°C(!) extends right along the south Arabian shore.

These examples should emphasize that only a careful study of local conditions can lead to an interpretation of the details of the temperature climate near the equator. Many local factors come into the foreground and one cannot lay down any rules on the relations between temperature, wind and rainfall, which are valid everywhere.

Upper air climate

Mean vertical structure

Over the tropical oceans, there is a large difference in thermal structure between the eastern and western portions. Over the eastern parts we encounter a temperature inversion in the lower atmosphere, the trade wind inversion, generally situated below three kilometres. In the equatorial trough zone and over the western parts of the trade wind belts an inversion does not exist as a mean condition, though stable layers appear in specific weather patterns and may be a mean condition in the dry seasons. During the rainy seasons the lapse rate is nearly moist-adiabatic through the bulk of the troposphere, including the land areas, above the convective condensation level, indicative that deep convection is the mechanism that has produced this vertical structure. While regional variations in temperature do occur, they are sufficiently small, apart from northern India and the mountainous plateau to its north, for it to have proved useful to define an oceanic tropical standard atmosphere, as distinct from the international standard atmosphere based on data near 50°N in Europe. Even though several mean atmospheres have been defined (6, 25, 30) and none has achieved official blessing, the widely-used mean atmosphere for the Caribbean rainy season computed by Jordan (16) will be presented as a good model guide (Fig. 2.18, Table 2.1). The temperature lapse rate exceeds the moist-adiabatic rate to about 600 mb close to the freezing level; a moist-adiabatic layer follows to 450 mb and then stability gradually increases upward, far below the tropopause near 100 mb. The pressures where the stability increases and where the return current of the meridional circulation begins (Fig. 1.18) are nearly identical, a fact that is more than coincidental. Defant and van de Boogaard (10) find a more sharply defined break in lapse rate sloping upward from around 240 mb at latitude 30° to 120 mb at the equator. This division is suggestive of what has been called a marked stable layer or secondary tropopause generally found in synoptic rain systems (14).

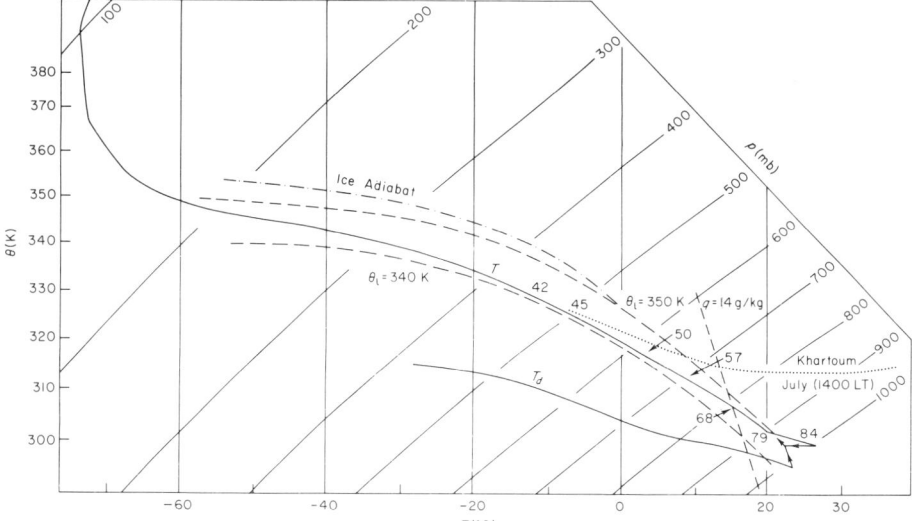

Fig. 2.18. Tephigram of the mean atmosphere for the Caribbean during the rainy season. Temperature and dewpoint solid, θ_e at 340 K and 350 K dashed; also mean July noon sounding at Khartoum to 500 mb. Numbers along the T-curve are relative humidity (%).

Table 2.1. Mean tropical atmosphere during the West Indies rainy season (based on night-time soundings only (16))

p (mb)	H (m)	T (°C)	θ (K)	q (g/kg)	RH (%)	θ_e (K)	ϱ (kg/m³)
80	17 883	−69·8	418			418	0·137
100	16 568	−73·5	386			386	0·174
125	15 260	−72·2	364			364	0·217
150	14 177	−67·6	354			354	0·254
175	13 236	−61·5	348			348	0·288
200	12 396	−55·2	345			345	0·320
250	10 935	−43·3	342			342	0·379
300	9 682	−33·2	338			338	0·434
350	8 581	−24·8	335			336	0·490
400	7 595	−17·7	332			335	0·545
450	6 703	−11·9	328	1·4	42	333	0·599
500	5 888	− 6·9	324	2·1	45	332	0·653
550	5 138	− 2·5	321	2·8	47	330	0·707
600	4 442	+ 1·4	318	3·6	50	329	0·760
650	3 792	+ 5·1	315	4·6	54	329	0·811
700	3 182	+ 8·6	312	5·8	57	330	0·862
750	2 609	+11·8	309	7·1	61	331	0·913
800	2 083	+14·6	307	8·4	68	333	0·964
850	1 547	+17·3	304	11·0	74	337	1·013
900	1 054	+19·8	302	13·0	79	340	1·062
950	583	+23·0	300	15·3	81	344	1·108
1000	132	+26·0	299	17·6	84	349	1·152

θ_e at 400 and 350 mb interpolated.

In the moisture profile, a sharp break between a lower moist and upper dry layer is evident on many individual days, even though the temperature lapse rate does not have any stable layer. In the mean sounding this distinction is obliterated, so that specific humidity decreases steadily upward, a little below and evidently controlled by the saturation specific humidity (Fig. 2.19). Over the layer from 900-600 mb $\bar{q}_s - \bar{q}$ remains fairly constant near 4 g/kg, then decreases slowly upward.

Structure of the equatorial trough zone

We now turn to coordinates following the equatorial trough and examine its structure for July following the positions given in Fig. 1.10. In this month, in the Northern Hemisphere summer, some differences are found between the "continental" half of the trough,

RADIATION, TEMPERATURE AND HUMIDITY

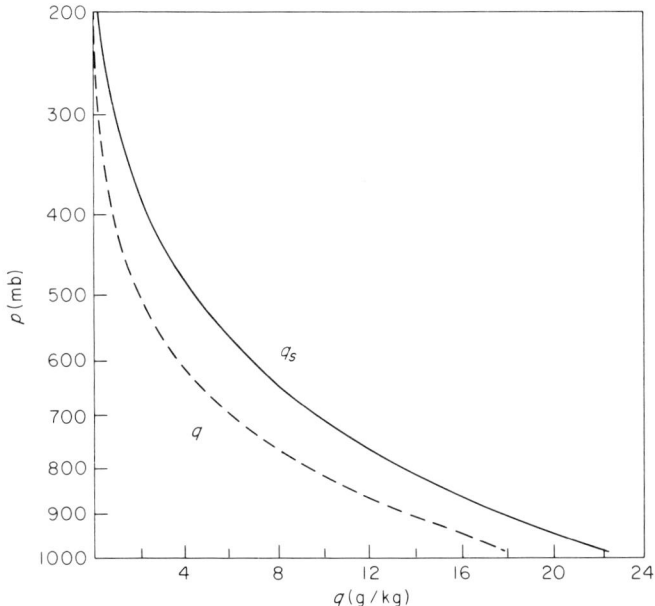

Fig. 2.19. Saturation specific humidity (q_s) and actual specific humidity (q) of the mean tropical atmosphere (Table 2.1).

taken from 30°W to 150°E, the interval with high tropospheric easterlies (Fig. 1.14), and the complimentary "oceanic" half over the Atlantic and Pacific Oceans. Therefore, cross-sections of temperature and specific humidity are presented separately in the form of deviations from the average from 20°N to 20°S of the troughline at each level (Fig. 2.20). Mean differences between the continental and oceanic halves are given in Table 2.2.

Temperature is warmest near the troughline at low levels, with displacement to the north over the continental section, related to the heat of the subtropical deserts and the elevated heat source across most of Asia (cf. also Figs 2.25 and 2.26). In the high troposphere and low stratosphere the well-known compensating layer is found, with cold temperatures above the equatorial trough zone and warmer temperatures toward the polar regions. On the south side, the horizontal temperature gradient is no larger than in the north, even though it is winter there. This is related to the position of the equator within the southern portion, where maintenance of a meridional temperature gradient is dynamically impossible.

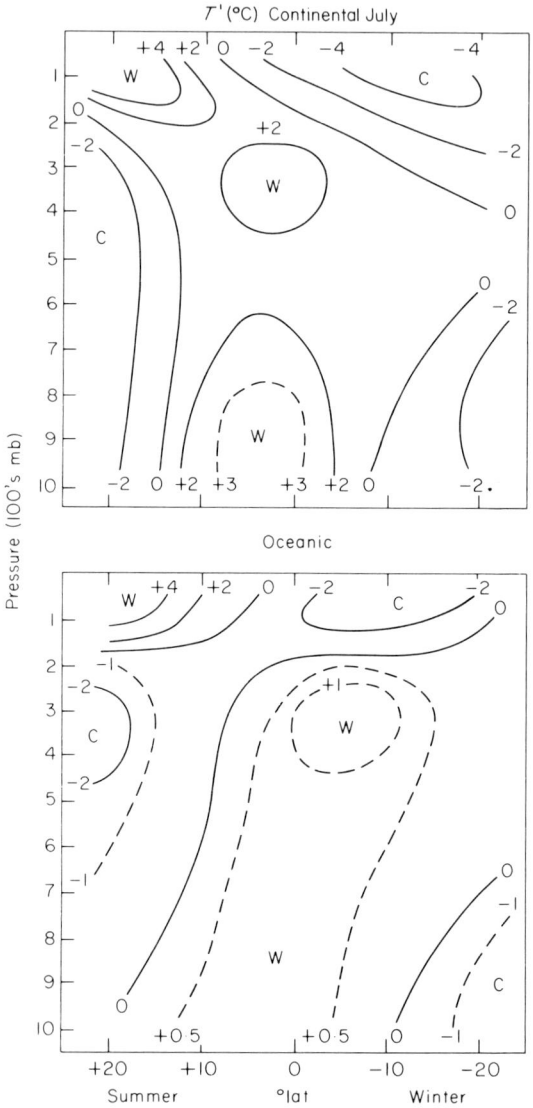

Fig. 2.20. Vertical cross-sections of temperature anomaly (°C) and of specific humidity anomaly (g/kg) across the equatorial trough in July for land and ocean halves. Data from van de Boogaard (A-23).

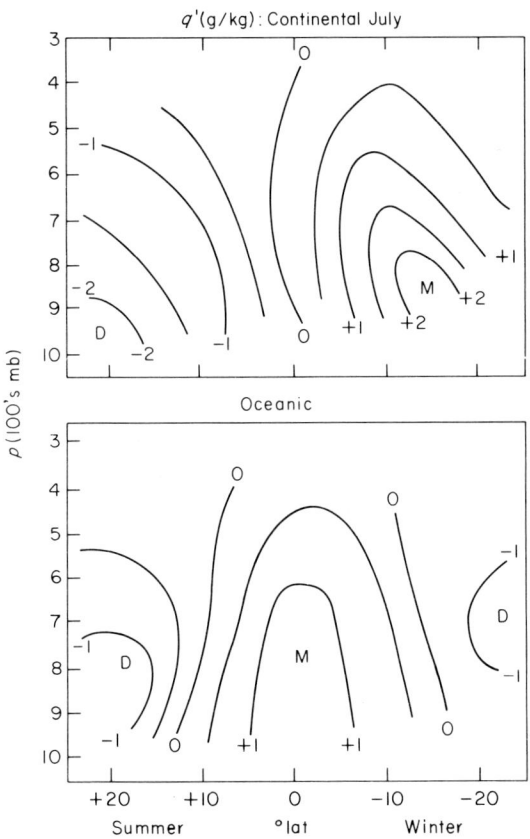

Fig. 2.20 continued.

The continental half is warmer throughout the troposphere, but, as noted from Table 2.2, the large surface difference decreases upward to 500 mb where it is almost zero, succeeded by another belt with 3°C differences in the high troposphere. Most interesting is the fact that both continental and oceanic halves have a "warm core" structure of the trough, though the locations of warmest air and heaviest rain need not be identical. The warm core is also found when averages are taken over smaller longitude intervals, say from 30-100°W: the Atlantic Ocean. Nowhere is there an indication of an average cold core for the season, though on individual synoptic maps cold upper Lows do occur just above the surface convergence zone. In limited areas where a strong trade wind inversion borders on the equatorial trough zone— eastern Atlantic off Africa in summer (31), eastern equatorial Pacific (32)—temperatures in the inversion zone or zones are warmer than those of the equatorial trough zone for a layer of limited depth (Fig. 2.21). These narrowly confined anomalies are washed out in the

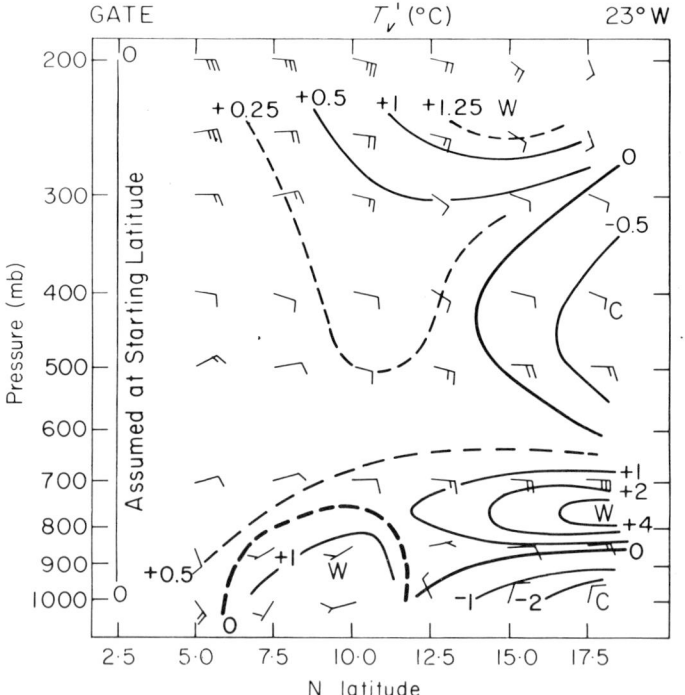

Fig. 2.21. Vertical cross-section of temperature (°C) and wind (knots) along longitude 23·5°W August-September 1974. Temperature deviations from latitude 2·5°N. Based on actual temperature fields and isotherms in (31) surface—850 mb. Computed geostrophically 850-200 mb.

integration over large longitude intervals. In the GATE area the meridional temperature gradient was surprisingly small (11), even though the temperature profile is entirely normal in the subcloud layer.

Table 2.2. Mean soundings from 20°S to 20°N across equatorial trough

p (mb)	Land T(°C)	Land q (g/kg)	Ocean T(°C)	Ocean q (g/kg)	Δ (land-ocean) T(°C)	Δ (land-ocean) q (g/kg)
100	−83·1		−83·8		+0·7	
150	−63·7		−67·1		+3·4	
200	−51·2		−54·6		+3·4	
300	−30·9		−32·8		+1·9	
500	− 6·2	2·2	− 6·3	1·7	+0·1	+0·5
700	+10·3	5·4	+ 9·1	4·8	+1·2	+0·6
850	+19·9	9·3	+16·6	9·6	+3·3	−0·3
Surface	+28·9	15·7	+25·9	17·2	+3·0	−1·5

	Precipitable water (cm) Land	Precipitable water (cm) Ocean	Precipitable water (cm) Δ (land-ocean)	Mean
20°N of Trough	5·4	5·8	−0·4	5·6
10°N of Trough	5·7	6·7	−1·0	6·2
Trough	6·7	7·5	−0·8	7·1
10°S of Trough	7·9	7·0	+0·9	7·4
20°S of Trough	7·2	5·6	+1·6	6·4

Specific humidity anomalies closely follow those of temperature over the oceanic section, while over the continental portion the centre of positive anomalies is displaced well to the south of the troughline, as is rainfall also (Chapter 3). In this broad band, also characterized by the largest expanse of surface equatorial west winds (Fig. 1.7), most moisture convergence and the track of synoptic weather systems is located asymmetrically in the belt with low-level equatorial westerlies (13, 28).

From the surface pressure profile (Fig. 1.11) and the preceding cross-sections we should expect hydrostatically that high pressure must overlie the equatorial trough zone in the upper troposphere. That this is the case is confirmed in Fig. 2.22 where the deviation pattern of the height of isobaric surfaces is given for the whole equatorial trough. The level of reversal from lower to upper pressure gradient field lies near 500 mb, which is also in the middle of the layer of non-divergence (Fig. 1.18) separating lower inflow and upper outflow.

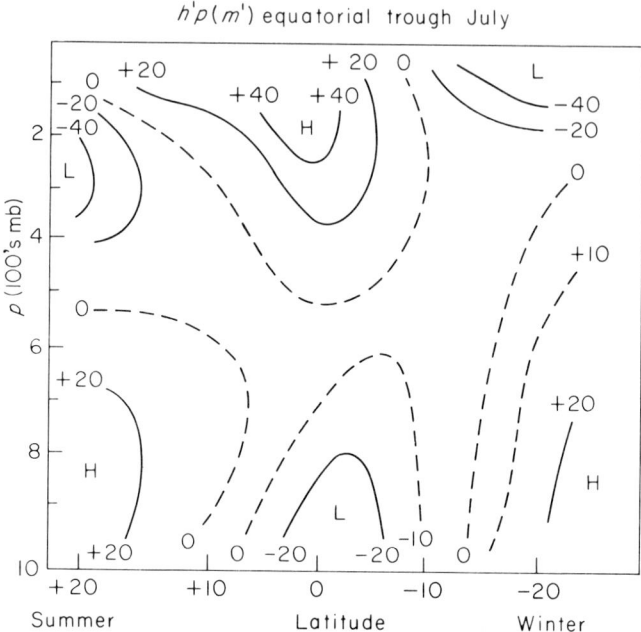

Fig. 2.22. Height anomalies (m) of constant pressure surfaces for the whole equatorial trough in July.

Thus, these mean fields are quite consistent and indicative of a "simple heat engine" type of general circulation for the tropics. Difficulties encountered with this classical picture will be brought up in subsequent chapters. That there is a problem is immediately evident from Figs 2.23 and 2.24 which depict the vertical distribution of thermodynamic energy, sometimes called "static" because kinetic energy is omitted. We define this energy

$$Q = gz + c_p T + Lq ,\qquad(2.2)$$

where g is acceleration of gravity, z height, c_p specific heat of air at constant pressure, L latent heat of condensation or vaporization and q specific humidity. Then gz is potential energy, $c_p T$ specific enthalpy and Lq latent energy. In the absence of outside heat sources and sinks, mainly radiation and ground-air energy transfer, Q is conserved by individual masses of air which, during vertical motion, act to establish the nearly moist-adiabatic structure of the mean atmosphere. However, from Figs 2.23, 2.24, Table 2.3 and the distribution of equivalent potential temperature (see Appendix to chapter) in Table

RADIATION, TEMPERATURE AND HUMIDITY

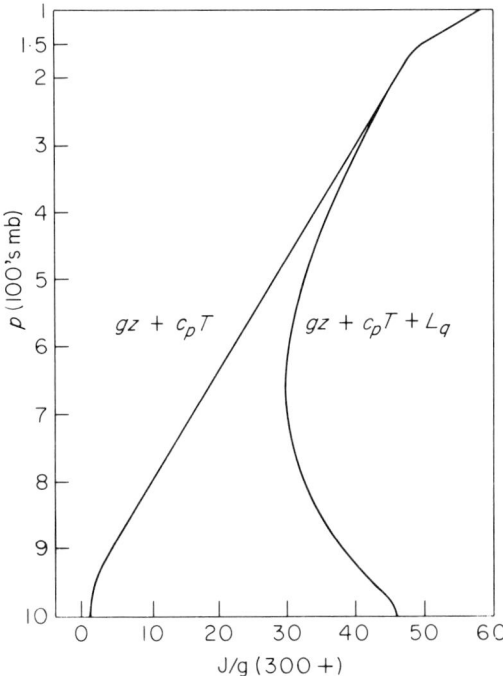

Fig. 2.23. Vertical profiles of thermodynamic energy (J/g) at the equatorial trough.

2.1, a remarkable fact is apparent: the thermodynamic energy has a mid-tropospheric minimum which holds not only for charts and tables, but also for nearly all of the well over a million individual radiosonde observations taken in the tropics. The mean ascent inferred from Fig. 1.18 for the inner equatorial trough zone obviously is in conflict with the Q-profiles. Further, the shape of the profiles above 600 mb runs entirely counter to the concept of energy diffusion from the energy source which is at the earth's surface. If a "normal" diffusion process away from the source is assumed, then total thermodynamic energy should decrease throughout the whole troposphere, proportional at least to net radiation cooling. The actual profiles foreshadow the necessity to introduce a different concept for heat transfer in the equatorial trough zone, transfer through "negative viscosity" (Chapter 4). In the cross-section of Fig. 2.24 the mid-tropospheric minimum is in evidence across the entire belt; it sinks and weakens toward the outskirts. At higher latitudes, thermodynamic energy increases upward all the way from the surface on average.

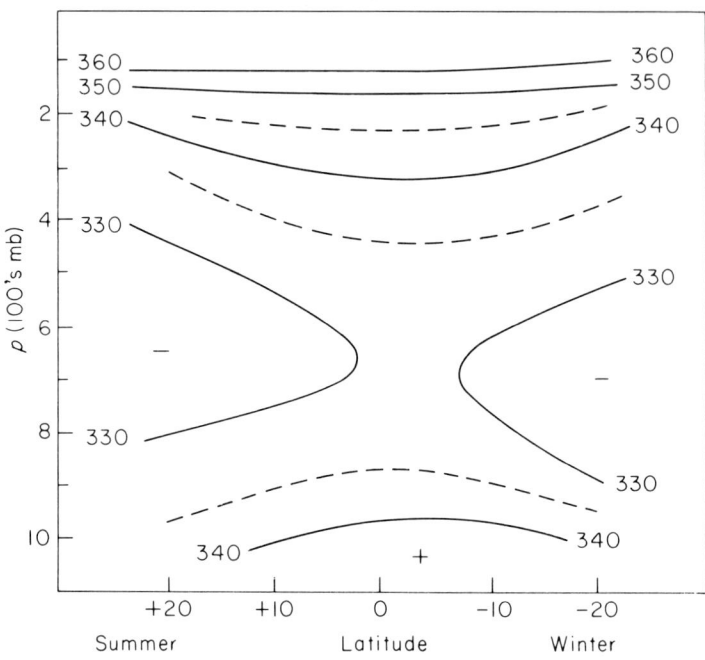

Fig. 2.24. Cross-section of thermodynamic energy (J/g) across the whole equatorial trough in July.

Table 2.3. Mean energy content of air (j/g) above equatorial trough

p (mb)	gz	$c_p T$	$gz + c_p T$	Lq	$gz + c_p T + Lq$
100	160	197	357		357
150	142	206	348		348
200	124	221	345		345
300	97	242	339	1	340
400	76	257	333	3	336
500	59	268	327	5	332
600	44	278	322	9	331
700	31	285	316	15	331
800	20	290	310	22	332
850	15	292	307	26	333
900	10	294	304	33	337
950	5	296	301	40	341
1000	1	300	301	45	346

RADIATION, TEMPERATURE AND HUMIDITY

Regional temperature contrasts

Examination of regional temperature differences on a latitude basis further emphasizes the different structure of the continental and maritime halves of the Northern Hemisphere tropics in summer, made very evident by the meridional profile of 200 mb temperature within each region (Fig. 2.25). Over the western Atlantic temperature is almost uniform indicating the level of temperature reversal; there is a region of slightly cooler temperature between 20° and 30°N, where the mean upper tropospheric trough is situated in Fig. 1.13. In contrast, we see a huge rise of 200 mb temperature along 80°E from the equator to latitude 35° where the highest, seasonal, high-tropospheric temperatures of the world are found. The same contrast appears along the vertical comparing, for instance, Gan, just south of the equator, and New Delhi (Fig. 2.26). The mean July temperature profile at Gan follows the oceanic tropical atmosphere (Table 2.1) almost precisely up to 200 mb. At New Delhi, temperature is relatively warm compared to the oceanic temperatures of the mean atmosphere in the low levels, as expected. In contrast to the mean picture (Table 2.2) where the ocean-continent temperature difference becomes zero at 500 mb,

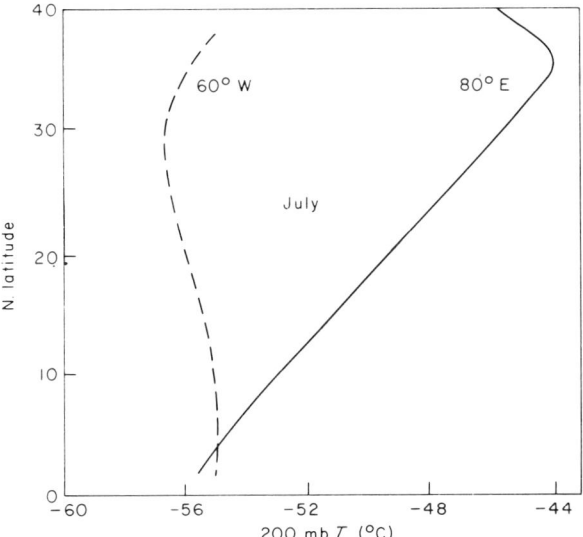

Fig. 2.25. Comparison of latitudinal 200 mb temperature profiles at 60°W and 80°E in July.

however, New Delhi temperatures depart increasingly with height from the oceanic atmosphere until a grand maximum of over 9°C is reached near 200 mb. This is the largest temperature contrast found in the tropical troposphere anywhere.

Let Khartoum represent the very large number of stations where the continent-ocean contrast does disappear in the middle troposphere. In Fig. 2.18, the mean July sounding at the time of maximum temperature is compared with the oceanic tropical atmosphere. In the desert, as we have seen, surface heat exchange is accomplished mainly through heating of air by sand. Thus, the temperature lapse rate, even of the monthly sounding, is superadiabatic in the lowest 500-1000 m, then dry-adiabatic to about 750 mb. Individual soundings show more extreme conditions when surface temperature rises well above 40°C.

The outstanding fact of Table 2.2 and Fig. 2.26 now is that, in contrast to winter, a second layer with positive temperature anomaly of the land area reappears above 500 mb with maximum at 200 mb. In Fig. 2.26, the Khartoum and New Delhi profiles are quite parallel; only the Khartoum departure from the oceanic sounding is smaller.

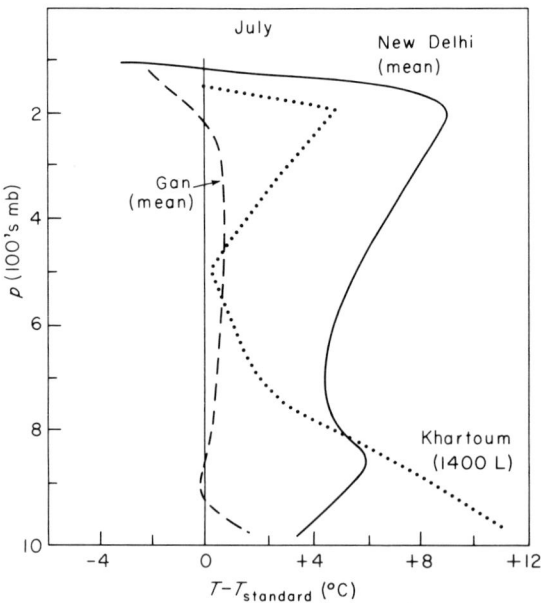

Fig. 2.26. Vertical profiles of difference between observed and tropical standard atmosphere in July at Gan (1°S, 73°E), New Delhi (29°N, 77°E), and Khartoum (15°N, 32°E). (New Delhi data 1961-1974.)

The high temperatures at Khartoum, often —48 to —46°C at 200 mb, are strongly indicative of upper subsidence. Over the bulk of the Asiatic highlands, subsidence on the south side of the summer subtropical westerly jet stream is also suggested as the dominant mechanism since monthly 200 mb streamlines cross from temperatures of —54°C over the Mediterranean to about —44°C over the plateau, a very large difference. For the region farther south, condensation heating from monsoon clouds has been advocated as the responsible mechanism, but doubts concerning this hypothesis, at least in its simplest form, are expressed in Chapter 8. It is more likely that dry convection from high elevations as well as cumulonimbus clouds carry large amounts of dust toward the tropopause. The dust appears as a thick layer with poor visibility; it may well be a means for long-wave radiational heating of the very high troposphere (Chapter 12).

Seasonal changes of temperature and humidity

Mean seasonal cross-sections of temperatures across the tropics conform, by and large, to Figs 2.7 and 2.20. Warmest air and equatorial trough migrate together; temperature differences between warmest latitude and the outskirts of the tropics are about 2°C, reversed above a level of zero meridional temperature gradient. As a single example, we see the 500 mb temperature profile for summer and winter of each hemisphere in Fig. 2.27. On the winter side the profiles are identical, while in summer, poleward of latitude 20°N, the Northern Hemi-

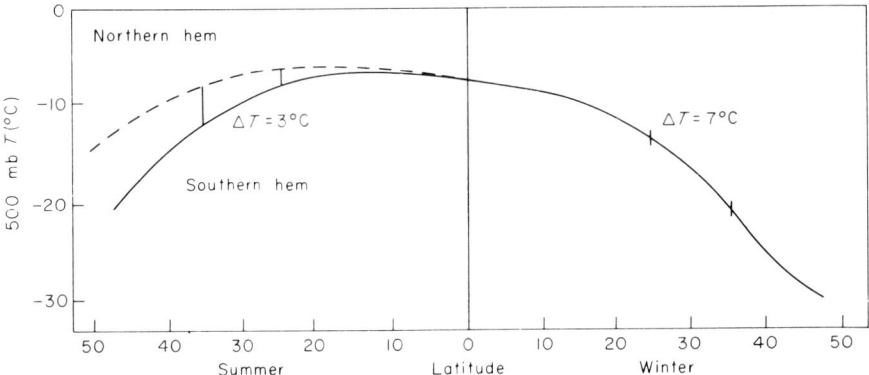

Fig. 2.27. Mean latitudinal profiles of 500 mb temperatures (°C) for January and July, arranged for summer and winter profiles to be shown together.

sphere is warmer than the Southern Hemisphere, again illustrating the difference between continental and oceanic climates. In terms of vertical profiles of geostrophic zonal wind, the east wind decreases or the west wind increases upward through 500 mb everywhere on the winter side. In summer, this increase only starts poleward of latitudes 20°. Temperature decreases slightly toward the equator from there, indicating an increase of the geostrophic east wind with height, often coupled with an increase of the actual east wind as in Fig. 1.14. Figure 2.27 permits computations of the meridional temperature gradient, a quantity which is vital not only for determining thermal zonal wind shear, but also for its relation to meridional heat transport. The gradient is 7°C/10° across 30° in winter, whereas the summer gradient of two hemispheres averages near 3°C/10° latitude. Thus, the gradient doubles from summer to winter as does the meridional heat transport. Further towards the equator, seasonal temperature variations become very small and often unsystematic at individual stations, depending on a variety of local factors. Year-to-year changes of temperature in one season may be as large or larger than changes between seasons. Reliable statistics of seasonal temperature differences in many instances are very difficult to establish for the troposphere, much more so than at the surface itself. One may well ask why temperatures remain so steady when substantial variations in radiative heating or cooling, and energy transfer from earth to atmosphere do occur. The answer must lie in the fact that the occurrence of isotherms on seasonal mean charts at a given pressure surface indicates a level of potential energy in the atmosphere which is raised above a zero reference furnished by complete coincidence of isobaric and isothermal surfaces; the term "available potential energy" has been coined by Lorenz (21) for this elevated condition. If such available potential energy exists on mean charts (i.e. persists), a mechanism must exist so that the cold air does not sink readily below the warm air. Such a mechanism can be provided by the earth's rotation through the thermal wind which varies inversely with the Coriolis parameter. Let us assume that a meridional circulation with conservation of absolute angular momentum gives the right order of magnitude for "baroclinity". Then, the typical vertical wind shear at latitude 15° will be 20 m/s/12 km given low-level easterlies of 5 m/s and upper westerlies of 15 m/s in the 200 mb return current. This shear, if it is assumed to be equal to the geostrophic shear, must be balanced by a mean tropospheric temperature gradient of about 1·5°C/1000 km in order to be seasonally maintained. This is exactly the order of magnitude observed with the best possible means.

RADIATION, TEMPERATURE AND HUMIDITY

We may conclude that the marked decrease of fields of available potential energy in the tropics, and their disappearance close to the equator, indicates that the dynamical factor is largely responsible for the very small ranges of temperature and temperature gradient in the tropics. The adjustment must be made by other variables, notably radiation, ground-air energy exchange and energy conversion within the atmosphere. These are all readily capable of adjustment, whereas the rotation of the earth at any latitude is an invariant.

Under these circumstances some other question about seasonal changes must be asked, if their computation is to provide something useful for the understanding of the large seasonal rainfall fluctuations shown in Chapter 3. Since tropical precipitation is largely convective, the right question may be whether one can find seasonal changes in average buoyancy which will favour or inhibit development of convective rain clouds. The average buoyancy would be determined from the difference in (virtual) temperature between a "parcel ascent" from the surface layer and mean seasonal temperatures at different upper levels. As just stated, the latter may be nearly invariant throughout the year, so the main element which changes seasonally must be the specific humidity, especially in the lower levels. However, slight changes in thermal stability must not be neglected either, induced, for instance, by the winter surface cooling indicated in Figs 2.12-2.17, especially when coupled with a little upper warming related to subsidence. The thermal factor in the low levels will become large where a trade wind inversion becomes established.

This approach is a difficult and rather dangerous one, especially since occasional synoptic weather systems in the tropics can force a very rapid, though temporary, interruption of the dry season structure. Even for a country as far north as Britain, Brooks (3) has made this remarkable statement: "The seasons in the British Isles are only slightly accentuated, and all through the year there are days which might belong to any month." Nevertheless, the approach as outlined has produced some positive results (24) and will be pursued briefly with a survey of a few stations, widely different in geographic setting (Table 2.4). The reader will gain considerable insight if he plots these and other examples on a thermodynamic diagram and makes a few comparisons of his own.

The seasonal comparison is most successful for oceanic and island stations without complicating topography. Barbados in the West Indies is such an island; examples from the Pacific atolls could also have been given. From 800 mb down, Barbados winter temperatures are 2°C colder than summer temperatures, a rather large change, and

Table 2.4. Seasonal variation of temperature, humidity and buoyancy

	Barbados (13°N, 60°W)						Mexico (19°N, 99°W) $p_0 = 782$ mb						Bogotá (5°N, 74°W) $p_0 = 751$ mb						Singapore (1°N, 100°E)						Khartoum (16°N, 32°E) $p_0 = 967$ mb					
	Feb.			Aug.			Feb.			Aug.			Feb.			Aug.			Jan.			July			Jan.			July		
p(mb)	T	q	ΔT_b	T	q	ΔT_b	T	q	ΔT_b	T	q	ΔT_b	T	q	ΔT_b	T	q	ΔT_b	T	q	ΔT_b	T	q	ΔT_b	T	q	ΔT_b	T	q	ΔT_b
Surface	25	17		27	18		6	6·6		13	11		8	8		10	9		23	16		24	18		30	6·0		37	15	
1000	24	16	0	26	17																									
950	20	14	0	22	16	0																								
900	17	11	0	19	13	+1																								
850	15	10	+1	17	11	+1													16	9·0	−1	18	11	−1	20	4·0		25	10	
800	13	7·2	0	15	8·0	+1	10	6·2		12	11								8	6·0	0	8	6·0	+1						
750							9	5·4		8	9·1	+1	9	8·4		8	8·4	−1												
700	9	2·6	−2	9	4·5	+1																			11	2·0		11	7·0	
600	3	1·4	−2	2	2·8	+2	2	2·3	−6	2	5·8	+2	2	4·5	−1	1	5·6	−1	−8	2·5	+1	−8	2·5	+2						
500	−6	0·8	−1	−6	1·5	+2	−7	1·1		−7	3·0	+2	−5	1·6	−2	−6	2·8	−1							−7	0·6		−7	3·5	
400	−17		−1	−17	0·7	+3	−18	0·7		−17	1·2	+1	−15	0·5	−3	−16	1·2	−1												
300	−32		−3	−33		+4	−34			−32		0	−31		−4	−32		−2	−34		0	−32		−1	−33					
	0800 LT						0500 LT						0700 L						0700 + 1900 LT						1400 L					

T(°C); q(g/kg); $\Delta T_b = T_{\text{parcel ascent}} - T$

specific humidity is 1-2 g/kg lower. From 700 mb upward, temperature is identical but specific humidity remains low throughout the troposphere in winter, in comparison with summer. The rather small, low-level changes (small compared to middle latitudes) are quite enough to inhibit buoyant ascent in winter and to favour it in summer. The nearby Trinidad precipitation diagram (Fig. 3.9) exemplifies the associated seasonal rainfall change, which is indeed large.

Above the high altitude of Mexico City, temperatures are even more uniform than at Barbados while specific humidity doubles from winter to summer. At the early morning hour of the sounding, stability is large in winter from 750 to 700 mb. However, even if a parcel ascent is started at 700 mb, with relative humidity near 50%, there is no hope for a buoyant continuation; Christmas is a good time to climb the high Mexican volcanoes. In summer, buoyant ascent is strongly favoured by the humidity increase; on average the very pronounced rainy season starts in June.

Comparison of Mexico City and Bogotá, with a similar rainfall regime, illustrates some of the problems with average soundings. At Bogotá, the low-level moisture change is small compared to Mexico, but middle-level temperatures, especially around 500 mb, are quite warm and would inhibit winter convection. In spite of relative humidity in the range of 80-90% on the summer sounding, the parcel ascent remains colder than the mean sounding and even the virtual temperature correction will not help. Here one must postulate a vertical structure different from the mean on days with deep convective clouds and heavy rains.

Turning next to Singapore, there is substantial rain in every month of the year (Fig. 3.13) with a December-January maximum. One would expect uniform soundings in both seasons shown and this is true from 700 mb upward. Lower down, temperature and humidity are slightly lower in the north (January) compared to the south (July) monsoon, so that computed buoyancy is reduced. Unfortunately, there are no readily available averages for the important low troposphere from international climatic data publications, but there is an indication that the heavier January precipitation is related mostly to convergence in the large-scale wind field at this time of year. Given a mean temperature structure neutral for convection—as in the case of the GATE equatorial trough zone—individual cumulonimbus towers can still ascend (Chapter 4).

Finally, there is Khartoum with very large humidity increases from winter to summer but identical temperatures at 700 and 500 mb. The rainy season is brief and rather unproductive (Fig. 3.8); on landing by aircraft one finds oneself in the desert with extensive green fields maintained only through irrigation from the Nile river. The afternoon sounding has been chosen for presentation in Table 2.4; on computation the reader will find positive buoyancy throughout the troposphere, but, nevertheless, it would be wrong to make the computation. At this continental location there are real air mass contrasts in summer between continental tropical and maritime tropical air. The mean sounding is a statistical mixture. Rain comes with the maritime tropical air from southwest with structure rather similar to the mean oceanic atmosphere so far inland. For a realistic computation, the two separate air mass structures are required. Such a separation was made by Solot (34) with soundings from the 1940s. No investigation is known in which his method has been applied with more modern rawinsonde instrumentation.

To sum up, the idea of viewing seasonal thermodynamic changes in the deep tropics mainly in terms of their effect on buoyancy has considerable merit and promise; but the method must be applied with great caution. If all tropical stations were analysed, as in the short sample presented, a whole new dimension would be added to tropical climatology.

References

(1) Betts, A. K. and Dugan, F. J. (1973). Empirical formula for saturation equivalent potential temperature. *J. Appl. Meteor.* **12:** 731-732.
(2) Betts, A. K. (1974). Further comments on "A Comparison of the Equivalent Potential Temperature and the Static Energy." *J. Atmos. Sci.* **31:** 1713-1715.
(3) Brooks, C. E. P. (1930). "Climate." Chas. Scribner's Sons, New York. 199 pp.
(4) Budyko, M. I. (1956). "The Heat Balance of the Earth's Surface." Reprinted in English by Office Tech. Svcs., US Dept. Commerce, Washington D.C. (1958).
(5) Budyko, M. I. (1963). "Atlas of Heat Balance of the Earth's Surface." Moscow Geophysical Observatory. 69 pp.
(6) Colón, J. A. (1953). Mean summer atmosphere of the rainy season over the western tropical Pacific Ocean. *Bull. Amer. Meteor. Soc.* **34:** 332-334.
(7) Deacon, E. L. (1969). Physical processes near the surface of the earth. *In* "World Survey of Climatology, 2." (H. Flohn, Ed.) Chapter 2, p. 86. Elsevier Publishing Co.
(8) Defant, A. (1921). Die Zirkulation der Atmosphäre in den Gemässigten Breiten. *Geografiska Ann.* **3:** 209-266.
(9) Defant, A. (1961). "Physical Oceanography, 1 (Charts)." Pergamon Press, Oxford.
(10) Defant, F. and van de Boogaard, H. Ref. 7, Chapter 1.
(11) Estoque, M. A. and Douglas, M. (1978). Structure of the intertropical convergence zone over the GATE area. *Tellus* **30:** 55-62.

(12) Fleming, J. R. and Cox, S. K. (1974). Radiative effects of cirrus clouds. *J. Atmos. Sci.* **31:** 2182-2188.
(13) Flohn, H. (1957). Studien zur Dynamik der äquatorialen Atmosphäre. *Beitr. Phys. Atmos.* **30:** 18-46.
(14) Graves, M. (1951). The relation between the tropopause and convective activity in the subtropics. *Bull. Amer. Meteor. Soc.* **32:** 54-60.
(15) Gruber, A. (1972). Fluctuations in the position of the ITCZ in the Atlantic and Pacific Oceans. *J. Atmos. Sci.* **29:** 193-197.
(16) Kordan, C. L. (1958). Mean soundings for the West Indies area. *J. Meteor.* **15:** 92-93.
(17) Kraus, E. B. (1977). Subtropical droughts and cross-equatorial energy transports. *Mon. Wea. Rev.* **105:** 1009-1018.
(18) Kraus, E. B. (1977). The seasonal excursion of the intertropical convergence zone. *Mon. Wea. Rev.* **105:** 1052-1055.
(19) Levine, J. (1972) Comments on "A comparison of the equivalent potential temperature and the static energy." *J. Atmos. Sci.* **29:** 201-202. Reply by R. A. Madden and J. F. Robitaille, pp. 202-203.
(20) London, J. and Sasamori, T. (1971). Radiative Energy Budget of the Atmosphere. *Space Research* **11:** 639-649.
(21) Lorenz, E. (1955). Available potential energy and the maintenance of the general circulation. *Tellus* **7:** 157-167.
(22) Madden, R. A. and Robitaille, F. E. (1970). A comparison of the equivalent potential temperature and the static energy. *J. Atmos. Sci.* **27:** 327-329.
(23) Oort, A. H. and Vonder Haar, T. H. (1976). On the observed annual cycle in the ocean-atmosphere heat balance over the Northern Hemisphere. *J. Phys. Oceanog.* **6:** 781-800.
(24) Palmén, E. (1948). On the formation and structure of tropical hurricanes. *Geophysica* (Helsinki) **3:** 26-38.
(25) Pisharoty, P. R. (1959). A standard atmosphere for the tropics. *Indian J. Meteor. Geoph.* **10:** 243-254.
(26) Raschke, E., Vonder Haar, T. H., Bandeen, W. R. and Pasternak, M. (1973). The annual radiation balance of the earth-atmosphere system during 1969-70 from *Nimbus* 3 measurements. *J. Atmos. Sci.* **30:** 341-364. (Many references.)
(27) Riehl, H. and Malkus, J. S. (1958). On the heat balance in the equatorial trough zone. *Geophysica* (Helsinki) **6:** 503-538.
(28) Riehl, H. (1977). Venezuelan rain systems and the general circulation of the summer tropics. Part II: Relations between high and low latitude circulation. *Mon. Wea. Rev.* **105:** 1421-1433.
(29) Rossby, C.-G. (1938). Thermodynamics applied to airmass analysis. *Quart. J. Roy. Meteor. Soc.* **83:** 342-350.
(30) Schacht, E. J. (1946). A mean hurricane sounding for the Caribbean area. *Bull. Amer. Meteor. Soc.* **27:** 324-327.
(31) Schnapauff, W. (1937). *Veröff. Meteor. Inst. Univ. Berl.* **2:** No. 4.
(32) Simpson, R. H. (1947). Synoptic aspects of the intertropical convergence near Central and South America. *Bull. Amer. Meteor. Soc.* **28:** 335-346.
(33) Simpson, R. H. (1978). On the computation of equivalent potential temperature. *Mon. Wea. Rev.* **106:** 124-130.
(34) Solot, S. B. (1950). General circulation over the Anglo-Egyptian Sudan and adjacent regions. *Bull. Amer. Meteor. Soc.* **31;** 85-94.
(35) Sverdrup, H. U. *et al.* (1942). "The Oceans (Charts)." Prentice-Hall, Inc., New York.

(36) Vonder Haar, T. H. and Hanson, K. J. (1969). Absorption of solar radiation in tropical regions. *J. Atmos. Sci.* **26:** 652-655.
(37) Vonder Haar, T. H. and Suomi, V. (1971). Measurements of the earth's radiation budget from satellites during a five-year period. Part I: Extended time and space means. *J. Atmos. Sci.* **28:** 305-314.
(38) Vonder Haar, T. H. and Oort, A. H. (1973). New estimates of poleward energy transport by the Northern Hemisphere oceans. *J. Phys. Oceanog.* **3:** 169-172.

Appendix

The relation between static energy and equivalent potential temperature (θ_e) is based on Rossby's (29) original derivation, and may be stated in the form

$$c_p \frac{T}{\theta_e} \frac{\mathrm{d}\theta_e}{\mathrm{d}t} = \frac{\mathrm{d}}{\mathrm{d}t}(c_p T + gz + Lq) \quad , \qquad (2.3)$$

where T on the left side is the temperature at the condensation level in case of unsaturated air. Several assumptions are made, especially that the condensate falls out immediately, that no provisions should be made for ice processes and that density differences between a rising air column and its surroundings can be neglected. The freezing process could be included by writing $(L + F)q$ instead of Lq where F is latent heat of fusion, but water drops freeze at variable temperatures, so that the effect cannot be included readily on thermodynamic diagrams. As defined above, θ_e is sometimes called "pseudo"-equivalent potential temperature. Various more complex formula have been discussed (2, 3, 19, 22, 33). The approximations of Eq. (2.3) are felt most strongly in the high tropical troposphere where an ascending air column may be 2-3°C warmer and θ_e 3-5°C higher than calculated, a substantial difference for tropical convection (Fig. 2.18).

3
Precipitation and Evaporation

Precipitation and water balance are by far the most important and sensitive ingredients of climate in the tropics. The whole range of human activities depends on an adequate ("normal") supply of water. Yet this is exactly what is not realized over a very large fraction of the tropical continents and islands. Seen in the broadest terms, heavy annual precipitation is confined to portions of the equatorial zone, while the subtropical margins contain stark deserts. Seasonal rainfall migrates with the equatorial trough zone (Fig. 1.10). Thus, the inner tropical zone, by and large, has two rainfall maxima per year, the outer envelope only one. Since rainy seasons are known to "fail" quite often, two chances per year are always better than one, especially when the interval between rainy seasons may be marked by just plain, clear skies without any rain.

Very large regional deviations are superimposed on this broad climatic picture. For instance, it is particularly difficult for a teacher of atmospheric science in Florida to convince his students that the subtropics are arid and for one in Brazil to insist that all of the equatorial belt has abundant rain. Around the equator, and even within Brazil itself, regions with hundreds of centimetres of rain a year alternate with virtually completely dry zones. The meaning of precipitation for water supply also intrudes. In high latitudes, dense forests grow happily with annual precipitation of 60 cm, while in the trade wind belts only bush will grow with such rainfall. Evaporation, and the biological need for evaporation, rises with temperature. Much effort has been expended to represent precipitation in terms of indices which take this factor into account.

In contrast to climatological elements which "occur" all the time, such as temperature, precipitation is intermittent. In fact, most precipitation at tropical stations occurs in just a small fraction of the time in a year; rain comes in spurts. Much of the water may run off and never benefit the area where it falls, unless captured in reservoirs. Sometimes, rain episodes are so heavy that they result in disastrous floods, and these may come right on top of a long drought. To this must be added the fact that precipitation is highly sensitive to even small and temporary variations of the general circulation, much more so than, say, temperature, and wide swings of annual rainfall from one year to the next are the result. Water management is called for to provide means for balancing a highly erratic water source to a stable supply. Beyond the interannual changes are longer period variations, decades, centuries and longer with high or low precipitation compared to, for instance, 40-year averages. These variations do not approach zero with increasing length of time over which records are examined, as attested to by historical events such as the rise and decay of nations and migration of populations as a result of changing water supplies. Many attempts have been made to find "cycles" of abundant and deficient precipitation, but the caprices of the general circulation are such that one can find only frequency distributions of, say, the duration of drought periods. Occasionally cycles (for instance, attached to sunspots which have a demonstrable cycle) have been discovered, but they invariably failed sooner or later.

Annual rainfall in the tropics

Observations

Of all the data so far considered, those on rainfall are by far the least satisfactory. There are no stations that record rainfall on a routine basis over any of the oceans. Over land, especially in mountainous areas, the record at any station is notorious for representing only the immediate environment. This situation is accentuated in the tropics compared with higher latitudes since precipitation is cumuliform to a large extent, and any heavy shower may just hit or miss any given station. One would think that, with time, the accidental part of records would even out, but, statistically, this is not so and experience with accumulating rainfall over periods of increasing length in small regions with dense precipitation networks has borne out the fact that isohyetal patterns do not simplify as time is increased.

In spite of all difficulties, mean rainfall charts have been prepared for the continents from records of often quite variable duration. Supan (38) first attacked the problem of mean annual precipitation over the oceans. He thought that ship reports gave a good rainfall frequency distribution and if, in addition, he could obtain some idea as to the average amount of rain per rainy day, he could by multiplication construct isolines of mean annual rainfall. A few ships have attempted rain measurements and the number of such observations has increased greatly since Supan's day. This technique is preferable to extrapolating precipitation even of small atolls out over vast ocean areas. Satellite observations now give improved precipitation frequencies over the oceans (18).

From the oceanographic side, Wuest (41) showed that salinity and the difference between precipitation and evaporation are related linearly. Herewith another approach to rainfall estimates over the oceans is opened, provided that the evaporation is known, and in Chapter 4 means for determining evaporation will be described. Lacking direct measurements, evaporation has been estimated from evaporation pans, always a dubious procedure; from turbulent interaction between sea and air, described in Chapter 4; and through computation of oceanic heat balances. The latter technique can only be applied on a very large scale when the mobility of ocean currents can be at least partly excluded, or in confined basins such as the Red Sea (24). A potent method, no doubt destined to supersede all others, is the estimation of precipitation through satellite cloud observations (2, 20). These are based on relationships between satellite-borne Electrically Scanning Microwave Radiometer brightness, temperature and rain rates over the oceans (30). Flohn (11, 12) has discussed the particular problems of estimating water budgets in tropical and subtropical, mountainous areas.

Annual precipitation and evaporation

Profiles of annual precipitation have been plotted according to estimates by Jaeger (A-7) who has compared his results with climate models by Wetherald and Manabe (40) and other investigators with fair results. The evaporation curve was taken by him from a study by Kessler (17). As there is ever increasing evidence that rainfall on the equator itself is depressed compared to the belts just north and south, the author has entered a small depression there. Such a depression also agrees with the results of modelling. Largest precipitation generally is agreed to be found near latitude 5°N, the meteorological equator; in the trade wind belts precipitation of both hemispheres is about equal. How reliable the profile is, remains an open question, but a satellite-derived atlas (30) agrees quite well over the latitude range of Fig. 3.1, though rainfall is estimated at about 10% lower.

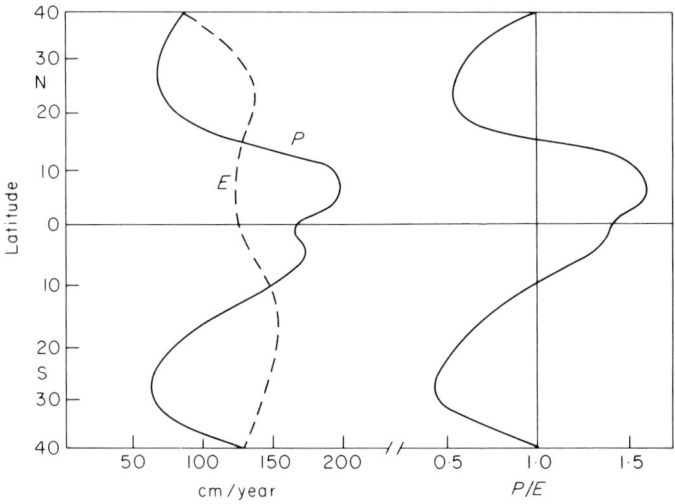

Fig. 3.1. Latitudinal profiles of annual evaporation, precipitation, and of the precipitation/evaporation ratio (after A-7, 17).

The evaporation curve varies less drastically than that of precipitation. Maxima occur in the trade wind belts, where, with subsiding motion, a high percentage of sunshine is absorbed in the ocean and evaporation is facilitated by steady strong winds. Cloud cover reduces evaporation in the region of the equatorial trough zone itself. Thus, maxima of evaporation coincide with minima of precipitation, and vice versa. Excess energy in form of latent heat is exported from the trades to the equatorial zone and also to extratropical latitudes where the bulk of latent heat is released (Fig. 3.2). The ratio P/E is greater than one where precipitation is heavy and less than one where precipitation is small. Comparison of the evaporation profile with computed potential evapotranspiration over the land areas (Fig. 3.3) shows remarkable agreement in many areas, e.g. the Amazon and Congo basins, Indonesia and the Gulf of Mexico and Caribbean areas. In contrast, potential evapotranspiration is higher over the subtropical land areas than over the oceans at comparable latitudes; about 150 cm/year (largest, 250 cm/year) on average according to estimates by Budyko (6) and others. Potential evapotranspiration is defined as the

PRECIPITATION AND EVAPORATION

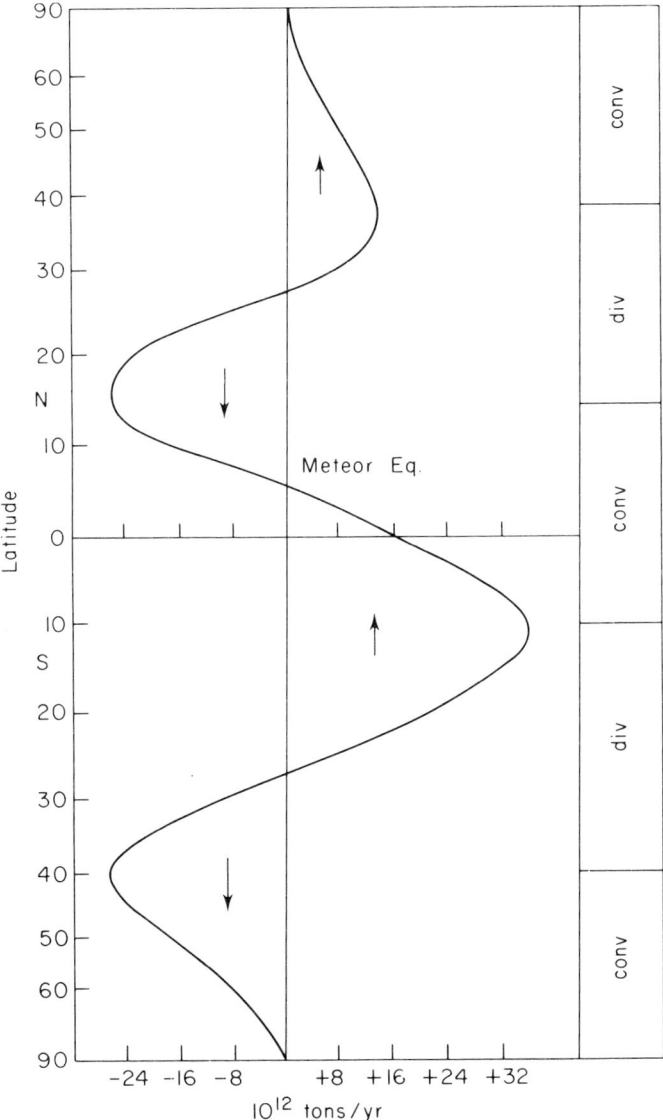

Fig. 3.2. Annual global transport of water vapour in the atmosphere toward north or south as indicated by arrows. Unit: 10^{12} tons/year.

evaporation that would occur if all solar radiation absorbed in the ground were given off to the atmosphere through evaporation. In view of the high albedo over arid land areas (35% and more compared to 4% over the oceans) one would expect to find lower potential evapotranspiration there. Since this is not the case, cloud albedo over sea must balance the land albedo at least partially. Further, evaporation and potential evaporation over sea are not identical; ocean currents transport away much of the stored energy. The sea is not dependent on local exchange with the atmosphere for energy balance.

Glancing at a map of annual precipitation (Fig. 3.4), we observe a marked crowding of isohyets (lines of equal precipitation) near the equatorial zone. The belt of heaviest rain closely coincides with the average position of the equatorial trough. Over the oceans there is a definite tendency for precipitation maxima to avoid the equator. In distinction, largest precipitation is shown centred on the equator over the Amazon basin and the "maritime continent" of Indonesia, which Ramage (29) regards as the greatest single source of energy for the extratropical regions of the Northern Hemisphere in winter. Farther north and south general dryness prevails in the trade wind belts except at the western ends of the oceans where, in the Northern Hemisphere, a clockwise-turning tongue of heavy rainfall extends poleward east of Asia and North America. Similar tongues are found over the southern oceans, though not as distinctly.

The crowding of isohyets at the poleward boundaries of the equatorial trough zone over the oceans lets these areas appear as maritime extensions of the subtropical deserts. Herewith, the assertion is refuted that deserts can be changed to lush tropical forest and agricultural land merely by creating substantial inland seas artificially (see also 22). Because of the sharp climatic gradients, equatorial trough shifts of as little as 2-3° latitude from one year to the next can produce extreme rainfall differences in the marginal zone. Glancing along the trough line, we find one dry regime over the East African shore, especially north of the equator, and over the Arabian Sea. Another overlies the trade winds of the South Atlantic and a large part of northeastern Brazil. The most famous dry zone on the equator is situated in the central part of the Pacific Ocean. Especially during the Northern Hemisphere summer, heavy rainfall along the trough gives way on its south side to almost complete aridity (39), though the width of this dry zone is only a few degrees. Still farther south a moist belt carrying the trade wind synoptic weather systems extends along the island chains of the South Pacific.

The arid belt coincides with the tongue of cold water protruding across the Pacific from the South American coast (Fig. 2.11) and correspondingly low vapour pressure difference between sea and air, which even becomes negative. At one time the explanation offered for

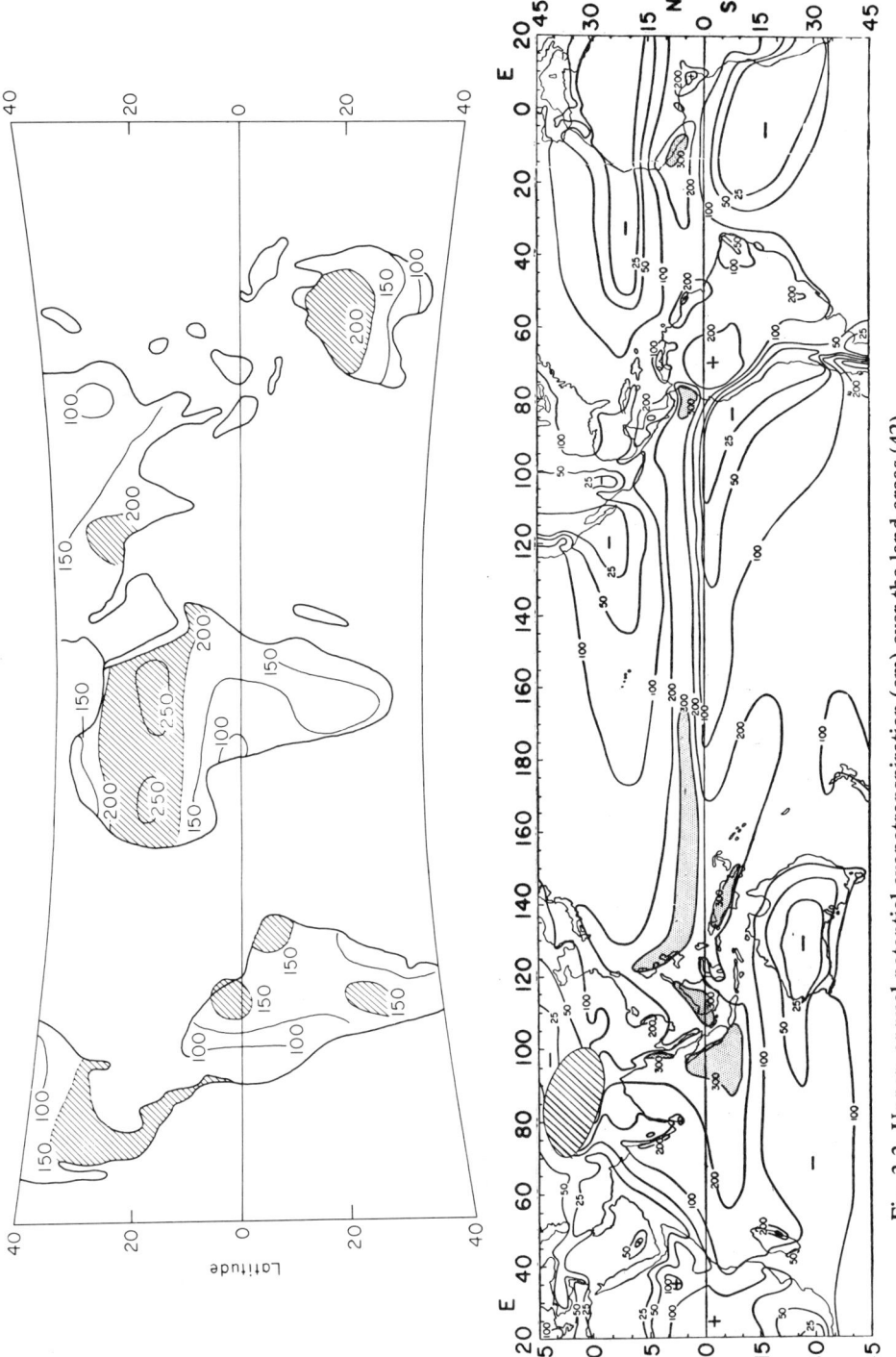

Fig. 3.3. Upper: annual potential evapotranspiration (cm) over the land areas (42).
Fig. 3.4. Lower: mean annual precipitation (cm) (23). Shaded: regions with more than 300 cm/year.

the dryness was simply that the cold water stabilized air passing over it and so prevented convection. This hypothesis became difficult to maintain when oceanographic expeditions discovered that ocean temperatures often are lower in the middle of the Pacific than upstream toward South America and that a dry inversion in the air overlies the area. Like the trade inversion, this stable layer cannot be accounted for by cooling from below, but only by considerable descent with adiabatic compression. Such descent implies horizontal divergence in the low levels. Actually the surface winds diverge in the mean (Fig. 1.12); rainfall and divergence patterns are remarkably well correlated (for further discussion see Chapter 12).

The role of tropical precipitation on the general circulation scale becomes particularly evident if precipitation is averaged following the seasonal equatorial trough positions, as was done for divergence and other parameters in Fig. 1.11. High correlation between mean precipitation and the mean vertical motion derived from the divergence profiles is evident, especially over the oceans (Fig. 3.5). Over the continents we observe the deviation of the precipitation maximum to the equatorial side of the trough with low-level equatorial west winds already noted in connection with temperature and humidity distribution (Fig. 2.20).

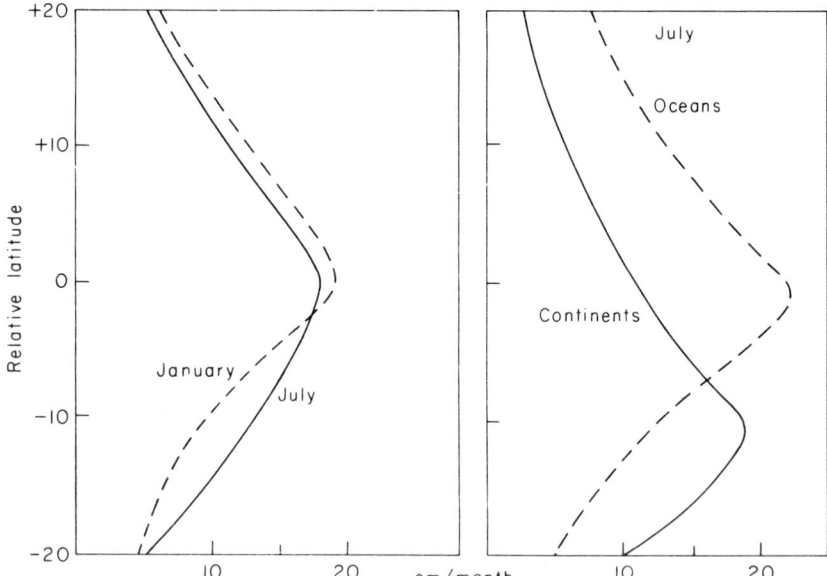

Fig. 3.5. Latitudinal profiles of precipitation relative to the equatorial trough. Left: January and July. Right: July, oceans and continents separate.

PRECIPITATION AND EVAPORATION

Precipitation weakens but does not vanish altogether where climatic surface convergence changes to divergence (Figs 1.5 and 1.11). Occasional synoptic weather systems make their appearance everywhere; it is their frequency with respect to the equatorial trough which determines the outer portion of the rainfall curves in Fig. 3.5. The relation may be expressed quantitatively by a regression equation between average precipitation and surface divergence, determined for July (Fig. 3.6) from Fig. 1.12 and Jaeger's July rainfall map. The regression is based on values in 10° latitude-longitude squares between latitudes 20°N and S relative to the trough, i.e. on well over 100 points. Where divergence is zero, mean July rainfall is still 10 cm.

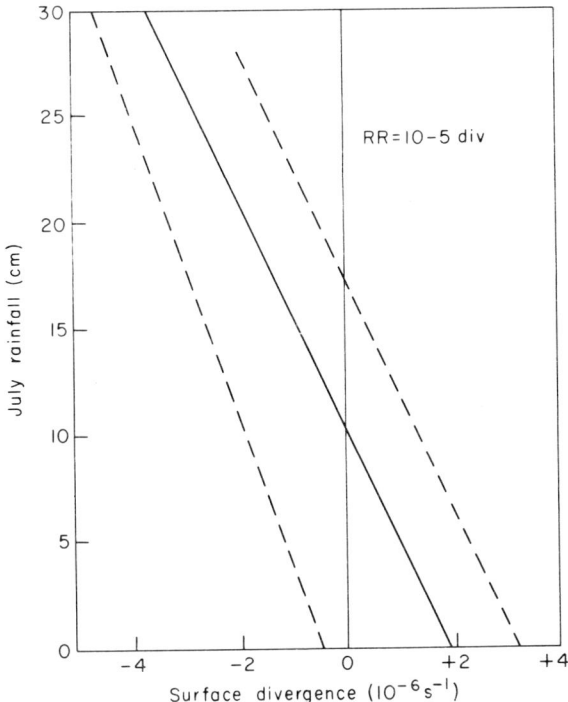

Fig. 3.6. Correlation diagram between divergence and rainfall relative to the equatorial trough in July, with 90% scatter band.

Seasonal march of rainfall

Ideally, seasonal-rainfall maxima should migrate with the equatorial trough, not only in the averages around the globe, but also at individual stations. As the trough reaches its extreme latitudes in

February and August, the equatorial margins of the trade-wind belts should experience a single rainfall peak in those months. Where the trough passes any latitude twice, double rainfall maxima should appear. If the trough oscillates symmetrically about the equator, rainfall should be heaviest on the equator in May and November, allowing for the lag of trough movement with respect to the sun.

This simple scheme cannot hold everywhere because of great differences in structure and behaviour of the trough itself at different longitudes. In addition, the controls on rainfall are not of a simple thermal kind. Dynamic effects in the upper troposphere, including close linkage with events outside the tropics, exercise a major control. Therefore, we must expect considerable variations of the seasonal march of rainfall from one part of the tropics to another, quite apart from extreme local influences.

As in the treatment of temperature, only a few basic rainfall types will be presented as a guide in this book. For detailed information, the reader should consult one of the numerous texts on regional climatology and also the "World Survey of Climatology" (A-11).

West African shore (Fig. 3.7)

Simultaneous with the exploration of Africa in the second half of the nineteenth century, the early climatologists evolved the picture of the seasonal variation of tropical rain, described above, which has come to be regarded as "standard". Small wonder, therefore, that we are able to find the classical model with greatest perfection on this continent.

The first profile extends along the West African shore from 15°N to 9°S. At Dakar, seven months of the year are completely dry and the brief rainy season correlates well with the summer excursion of the equatorial trough. Here, as in all of the following, mean rainfall does not denote steady rain, produced only by the presence of the trough. A few potent weather systems—seen in detail during the 1974 research project GATE (Chapter 8)—are responsible for a few days with heavy rain that make up most of the seasonal variation. Moreover, the fluctuations from year to year and over decades are large (Chapter 12).

At Freetown, the rainy season is longer; we also see the overpowering effect of a mountain range downstream, the Sierra Leone. Around the southern coast of West Africa, at some distance from the most marginal rainfall belt of the Sahel zone, much of the precipitation falls in form of rain rather than squall lines with thunderstorms. There is an interesting observation that thunderstorm frequency tends to rise at rainy season monthly precipitation up to,

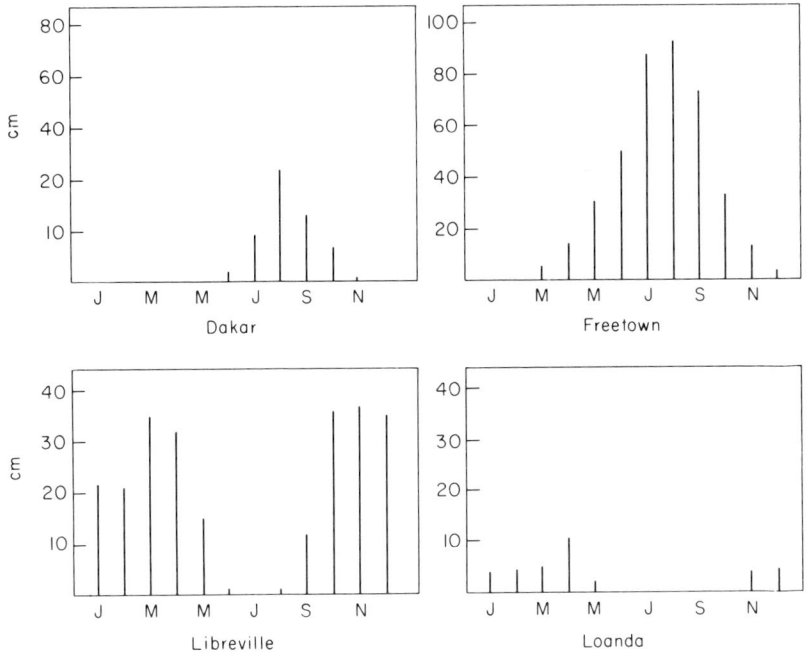

Fig. 3.7. Monthly rainfall along the coast of West Africa: Dakar (15°N, 17°W), Freetown (8°N, 13°W), Libreville (1°N, 10°E), Loanda (9°S, 13°E).

roughly, 30 cm (27). At higher precipitation values it decreases again markedly. Nevertheless, most precipitation is still derived from brief, intense showers, like everywhere else (16).

A double equatorial rainfall maximum appears south of Freetown. At Libreville the rainfall curve is very "regular", with two peaks of equal intensity. Since the equatorial trough oscillates about a mean position well to the north of the geographical equator, the dry season is severe from June to August; the trough never passes far enough south to permit complete cessation of rain, say from January to February.

Luanda is the Southern Hemisphere counterpart to Dakar, even though its latitude is only 9°S, further evidence of the difference between geographical and meteorological equator. In the mean, the trough does not reach latitude 9°S. According to the best available descriptions, the scanty rainfall comes during situations when the equatorial trough zone takes on a complicated structure with two branches, such as found in the Indian Ocean in July (Fig. 1.12).

Central Africa (Fig. 3.8; cf. also Fig. 2.16)

The rainfall at Khartoum takes the same seasonal course as at Dakar. Continental and shore regimes do not differ in timing, but Khartoum precipitation is very low compared even to Dakar which itself has no more than marginal rain. At Entebbe the two peaks coincide well with the two passages of the equatorial trough; again, precipitation is rather low at 150 cm/year, in contrast to the heavy snow loading of the Ruwenzori Mountains with permanent snow line below 5000 m altitude. There the rain-bearing winds come up from the Congo basin. Over East Africa the southward migration of the trough is very large. Thus, Bulawayo, even though located at 20°S, has considerable rain from summertime synoptic weather systems. The curve is out of phase with Khartoum as should be; the precipitation is larger and corresponds to a Northern Hemisphere location a little south of Khartoum.

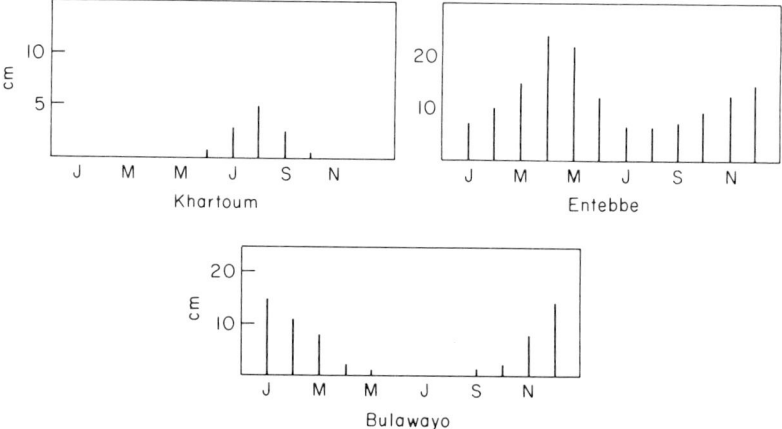

Fig. 3.8. Monthly rainfall in Central Africa: Khartoum (16°N, 32°E), Entebbe (0°, 32°E), Bulawayo (20°S, 29°E).

South American east coast and Caribbean (Fig. 3.9; cf. also Fig. 2.14)

The regime of Rio de Janeiro is typical for the eastern shore of continents in the subtropics including, for instance, Australia; some rain falls in all months, but summer has the most. At Rio the summer maximum is rather flat; in individual years the highest rainfall may be recorded anytime between November and April.

PRECIPITATION AND EVAPORATION

At Recife, on the "bulge" of Brazil, the rainfall curve is very simple and symmetrical. Average annual rain (165 cm) is more than at Rio (110 cm) and three times what is found in the arid interior to the south. Stations on the bulge generally report highest rainfall between March and June. The Recife maximum does not agree with the timing expected from equatorial trough movement. Nor do satellite photos often depict a cloud pattern reminiscent of equatorial trough.

An equatorial trough type of rainfall first appears at Belem. Since the trough, in the mean, hardly penetrates south of the equator in

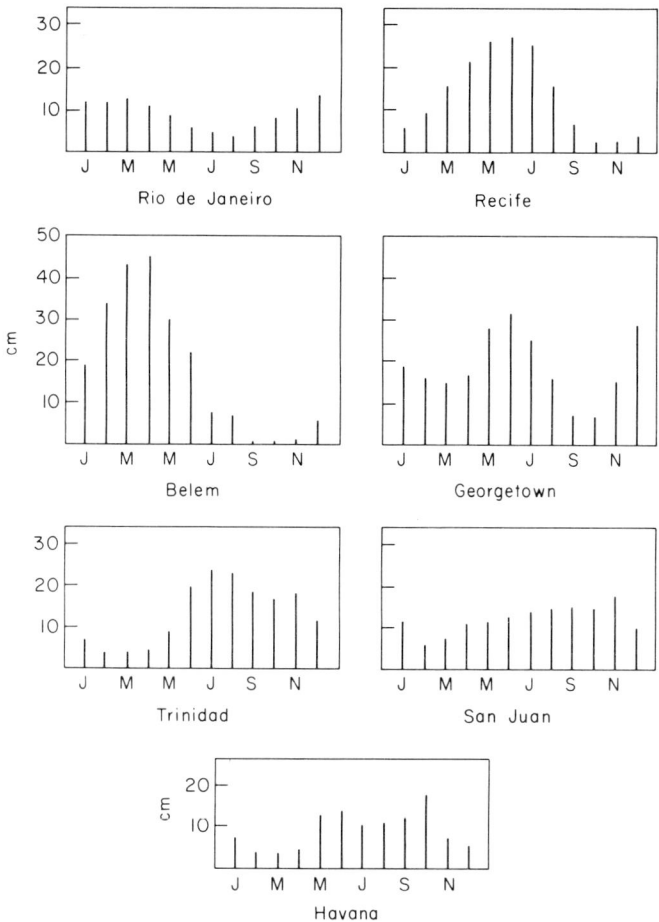

Fig. 3.9. Monthly rainfall along the South American east coast and Caribbean: Rio de Janeiro (23°S, 43°W), Recife (8°S, 35°W), Belem (1°S, 48°W), Georgetown (7°N, 58°W), Trinidad (10°N, 61°W), San Juan (18°N, 66°W), Havana (23°N, 82°W).

South America, rainfall is heavy from February to May and negligible from August to November at the mouth of the Amazon River. The lag of the rainfall with respect to the sun's migration is very large—fully three months.

The trough movement can be traced readily from Belem to Georgetown, where annual rainfall is heaviest (223 cm) along the whole section from Rio to Havana. The Belem maximum occurs in March-April, and shifts to June at Georgetown, indicating the trough's northward progression. The double peak is one of the most regular ones along the South American coast.

The Trinidad curve appears to be a logical continuation of the northward shift of equatorial trough rainfall from Belem and Georgetown. An accumulation of many satellite photos has indeed shown that weather systems along the trough, often narrow north-south and elongated east-west, frequently reach the South American coast not far from Trinidad. However, weather systems of the trade also influence Trinidad and the Lesser Antilles with increasing latitude.

Over Puerto Rico an equatorial trough influence is seldom felt. Weather systems of the trade wind belt combine to give the long, flat peak lasting half a year. Hurricane influences are much weaker than is often assumed, for the number of hurricanes actually striking the island is small. Some very heavy rains have occurred at the far tail end of hurricanes, however (up to 50 cm in 24 hours, three times the monthly average), especially in and around the eastern mountains of the island which catch the strong southeast winds fully (1, 14).

Havana should correspond to Rio and it does so, but only in a broad sense. Rainfall is highest in summer and fall, but there is a good chance for rain in every month on this profile with a rare triple maximum.

A few words may be appended on the drought region near the tip of Brazil (Fig. 3.10) which has been labelled a "problem climate" (A-22) not without good reason. Average annual precipitation sinks to values below 60 cm over a large area where the evaporating power may be three or four times as much. Flying from Recife, one encounters at first green fields, many villages and forested hillsides as the plane moves upward over the coastal mountain range with mean height of the divide near 600 m. Even before this divide is reached, the cumulus clouds driven inland by the easterly winds become sparse and then disappear, while underneath life gradually dies out and the countryside becomes brown. Eroded desert forms of peaks and ridges mark the divide; to the west the land flattens out into a stark steppe with sparse growth of grass and only an occasional village. The São

PRECIPITATION AND EVAPORATION 95

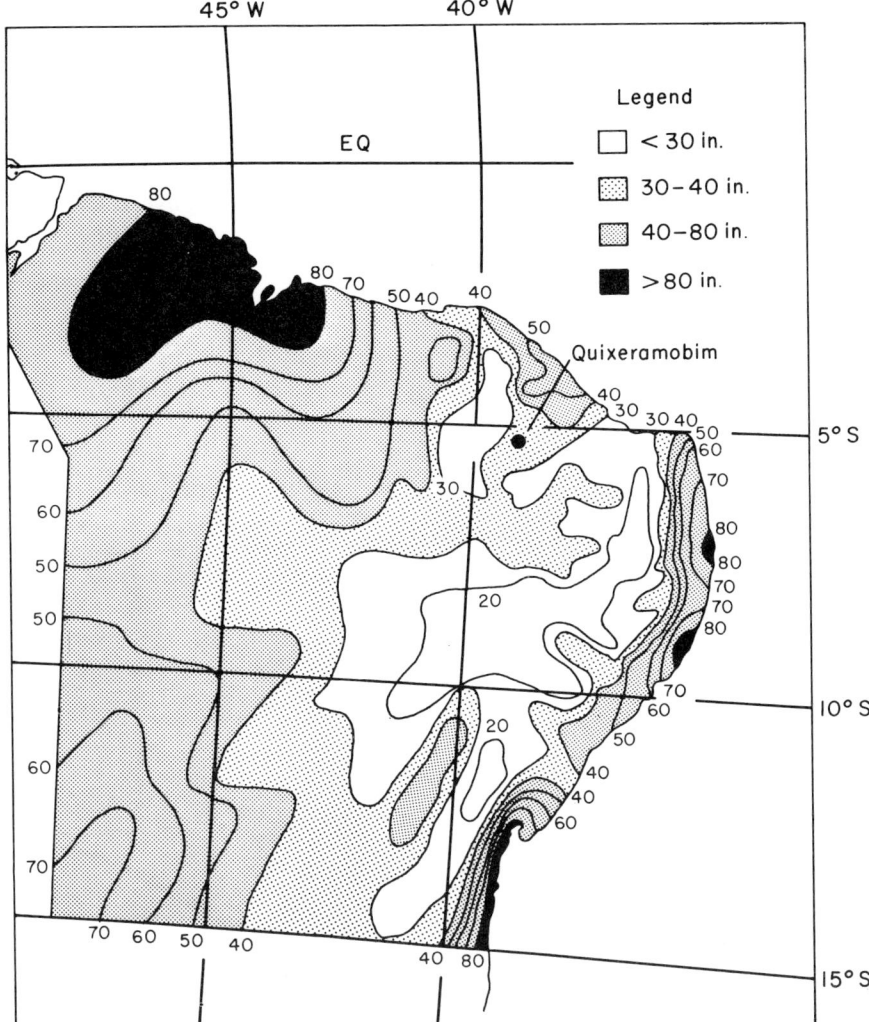

Fig. 3.10. Annual precipitation of the dry zone in northeastern Brazil.

Francisco River flows through the centre of the low population area. Here and there, some green irrigated fields break the brown monotony and these efforts no doubt will be increased. There is also some hope for cloud modification as many cumuli grow to the cumulus congestus stage but, with a continental nuclei spectrum, rarely manage to reach the ice phase.

At Quixeramobim, a station with a 65-year climatic record, precipitation less than half of the scanty annual average of 69 cm has occurred 13% of the time (Fig. 3.11). The average is upheld by a few years with heavy rains (125-150 cm), still marginal for the latitude. Wholesale migration of population from the area has been set in motion when low precipitation has persisted for more than one season. During the record, below average precipitation has occurred twice for three successive years, once each for four and six years and once for no less than nine years. A total of 25 out of 65 years thus has had prolonged drought.

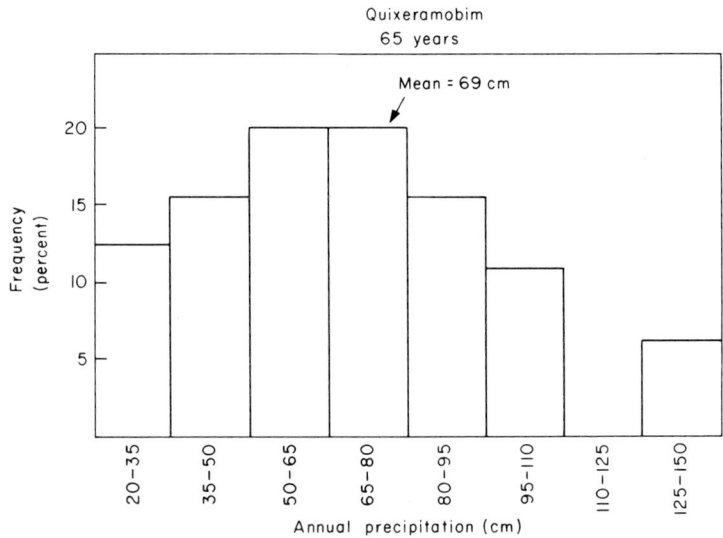

Fig. 3.11. Frequency distribution of annual precipitation at Quixeramobim (5°S, 39°W).

Pacific Ocean (Fig. 3.12, see also Fig. 2.15)

Midway and Norfolk Islands, at the northern and southern margins of the tropics, should provide a match, and to a great extent they do. Precipitation, mainly influenced from higher latitudes, is least in spring and rises in late summer; maxima fall in autumn and winter. Honolulu and Papeete, Tahiti, should also match, but they do not. Upper Lows during winter produce the Hawaiian rainy season, where summer is dominated by a strong trade wind inversion, which is weakened only slightly in August. A triple maximum is encountered. In contrast, Papeete has a straight summer maximum and winter minimum; far more typical for its position in the southern trades and well in accord with corresponding stations in the American section. In the "problem climate" area near the equator, Canton and Fanning Islands have irregular courses of rainfall, but Canton is in the divergent and cold water regime most of the time with large interannual variations (Chapter 12), whereas Fanning lies close to the equatorial trough zone which, in the mean, moves very little during the year in this area.

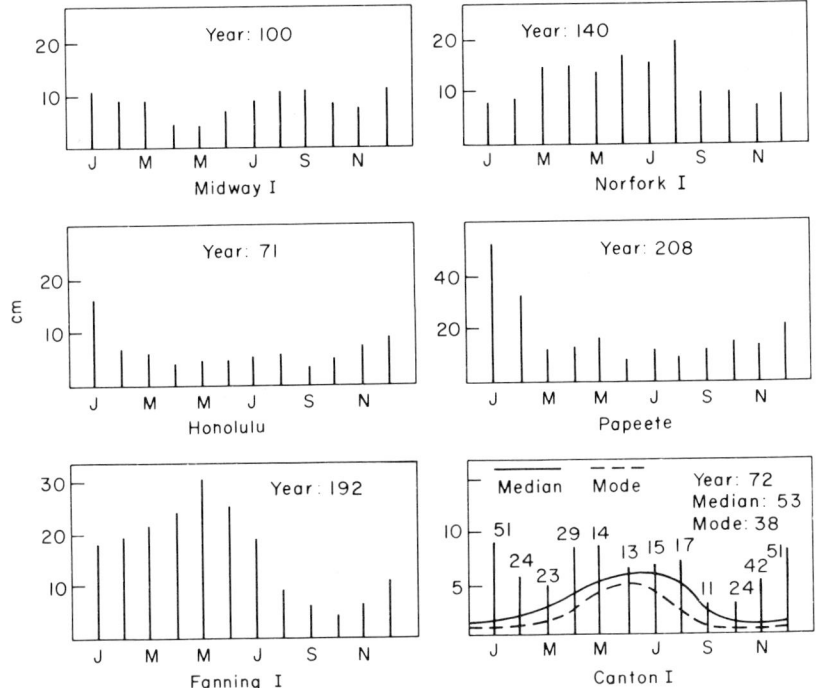

Fig. 3.12. Monthly rainfall in the Central Pacific: Midway (27°N, 177°W), Honolulu (21°N, 158°W), Fanning (4°N, 159°W), Canton (3°S, 172°W), Papeete (18°S, 148°W), Norfolk (30°S, 169°E). Also largest monthly values at Canton (n = 25).

The maritime continent of the Far East (Fig. 3.13)

The stations chosen lie in the monsoon regime of the Malay peninsula and Indonesia. At Djakarta, a single, large rainfall peak coincides with the season of prevailing north to northwest winds and the proximity of the equatorial trough. The season of southeast trades, in contrast, brings marked aridity. Altogether, the regime is quite "regular". The Indonesian region as a whole stands out for being the region with largest rainfall on the globe. Dhakarta's annual precipitation actually is not large for the area, considering that Buitenzorg in the nearby hills reports over 300 thunderstorm days per year, the highest recorded frequency on earth.

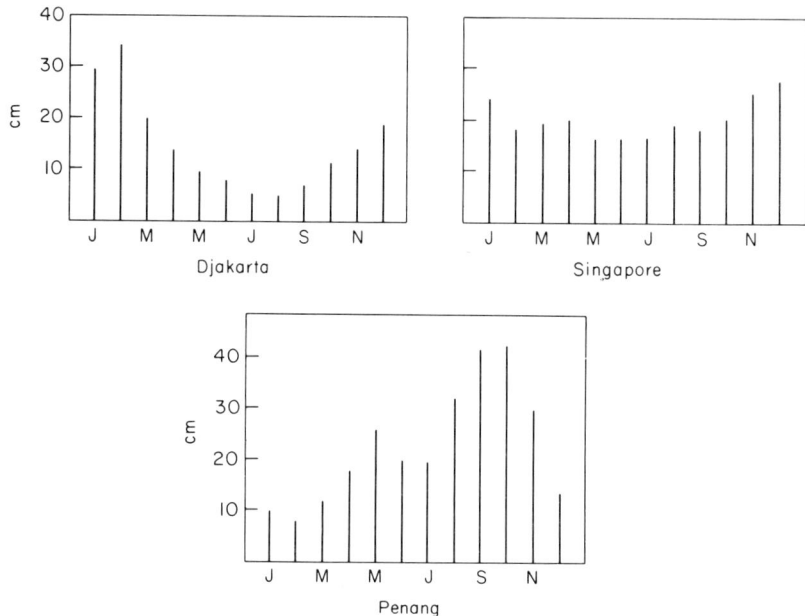

Fig. 3.13. Monthly rainfall in the Far East: Djakarta (6°S, 107°E), Singapore (1°N, 100°E), Penang (5°N, 100°E).

The transition from Southern to Northern Hemisphere differs greatly from that seen on both sides of the Atlantic. There, as well as in the interior of Africa, dry and rainy seasons alternate near the equator. At Singapore, however, rainfall is heavy and nearly uniform in each month in that part of the world. Penang, on the Indian Ocean side of Malaysia, has the large late autumn maximum typical of the southern coast of India, notably Madras. Stations along the east shore of Malaysia have their highest rainfall in November and December, coincident with an average convergence centre, a result of weather systems travelling westward to southwestward. Close to the equator these at times become organized as weak cyclones after a large monsoon outbreak from China (5). Record floods may occur even at Singapore, almost on the equator. The average monthly precipitation of 30 cm has come down in just two days, an event rated with a probability of return period of 100 years (35). Such an occurrence is usually reserved for dry climates; in deserts it becomes the rule rather than the exception.

Most of the winter monsoon air from the Asian continent which passes through the South China Sea does not cross the equator but converges and gives rise to the heavy Indonesian rainfall. In this way the secondary equatorial trough zone becomes manifest, in marked contrast to the divergence experienced elsewhere along trajectories toward the equator.

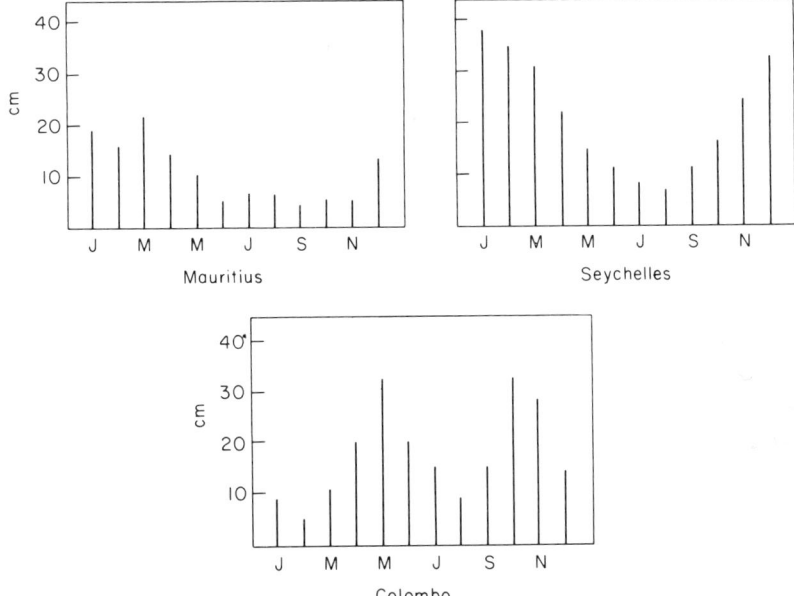

Fig. 3.14. Monthly rainfall in the Indian Ocean: Mauritius (20°S, 58°E), Seychelles (5°S, 55°E), Colombo (7°N, 80°E).

Indian Ocean (Fig. 3.14)

The last section across the equator also contains features that distinguish it from all other regions. At Mauritius, the course of rainfall is similar in general outline to that of Rio de Janeiro, Havana and Papeete, but here it is the equatorial trough of the South Indian Ocean which migrates close to the Mauritius area with an attendant tropical storm season that is one of the world's largest.

The Seychelles Island trace is most surprising. One would expect to find a double peak, but one does not. In contrast, Colombo shows one of the most regular double peaks found anywhere and just at the right time.

In summary, tropical rainfall regimes are far more complicated and less amenable to simple modelling than one might wish. In the predominantly marine environment, only a limited amount of the blame can be put on continental and orographic features. The model of a regular north-south migration of a single equatorial trough zone following the sun and yielding the rainfall model sketched at the beginning of this section occurs only in some areas. Facts such as the presence of high-tropospheric westerlies over parts of the tropical belt and easterlies over other parts, must be taken into account. While some modelling of the global circulation has produced a fair average structure of the tropical belt, further great improvement in the understanding of atmospheric dynamics is awaited to resolve the many intricate seasonal courses of precipitation.

Extreme precipitation

A very extensive literature is concerned with the subject of frequency distributions of rainfall and especially the occurrence of extreme precipitation for different time intervals at any location and also in global overviews. The subject is one of great importance to hydrologic engineers and the reader is referred to texts such as (A-3); it cannot be covered here in the detail it deserves. There are rainstorm probability curves, usually plotted on logarithmic probability paper as straight lines, at least outside the high and low extreme values. The Singapore curve mentioned above shows the average "return period" for a 5 cm rain as two years, 10 cm as 50 years and 30 cm as 100 years. Beyond this point the extrapolation is purely speculative; it is carried to 50 cm in 10^4 years. Such curves have been produced all over the world and they must constantly be revised as heavier rainfalls occur.

Another method consists in drawing depth-duration curves, i.e. the envelope giving the largest rainfall in 1, 10, 60 minutes, days, years and centuries. Many authors (for instance 21, 26) have produced global envelopes. The highest values at long time intervals are all found on Cherrapunji, N.E. India, though it is believed now that the world's rainfall peak is on the island of Kauai in the Hawaiian group (22°N, 159°W). Fletcher (10) and Marx (21) have offered approximation formulae for the envelope.

Rainfall analysis

A knowledge of annual rainfall values and the seasonal course in different areas is the first step in becoming acquainted with precipitation in the tropics. In the following section we shall be concerned with (a) variability of rainfall and (b) daily rainfall and rainstorms.

Variability of annual and monthly rainfall

Glancing at the world precipitation (Fig. 3.4) it is very crucial to know to what extent the mean annual precipitation can be expected to occur in any one year, i.e. how reliable it is. This is true especially for marginal regions where the average annual amount barely suffices for a wide range of activities undertaken there, based on the belief that the average value may be expected to occur. Unfortunately, this is far from the truth, as is brought out by a comparison of Figs 3.4 and 3.15. In Fig. 3.15 the average deviation from the annual mean is plotted, expressed as a percentage of this mean. As can be seen at once, variability is highest within any latitude belt where mean annual rainfall is lowest, and vice versa. This is of course the exact opposite from what is wanted, a very unfavourable outcome for all dry areas. Regions with variability of less than 15% are restricted in the main to the Congo Basin, the Amazon Basin, and parts of Indonesia in the tropics. From there, variability increases toward the subtropical deserts and also toward the bulge of Brazil in South America, where variability is very high compared to mean rainfall (Fig. 3.10).

As an example of the frequency distribution of annual rainfalls leading to the overall variability picture, Darwin (Fig. 3.16) has a very regular or Gaussian envelope. In 50% of the years, the deviation remains within 10% of the mean; about 20% of the years lie between each + 10-30% and —10-30% and only a few years are outside this limit. Darwin's mean annual rainfall is 165 cm which is mostly produced in the five month summer monsoon season. With a negative deviation of 30%, Darwin still has 115 cm in a year.

At Bombay, in a monsoon regime corresponding to that of Darwin in the Northern Hemisphere, both mean annual precipitation (185 cm) and frequency distribution are very similar. The mode, in the class interval of ±10%, is even more strongly pronounced, but the "wings" of the distribution are flatter, indicating more years with very high or very low rainfall. In the 110 years of recordings, highest precipitation was 350 cm and lowest was 84 cm. The old rule is that, as a first approximation, one should estimate highest precipitation as

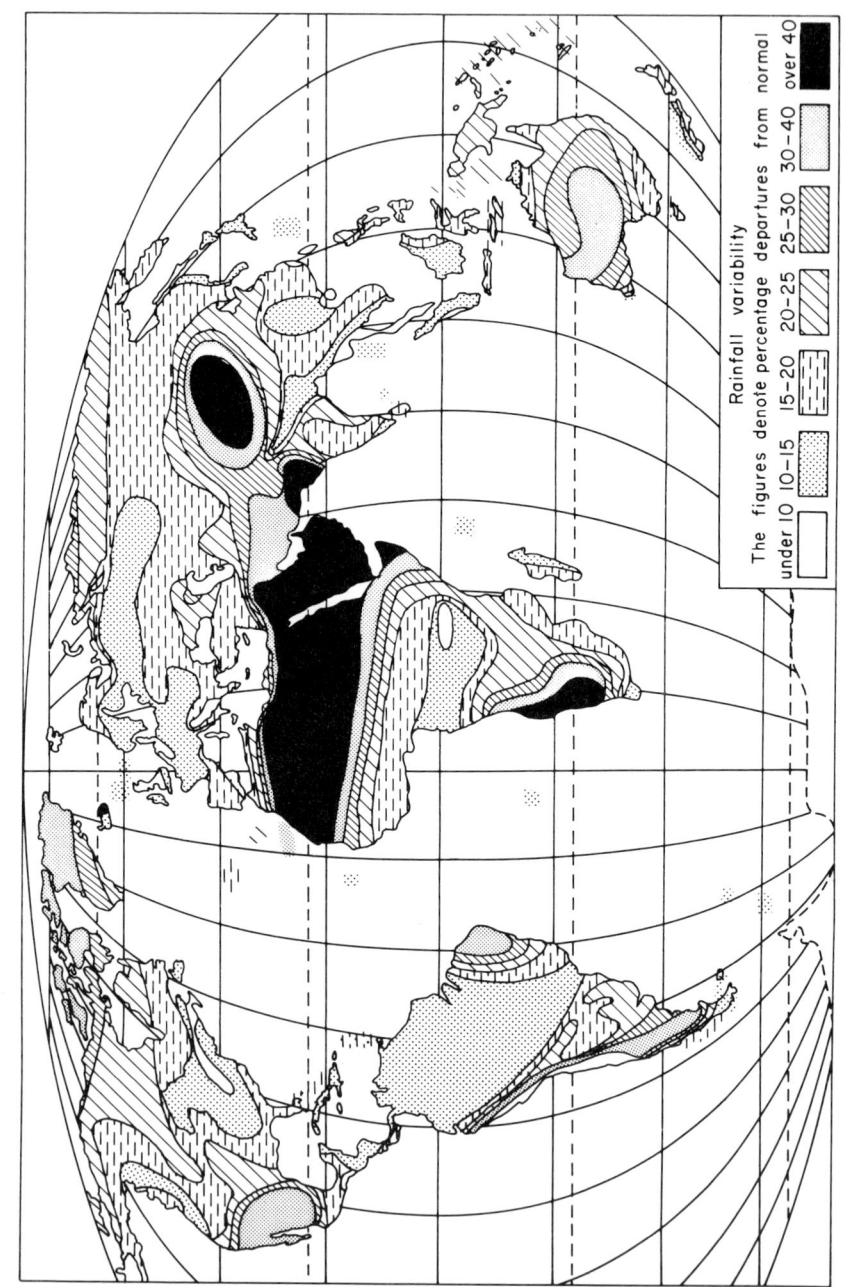

Fig. 3.15. Annual rainfall variability of the globe (per cent). From Erwin Biel (4).

PRECIPITATION AND EVAPORATION

double the mean, and lowest as half of the mean when a frequency distribution is not available. The upper limit is very well maintained by Bombay, as at most stations. The lower limit is exceeded, and that is also encountered at many other stations, especially when average rainfall is low!

The third curve of Fig. 3.16 is for Honolulu, where the frequency distribution is quite flat and the mode in the middle contains only 30% of the years. Correspondingly, the wings are much broader, just the inverse of what is needed for a balanced water economy in a climate with marginal rain. At Honolulu, for 85 years the average precipitation has been 71 cm, quite low for the tropics and even the trades. Highest rainfall was 142 cm, exactly double the mean, while the lowest was 25 cm, substantially below the threshold of half the average. Since the probability of precipitation within 10% of the mean

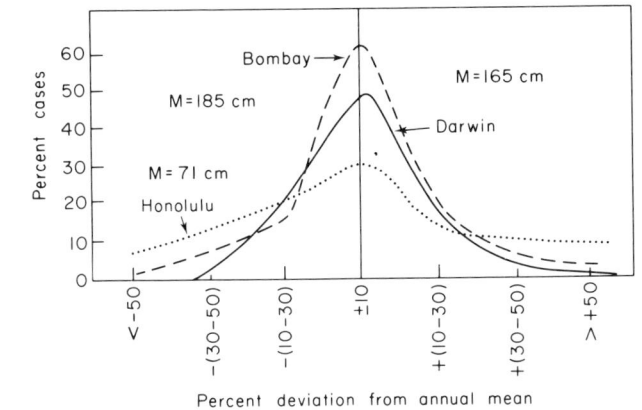

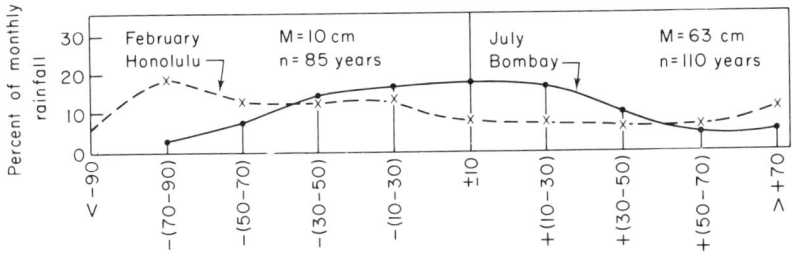

Fig. 3.16. Upper: frequency distribution of annual precipitation in class intervals, per cent deviation from the long term mean: Honolulu (21°N, 158°W), Darwin (12°S, 131°E), Bombay (19°N, 73°E).
Lower: Same for July precipitation at Bombay and February at Honolulu.

is only 30%, one should always speak of "average" and not of "normal" precipitation.

On a monthly basis, the rather nice annual frequency distributions fall completely apart (lower diagram of Fig. 3.16), even at Bombay. Essentially, the frequency distribution becomes flat over a large range of percentages; a difference between Bombay and Honolulu is no longer discernible. There were months with more than double the average monthly precipitation (highest 150 cm) and an almost rainless month (10 cm). Considerable skewness is evident at Honolulu, where rainless months occur, and skewness must appear in such cases since precipitation as a climatic element has one fixed limit which is zero. These records, compared to the annual, of course indicate that one must think in terms of periods longer than a month for water supply estimates. Within a rainy season there is marked compensation and stabilization.

A single drought year, or even two, usually can be overcome without extreme water problems. A succession of three or more drought years, in contrast, becomes serious almost everywhere in modern times with our great and ever increasing demands for water. Bombay, in 110 years, had three instances with precipitation below average in three successive years and one instance of a four-year drought. The three-year drought of 1940-1906 should have been the worst, since two successive years had less than half of the mean annual precipitation; in fact, they were the two lowest years of the whole record. Altogether, however, only 13% of all years were subjected to more than a two-year drought, and there were periods of as much as eight years with above average precipitation. The water management problem appears confined to guarding against a three-year deficiency.

At Honolulu, the record shows two instances of three-year drought, one of four and one of five years duration. In addition, a severe two-year drought occurred from 1952 to 1953 when precipitation was below 50% of the average in both years. Altogether there were 17 bad years, or 20% of all years, aggravated, of course, by the fact that the average precipitation is so low to begin with. Provision against drought must be far more substantial than at a location like Bombay, if it can be achieved.

The variable contribution to total precipitation can be brought out in another way, namely by writing the individual numbers of a time series in rank order and adding them cumulatively. A curve of cumulative per cent years versus per cent rain at Bombay (Fig. 3.17) brings out that 30% of the years deliver half of the rain. This, of course, is due to some years having very heavy rain. In general, one should find on such a diagram that the curves become steeper as area is decreased and, conversely, flatter as area is increased. For the globe as a whole, one would imagine the curve to become very nearly a straight line. One of the interesting questions of meteorology concerns global precipitation. Does the same amount of rain fall in each year, or perhaps even on each day, and, if not, are the deviations large enough for it to matter? Observations are as yet grossly insufficient to answer questions of this type; our hope must lie in refinements of satellite estimation techniques.

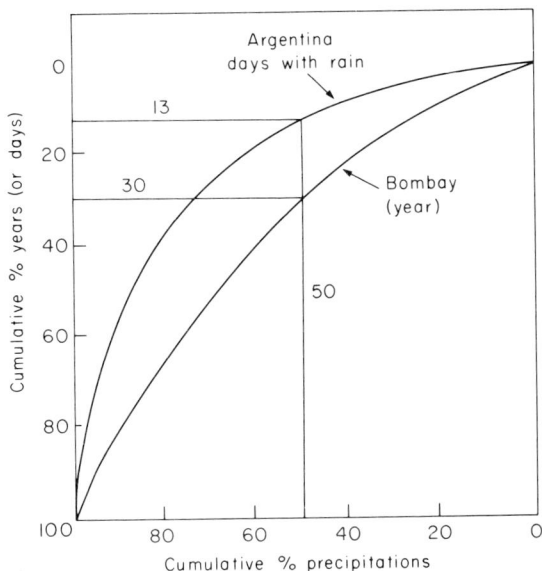

Fig. 3.17. Cumulative annual precipitation by rank order at Bombay versus cumulative years of record, in per cent. Also the curve for daily precipitation from Olascoaga (25).

One should also see increasing curvature, or skewness, of the rainfall distribution when comparing long records made up from the frequency distribution of, say, annual, monthly and daily values. The shorter the time interval, the greater should be the chance for extreme values to occur. Fig. 3.17 verifies such an increase in skewness. At Bombay, 50% of the annual rainfall is delivered by 30% of the years. Passing to daily values encountered in a variety of South American climates in small station networks (25), it turned out that 12% of the days with rain deliver half of the precipitation, whereas the lowest 50% of the days by rank produce only 12%.

Daily rainfall and rainstorms

Investigation of the short-period composition of tropical rainfall can be carried further by considering days, hours and even minutes with precipitation. As we take up these time scales, we leave the broad climatological aspects and become more concerned with suggestions inherent in the data as to the mechanisms that produce rain.

The bulk of low-latitude precipitation falls in the form of showers. It has been the custom to portray convection as occurring predominantly with a random spatial distribution, limited only by climatic boundaries. This picture follows from the conception of a steady tropical circulation. Slight changes in rainfall should accompany the minute fluctuations of surface temperature, wind and pressure that are often observed. As just seen, this does not hold from longer time viewpoints. In any given area, excess rainfall may flood the fields and fill the reservoirs in one year, and drought may strike in the same season of the next year. When we determine the manner in which individual days contribute to rainfall totals, the discrepancy between the classical description and reality becomes still more striking, even though we hear that in many localities showers begin every day at a certain hour with clockwork regularity, and even though the Hawaiian tongue contains no word for "weather". In any area, clear and rainy spells alternate, and this alternation occurs in definite relation to changes in the upper wind field (Chapter 8). Satellite cloud and radiation maps continually produce evidence of great concentration of heavy cloud masses in very small areas. Nevertheless, the convenient "classical" viewpoint has been slow in dying.

Concentration of precipitation

Because of the cumuliform character of tropical rain, it is always better to deal with small networks of precipitation than with single stations, with radar area-integrated rain, or with runoff from small watersheds (area order of 1000 km^2). Olascoaga (25) made small network analyses for various climatic regimes of Argentina, also Sircar (36) for India, and found the following: essentially, 10-15% of the days with rain account for 50% of the rainfall; 25-30% of the days account for 75% of the rain. Conversely, the water derived from the 50% of the days with smallest rain amounts to only 10% of the total. This contribution, which tends to be quite steady on account of the many days involved, often evaporates quickly and is dismissed as "noise".

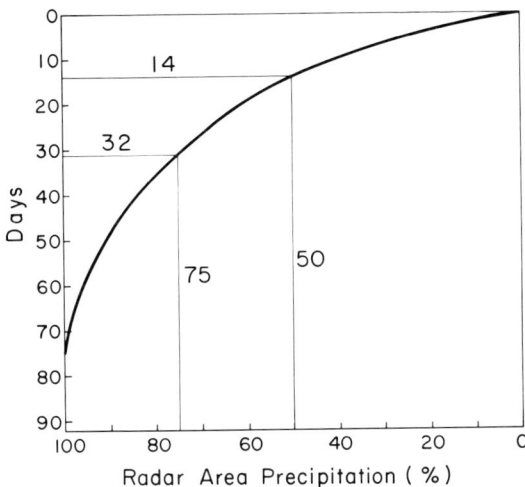

Fig. 3.18. Cumulative per cent precipitation in radar area (radius 80 km) versus cumulative days during Venezuelan research experiment in 1969 (33).

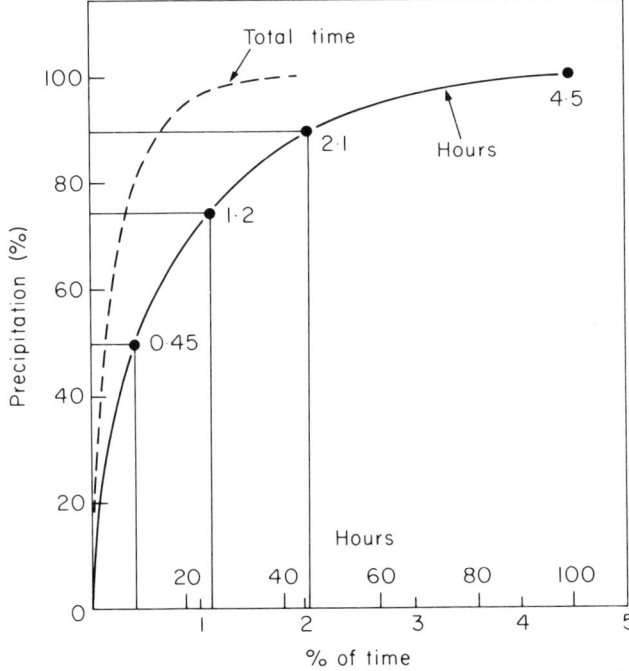

Fig. 3.19. Cumulative per cent frequency distribution showing precipitation at 11 weighing rain gauges versus hours with precipitation, in per cent of total hours and in actual hours (solid curve); and total time with precipitation determined from minutes of precipitation, in per cent of total time (dashed curve) (33). Venezuela experiment of 1969.

In the two Venezuela experiments during the rainy seasons of 1969 and 1972, networks of recording rain gauges were maintained in connection with a 10 cm radar covering about $2 \times 10^4 km^2$, a rather large area (3, 8, 33). Nevertheless, the result (Fig. 3.18) conveys the identical picture of concentration of daily precipitation. More detailed analysis of the recording rain gauge records brought out that, at individual stations, half of the precipitation in a 100 day sample fell in 10 hours and even less in time when minutes were counted (Fig. 3.19). Thus, it almost never rains in a cumulus regime, in contrast to the climates with stratus precipitation in higher latitudes!

Rainstorms

In passing from the statistical analysis of time to the events that produce the short bursts of heavy rain, it is useful to introduce the concept of "rainstorm" or rain episode with variable duration for a rainfall network. It has been employed, for instance, by Fletcher (9) and by Riehl and Byers (32) in Venezuela in analysing occurrences of high river streamflow. While there is no uniform definition, it has been found useful to define a rainstorm day as one in which the area-averaged precipitation exceeded the daily mean precipitation for the whole rainy season. For example, in the area of the first Venezuela experiment in the northeast part of that country, average daily precipitation was 4·4 mm. A day with higher area precipitation would be classified as a rainstorm day if more than half of the stations in the network reported rain. Duration of rainstorms on this basis ranged from one to nine days, average two days. The threshold of 4·4 mm per day coincided with the accumulation of 75% of total precipitation, i.e. days with precipitation above 4·4 mm accounted for 75% and days with less than 4·4 mm accounted for the residual 25%.

Thus, one could define the upper 75% as produced by organized synoptic systems passing across the experimental area; the other 25% include local precipitation at isolated places and also synoptic systems which merely brush the area with their fringes. With this definition, 14 rainstorms occurred in the 100-day season of 1969 (Fig. 3.20). The picture is rather similar to that of the time analysis. Two episodes accounted for 50% and seven episodes accounted for 75% of the organized precipitation; the other seven events produced rather little, a type of statistic valid also in many middle latitude climates where one cyclone can produce the average monthly precipitation in one day.

June 1969 provided a typical example of heavy rainfall month for a small network of rain gauges around the Lago de Valencia in north-

PRECIPITATION AND EVAPORATION

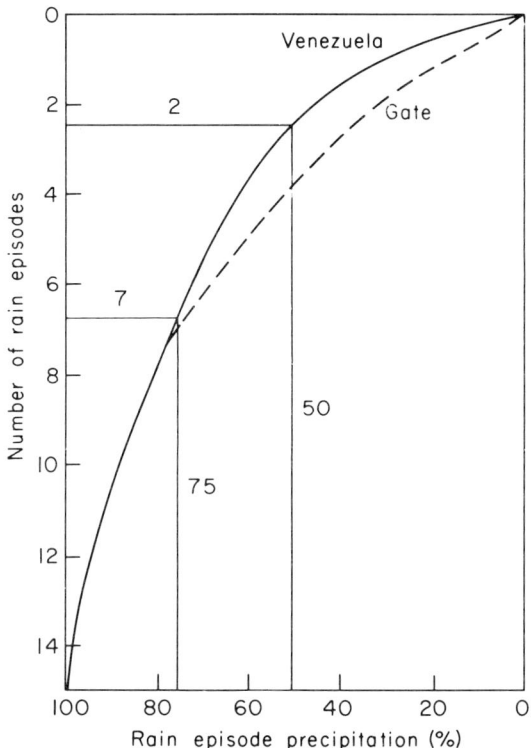

Fig. 3.20. Per cent cumulative precipitation due to synoptic episodes versus cumulative number of episodes. Solid: Venezuela 1969 (33), dashed: GATE (1974). Both curves coincide in lower part. Rainstorm precipitation was 75% of total precipitation in both cases.

central Venezuela (Fig. 3.21). Four rainstorms occurred in this month, three of one day's duration and one overlapping the hour of 0800L when the daily raingauge measurement is made (Fig. 3.22). Total precipitation for the month was 200 mm, of which 26 days produced 50 mm or 25%; the four rainstorm days delivered 150 mm or 75% (Fig. 3.23). If the two largest storms with 87 mm had not occurred, precipitation for the month would have been 113 mm or 3·8 mm/day, almost average. Thus, the two large storms lifted the month's precipitation for the network to almost twice the average value.

In a parallel study for India, about 20% of rainstorms accounted for 50% of precipitation in a rain gauge network in the Delhi area of India; 28% of the rainstorms in four areas of India accounted for 50%

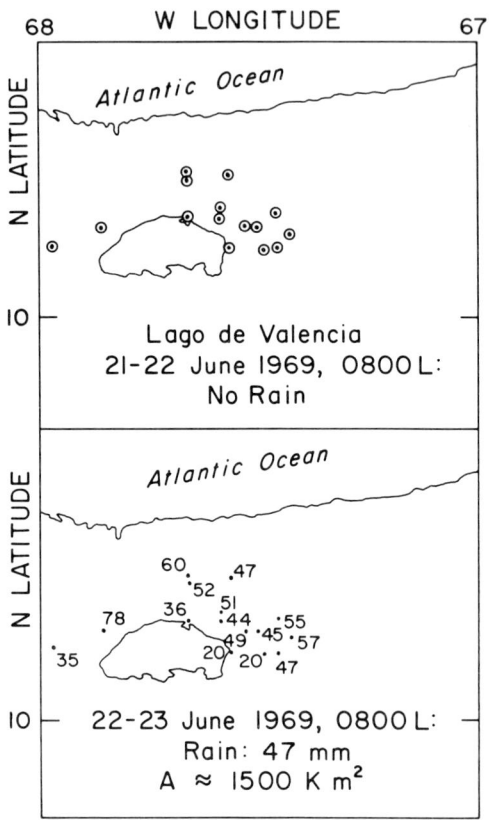

Fig. 3.21. The precipitation network north and east of Lago de Valencia in northern Venezuela. Precipitation is shown for 21-22 June, 1969, followed by the largest event in the month (mm) (Riehl A-19).

of the whole summer monsoon, an excellent corroboration of the uniform behaviour of storms in widely different regions (36). The precipitation record for the whole GATE, composed from (unpublished) ship and radar measurements, furnishes another telling example of the typical course of rainfall during a rainy season—45 cm on 60 experimental days, peak latitude 68 cm (Fig. 3.24). This is almost identical with the 0·7 cm/day indicated for the oceans along the equatorial trough line in July (Fig. 3.5), even though 1974 was still part of the great drought period of the early 1970s over Africa. Precipitation decreases very sharply both toward north and south; the area is bordered by the northern trade inversion and its cold water

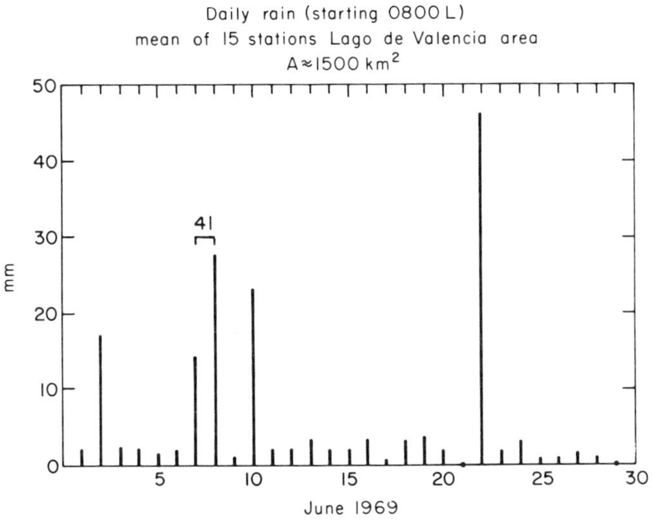

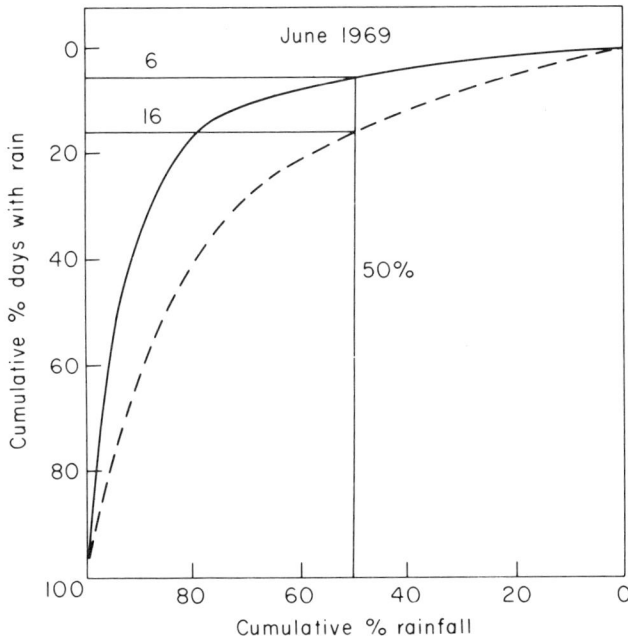

Fig. 3.22. Daily precipitation averaged over the network of gauges in Fig. 3.21. The rain of 7-8 June, 1969 was a single event (A-19).

Fig. 3.23. Cumulative per cent daily precipitation versus cumulative per cent days with rain for extreme month of June 1969 at Lago de Valencia compared with average distribution in small network with about 1500 km² area.

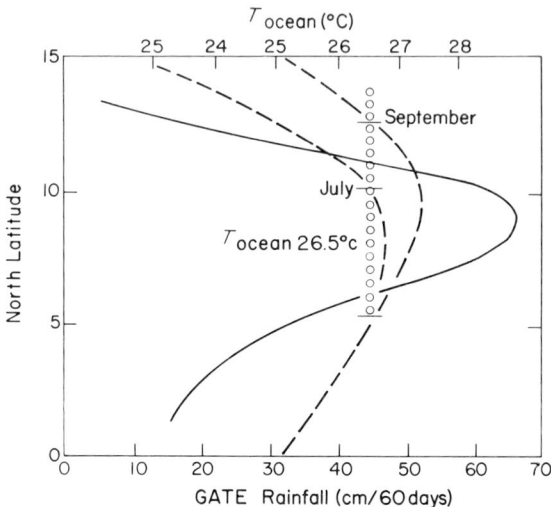

Fig. 3.24. Latitudinal profile of GATE precipitation (lower scale); also mean ocean temperature at longitude 23°W in July and September (upper scale). Circles: ocean temperature of 26·5°C, the "threshold" of deep convection.

regime on one side and the southern trade inversion plus equatorial upwelling of cold water in the south. A small northward migration of heaviest rain occurred along with the rise of ocean temperatures in the north. The line indicating T_w of 26·5°C denotes the approximate latitudinal limits within which free convection with $\theta_e = 346$ K is barely possible (7). This limit defines the small latitude belt with heavy rainfall quite well. Substantial ocean temperature gradients such as are found here generally act as a decided constraint on tropical precipitation. It is interesting that, if only the inner network close to the equatorial trough position (dashed boundary in Fig. 5.2) is considered, rainfall was 60 cm/season or about 1 cm/day rather than 45 cm/season for the larger network (solid boundary).

Fig. 3.25 shows the daily time history of precipitation for the larger GATE network and the periods classified as rainstorms. The line marking 75% of precipitation, separating rainstorm from non-rainstorm days, lies at the mean value of 0·7 cm/day as already indicated. The time histories of Figs 3.22 and 3.25 are quite similar, although the GATE network extended over a much larger area and, therefore, is somewhat less extreme with more rainstorm events.

PRECIPITATION AND EVAPORATION   113

Nevertheless, the role of one or two large storms is again emphasized. If the two biggest events of GATE had not materialized, rainstorm precipitation would have totalled 235 instead of 353 mm, or one-third less, marking the difference between drought and no drought.

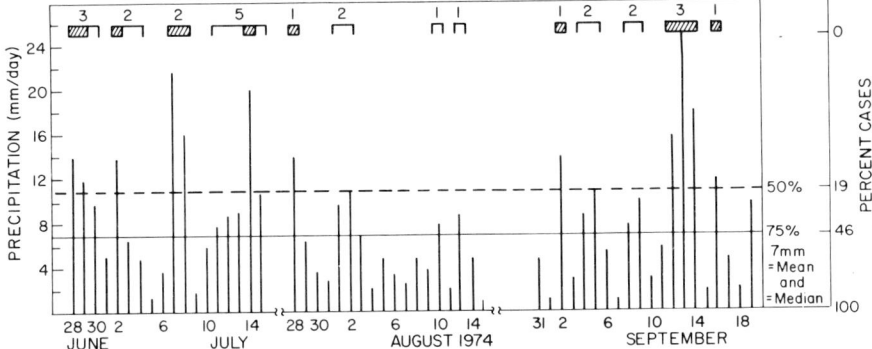

Fig. 3.25. Daily precipitation in central GATE area (7-10°N, 22-25°W), in mm from ship observations. Days extending above 50% contributed to the upper 50% of precipitation; days between the 50% and 75% lines contributed the next 25%, and days remaining below the 75% line (the daily mean precipitation) contributed to the lowest 25% of precipitation arranged in rank order from highest to lowest. Upper markings: rainstorms and their duration in days. Shaded: days in upper 50%.

When the rainstorms of Fig. 3.25 are rank ordered, it turns out that 25% of storms deliver 50% of precipitation due to such storms and half of them produce 75% of rainstorm precipitation (Figs 3.20 and 3.26). The curve of Fig. 3.26 is almost identical with that for the Indian monsoon composite (36). An individual small network, such as Delhi, of course stands out with greater skewness. Here, 20% of the storms account for half the precipitation. For the Indian composite, therefore, also GATE and Venezuela, the curve may be expressed by

$$x = aye^{by}, \qquad (3.1)$$

where x is the rainfall expressed as a cumulative percentage of the total, and y the number of rainstorms, also expressed as a cumulative percentage of the total number. For the monsoon composite, Sircar (36) finds $a = 16 \cdot 20 \times 10^{-2}$, $b = 1 \cdot 82 \times 10^{-2}$, in close agreement of the original values derived for Argentina (25). A similar expression holds for total precipitation against days with rain (excluding days without rain or trace); there the constants have been found to be in the range $a = 2 \cdot 30$ to $2 \cdot 87 \times 10^{-2}$, $b = 3 \cdot 55$ to $3 \cdot 78 \times 10^{-2}$. As also suggested by Sircar, Eq. (3.1) roughly represents a statistical law representative for most regions of the globe evaluated, for example at Hong Kong (28) and the island of Oahu in the Hawaiian group (31). There, ten rainstorms (out of 24) account for over two-thirds of the annual rainstorm precipitation, which is found to be about 90% there in the otherwise dry climate, apart from windward mountain sides (Chapter 6). These values fit very well on Figs 3.20 and 3.26.

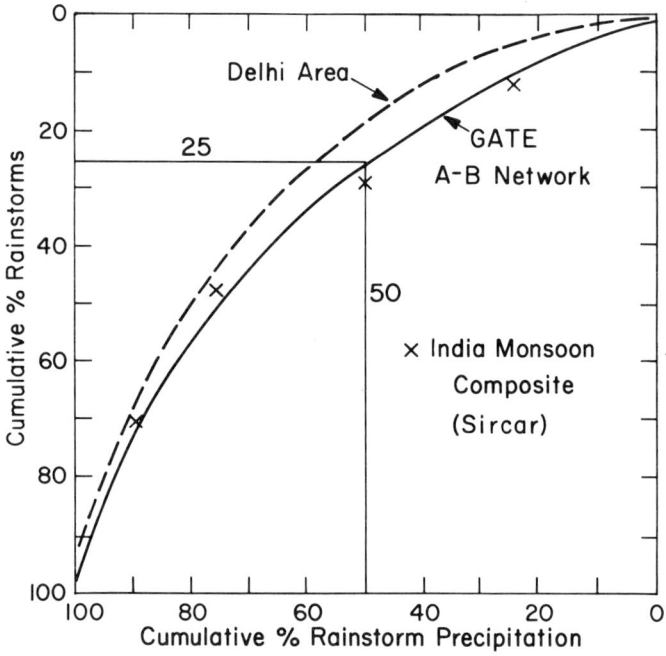

Fig. 3.26. Cumulative per cent precipitation versus per cent rainstorms for GATE, the Delhi area of India and the monsoon composite (crosses) (36).

Mesoscale rain systems

Mesoscale areas

When the passage of rainstorms is examined at the individual stations of a network, one curiosity stands out. In over 70% of the cases the passage is marked by a single shower, rarely by two showers, and most rain falls at the forward edge of the cloud in 15-30 min (13, 34). This characteristic of concentration of rain inside a synoptic rainstorm envelope to about 10% of its area is readily observed on radar. There, these smaller concentrated areas may appear as blobs or lines oriented along or perpendicular to the wind. Their average area tends to be approximately 2000 km^2, or 3000 km^2 (13), with range from 1000-10 000 km^2 and sometimes even larger. Because of their intermediate size between synoptic and cloud scale they have been named "mesoscale systems" in analogy to the terminology employed for middle latitude severe weather events. It should be emphasized that the radar echoes show the location of raindrops and not necessarily of cloud matter with much smaller drops. Much of the volume through which rain falls is actually cloud free.

Most precipitation of rainstorms, and therewith most precipitation altogether, comes from the cloud masses organized on the mesoscale with dimension of 25-100 km. There is a rather definite lower limit to the area of a mesoscale system for producing daily rain above the seasonal average, i.e. rainstorm precipitation. In Table 3.1 maximum radar echo size is related to daily rainfall in the first Venezuela experiment. When the greatest echo area observed on a particular day was in the range of 200-500 km^2, rainfall remained at or below the seasonal average. Some large echoes, mainly from middle clouds, also produced little rain, but when rain did exceed the threshold for rainstorm precipitation, it was associated with echoes starting around 1000 km^2. This relation was found to hold in the second Venezuela experiment and also in GATE (15). Apparently a certain minimum size of radar echo area is required to permit the system to collect enough water for a rainstorm to materialize. In contrast, echo height correlated poorly with precipitation (Table 3.2), a result that may be valid only over land. At low echo heights, precipitation does not occur since there is little warm precipitation over land. The greatest heights are attained on otherwise dry days, when it is very hot. Then, isolated cumulonimbi which break through during the afternoon can reach 16-18 km.

Table 3.1. Rainfall for 24 hours in Anaco radar circle, and greatest echo area for corresponding days (33)

Radar	Rainfall, mm/day			
A_{max} (10^2 km^2)	1-5	6-10	11-15	>15
2-5	13	—	—	—
8-11	15	6	—	—
14-17	2	8	—	3
>17	—	—	4	1

Numbers indicate frequency of days in indicated rainfall classes.

Table 3.2. Rainfall for 24 hours in Anaco radar circle, and greatest echo height on corresponding days (33)

Radar H_{max} (km)	Rainfall, mm/day			
	1-5	6-10	11-15	>15
< 9	1	—	—	—
9	3	3	—	1
11	7	5	4	3
13	11	3	—	3
15	4	1	—	—
17	1	—	—	—

After a rain area has built up to maximum size, heavy rainfall ceases within 15-30 minutes. "Late precipitation" falling at a light rate may continue for an hour or more; it is at least partly derived from cumulonimbus anvils when they are properly oriented. An average plot of echo area versus time shows a symmetrical picture of two hours increase, followed by two hours of decrease (Fig. 3.27) (3).

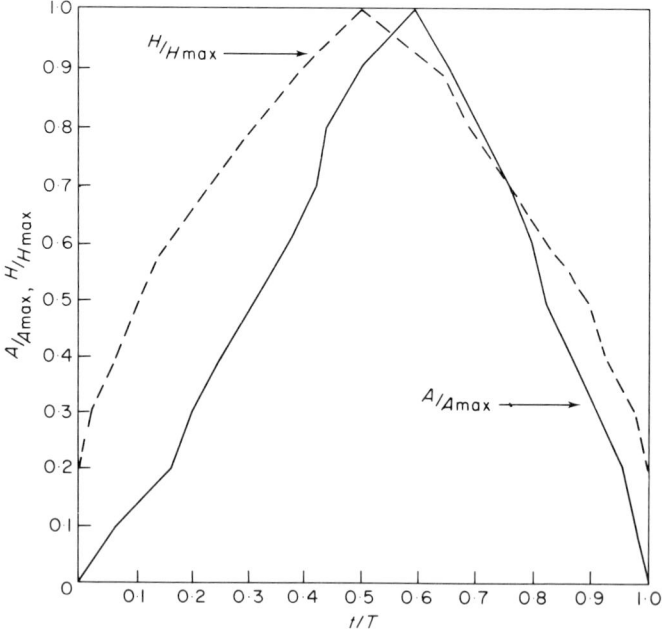

Fig. 3.27. Growth and decay of radar echo areas and heights versus time, non-dimensional (3). Total time was four hours.

Mesoscale ensembles

Out of the total radar echo population in the Venezuela experiments, 20% attained a maximum echo area of 1000 km² at radar antenna tilt of 6°, where the echoes are usually smaller than at lower elevation angles (Fig. 3.28). These 20%, or 80 echoes out of a total of 400, were contained in the rainstorm envelopes. As is widely found (19), the frequency distribution of radar areas is logarithmically normal over the largest portion of the sample. A similar distribution was obtained in GATE (15). There is a difference in the intercept at, say, 50%, which ranged from 20-100 km² in GATE compared to 300-400 km² in Venezuela. The difference is considered attributable to the very large number of small, warm rainclouds over the tropical ocean. As is well known, these showers do not occur over land with the continental condensation nuclei distribution.

The rain productivity of different mesoscale systems, a quantity of much importance also for hydrology, should be related to area (A) and duration (D) of mesoscale systems. One finds that duration increases with A_{max} (8); Fig. 3.29 shows the relation for the Venezuela experiments—the crosses represent groups of cases, the circles large individual storms. The value of $A_{max} \times D$, when given in 10^5 km²min,

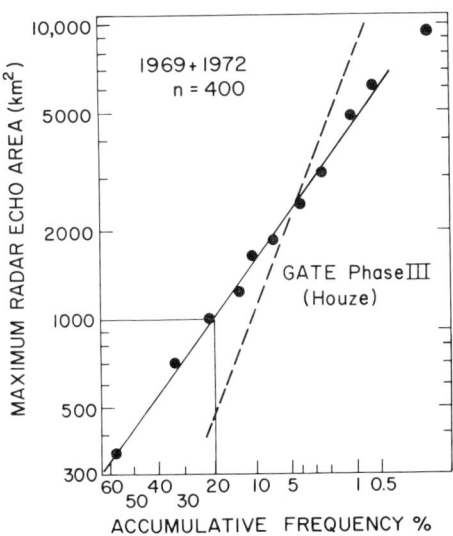

Fig. 3.28. Logarithmic probability distribution of maximum radar echo area during 1969 and 1972 Venezuela experiments (A-19) and Gate (15).

ranges from 1 at $A_{max} = 10^3$ km² to 55 at $A = 11 \times 10^3$ km². Almost the same relation was obtained during GATE (15), which reveals how independent the size of convective elements is from other factors, such as thermal and wind structure of the atmosphere, types of synoptic systems and even the drastically different nature of the underlying surface.

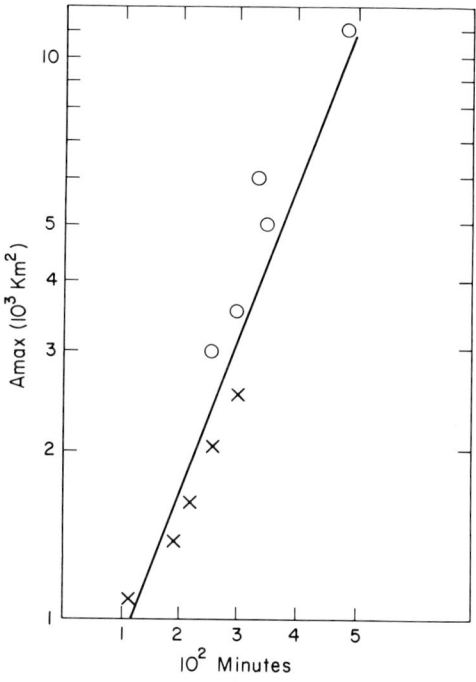

Fig. 3.29. Maximum area of mesoscale systems versus duration as measured in the Venezuela experiments (33).

While $A \times D$ is proportional to precipitation for an individual echo, its role in the ensemble of echoes of different size in a synoptic envelope cannot be assessed without taking frequency into account. The contribution of any echo size to total precipitation of the synoptic system may be defined as $Y = A \times D \times n$ where n is the frequency which can be deduced from Fig. 3.28. Comparison of the Venezuela experiments and GATE is not immediately possible, since there was a large population of small, echo-producing showers over sea not found over land. A continental spectrum of condensation nuclei may be assumed to exist well inside Venezuela. In order to eliminate this extraneous feature, only echoes of 500 km² and greater were considered for both distributions. Since $A \times D$ was equal, only the frequency distribution n can produce a difference. In the following table the quantity Y is compared for both projects, normalized through dividing by the contribution made by average mesoscale systems of 2000 km² (Y_0).

Table 3.3. Comparison of contribution of Venezuela and GATE radar echo distributions to precipitation of synoptic envelopes

A_{max} (10³ km²)	Y/Y_0 GATE	Y/Y_0 Venezuela
10	2·0	0·4
5	1·8	0·8
3		1·1
2	1·0	1·0
1	0·3	0·7
0·5	0	0

The distribution $Y = $ const, originally expected, is not found. Instead the contribution of mesoscale echoes to ensemble precipitation increases in GATE to $A_{max} = 5 \times 10^3$ km² and very slightly from there. In Venezuela, on the contrary, medium-sized echoes made the largest contribution. This difference is most likely due to the character of the rainy season as a whole. In GATE, rainfall was at or above average, especially in the last phase of observations. In Venezuela, the first project had about average rain (and $Y = $ const), while severe drought marked the second project. If interpreted in this sense, the result agrees exactly with what has been said before: the raininess of a particular season depends on whether or not a few events with very large precipitation will occur. Thus, the relative contribution shifts to medium-small echo sizes in Venezuela compared to GATE.

Conclusion

At the end of this exploration of precipitation we must re-examine the excellent correlations found in Figs 3.5 and 3.6 between mean seasonal fields of divergence at the surface and rainfall. Evidently, we cannot relate rainfall to the mean vertical motion in a simple way, if most rainfall is concentrated in a few days. In an extensive study, Solot (37) noted certain correlations between Hawaiian rainfall and monthly sea-level pressure anomalies over the Northern Hemisphere. On the basis of the above discussion these denote a basic state of the general circulation favourable or unfavourable for generating a few potent rainstorms. They do not indicate deviations of the properties of the large-scale flow from average that would tend to raise or suppress daily rainfall throughout a month or season.

In Venezuela, and the Caribbean area as a whole, it has become well known that certain states of the high-tropospheric wind circulation will promote the growth of synoptic weather systems and that other states will suppress them, notably hurricanes. As far as one can judge, similar considerations apply to all tropical regimes. Thus, the climatic rainfall curves (Figs 3.7-3.14) indicate a tendency of the atmosphere to produce two or three major storms on average in some months and suppress them in others. When we speak of "equatorial trough rainfall", we do not mean rainfall on all days when the trough is present, but heavy rain on a few days with passage of some wind-field anomaly such as a vorticity concentration (Chapter 8).

The vertical motion from the divergence charts is 100-500 m/day assuming the divergence is the mean for the average lowest 100 mb. In contrast, the vertical motion in cumulonimbi has the order of metres per second, three orders of magnitude larger than the climatic vertical motion. Thus, cumulonimbi will occupy only one-tenth of one per cent of the tropical space at any time, and a complex interpretation of the mean divergence charts must be made. Average convergence and mesoscale cloud concentration will follow similar paths. The cumulative effect of these systems, and not persistent convergence through the month, produces the mean convergence belts in Fig. 1.12. Most rainfall occurs in consequence of organized, not unorganized, convection. Within only a few relatively narrow zones of active weather does the tropical atmosphere obtain the largest portion of its heat from condensation and precipitation.

References

(1) Barnes, H. H., Jr. and Bogart, D. B. (1961). Floods of September 6, 1960, in Eastern Puerto Rico. US Geological Survey, Circular 451. 13 pp.
(2) Barrett, E. C. (1970). The estimation of monthly rainfall from satellite data. *Mon. Wea. Rev.* **98**: 322-327.
(3) Betts, A. K. and Stevens, M. A. (1974). Rainfall and Radar Echo Statistics. Tech. Rep. Dept. Atmos. Sci., Colorado State Univ., Fort Collins. 150 pp.
(4) Biel, E. R. (1943). Figure 137 *in* "Introduction to Weather and Climate." 2nd ed., by G. T. Trewartha. McGraw-Hill Book Co., New York.
(5) Bryant, K. (1958). Comparison of months giving extremes of rainfall during northeast monsoon at Changi, Singapore Island. *Meteor. Mag.* 87: 307-312.

(6) Budyko, M. I. Refs 4, 5, Chapter 2.
(7) Carlson, T. N. (1969). Some remarks on African disturbances and their progress over the tropical Atlantic. *Mon. Wea. Rev.* **97**: 716-726.
(8) Cruz, L. A. (1973). Venezuelan rainstorms as seen by radar. *J. Appl. Meteor.* **12**: 119-126.
(9) Fletcher, R. D. (1949). A hydrometeorological analysis of Venezuelan rainfall. *Bull. Amer. Meteor. Soc.* **30**: 1-9.
(10) Fletcher, R. D. (1950). A relation between maximum observed point and areal rainfall values. *Trans. Amer. Geoph. Union* **31**: 344-348.
(11) Flohn, H. (1969). Zum Klima and Wasserhaushalt des Hindukuschs und der benachbarten Hochgebirge. *Erdkunde, Arch. Wiss. Geogr.* **23**: 205-215.
(12) Flohn, H. (1970). Comments on water budget investigations, especially in tropical and subtropical mountain regions. Intern. Assoc. Scientif. Hydrology; Proc. Reading Symposium on World Water Balance, Reading, England.
(13) Henry, W. K. (1974). The tropical rainstorm. *Mon. Wea. Rev.* **102**: 717-725.
(14) Higgs, Ralph (1955). Severe floods of October 12-15, 1954, in Puerto Rico. *Mon. Wea. Rev.* **85**: 251-253.
(15) Houze, R. A., Jr. and C.-P. Cheng (1977). Radar characteristics of tropical convection observed during GATE: mean properties and trends over the summer season. *Mon. Wea. Rev.* **105**: 964-980.
(16) Hubbard, J. R. (1954). A note on the rainfall of Accra Gold Coast. *Geogr. Studies* (London) **1**: 69-75.
(17) Kessler, A. (1968). Globalbilanzen von Klimaelementen. Ber. Inst. Meteor. Klimatol., Technische Hochschule Hannover. No. 3.
(18) Kidder, S. Q. and Vonder Haar, T. H. (1977). Seasonal oceanic precipitation frequencies from NIMBUS 5 microwave data. *J. Geoph. Res.* **82**: 2083-2086.
(19) López, R. E. (1977). The lognormal distribution and cumulus cloud populations. *Mon. Wea. Rev.* **105**: 865-872.
(20) Martin, D. W. and Scherer, W. D. (1973). Review of satellite rainfall estimation methods. *Bull. Amer. Meteor. Soc.* **54**: 661-674.
(21) Marx, S. (1969). Über die extremsten Niederschlagsmengen auf der Erde. *Z. Meteor.* **21**: 118-119.
(22) McDonald, J. E. (1962). The evaporation-precipitation fallacy. *Weather* **17**: 168-177.
(23) Meinardus, W. (1934). Die Niederschlagsverteilung auf der Erde.' *Z. Meteor.* **21**: 345-350.
(24) Neumann, J. (1952). Evaporation from the Red Sea. *Israel Explor. J.* **2**: 153-162.
(25) Olascoaga, M. J. (1950). Some aspects of Argentine rainfall. *Tellus* **2**: 312-318.
(26) Paulhus, J. L. H. (1965). Indian Ocean and Taiwan rainfalls set new records. *Mon. Wea. Rev.* **93**: 331-335.
(27) Portig, W. H. (1964). Thunderstorm frequency and amount of precipitation in the Tropics, especially in the African and Indian monsoon regions. *Arch. Meteor. Geoph. Bioklim.* (B) **13**: 21-35.
(28) Ramage, C. S. (1951). Analysis and Forecasting of summer weather over and in the neighborhood of South China. *J. Meteor.* **8**: 289-299.
(29) Ramage, C. S. (1968). Role of a tropical "Maritime Continent" in the atmospheric circulation. *Mon. Wea. Rev.* **96**: 365-370.
(30) Roa, M. S. V. and Theon, J. S. (1977). New features of global climatology revealed by satellite-derived oceanic rainfall maps. *Bull. Amer. Meteor. Soc.* **58**: 1285-1289.

(31) Riehl, H. (1949). Some aspects of Hawaii rainfall. *Bull. Amer. Meteor. Soc.* **30:** 176-187.
(32) Riehl, H. and Byers, H. R. (1960). Computing a design flood in the absence of historical records. *Geof. Pura Appl.* **45:** 215-226.
(33) Riehl, H., Cruz, L., Mata, M. and Muster, C. (1973). Precipitation characteristics of the Venezuela rainy season. *Quart. J. Roy. Meteor. Soc.* **99:** 746-757.
(34) Riehl, H. and Lueckefedt, W. (1976). Precipitation and thermodynamic structure of rain events in Venezuela. *Mon. Wea. Rev.* **104:** 1162-1166.
(35) Sien, Chia Lin and Chang Kin Koon (1971). The record floods of 10 December 1969 in Singapore. *J. Trop. Geogr.* (Univ. Singapore and Malaya) **33:** 9-19.
(36) Sircar, N. C. R. (1955). Some aspects of monsoon rainfall in India. *Indian J. Meteor. Geoph.* **6:** 217-224.
(37) Solot, S. B. and Haggard, W. H. (1948). Relationships between large-scale atmospheric flow patterns and Hawaiian precipitation. *Trans. Amer. Geoph. Union,* **29:** 796-802.
(38) Supan, A. (1898). Die Jährlichen Niederschlagsmengen auf den Meeren. *Petermanns Geogr. Mitt.* H, 124.
(39) Taylor, R. C. (1973). "An Atlas of Pacific Islands Rainfall." Hawaii Institute of Geophysics, University of Hawaii, Honolulu.
(40) Wetherald, R. T. and Manabe, S. (1972). Response of the joint ocean-atmosphere model of the seasonal variation of the solar radiation. *Mon. Wea. Rev.* **100:** 42-59.
(41) Wuest, G. (1936). "Oberflächensalzgehalt, Verdunstung und Niederschlag auf dem Weltmeere nebst Bemerkungen zum Wasserhaushalt der Erde." p. 347. Festschrift Norbert Krebs, Stuttgart.
(42) Zubenok, L. I. (1965). World Maps of Evaporativity. *Soviet Hydrology* 274-289, *Trans. Voeykov Main Geoph. Obs.* **179:** 144-160. Translated by S. Molansky.

4
Vertical Energy Transfer

For the visitor to the tropics, the outstanding element of scenery is the cumulus cloud, set off against the blue sky or reflecting moonlight by night. Its graceful appearance on a travel poster may often have helped to lure him there. The cumulus cloud is the queen of beauty in the atmosphere, seen near to or far away; many astronaut photographs testify to the earth's varied colour display due to cumulus clouds. It is one of the amusements reserved for the meteorologist to stroll through a picture gallery and deduce the weather situation from the clouds appearing in landscapes, especially the cumuli. Since good painters are accurate observers, their clouds convey an excellent impression of the state of the atmosphere.

Cumuli are not all serenity; they act to produce turbulence and much needed rain, and their growth, movement and decay reveal secrets of the air circulation on varied scales. The structure and maintenance of great vortices, even hurricanes, is governed in part by the ability or inability of great cumuli to develop. Cumulus convection is the most important mechanism for funnelling heat upward in the tropics, and it provides a basic link in maintaining the general circulation through so-called *scale interaction*.

As he watches the tropical sky, the meteorologist soon realizes that the cloud sequence is not the same day after day. The average height, thickness and distribution of cumuli change. Sometimes they are erect, and sometimes they slant with height. Often they are arranged in neat rows parallel or perpendicular to the wind or vertical shear of the wind, and often no such pattern can be discerned. At any time, we find a great variety of cloud forms, sizes and heights in the sky. Cumuli go through a life-cycle of variable length and degree of cloud development observed most readily by time-lapse motion pictures projected rapidly, so that the events of one minute pass before our eyes in one second. If the clouds are trade cumuli of average vertical thickness, roughly 1500-2000 m at peak development, their life has a length of perhaps 30 minutes. The growing stage is brief, about 10 minutes; the decay lasts longer, 20 minutes or more. Therefore, the sky contains far more clouds that have reached maturity or are dying than clouds that are growing. Clouds rarely persist even for a few minutes in "steady state", i.e. without changing shape and structure.

When looking at a sky with tall clouds, perhaps including cumulonimbus, we can expect to find a longer life-span. The period of growth may last 30 minutes, and although the average life of a cumulonimbus is 60-90 minutes, it may also be protracted over many hours. While the naked eye sees one great cloud mass, sophisticated radars reveal active cells contained within it. Many such cells can grow and decay during the life of the whole cloud as the activity seen by the intensity of radar energy return keeps shifting to new levels. Cumulonimbi will also band together into mesoscale cloud masses where the extent of a radar echo covers an area of 2000-3000 km^2, and even more, as was described at the end of the last chapter.

These sketchy observations may serve to introduce the many interesting things found in a cumulus sky. The underlying theme is "vertical energy transfer". The prime energy source for the atmosphere is at the earth's surface, yet ways and means must exist for this energy to reach the vicinity of the tropopause. Only there do other sources for providing radiation balance independently make their appearance. This is still a wide field for investigation, a field in which many data have been collected, in which theory has advanced rapidly and which continues to be the object of intense research. We need to make a survey of vertical energy exchange mechanisms in order to understand the topics that follow. Of course, vertical energy transfer takes place at all latitudes, but, in contrast to dynamics where the tropics mainly provide a variant of general laws at low Coriolis parameter, the subject of energy transfer contains aspects which are of primary concern in the tropics. To a considerable extent, the general circulation and its synoptic weather systems utilize the latent heat of condensation as direct energy source. Yet the release of the latent heat takes place in minute fractions of the area of the tropics, as demonstrated with emphasis in Chapter 3. These minute areas, and all that takes place in them, are not caught readily by the grid networks used in modern weather and climate analysis and prediction with numerical means.

In order to acquire some understanding of convective energy transformations and also the many other manifestations of the tropics dependent on vertical overturning, we shall take up (a) the layer below the clouds, (b) the onset, structure and life-cycle of clouds and convective ensembles, and (c) convective modelling.

The subcloud layer

Release of latent heat of condensation is the major energy source for the building and driving of cumuli. To initiate convection, however, another energy source and a dynamic mechanism are required. Consider a normal situation with tall cumuli and precipitation, which complicate the picture. Initial energy will be furnished by conversion of some of the energy of the general air stream, especially through its vertical shear, into that of smaller "eddies" (64). A dynamic mechanism is contained in the vertical equation of motion which relates vertical acceleration to imbalance between pressure and gravitational forces, known as *buoyancy* (Eq. (4.15)). Heating of land in daytime by the sun produces an unstable thermal stratification along the vertical. Soon, overturning sets in and "air parcels" or "columns" become accelerated upward in "thermals". Over the ocean, such strong heating cannot occur and the sea-air temperature difference is small compared to land. Nevertheless, the ocean must be at least 0·5 to 1°C warmer than the air in order for convection to occur at all; when the reverse is observed all convective motion dies out.

The dynamic mechanism must also be capable of constraining air motion into convergent flow patterns to enable air columns to respond to buoyancy accelerations. The structure of wind eddies created from the mean wind energy must be such as to permit the occurrence of vertical currents up, and compensatory currents down as long as the subcloud layer is not acted upon from above through interference by clouds. That the wind eddies have such a structure for a good part of the time at least, is revealed by convection cell patterns in the subcloud layer made visible with chemical smoke. Figs 4.1 and 4.2 are aged, yet immortal photos of a very fine experiment conducted by Woodcock and Wyman (91) with ship and aircraft. The pattern of Fig. 4.1 drifted downstream with the wind. Whereas the direction of the axis changed more than 45° from one portion of the smoke plume to another in the picture, the surface wind on the ship fluctuated by no more than 5° in the hour prior to smoke release. If such angular departures of the wind direction from the mean occur in an ordered

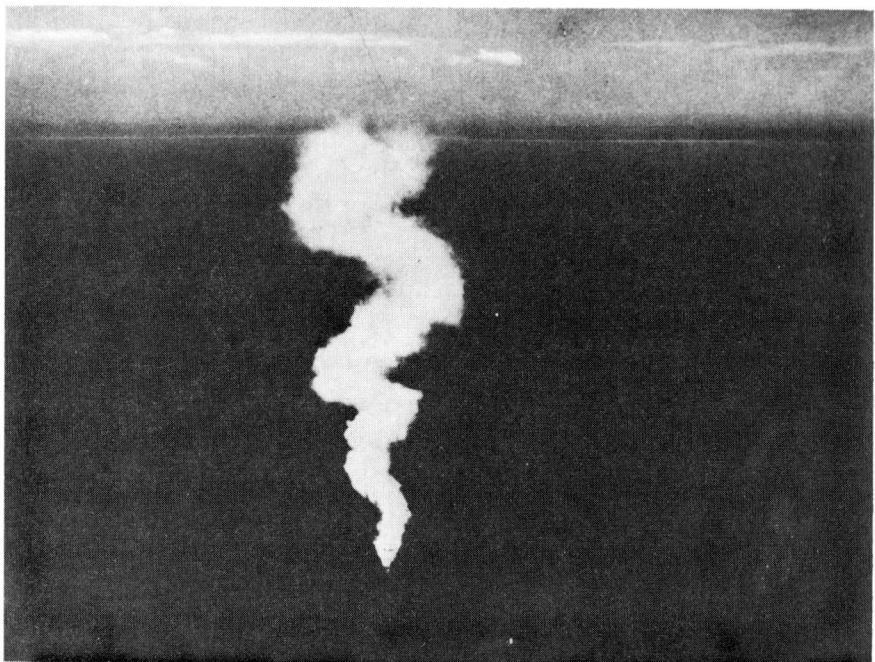

Fig. 4.1. Smoke from stationary source near Panama, seen from above (91). Courtesy of New York Academy of Sciences.

pattern of the Bénard cell type (6), the measured lateral displacement of the smoke can be explained in terms of hexagonal cells. When vertical wind shear is present the cells become elongated roll vortices oriented along the shear vector.

These displacements, denoting convergence and divergence patterns, must be accompanied by vertical motions, and these do exist (Fig. 4.2). In this case the plume had been laid parallel to the wind by aircraft flying near 100 m altitude. Vertical motion of the plume attained 1 mps which, over the short vertical distance involved, can only be produced by buoyancy accelerations. The average height of the cells appears to be about 300 m, half the length of the cell sides, in agreement with Bénard's experiments. The top of the cells coincides with the altitude at which the stratification of the subcloud layer becomes thermally stable.

Figures 4.1 and 4.2 essentially provide a model and setting for the occurrence of vertical energy fluxes expanded in model form by LeMone (41). With this structure in mind, we turn to energy transfer at the air-ocean interface.

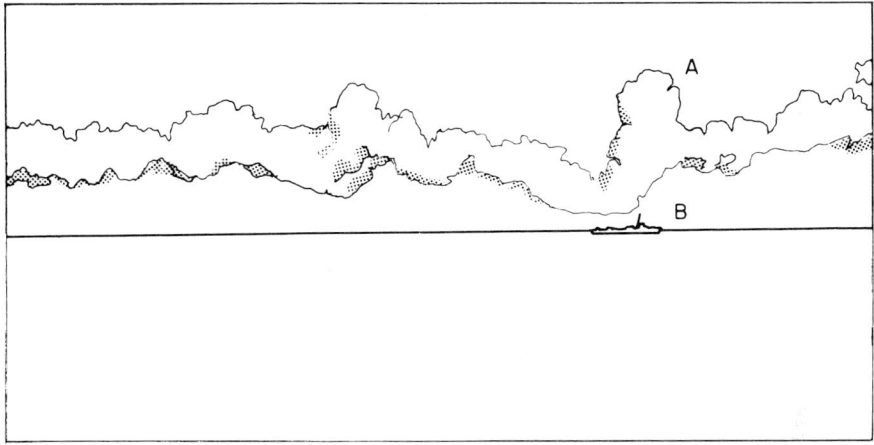

Fig. 4.2. Vertical distortion of smoke laid parallel to wind by aeroplane flying horizontally at 100 m altitude (91). Courtesy of New York Academy of Sciences.

Energy exchange near the sea surface

Energy is transferred from ocean to atmosphere when downward currents, as contained in the convection cell model, descend to the ocean surface where they are heated and absorb moisture through evaporation. The descent at point B in Fig. 4.2 illustrates this motion. For energy transfer to take place, the sea surface temperature must be higher than the temperature of the air coming in contact with the water, and the vapour pressure over the water must exceed that of the air. Since the vapour pressure over the ocean surface is usually assumed to be equal to that over a flat water surface at saturation, the water temperature determines the vapour pressure over the surface. The relative humidity of the air is nearly always less than 100% and the vapour pressure difference between sea and air remains substantial over most parts of the oceans with average values of 7-10 mb. Therefore, both latent and sensible heat will be transferred from ocean to air when the sea surface temperature exceeds air temperature. The ocean does the work required for the evaporation process. In the special case when air temperature equals water temperature, energy exchange at the interface can still take place. The ratio of molecular weight of water to that of air vapour is only 0·62. Thus, masses of air adding moisture but not sensible heat from the ocean still will be less dense than the surrounding atmosphere and become accelerated upward. This is called water convection, a potentially important process at least.

The energy exchange per parcel of air in unit time will increase with the temperature and vapour difference between sea and air. Strong wind will serve to sustain both gradients, since the turbulent wind energy gained from the wind shear will act to ventilate the sea surface rapidly. Thus, wind speed is one obvious parameter in computing surface fluxes in addition to the vertical differences of temperature and vapour pressure. These parameters are not fully independent, since wind speeds affect the vertical gradients. In touching the ocean, air particles also change their wind speed because of the frictional breaking action of the water. In the trades, this means that the east wind is reduced or, as it is frequently stated, that the air has received westerly momentum from the earth (Chapter 1). Ascending particles should, therefore, have moisture and temperature above average and wind speed below average when the 50-100 m distance scale of dry convection cells is the unit length of computation. For steady state in the air the mean wind speed is usually maintained by flow of air toward lower pressure (Chapter 5). The heat and moisture acquired from the ocean must be removed to higher levels or advected downstream.

A special case not really covered in the foregoing or in the following discussion is a spray-covered ocean at high wind speeds especially, of course, in hurricanes. The air-ocean boundary essentially vanishes and the area of contact between air and ocean increases vastly. All exchanges are greatly enhanced when referred to the path of innumerable water particles through the air. Information on spray, its density, height and residence time in the air has been too sketchy to construct an air-sea exchange model including spray.

Transfer formulae

We have now acquired a physical picture of the exchanges between ocean and atmosphere. Both differences in heat and moisture between ocean and atmosphere and wind speed in the air are major variables that have come to light, but, unless a model is formulated that is explicitly based on convection cell structure, we must still seek a method for actual transport calculations. Three approaches have been followed for this purpose.

(1) The *energy balance method* can be applied to land or ocean. The radiation surplus stored in the earth's surface is exported to the atmosphere through sensible and latent heat transfer. All heat stored over land must be given off to the atmosphere at every point, omitting very small storages in the ground and heat transfer within the earth.

VERTICAL ENERGY TRANSFER

Over sea, the portion of surplus energy received and not carried away by ocean currents or mixed downward inside the ocean must be given off by the same processes in steady state.

(2) The *turbulent flux method* is considered to be the most accurate atmospheric technique as long as the transfer from land or ocean takes place mainly in convection cells with length up to 1000 m and, also, longer cells with a wavelength of perhaps up to 50 km. Both of these scales are seen in Fig. 4.3, which depicts turbulent variations of temperature, density of water vapour and vertical motion over an aircraft leg of 33 km length. The mean values of these quantities over the leg have been subtracted, as have also linear changes from beginning to end which would indicate the presence of still longer waves. (For instance, for determining the deviation of a day's temperature from average, the mean annual temperature wave must be excluded, else all summer departures will be positive and winter departures negative—not very informative.) In so doing, the principal expectation is that the net mass flow averaged over the distance either vanishes or is unimportant. Updrafts and downdrafts are viewed as turbulent motions with zero net upward or downward transport. If indeed our observational distance has captured the scales, contributing all or nearly all of the variances in temperature, moisture and vertical motion, we are justified to write the flux formulae

$$Q_s = \overline{(\varrho w)'(c_p T)'}, \qquad (4.1)$$

$$Q_e = L\overline{\varrho_v' w'} = \bar{\varrho} L\overline{w' q'}. \qquad (4.2)$$

Q_s and Q_e are sensible and latent heat transfer, ϱ_v is density of water vapour and, here, the bar denotes averages of individual products $T'w'$ and $\varrho_v' w'$ over the length interval or time of measurement. The ratio Q_s/Q_e is widely referred to as the *Bowen ratio*.

For many years direct flux measurement with Eqs. (4.1) and (4.2) were considered the ultimate which, however, could not be attained on account of instrumental shortcomings. The lag of instruments sensing temperature and moisture was too great for measuring from an aeroplane flying at 100 m/s; a deficiency since remedied. It was the measurement of vertical air motion, however, that was most difficult to engineer. Only from 1972 onward have so-called "gust probes" satisfactorily determined vertical motion in air from aircraft. Given a 500-1000 m "wave length" for important small scale fluctuations contributing importantly to the fluxes, readings from a 100 m/s aircraft must be made at least once per second, and none of the

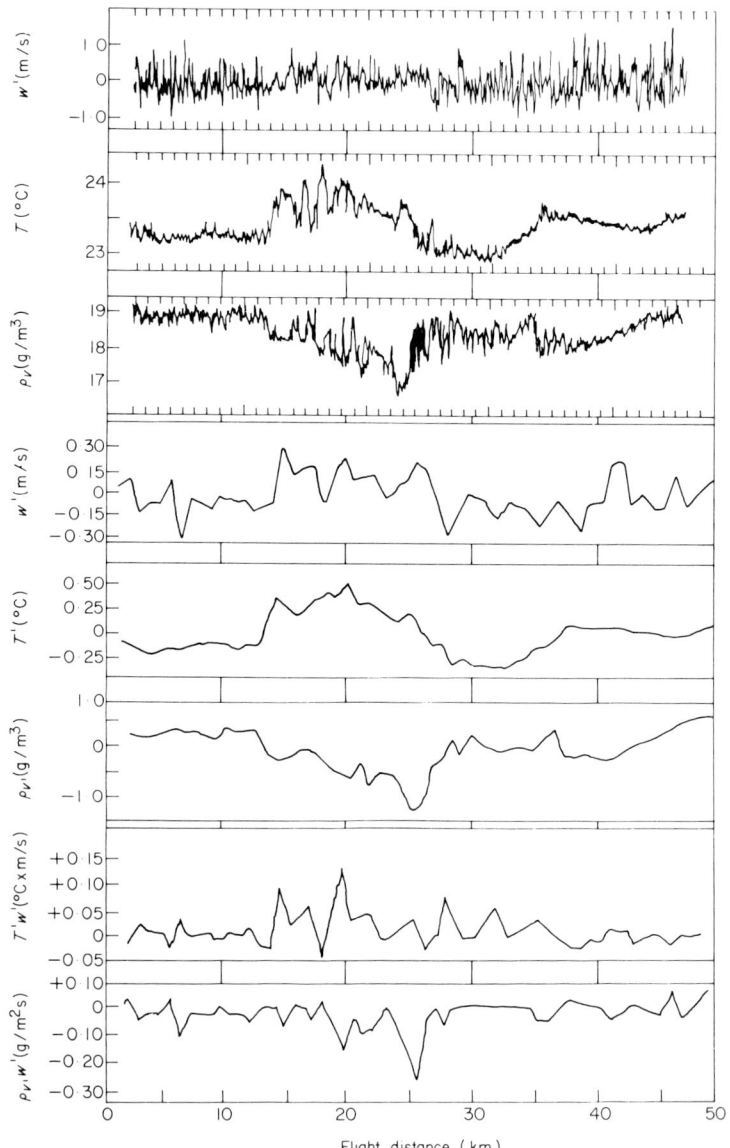

Fig. 4.3. Upper: fluctuations of vertical motion, temperature, and density of water vapour during GATE measured by DC-6 aircraft at 150 m height on approach to weather-active area near 8°N, 23°W in Atlantic off West Africa on 10 August, 1974. Middle: 10s averages. Lower: cross-products $T'w'$ and $\varrho'_v w'$ (71).

measuring instruments may lag the others. Computations were made at a rate of 20 per second in the experiment shown in Fig. 4.3. The fluxes were then added to 10 second values for a legible presentation; in such an addition the transport by individual convection cells, of course, is lost, as each summation covers 10 to 20 of these.

The sensible heat transfer has been related by Priestley (A-14) to the difference between actual and dry-adiabatic temperature lapse rates, which is important for vertical flux profiles when the atmospheric structure differs from adiabatic. Use of $\theta'w'$ also takes the lapse rate into account (43). Brook (14) points out that in evaluating Eq. (4.1) the difference in specific heat of water vapour (c_{pv}) and of dry air (c_{pd}) must be taken into account, since c_{pv} exceeds c_{pd} so that $c_{pv}/c_{pd} = 1.83$. The heat supply for warming one gram of water vapour by 1°C is almost double that required for heating dry air. Following the type of derivation found in textbooks for virtual temperature he finds that

$$c_p T = c_{pd}(1 + 0.83q) \cdot T = c_{pd} T_c. \tag{4.3}$$

Then Eq. (4.1) becomes

$$Q_s = \bar{\varrho}\, c_{pd} (\overline{w'T'} + 0.83\, \overline{w'q'T}) \tag{4.4}$$

omitting two small terms. Introducing Eq. (4.2), the second term becomes $(0.83\, \bar{T}\, c_{pd}/L)\, Q_e$. Given $\bar{T} = 300$ K near the surface in the tropics, the constant is 0.1 almost exactly so that

$$Q_s = \bar{\varrho}\, c_{pd}\, \overline{w'T'} + 0.1\, Q_e. \tag{4.5}$$

Thus, Q_s is augmented by 0.1 of the latent heat flux. The correction matters little at large values of the Bowen ratio such as are found over land. Over the tropical oceans, however, outside of zones of active convection, Q_s/Q_e is very small, averaging perhaps no more than 5% and in many areas less. This means that the correlation $\overline{w'T'}$ is very small for reasons to be explored shortly. Therefore, over the tropical oceans at large the expansion of Eq. (4.1) into (4.5) is highly important. In particular, Q_s may be positive even when $\overline{w'T'}$ is negative.

(3) *Modelling* of surface-air transfer was a long-standing substitute for direct flux measurements. Of course, models using mean quantities read from synoptic and even climatic charts will always have to be used for transfer determinations since the globe cannot be covered constantly with fleets of aircraft and ships

measuring fluxes directly. A vast amount of literature on modelling the lowest part of the "boundary layer", often referred to as "constant flux layer" has accumulated. Whole textbooks have been written on air-sea exchange (A-10, A-20), to which the reader is referred for the derivations of "bulk transfer formulae". Only their application to the tropical oceans will be considered here.

The background for the transport formulae comes from molecular physics; a transfer coefficient is supposed to exist so that the ratio of flux to gradient is constant, as introduced in Chapter 2. In applying this concept to the turbulent atmosphere, the molecular transfer coefficient is replaced by an "eddy coefficient" K_h or K_v which is greater than the molecular one by several orders of magnitude. We may then consider the fluxes across a ship's deck level (6-10 m) which is certainly in the constant flux layer, i.e. any small storage in the one thousandth of the atmosphere's mass between the surface and 10 m height may be safely neglected. The bulk transfer formulae then are

$$\frac{Q_s}{-d\overline{T}_c/dz + \Gamma} = c_p K_h, \qquad (4.6)$$

$$\frac{Q_e}{-d\overline{e}/dz} = L K_v, \qquad (4.7)$$

where e is vapour pressure, Γ dry-adiabatic lapse rate and the bar denotes averaging carried out over ranges of space or time, from minutes or a day in small areas ranging to mean annual values. Of course, not all of these averages can give a correct result.

In principle, the relations are clear enough. When a warmer and a colder body come into contact, heat flows from the warmer body (in our case mainly the ocean) to the colder body (the air) and this is well portrayed by Eq. (4.6). The vapour pressure difference between ocean and air certainly denotes a clear mechanism for moisture transport. Because of the well-known approximation $q = 0.62 \, e/p$, the quantity dq/dz is often introduced in Eq. (4.7) for convenience of computation. Then K_v must be multiplied by $p/0.62$.

Momentum is exchanged between ground and atmosphere through frictional stresses. The third transfer equation is:

$$\frac{\tau_s}{-d\hat{V}/dz} = K_M \text{ (dimensions m l}^{-1}\text{ t}^{-1}\text{).} \qquad (4.8)$$

VERTICAL ENERGY TRANSFER

On introduction of aerodynamic similarity considerations and the concept of a laminar boundary layer with a "roughness length" (Priestly and others) $K_M = \varrho_s C_d \overline{V} H$, where C_d is an aerodynamic drag coefficient and V the velocity at 10 m height if $H = 10$ m. After integration,

$$\overline{\tau_s} = \varrho_{10} C_d \overline{V_{10}^2}. \tag{4.9}$$

Sometimes $\overline{V^2}$ from climatic charts is used instead of $\overline{V}^2$, a dubious assumption since a surface wind observation averaged over one to five minutes is already a mean value with respect to dry convection cells.

Since the temperature lapse rate governs thermal stability, one would expect the coefficient C_d in Eq. (4.9) to depend on the thermal stratification. Actually, the equation is fully valid only for adiabatic and near-adiabatic stratification. For this "neutral" case, widely found over the tropical oceans, we may introduce K_M as defined above into the expressions for sensible and latent heat transport. Then

$$K_H = K_M,$$

$$\frac{p}{0 \cdot 62} K_v = K_M.$$

After integration and introduction of finite differences,

$$Q_s = \overline{\varrho_s} c_p C_d (\overline{T_w} - \overline{T_{as}}) \overline{V_s} + 0 \cdot 1 Q_e, \tag{4.10}$$

$$Q_e = \overline{\varrho_s} L C_d (\overline{q_w} - \overline{q_{as}}) \overline{V_s}. \tag{4.11}$$

The subscript s denotes surface values at about 10 m height, T_w and q_w are temperature and specific humidity at the water surface. These equations should give the same results as Eqs. (4.1) and (4.2) evaluated with, say, the turbulent flux technique. Kraus (A-10, pp. 147-151) has treated modifications of the equations for stable and unstable stratification. The drag coefficient becomes smaller for stable and larger for unstable air. When $T_w - T_{as}$ is much in excess of 1°C, modified formulae must be used. For instance, with $T_w - T_{as} = 6$°C during cold outbreaks from China over the South China Sea, Eqs. (4.10) and (4.11) underestimated the energy transfer from sea to air, determined from an energy budget, by a factor of two which, of course, is a gross error (69).

Even under neutral conditions, computations with Eqs. (4.10) and

(4.11) may be as much as 50% off. In spite of concentrated efforts, this range has not narrowed over the years. Thus some important factors must be missing in the equations, the sea state, for instance, which does not depend on instant local conditions most of the time. (The best determination of C_d often can be made from observations of the ocean.) Wind speed over the tropical oceans averages 6-7 m/s; there is always a tendency to return to this mean which, from Eq. (4.9), suggests a threshold value. At one time the ocean was considered "smooth" below the threshold and "rough" above it. This concept has not held (13); transition to increasing roughness appears to be continuous. Functions of $C_d(V)$, if known, can be readily incorporated into computer models (75). Nevertheless, the thought that the roughness of the sea surface changes most rapidly at wind speed of 7 m/s is attractive and continues to be kept alive (89). Because the speed of the trade winds oscillates about the mean in a semi-cyclic fashion, the total evaporation over broad ocean areas varies by large percentages from one week to the next, affecting the general circulation. Thus, the energy exchange at the sea surface is not only important for the production of cumuli, but is also a factor in large-scale and longer period circulation changes (50).

Structure of the subcloud layer

Temperature and humidity

An early notion of subcloud layer structure was that of a fully mixed layer with potential temperature, specific humidity and also momentum constant above the lowest part of the boundary layer, a state that could be brought about by small-scale fluctuations "diffusing" heat, moisture and momentum upward. In reality, the layer is fully mixed only in rare instances, so that the "mixed layer concept" will hold only in the first approximation. Temperature, or potential temperature decreases by, perhaps, 1°C very close to the sea surface in the trades at large. Then potential temperature is constant to about 200-300 m, followed by a slow upward increase at a rate of about 0·2°C/100 m to the top of the subcloud layer. Such a profile is seen in Fig. 4.4(a), a sounding from the research vessel *Meteor* in a general rain area. Another profile (Fig. 4.4(b)), taken in the northeast trades two days earlier, is slightly more unstable until a sharp decrease in temperature lapse rate is encountered near 950 mb, called the "transition layer". In contrast, over land in daytime (Fig. 4.4(c)) the lapse rate is superadiabatic almost to 950 mb and then dry-adiabatic through a deep layer almost to 850 mb.

Mixing ratio is more variable than temperature. In the mean oceanic atmosphere, for instance, specific humidity decreases from the ground upward (Fig. 2.18). Such a decrease is typical for the great majority of radiosonde ascents of all types of manufacture. However, in the convergent atmosphere in which the sounding of Fig. 4.4(a) was launched, specific humidity is constant above a sharp drop just above the ocean, very favourable for inception and growth of convective clouds. In Fig. 4.4(b) specific humidity decreases through the lower half of the subcloud layer and then becomes nearly constant in the upper half, a much less favourable structure. Very often moisture just

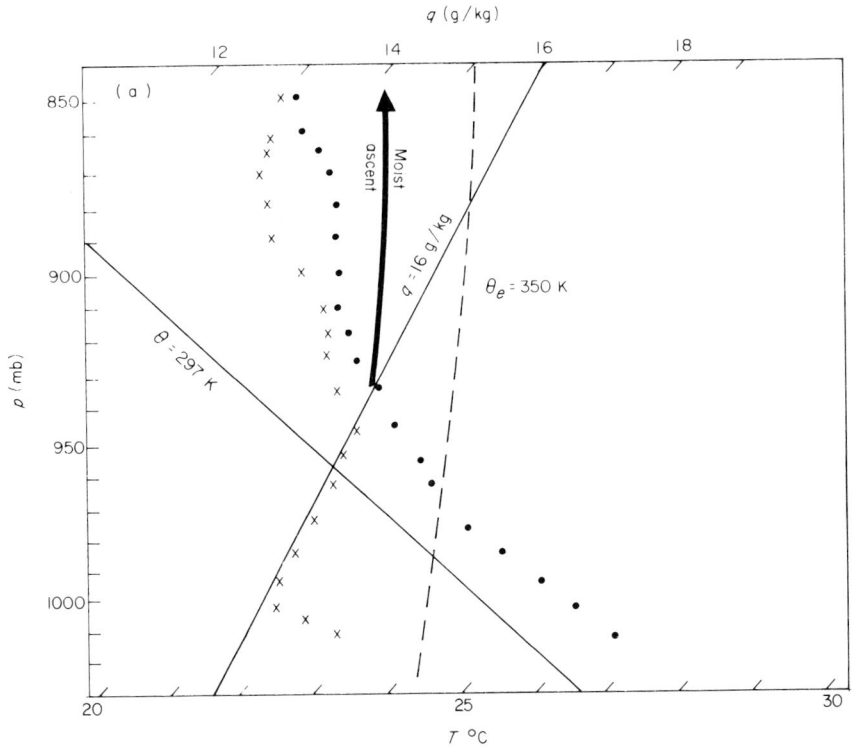

Fig. 4.4. Tephigrams showing structure of subcloud and lower cloud layers for (a) equatorial convergence. R.V. *Meteor* in tropical Atlantic 15 February, 1969; (b) northeast trades, R.V. *Meteor* 12 February, 1969; (c) over land at Carrizal, Venezuela (9·5°N, 67°W) 24 July, 1972, 1730 LT. Dots: temperature; crosses: dewpoint.

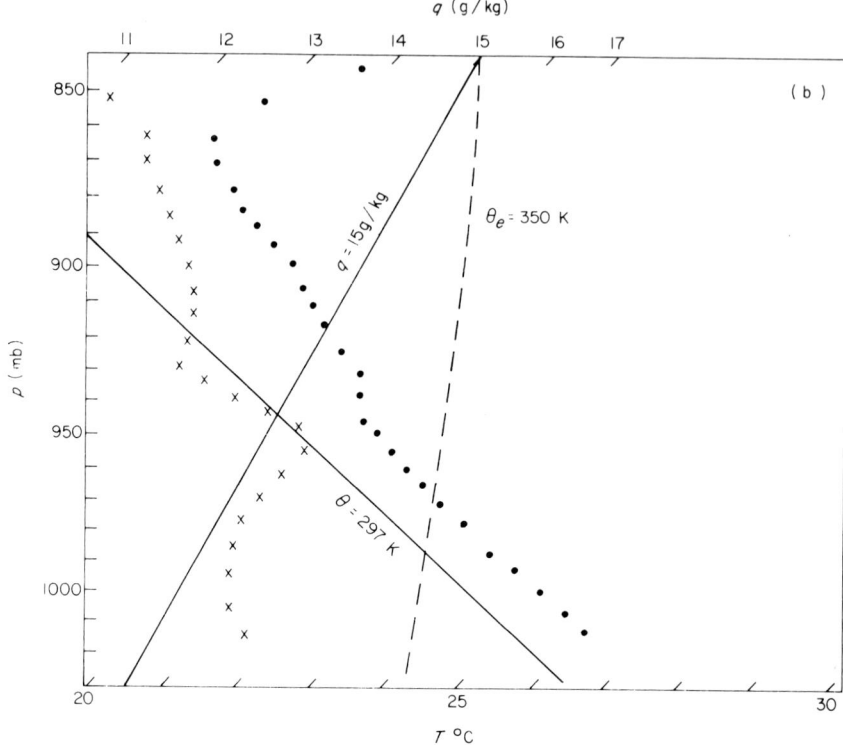
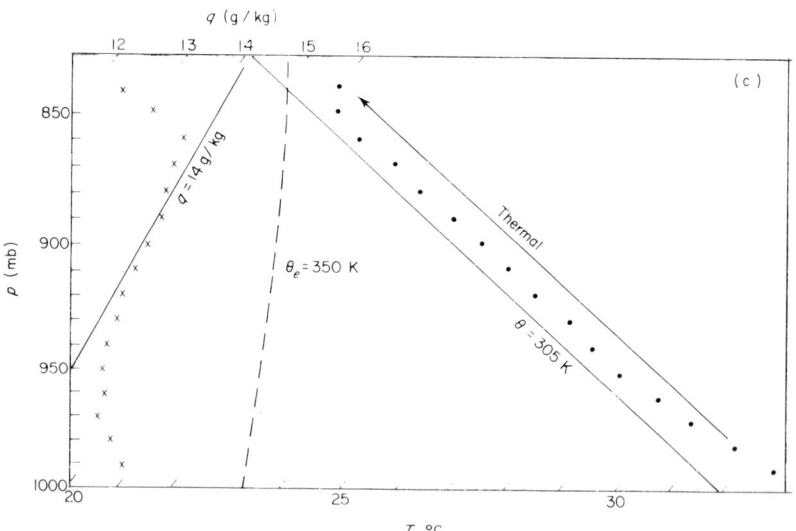

Fig. 4.4 continued.

keeps decreasing to the top of the subcloud layer. Over land (Fig. 4.4(c)) the moisture profile has a rather similar shape. Above a marked decrease to 950 mb, specific humidity decreases only very slowly to 860 mb where the base of a dry and stable layer is situated.

Betts (9) has computed composite soundings for ascents into clouds and between clouds for Venezuela. A plot of the mean of 14 radiosonde ascents followed through cloud base indicates the top of the superadiabatic layer at 0·3 of the pressure difference (88 mb) between cloud base and the surface (p^*) or near 970 mb (Fig. 4.5). From there potential temperature is practically constant to cloud base. Specific humidity also decreases near the ground, then rises slowly and starts to drop again below cloud base. Evidently, the balloons did not ascend in a "thermal". Nevertheless, on comparing this mean sounding with another one taken between clouds, the subcloud layer is distinctly more humid underneath clouds while potential temperature is nearly constant up to about $p^* = 0.6$. From there potential temperature increases between clouds, and the increase becomes most rapid in a shallow layer near cloud base marked "transition layer" (16). Specific humidity decreases markedly, starting already below cloud base. This transition layer is also

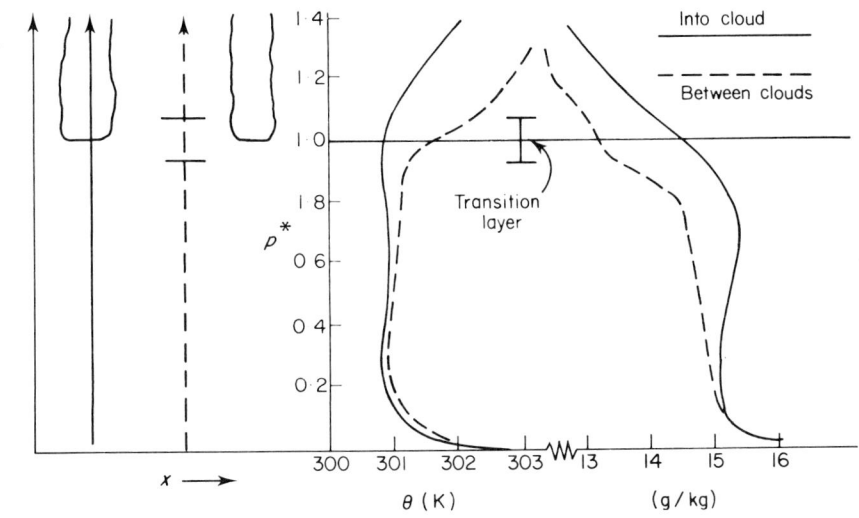

Fig. 4.5. Profiles of potential temperature and mixing ratio (r) for the subcloud and lowest part of cloud layer at Carrizal, Venezuela (9·5°N, 67°W) for balloons entering cloud base (solid) and for balloons bypassing cloud (9).

observed over sea (Fig. 4.4(b)); land and sea structures are remarkably similar. Combination of thermal stability with drying always denotes subsidence, so that the "transition layer" is one of compensation against the upward mass flow in clouds.

Turbulent mass exchange

Thermals with undilute ascent can be found readily over land and at least occasionally over sea. During a typical ascent of research aircraft in Venezuela close to the equator, moisture decreased in general upward through the subcloud layer (Fig. 4.6), but "spikes" of humidity with values equalling surface specific humidity continued to be encountered by the aircraft all the way to 850 mb, the cloud base height. The same aircraft encountered moist spikes over the Caribbean, off Venezuela (Fig. 4.7). There the condensation level was close to flight altitude near 900 mb. Again, moisture values equalled those at the surface and this demonstrates that masses of air from the lowest levels can pass through the whole subcloud layer without dilution and that the roots of cumuli can be found well inside the subcloud layer (23). As in Fig. 4.5, temperature at cloud base was below ambient air temperature, about 0·5°C.

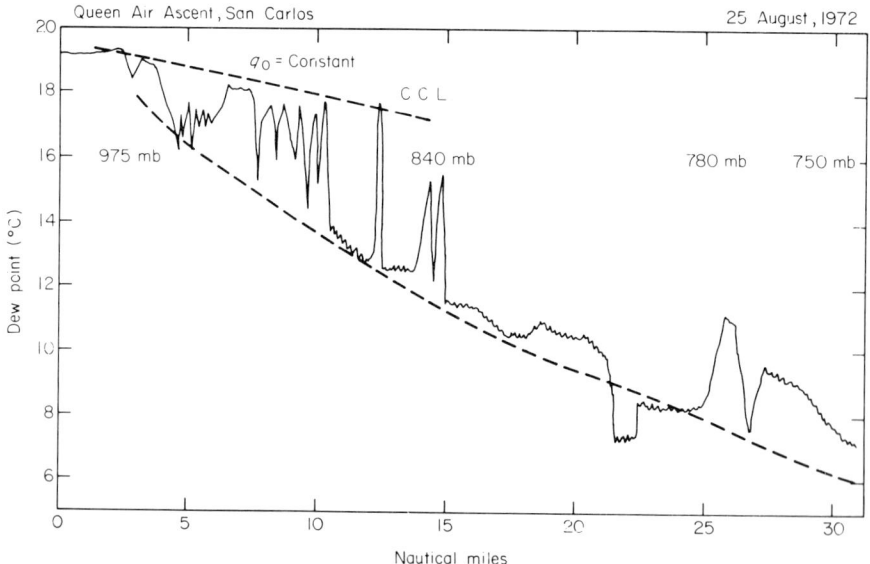

Fig. 4.6. Dewpoint fluctuations during ascent of research aircraft near San Carlos, Venezuela (2°N, 67°W), 25 August, 1972.

VERTICAL ENERGY TRANSFER

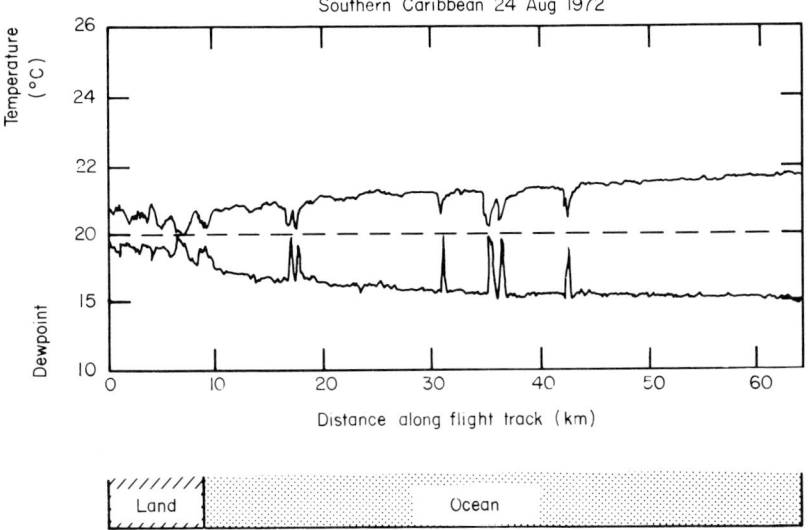

Fig. 4.7. Spikes of moisture as research aircraft passes through the base of small undilute cumuli just off central Venezuela over the Caribbean on 24 August, 1972. Range of dewpoint corresponds to range of 5 g/kg specific humidity (71).

How do these turbulent motions occur? The classical answer is drawn from the vertical equation of motion which, omitting frictional retardation, is

$$\frac{dw}{dt} = -\frac{1}{\varrho'}\frac{\partial p'}{\partial z} - g, \qquad (4.12)$$

where the primes refer to the properties of a particular mass followed during upward or downward acceleration. Now one postulates that the surrounding mass in the environment is accelerated neither upward nor downward so that $-\frac{1}{\varrho}\frac{\partial p}{\partial z} - g = 0$; the balance is hydrostatic. Although the pressure force in Eq. (4.12) must contain a non-hydrostatic part, it is a premise in nearly all convection work that the accelerated mass immediately assumes the pressure of the surroundings so that $\partial p'/\partial z = \partial p/\partial z$. If this assumption is accepted, the pressure gradient is eliminated from Eq. (4.12), so that

$$\frac{dw}{dt} = g\frac{\varrho - \varrho'}{\varrho'}. \qquad (4.13)$$

Now density is not measured directly; it can be eliminated through the equation of state for air. In the tropics and in higher latitudes during summer it is important to use the equation of state for moist air which includes water vapour with lower density than dry air ($m_v/m_d = 0\cdot62$, where m_v and m_d are molecular weights of water vapour and dry air). The inclusion of water vapour is conveniently carried out by the definition of virtual temperature (T_v),

$$T_v = T(1 + 0\cdot61 q). \tag{4.14}$$

Since both dry temperature and relative humidity (convertible to q given $T(p)$) are routinely measured at the surface and in radiosonde ascents, T_v is readily available. Approximately, T_v increases 1°C for every 6 g/kg of water vapour added; this is not a small quantity when density differences are desired for evaluating Eq. (4.13) in the tropics! For the dewpoint difference of 20-16°C at 900 mb in Fig. 4.7, the specific humidity difference is 3·5 g/kg, so that the virtual temperature correction is slightly more than 0·5°C cancelling the negative dry temperature difference. Thus, the case of Fig. 4.7 essentially represents "water convection".

With use of Eq. (4.14), the "buoyancy" equation (4.13) becomes

$$\frac{dw}{dt} = g\,\frac{T'_v - T_v}{T_v} \equiv g', \tag{4.15}$$

a widely used expression. Masses with higher T_v than their surroundings will be accelerated upward. Masses with lower T_v are accelerated downward. In the subcloud layer the process described by Eq. (4.15) will be most effective if air that has descended to the surface picks up both heat and moisture; then T_v of the ascent will be definitely higher than T_v of descent in the subcloud layer. In Fig. 4.7, $T'_v - T_v = 0$, i.e. the situation is neutral, but a new element, not yet introduced, arises due to cloud formation with release of latent heat of condensation.

One can determine from aircraft observations how far Eq. (4.15) is valid by computing the quantity $\overline{T'_v w'}$, the vertical density transport, not to be confused with heat transport. As long as this correlation is positive, convective elements are buoyant up and down, at least in the average over the sample distance chosen. In measurements in the subcloud layer of the trades (42) and the equatorial trough $\overline{T'_v w'}$ turned out to be larger and positive to higher elevations in the subcloud layer than $\overline{T' w'}$. Herewith, it has been demonstrated that the water vapour picked up by the air from the oceans plays an important role for positive buoyancy itself.

We now return to the problem of Fig. 4.5. There, the upper part of the subcloud layer is warmer and drier than the ascent paths of air forming clouds higher up. The (potential) temperature difference at cloud base is $-0.5°C$ seen from the viewpoint of the rising mass; the moisture excess is only 1 g/kg, so that $T'_v - T_v$ is negative, therewith also the buoyancy acceleration $g' = g\,(T'_v - T_v)/T_v$. One can hypothesize that acceleration in the lowest part of the subcloud layer is large enough so that an ascending plume, though decelerating, will still reach the condensation level where release of latent heat stretches out a helping hand. Newton and Newton (60) have proposed an alternate mechanism to overcome the difficulty. The buoyancy acceleration g' in Eq. (4.15) acts to increase the potential energy of the air during ascent, or lower it during descent; but thermal buoyancy is not the only means to bring about potential energy changes. From classical mechanics $P + K$ = constant, when there are no other energy transformations. In the atmosphere, variations in K are generally neglected compared to those in P, since they average two orders of magnitude smaller. However, changes in kinetic energy are not to be compared properly to changes in gz during free fall but to changes in $g'z$, since only the small departures from hydrostatic balance need be overcome. A small amount of "available kinetic energy" may suffice to supply the required increase in potential energy for air to reach the condensation level, which in this case would be called "lifting condensation level". If the downward acceleration is fully balanced, the kinetic energy ($K = V^2/2$) following the moving mass along the vertical (d/dz) will change according to

$$\frac{dK}{dz} = -g'$$

and

$$dz = -\frac{dK}{g'} = \frac{-d\left(\frac{V^2}{2}\right)}{g\frac{T'_v - T_v}{T_v}}. \qquad (4.16)$$

For $dK = (15 \text{ m/s})^2/2$ and $g' = -10^{-2}g$, $dz = 1$ km. In this formulation, the order of magnitude difference between K and P has disappeared, since hydrostatic balance is chosen as a reference state of rest, and rising parcels or masses of air will readily overcome negative values of $10^{-2}g$, or $T'_v - T_v = -3°C$ in the tropical atmosphere, if the kinetic energy can be reduced by the stated amount over the vertical distance dz. Evidently, this mechanism will be more effective at strong than at light general winds.

The foregoing has shown two ways by which turbulence in the subcloud layer may be sustained in the face of very weak or even negative values of $\overline{w'T'}$. In a series of 12 traverses near 300 m altitude made by Bean et al. (unpublished) between the equator and 10°N on the route between Hawaii and Tahiti between November 1977 and January 1978 $\overline{T'_v w'}$ was consistently positive at that height whereas $\overline{T'w'}$ was small, zero or negative. Outside a narrow equatorial trough zone this route leads over undisturbed or even suppressed subcloud layer conditions. In and near heavy convection, of course, other factors may enter, clearly indicated in Fig. 4.3, where temperature and specific humidity are out of phase on the 20 km scale which dominates vertical eddy transport. Vertical motion was positively correlated with dry temperature. Thus, sensible heat flux was up, but latent heat flux was downward and it exceeded the heat flow in magnitude. It is expected that wide experimentation with gust probe measurements will reveal many of the intricate details of ground-cloud interaction across the subcloud layer.

It may be mentioned here that no temperature sensor has as yet been developed which does not act as a "wetbulb" on entering cloud, whether the carrier be aircraft or radiosonde. Temperature drops regularly on entering cloud; one never knows in cases such as Fig. 4.7 whether the drop is real or whether the temperature element is "wetbulbing". For this reason the resistance against cumulus growth near cloud base often appears exaggerated. It seems that there is only a "lifting condensation level" and no "level of free convection". Probably at least a good portion of such "evidence" rests on inadequate instrumentation.

Turbulent energy transport

Measurements made using Eqs. (4.1)-(4.2) should yield upward flow of moisture throughout the subcloud layer—ascent coupled with high moisture, descent with low moisture. Heat transport should be downward at cloud base where potential temperature between clouds increases upward sharply. The transport $\overline{T'w'}$ should remain negative until far down in the subcloud layer where θ becomes constant, but, on account of the correction term in Eq. (4.5), sensible heat transport

VERTICAL ENERGY TRANSFER

should be positive to much higher elevations. Exactly this result was obtained in a summary of aircraft measurements made during the rainy season of 1974 in GATE (Fig. 4.8) of which Fig. 4.3 is one example (71).

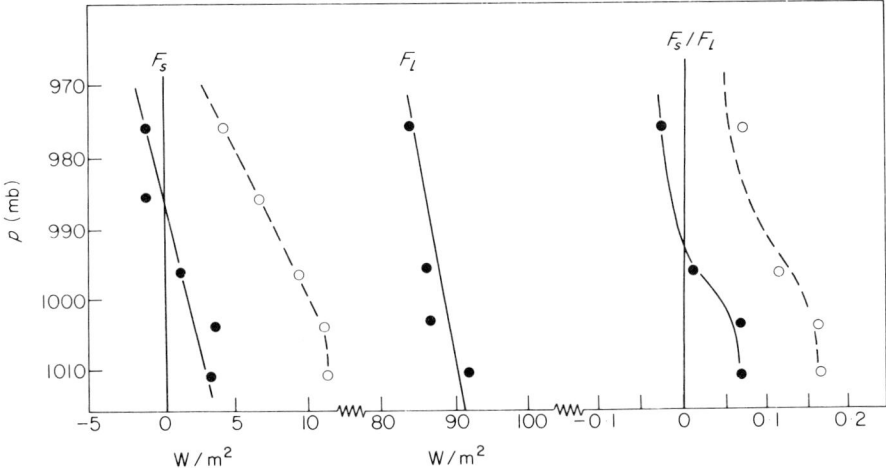

Fig. 4.8. Average fluxes of sensible (F_s) and latent (F_l) heat energy during GATE project outside convective areas, measured by DC-6 aircraft. The ratio F_s/F_l is also shown (71). Open circles denote F_s computed with Eq. 4.5.

The sinking motion between the clouds in Fig. 4.5 is the mechanism for importing thermal energy downward so that a good upward thermal gradient can never develop; therefore, sea and air temperature are very nearly in equilibrium. The same sinking motion takes care that, with import of dry air from above, a large vapour pressure or specific humidity difference between sea and air is sustained. For these reasons heat flow from the undisturbed oceans is always small in the tropics, and moisture flow large. Quite a different situation prevails over land where heat flux may attain 60% and more of the land evaporation, when the water supply is not so ample as to equal the oceanic supply, i.e. attain potential evapotranspiration. As available moisture sinks to lower levels, sensible heat transport inevitably must rise with increasing $T_o - T_a$ and eventually lead to the desert conditions of Fig. 2.9, where, in southern Australia and other deserts, the sensible heat flow alone must accomplish all the exchange between ground and atmosphere.

The subcloud layer in relation to moist convection

Whenever rain falls from large cumulus clouds, downdrafts tend to be initiated which transport low-energy air from the middle troposphere toward the surface. The intrusion from above has the same effect as the arrival of a strong cold front from higher latitudes. A complete upheaval of the subcloud layer takes place; in fact, one cannot speak of a subcloud layer as presented here for some hours or even longer until the intruding air has been modified through interaction at the surface (8). The intrusions will be discussed in the following pages on the cloud layer.

Surprisingly enough, a rainfall experiment in the Florida Everglades has revealed that convergence starts long before the onset of thunderstorms, and that there is a good correlation between rain amount and the time before rain that convergence was detected in the subcloud layer (86). At rainfall above 10 cm, convergence was observed to start as much as 1-2 hours earlier. The subcloud layer increases in depth with rising temperatures, until onset of rain drives surface temperatures down again (30).

The intensity of turbulence itself was measured during experiments by Woods Hole Oceanographic Institution in the trade winds near Puerto Rico (40). Vertical acceleration in the subcloud layer was on the other of $0 \cdot 1$ to $0 \cdot 2$ g in the subcloud layer, strongest near 400-500 m height. Above cloud base, turbulence in the air between cloud was well below $0 \cdot 1$ g while it rose to $0 \cdot 4$-$0 \cdot 5$ g in trade cumuli.

The cloud layer

The tropical sky

The factors that govern the development of cumuliform clouds in the tropics are (a) horizontal convergence of mass, (b) the vertical wind structure, (c) the depth and structure of the subcloud layer, (d) vertical stability and (e) orography and soil characteristics over land. Of these, the first factor is one of scale interaction, notably the appearance of synoptic scale convergence areas and also wind discontinuities along coast lines. Mean vertical wind structure is partly a climatic property of any area but subject to very strong synoptic, seasonal and interannual variations. Depth and structure of the subcloud layer again are partly climatic properties, such as are furnished by the trade wind inversion, but in a convergence zone the subcloud layer normally will become more moist so that its depth decreases with lowering cloud base. Vertical stability diminishes slightly in the upper part of the subcloud layer to yield the temperature lapse associated with balloon ascents into clouds (Fig. 4.5). Stability also is subject to vertical differences of net radiation. Finally, the arrival of a convergence zone is usually made evident at first by increased cloud growth over mountain ranges and convection points. Thus, "scale interaction" pervades all criteria named above. Nevertheless, on many days the properties of purely diurnal convection can be identified and discussed; in this case there is no net mass flow up and down and over areas with size perhaps of a 1° square ascending and descending currents cancel.

Depending on the relative importance and distribution of the parameters listed, various cloud types develop, ranging from cumulus humilis to cumulonimbus as in the higher latitudes. In fact, since the Coriolis parameter normally does not enter into analyses of convection, at least on the cumulus scale, the criteria are quite general and apply equally well at any latitude. However, dissatisfaction of meteorologists working in the tropics with the international cloud code has been continuous, since the latter in no way makes allowance for the nuances of cumuli and for sky composites, which are so revealing and important for judging the state of weather in the tropics. After development of a tropical "whole sky code" in 16 categories based on extensive photographic surveys (52), Holle (28) made a detailed analysis of cloud populations observed aboard ship in August and September 1963. He then condensed the "whole sky code" to one with only ten entries to fit into synoptic data reporting schedules (29). Table 4.1 reproduces this classification; Figs 4.9 and 4.10 show some photographic examples. It is evident that the classification is restricted to cumulus clouds; the various groupings with cloud leaning upstream or downstream, necks evaporating rapidly, isolated cumulonimbi and cumulonimbi consolidating with sheets of layer clouds are all significant and will be discussed at the appropriate place in this and later chapters.

The arrival of satellite photography has greatly extended the view of cloud ensembles beyond that available to the individual observer on earth; a considerable literature has developed and many spectacular views can be found in the column entitled "Picture of the Month" in the *Mon. Wea. Rev.* Among early analyses, those of Oliver and Oliver (61) and Fraedrich (24) on cloud groups in South East Africa from

Table 4.1. State of sky in the tropics (29)

Code number	Description
0	Cumuli, if any, are quite small; generally less than 2/8 coverage except on windward slopes of elevated terrain; average width of cloud is at least as great as its vertical thickness.
1	Cumuli of intermediate size with cloud cover less than 5/8; average cloud width is more than its vertical thickness; towers are vertical with little or no evidence of precipitation except along slopes of elevated terrain; a general absence of middle and upper clouds.
2	Swelling cumuli with rapidly growing tall turrets which decrease in size with height, and whose tops tend to separate from the lower cloud body and evaporate within minutes of the separation.
3	Swelling cumuli with towers having a pronounced tilt in a down wind direction; vertical cloud thickness is more than 1·5 times that of its average width.
4	Swelling cumuli with towers having a pronounced tilt in an upwind direction; vertical cloud thickness is more than 1·5 times that of its average width.
5	Tall cumulus congestus clouds with vertical thickness more than twice the average width; not organized in clusters or lines; one or more layers of clouds extend out from the cloud towers, although no continuous cloud layers exist.
6	Isolated cumulonimbi or large clusters of cumulus turrets separated by wide areas in which clouds are absent; cloud bases are generally dark with showers observed in most cells; some scattered middle and upper clouds may be present; individual cumulus cells are 1-2 times higher than wide.
7	Numerous cumuli extending through the middle troposphere with broken to overcast sheets of middle clouds and/or cirrostratus; cumulus towers do not decrease generally in size with height; ragged bases with some showers present.
8	Continuous dense middle clouds and/or cirrostratus cloud sheets with some large isolated cumulonimbus or cumulus congestus clouds penetrating these sheets; light rain occasionally observed from the altostratus; cumulonimbus bases ragged and dark with showers visible.
9	Continuous sheets of middle clouds and/or cirrostratus with cumulonimbus and cumulus congestus in organized lines or cloud bands; rain is generally observed from altostratus sheets and heavy showers from cumulonimbus; wind has a squally character.
/	State of sky unknown or not described by any of the above.

In the event of obscuration of clouds due to heavy rain, the observer should use classification 5 or 8. He should use 5 if the rain is localized or is brief in duration. He should use 8 if the rain is widespread or lasts for longer periods of time.

Nimbus I photos may be mentioned. Most astounding, perhaps, was the discovery of cloud streets of a very great extent. After Woodcock's classical paper on soaring of herring gulls related to thermal instability and wind shear off New England (90), Kuettner discussed "the band structure" of the atmosphere just before the satellite age (34) and then again some years later (35) with applications for soaring. He gives the following characteristics of low-level tropical cloud streets as typical (36, 37): length, 20-200 km; spacing, 2-8 km; height 0·8-2 km; width to height ration 2:4. He also indicates that the orientation of lines is along the mean wind of the convective layer and that little turning of wind direction with height is found in cloud street situations; the latter phenomena can be regularly noted. In an extensive theoretical analysis Kuettner discards the linear wind profile normally thought conducive to formation of streets. Instead he introduces the curvature of the wind profile (magnitude 10^{-7} to $10^{-6} cm^{-1} s^{-1}$) which, he says, enforces alignment of convective cells with flow direction. The cloud roll analysis of LeMone (41) may be mentioned again in this context.

This section closes with a brief mention of severe weather. About 2000 thunderstorms are present continually on earth, mostly in equatorial Africa and America (15, 65). They introduce a diurnal variation of earth's electric charge. Over sea, intensity of the thunderstorms is normally moderate, but over land they can be violent, as in middle latitudes, especially during periods of advance and retreat of monsoons. Hail formation is most uncommon near sea level, but in springtime severe thunderstorms with hail can develop in northwestern India and Pakistan along the subtropical westerly jet stream with passage of a cold upper-air trough (66), reminiscent of descriptions for the western plains of the United States.

In southern Florida around Miami, occasional severe hail is encountered at times under circumstances resembling those of severe storms in the midwestern United States (59). In a particular situation in March 1963, intense hail developed as an upper, cold-core Low travelled far south of the normal path toward Miami from northwest leading to large vertical instability. The hail incidents are remarkable in that instability appears to play its well-known part of setting off violent overturning as observed in middle latitudes. In contrast, on all other occasions tropical rain is most intense when the lapse rate is very close to moist adiabatic and just short of being absolutely stable, as also confirmed for the Bombay area (58). A large amount of convective instability—lapse rate much steeper than moist adiabatic— is generally indicative of upper subsidence and a dry atmosphere so that any rising towers rapidly evaporate. Therefore, the correlation

Fig. 4.9. Illustrating trade wind cumuli, cumulus skies and convection bands. Most pictures Figs. 4.9-4.10 courtesy Ron Holle (NOAA).

Fig. 4.9 continued.

Fig. 4.9 continued.

Fig. 4.10. Illustrating cumulonimbi and rain-producing skies.

Fig. 4.10 continued.

between precipitation and lapse rate is almost always strongly negative—the hail situations are a marked exception.

Hail is also observed in some areas on the equator, notably on the eastern shore of Lake Victoria in Kenya, where extensive tea plantations are situated at an altitude of about 2000 m. The hail may occur on as many as 200 days in the year; it is at least partly related to a lake breeze moving uphill from the large lake (Chapter 6). Circumstances attending formation are not altogether clear, nor is it obvious why the precipitate should reach the tea plantations in hail form. Of course, the distance between the altitude of the plantations and cloud base is much less than for locations at sea level. Also, temperatures are lower and especially depressed temperatures are created during hail occurrence (wetbulb effect); but similar conditions exist in many other parts of the tropics around the globe, where hail is not experienced. In western Kenya the size and intensity of the hail is sufficient to create a marked economic problem.

The growth of cumuli

We have already followed the onset of convection in the subcloud layer as far as cloud base, so that we can now take up the growth of cumuli from the time of first formation of cloud. It is assumed that an adequate amount of condensation nuclei is present, whether over land or sea. The new element to be considered is the release of latent heat of condensation.

Over land, cumulus inception in daytime clearly occurs through "thermals" which penetrate the subcloud layer "undilute", i.e. without mixing with subcloud layer air of lower humidity along the way (Figs 4.4(c), 4.6). This mechanism is clearly inoperative during night. As the years go by, nocturnal rain over land is reported in increasing amount; without detailed investigation, it is quite erroneous to assume that the bulk of precipitation is delivered by afternoon showers at any location, even when an afternoon build-up is observed regularly. Over sea, in contrast, slight instability develops at night when cloud tops may radiate strongly to space whereas surface temperatures are held constant by contact with the sea surface. Here the question has centred on whether cumuli have their "roots" at the surface just as over land, or whether they originate near the top of the subcloud layer as the depth of the latter undergoes vertical oscillations due to convergence and divergence associated with convection cell patterns (Figs 4.1, 4.2). When the subcloud layer is fully mixed, as in Fig. 4.4(a), the latter mechanism may indeed be operative. However, when moisture decreases throughout the subcloud layer (Fig. 4.7), clouds with specific humidity characteristic of the surface must have their roots low down, in particular through subcloud layer roll vortices. Both solutions, therefore, occur in nature; the more "mixed" the subcloud layer is (i.e. the higher the moisture content available for convection), the stronger should be the ensuing cumulus development.

Equations (4.12) to (4.15) are entirely valid for describing the inception of cumulus convection. When they are used as an onset criterion for determining the convective condensation level or the lifting condensation level, observations decisively support the validity for determining the approximate height of cumulus bases. It is remarkable to see how uniform these bases are when thousands of cumuli are in the sky.

Parcel ascent model

The "parcel technique" has been applied not only as onset criterion, but also as a model to predict the height of cloud tops and the probability of thunderstorm formation. After the air passes the condensation level, its path on a thermodynamic diagram should be given by the moist-adiabatic lapse rate. This path may intersect the sounding again after penetration of only a shallow layer, especially when a marked stable layer or inversion is present. When this happens, the parcel method gives an accurate prediction of cloud tops. The cumuli will be shallow, their bases will be broad and the sky may assume the character of stratocumulus. Over the oceans this occurs generally where the trade wind inversion is situated at only 500 to 1500 m above the convective condensation level.

On many summer days over land, and quite generally over the oceans away from subsidence inversions, the temperature lapse rate in the situations of interest lies between the dry and moist-adiabatic rates to great heights: 350 to 300 mb. The atmosphere then is called "conditionally unstable". Air ascending moist-adiabatically should continue to gain upward speed often to the high troposphere; tall clouds and thunderstorms, should form all the time, but predictions made on this basis very frequently fail, proving that the simply buoyancy model does not suffice to explain the observed clouds except for their beginnings.

Equation (4.15) describes a very powerful mechanism. Say, for example, that $g' = 10^{-2}g$; a value often encountered. Air starting from rest near the ground will reach the 3 km level after four minutes under such conditions and the tropopause after nine or ten minutes. The cloud growth predicted by this calculation exceeds by a large factor that usually observed. We must conclude that resistive forces not considered by the parcel model operate normally against the buoyancy force. In particular, we have neglected (a) the mass of liquid water carried in updrafts, (b) the continuity of mass and (c) interactions between the rising air and its environment—mixing as well as friction.

Inclusion of weight of water

In many theoretical derivations, all condensed water is assumed to drop out immediately as rain. This may be acceptable for some purposes, but a powerful mechanism opposing buoyancy of updrafts and enhancing buoyancy of downdrafts has thereby been eliminated.

Assuming that liquid water is carried in an updraft, its density, ϱ'', becomes $\varrho'(1 + 10^{-3}n)$, where n is the mass of water per kilogram of air. With this inclusion the buoyancy formula becomes

$$\frac{dw}{dt} = g\frac{\varrho - \varrho'(1 + 10^{-3} n)}{\varrho'(1 + 10^{-3} n)} \approx g\frac{T'_v(1 - 10^{-3} n) - T_v}{T_v}. \quad (4.17)$$

Consider $T_v = 300$ K, $T'_v - T_v = 3°C$. The additional term will be $-T'_v \cdot 10^{-3} n$. For $n = 3.3$ g/kg, we then have $-T'_v \cdot 10^{-3} n = -3 \cdot 10^2 \times 3.3 \cdot 10^{-3} = -1°C$. Thus, the buoyancy acceleration is reduced by no less than one-third; with $n = 9$, it will vanish entirely. We have therewith found a large modification of Eq. (4.15); one that will always be present to a greater or smaller degree, since no cloud can drop out all condensed water continuously and remain visible! Of course, the amount of water carried upward will vary greatly with circumstances, so that the number of variables in Eq. (4.17) cannot be reduced without further modelling (78).

Slice method

Bjerknes (10) was the first to recognize the omission of continuity of mass as a serious deficiency of the parcel ascent theory. He introduced a convection cell pattern with a scale of several kilometres to 30 km depending on cloud size. Since there is no net vertical mass transport over the area considered, this is perhaps the first convective turbulence model. Later, Cressman (21) extended Bjerknes' development by permitting net convergence or divergence to take place inside the area considered and so including synoptic effects.

Bjerknes permits his air columns to rise moist-adiabatically. After shedding all their water as rain they will descend dry-adiabatically (radiation was not considered on the time scale of cumulus life). Thus, the environment will warm up in relation to the ascending stream, whose temperature does not change. The initial temperature difference will decrease, become zero and reverse, terminating all convection. Herewith the concept of a life-cycle of cumuli is introduced and in many ways, the model is quite realistic. One always sees clear spaces between cumuli; often, cumuli are ordered into long "cloud streets" (Figs 4.9-4.10) where, as in case of the Bénard cells, the spacing of rows is about twice their height. Further, the spacing increases with cloud size and becomes 30 km for cumulonimbi. There still remains the problem of moist-adiabatic ascent that is much too rapid in the first stage of the cumulus evolution.

Entrainment

Let us suppose, for the opposite extreme of the assumption in the preceding discussion, that a rising column carries all condensed moisture with it. Then, an intercepting aircraft should record the "adiabatic moisture" content. For instance, consider that condensation started at $p = 900$ mb, $T = 20°C$, therefore, specific humidity of 17 g/kg, and that the cumulus cloud reaches the 710 mb level with $T = 11°C$ and $q = q_s = 12$ g/kg. Then the adiabatic water content is 5 g/kg or, roughly, 5 g/m^3. Various aircraft tests of the ratio of actual to adiabatic water content in non-precipitating cumuli (80, 87) (see A-4 for summary) have revealed that the moisture content of trade wind cumuli usually is nowhere near the adiabatic value. Even in hurricanes the adiabatic water content is only rarely encountered (2). Therefore, the water must have been removed from the cloud by some mechanism, and this mechanism is known not to be rain.

Already in 1946, Stommel (83, 84) was aware that aircraft measurements from Woods Hole Oceanographic Institution did not show a moist-adiabatic lapse rate upward from the convective condensation level in trade wind cumuli; rather, a lapse rate and a temperature that closely approximated the environmental atmosphere. From this fact he deduced that mixing between the ascending plumes and their surroundings must occur and he worked out the rate of mixing or "entrainment" on the basis of conservation of mass and energy. Although he did not prescribe a mechanism for entrainment, he suggested that buoyant air may be likened to an aerodynamic jet stream moving through an environment nearly at rest. On account of vertical wind shear, as depicted in Fig. 4.14, one can also take the position that the entrainment results from mixing inward of faster moving environmental mass with increasing height. This viewpoint permits modifying the assumption inherent in Stommel's treatment that the mass flow in the entraining jet increases right up to cloud top. Much effort has been expended in the literature to determine whether the outside air enters rising plumes directly or as a wake behind rising "bubbles" from below, or even, from the downwind side of the cloud. For our purposes we need not be concerned with the details of these distinctions.

Another constraint is that a rising cumulus tower must push the air above it out of its way, resulting in a "form" or "profile drag" accelerating the surrounding air back as a countercurrent (49). If the outside air enters (is entrained into) the rising top directly, a rapid decrease of buoyancy may occur. When a cloud breaks into the trade wind inversion (Fig. 5.17) or into a general, upper dry air layer above the lower moist layer, one can literally see the tops evaporate, a clear indication that direct mixing of cloud tops with the surroundings does occur and that cloud tower growth will only continue if the environment, mainly its humidity, remains favourable to sufficient heights (1, 5, 17).

Subsequently, the concept of "dynamic entrainment" through horizontal pressure forces was introduced (4). In Eq. (4.12) the non-hydrostatic pressure gradient dp'/dz appears. For vertical acceleration to occur p' must decrease with height. Thus, it is possible that $p_{outside} - p'$ at one level does not vanish but that it accelerates outside air inward. Austin does recognise that the assumption of horizontal pressure gradients around a cloud conflicts with the derivation of Eq. (4.15). On the other hand, he points out that the mass continuity requirement is violated if the cloud has a cylindrical shape, yet vertical motion increases upward, without lateral entrainment. The issue was taken up again by Newton and Newton (60) who attribute a major role in deep convection to the hydrodynamic pressure.

There are suggestions that computations based on single clouds may not be most relevant since, even in the absence of synoptic forcing, cumuli tend to occur in groups which may be located along convection bands. Over sea, they have been related to warm areas on the ocean with temperature anomaly of perhaps 0·1 to 0·3°C; the areas may have an extent of several kilometres only (51). Over land, around Salisbury (18°S, 31°E), almost all radar echoes have been observed associated with cumulus groups, very few with isolated clouds. The groups essentially form convection streets whose motion is due to cloud development and decay rather than simple cloud advection (79). This observation should hold over the oceans as well.

The widespread formation of rain showers in warm clouds, often very heavy, is largely a discovery of the early 1940s. Fletcher's text (A-4) contains a thorough description of the extensive literature. Here it may be mentioned only that marine showers appear to be closely related to salt nuclei partly released into the atmosphere through breaking ocean waves (92, 93). Inland, these frequent showers from small clouds with perhaps no more than 2-3 km thickness are not observed. We saw this in the difference between marine and continental radar echo area frequency distributions of Fig. 3.28. Ludlam (45) has given a comprehensive survey of the factors leading to cumulus and cumulonimbus convection.

The growth of cumulonimbus

The entrainment mechanism, whatever its details, is entirely capable of holding clouds, which adiabatically could attain 12 km, to only 2 km height. One may well ask, then, how the cumulonimbi which do occur ever come into existence? The entrainment rate is defined as $1/M \, dM/dz$, the percentage mass increase. Entrainment rates of about 100% in 500 mb for bulging cumuli have been measured (17), this is about five times smaller than that computed by Stommel for trade cumuli. Suppose that entrainment, as indicated in Fig. 4.14, acts on the boundaries of bulging and trade cumulus and that the intensity of mixing over a unit surface of the circumference is equal in both cases. Then the entrainment rate should depend on the ratio of the length of the boundary to the mass (for unit depth, the area) of the cloud, that is $2\pi r/\pi r^2 = 2/r$, and it should decrease by a factor of five as r increases by five. This result is reasonable and supports the lateral mixing hypothesis. It suggests that the effectiveness of entrainment in reducing the buoyant energy inside rising plumes will depend on the width of these plumes given a cylindrical shape. As the width increases, the core of the rising air becomes more and more protected from the influences of the environment which will not have enough time to penetrate inward as the column rises through a given layer. A narrow central portion may then have higher temperatures and larger updraft speeds than the bulk of the cloud, and this again is what has been observed (11, 12). An updraft area of 25 km^2 appears to be of satisfactory size to prevent destruction of the core except under conditions of extreme vertical wind shear, when cumulonimbus anvils are very low and have a decidedly eroded appearance. Further discussion of entrainment and other restrictive factors in relation to cumulus and its growth to cumulonimbus is given in (25, 26, 53, 54).

Undilute towers and the location of heat sources for the troposphere

For a convective element rising without external heat or cold sources, the first law of thermodynamics may be stated in the form

$$0 = \frac{L \, dq}{T} + c_p \frac{dT}{T} - \frac{R \, dp}{p}. \qquad (4.18)$$

Assuming steady state and neglecting horizontal pressure gradients (i.e. kinetic energy), the last term in Eq. (4.18) may be transformed to $+g \, dz/T$. When considering an ascent where inside and outside temperatures are the same, the temperature in the last term will be the same as that in the other terms of Eq. (4.18). For this special case the first law may be integrated and

$$gz + c_p T + Lq = \text{constant} . \qquad (4.19)$$

This equation should be applicable to synoptic situations, especially hurricanes, where ascent takes place over the whole inner core (73) (Ch. 9). Often we are interested in buoyant towers where $T'_v - T_v > 0$. Then the buoyancy term of Eq. (4.15) must be added for the vertical kinetic energy acquired during ascent. This energy is generally expended against friction in the high troposphere. The term will be very small if much liquid water and ice is carried in the updraft (Eq. (4.17)).

We now return to Figs 2.23 and 2.24 which present a mid-tropospheric minimum of energy in the tropics. This minimum was considered to be an extraordinary fact in nature. Diffusion from the surface heat source $Q_s + Q_e$ throughout the troposphere is impossible when much of the transport must take place up-gradient. The need for a countergradient transport mechanism, a phenomenon of *negative viscosity* (82), becomes apparent. This mechanism is now seen to be furnished by undilute towers which during ascent bypass the middle troposphere and deliver their energy through spreading anvils in the high troposphere.

There emerge two distinct means of vertical energy transport: cumulus and cumulonimbus. The matter is more significant than just noting that some clouds are bigger than others. Cumulus transport is bound to decrease upward because of the limiting saturation vapour pressure (Fig. 2.19). Another mode of transport arises for maintaining the temperature of the upper troposphere against radiation and for providing energy for poleward transport. Through the cumulonimbus "hot towers", the surface heat source is essentially transferred in part to the high troposphere. The whole picture is portrayed in Fig. 4.11. Mixing from cumuli extends to the level of minimum θ_e in mid-troposphere. Above this level, cumulonimbus undilute towers provide the energy source.

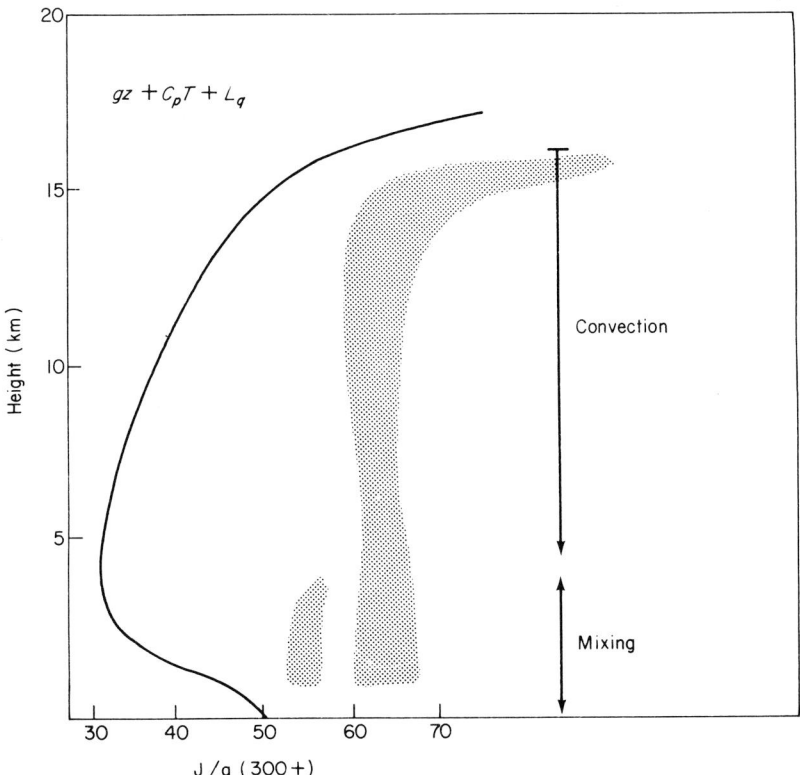

Fig. 4.11. Model of mean vertical distribution of $gz + c_p T + Lq$ in troposphere and stratosphere; mechanisms of upward heat flow in troposphere, and limit of upward penetration of heat gained by atmosphere from ground (73).

The undilute towers will occupy only a very small fraction of the whole tropical area. Given an area of 25 km² and upward motion of a few metres per second in the middle troposphere (i.e. carrying much water and ice) the whole mass converging into the tropics in the low troposphere (Figs 1.5, 1.18) will go up in these narrow tall chimneys occupying only one-tenth of one percent of the tropical area (see end of Chapter 3). It has, therefore, proved difficult to demonstrate their existence. Radiosonde balloons almost never enter the updrafts or are beaten down by heavy precipitation. While research aircraft have penetrated high-level anvils occasionally, the prime source proving the existence of undilute towers are radar and satellite observations.

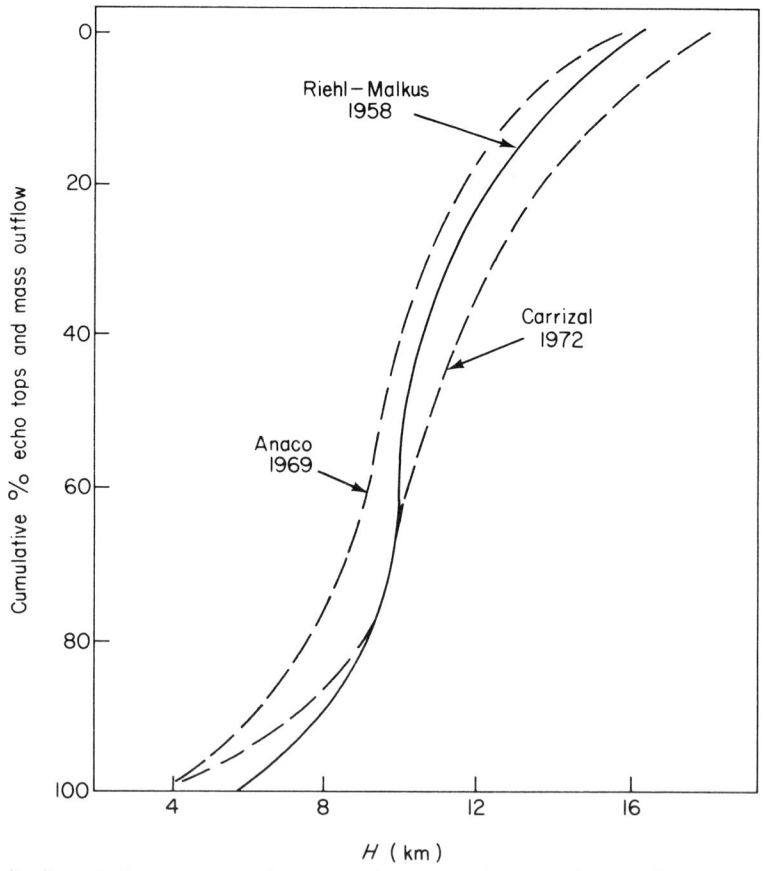

4.12(a). Cumulative per cent decrease of computed upward mass flow in tropical undilute towers (70) and of radar echo tops observed during Venezuela field experiments in 1969 and 1972, with radar stationed at Anaco (9·5°N, 64·5°W) and Carrizal (9·5°N, 67°W).

VERTICAL ENERGY TRANSFER

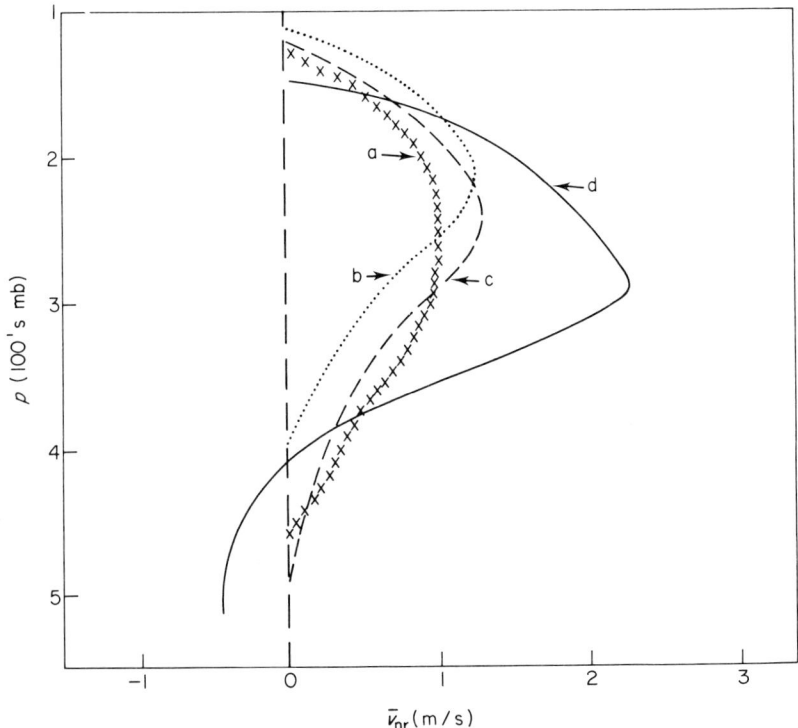

Fig. 4.12(b). High tropospheric mass outflow profiles: 1) GATE; b) from (77); c) from (73); d) from (74) is mesoscale, reduces to 1 m/s when reduced to synoptic scale of (a) to (c).

Radar echoes frequently extend to heights which clouds carrying precipitation particles visible on radar screens could never attain, if their energy was substantially reduced during their passage through the low and middle troposphere. Figure 4.12 illustrates the mass flow out of the equatorial belt in relation to the vertical distribution of radar echo tops during the Venezuela experiments.

It is seen that the median radar echo height (seeing 1 mm drops) lies above 10 km and that 16 km is attained occasionally. Satellites such as *Nimbus* and *Goes* now monitor the tropical distribution of cloud top temperatures continually. Even when average values over 1° latitude-longitude squares are taken, up to half of the synoptic cold centres with convective clouds may report radiation temperatures of —50°C and even —60°C (Fig. 4.13), i.e. pressures of 200 to 170 mb and corresponding heights from Table 2.1. Such cloud top pressures have also been observed by aircraft over India (22) and Malaya (96). Very strong evidence is furnished by aircraft ascents measuring high Aitken nuclei counts in very low and very high troposphere, with very few such nuclei in the deep intervening layer (pers. comm. Mel Shapiro, National Center for Atmospheric Research, on non-cloudy flights).

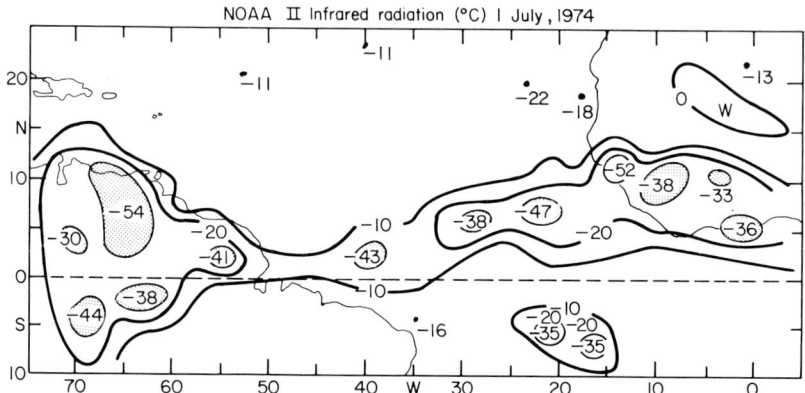

Fig. 4.13. Synoptic chart of effective long-wave radiation temperature (°C) for the Tropical Atlantic Ocean, 1 July, 1974, near 2100 local time (72).

Horizontal wind structure and life-cycle of cumuli

If a cumulus cloud builds in an atmosphere with uniform wind along the vertical, the wind inside the cloud is the same as the wind outside. The cloud moves with the speed of the wind and its axis is vertical. More often wind speed either increases or decreases with height. The clouds then no longer move with the mean speed of the wind in the surroundings, and their axes lean vertically in the direction of the shear (Fig. 4.14). In the layer through which trade cumuli usually

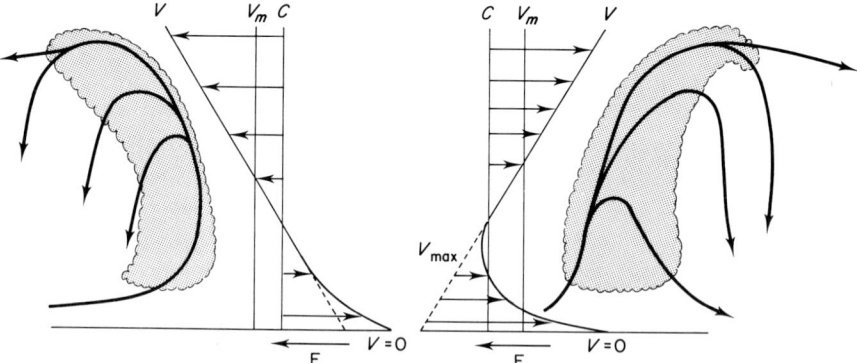

Fig. 4.14. Model of cumulus rotation, relative motion, and dynamic component of displacement for east wind increasing upward (left) and decreasing upward (right).

VERTICAL ENERGY TRANSFER

extend, wind direction seldom changes appreciably so that, in this case, one can also say that clouds and cloud streets are oriented along the direction of the wind (37). Cumulus congestus and cumulonimbus, however, commonly reach into strata with a wind structure quite different from that of the low levels. The cloud axis is still oriented along the shears (Fig. 4.15), and we observe the twisted structure characteristic of the upper portion of many tall clouds (53).

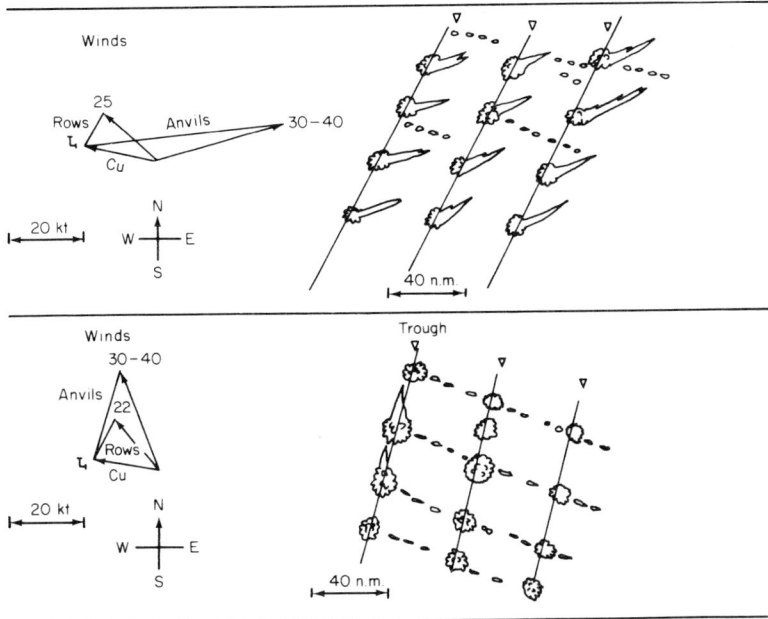

Fig. 4.15. Illustrating arrangement of cumulus rows and orientation of cumulonimbus anvils under the influence of external shear as observed in the western tropical Pacific Ocean between Kwajalein and Guam during the 1957 rainy season (52).

The effect of vertical wind shear on cumulus has been the subject of several studies (47, 17). These analyses have mainly developed two-dimensional models, i.e. changes of wind speed, but not of direction with height. Using the law of conservation of momentum, a continuity equation is set up for horizontal momentum. Given an upper and a lower pressure surface slicing through an entraining cumulus cloud, the flow of momentum through the upper surface must be equal to the flow through the lower surface plus momentum entrained into the cloud in the intermediate layer, if there are no accelerations by pressure forces. With suitable assumptions regarding the vertical shear in the environment and the entrainment rate the horizontal wind speed and the slope of the cloud axis can be determined as a function of altitude.

The external wind structure governs the "life-cycle" of cumuli to a large extent. The clouds should build into the wind, i.e. move slower than the surrounding air when the shear is positive, and they should move faster when it is negative (Fig. 4.14). This distinction appears in the cloud code of Table 4.1. Direct measurement has proved the reality of relative motion, the dynamic component of the motion vector (17). It is also substantiated by large quantities of time-lapse motion pictures which show that cloud matter commonly streams out on one side alone—the side toward which the cloud leans with height (47). Cloud towers build on the upshear side of the cloud and rain out and evaporate on the downshear side; hereby the "dynamic component" of cloud motion relative to the external wind arises. I have found such motion pictures highly illuminating and urge readers to view a sample themselves.

Precipitation derived from cumuli is also influenced substantially by the external shear. "If the wind shear is weak, cloud turrets will remain nearly vertical even as their updraft velocities decay and they dissipate. When, on the other hand, the external shear is strong, the dissipating towers become very strongly slanting (as in Fig. 4.14) and the new vigorous towers, which grow much more nearly vertically, are penetrating dry air." (47). We may add that rainfall becomes progressively inhibited at large cloud slopes. Raindrops grow mainly from collision with other drops. A minimum vertical cloud thickness is needed—2-3 km in the tropical cloud layer (17)—for drops to grow sufficiently to fall out as rain. A small slant of the cloud axis, as in Fig. 4.14, will be beneficial to cloud growth since the precipitation will fall at some distance from the updrafts and not interfere with them. As the slant of the cumulus axis becomes large, however, the vertical cloud thickness through which the drops fall becomes reduced diminishing the chances of precipitation. Such small drops as do form then evaporate before reaching the ground. Not infrequently do we see the spectacle of shearing trade cumuli completely "raining out" aloft. From computations of drop evaporation (summarized in A-4) it has become very evident that small drops evaporate much more rapidly than large ones which, in an atmosphere of, say, 70% relative humidity will lose only a fraction of their mass during a fall over several kilometres at their terminal speed. Very strong shears as encountered in the Line Islands expedition (46) and also the Pacific

Ocean (52) in middle and upper troposphere can even erode cumulonimbus towers completely and make the "hot tower" energy transfer mechanism inoperative.

Conservation of vorticity

The external shear dU/dz is a vorticity component about the horizontal axis perpendicular to the shear. When the rotation of cloud elements in the sense of this shear, so evident in time-lapse photography, is also considered there is a strong suggestion that a vorticity equation about the horizontal axis is a governing dynamical law for structure and life-cycle of cumuli. If the shear is that of a zonal basic wind, say, the trade winds, the vorticity equation for an incompressible fluid becomes

$$\frac{d\eta}{dt} + \eta \left(\frac{\partial u}{\partial x} + \frac{\partial w}{\partial z} \right) = 0. \tag{4.20}$$

where $\eta = \partial u/\partial z - \partial w/\partial x$. During the growth and decay phases of cumuli, and for tall cumuli with rain, a term involving density gradient must be added.

If wind and wind shear have the same orientation, i.e. the motion is two-dimensional in the x-z plane, the divergence $\partial u/\partial x + \partial w/\partial z = 0$ and η = constant for each moving column of air. The left side of Fig. 4.14 is a model for east wind increasing upward. The shear dU/dz may be linear. The cloud leans downwind and inside it the wind increases upward at a much slower rate in the updraft than outside. The vertical motion arrows are patterned on the time-lapse films.

The vorticity to be conserved is that of the basic current. Thus,

$$\eta = \frac{\partial u}{\partial z} - \frac{\partial w}{\partial x} = \frac{\partial U}{\partial z} \text{ or } \frac{\partial}{\partial z}(u - U) - \frac{\partial w}{\partial x} = 0.$$

Consider a trajectory entrained into the cloud from upstream. During the approach $\partial u/\partial z$, and, therefore, also $\partial/\partial z\,(u - U)$ increases. Thus, $\partial w/\partial x$ must be positive and also must increase for conservation of vorticity. This we observe. On the downwind side $\partial w/\partial x$ is negative. Therefore, $\partial/\partial z\,(u - U)$ should also be negative, again in agreement with observations. Thus, the illustration correctly portrays the motion field for the case of η = constant which appears to be a realistic dynamical law. Give $dU/dz = 2$ m/s/km in the cloud layer, the rotation rate is 2 × 10⁻³rad/s or 1 rad/30 min. This rate corresponds to the average life of trade wind cumuli. For verification of the conservation law, trade wind cumuli in Venezuela photographed with a time-lapse camera were examined through a projector modified so that it could follow a particular cloud on the screen and with a device by which the projector was rotated so that the cloud became stationary in the image (C. Kumitz, unpublished Diploma thesis at the Institute for Meteorology of the Free University of Berlin). The rotation rate so measured agreed closely with $\partial U/\partial z$ from balloon ascents at the same site over most of the cloud life. In one case, when an upper cloud in reverse shear could be followed without obstruction from low clouds, the result was also satisfactory.

Cumulus life-cycle in undisturbed trade winds

The foregoing comments on the life-cycle of cumuli in relation to basic current structure may be amplified from inspection of Fig. 4.14. Models for both increasing and decreasing east wind are shown. The mechanism for entrainment and detrainment is provided by the external shear in relation to the rotating clouds, inflow on the upshear side and detrainment downshear. The vertical lines V_m and C indicate the mean motion of the atmosphere over the vertical extent of the cloud (V_m) and the cloud speed C. The differences $C - V_m = C'$, the dynamic component of cloud propagation.

For positive shear the model corresponds to the left side of Fig. 4.14. The ascending towers are situated at the eastern cloud edge; water falls into the surrounding atmosphere on the downwind side and cloud base rises strongly downwind on account of the combined action of entrainment, precipitation and evaporation of cloud drops. Both inflow and outflow are situated on the western cloud side. Given the rotation rate of 1 rad/30 min, rain will fall out of the western part of the cloud within half an hour after cloud formation. Evaporation of water from falling drops plus the weight of the water create a downdraft (Eq. (4.17)) which, given enough water for evaporation, will penetrate into the subcloud layer. The air in the downdraft has low energy compared to the subcloud layer air originally forming the cloud—equivalent potential temperature of 330-340 K (Table 2.1), compared to 350 K in the initial subcloud layer. Therewith, the supply of high energy air for the cloud is cut off and the cloud dies. Thus, dU/dz determines the length of the life-cycle, even though the slant of the cloud may be much less than dU/dz (60).

In the case of negative shear (right side of Fig. 4.14), the surface layer with upward increase of wind will necessarily be deeper than for the atmosphere with positive shear. Subcloud inflow layer is still from west while the outflow is toward east. In the middle of the cloud the relative wind may be from east, often, however, the dynamic component of propagation is sufficiently strong, so that cloud and wind motion will be nearly equal in the middle and no ventilation from east occurs. Thus, the entrainment, if any, may be small with this arrangement in the lower portions, and the cloud, not subject to strong restraint by entrainment, can become more vigorous as $T'_v - T_v$ remains relatively large. Further, the downdraft is situated east of the cloud and does not enter the inflow. Theoretically, this cloud could go on forever given a moist atmosphere. A steady-state type of model has been advanced for the maintenance of long-lived squall lines (56, 57).

Life-cycle in convergent environment

When a synoptic convergence area is present, the life-cycle of cumuli along their path through the convergence zone must undergo some modification from the cycle just depicted. Cumulonimbus clouds appear. Their formation is favoured by low-level convergence plus small shear of the basic current so that, as noted earlier, rising towers are not exposed to large environmental ventilation and can have a full growing stage.

In Fig. 4.16 we see the cloud pattern through a wave trough in the easterlies in the western Caribbean (68). The vertical shear of the easterlies was very small, below 600 mb; higher up the easterlies decreased slowly to 200 mb. Relative to westward wave propagation of 9 m/s, the wind was from west at all heights. Thus, the area of low-level convergence was situated west of the wave trough, followed by divergence on the eastern side (Fig. 8.6).

Along the relative trajectory, depicted by the wind arrows in Fig. 4.16, skies were clear on the far western starting point of the trajectory, marking the limit of the convergence zone. Then, rows of small cumuli, followed by cumulus congestus, made their appearance, breaking into cumulonimbus just ahead of the wave trough. With wind structure similar to that of Fig. 4.15 from the middle troposphere upward, the anvils were oriented just that way and they also attained great length up to 100 km. East of the wave axis the lower clouds began to disappear under the influence of divergence near the ground, while the mass above 600 mb continued to ascend. The anvils merged into a huge altostratus sheet with area of at least $(300 \text{ km})^2$. Research

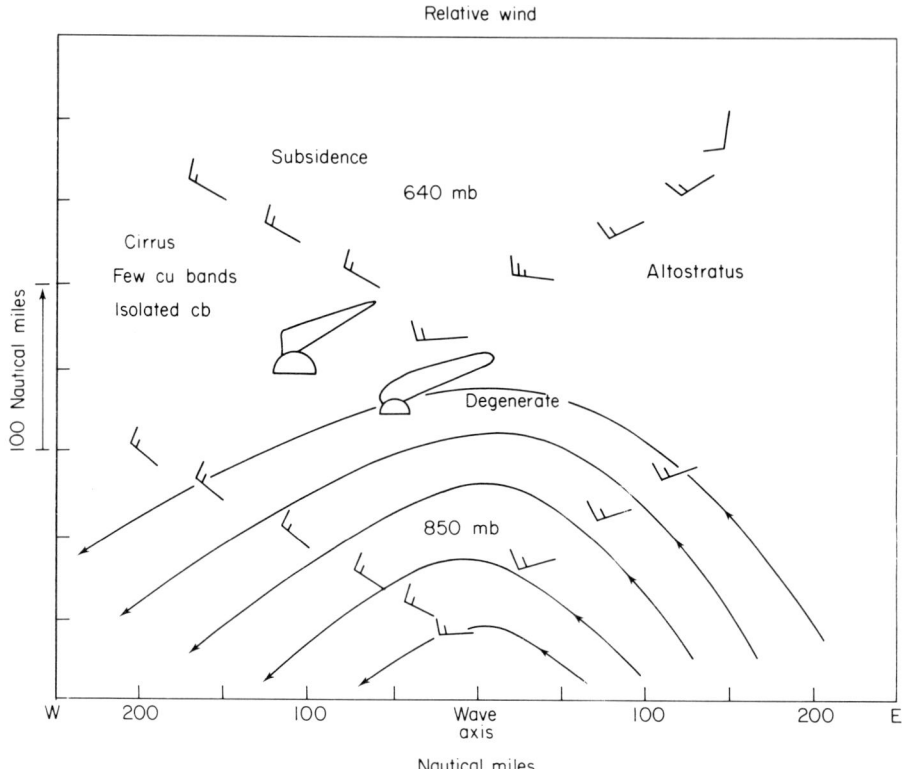

Fig. 4.16. Streamlines of wave in easterlies explored by research aircraft in western Caribbean on 8 August, 1964. Winds relative to the moving wave (relative trajectories) are shown for 850 and 640 mb, also distribution of cloud types encountered (68).

aircraft traversing along the base of this stratus layer near 600 mb encountered continuous light rain; no lower clouds were visible.

In principle, this description resembles that of Fig. 4.14 for negative vertical shear, but on a greatly expanded scale. An unstable stratification was changed to a stable one across the wave axis. On the forward side along the air trajectory are cumulus and cumulonimbus clouds with release of convective instability and overturning of the atmosphere. In the end phase, all clouds are layer clouds. The transformation of the cumuli has an important feedback on the synoptic scale structure. In Fig. 4.17, the 24-hour changes during wave passage at Gran Cayman (19°N, 81°W) are shown and compared with the findings of the two-plane research mission. On the left we see the vertical overturning in terms of θ_e profiles ahead and behind the wave

VERTICAL ENERGY TRANSFER

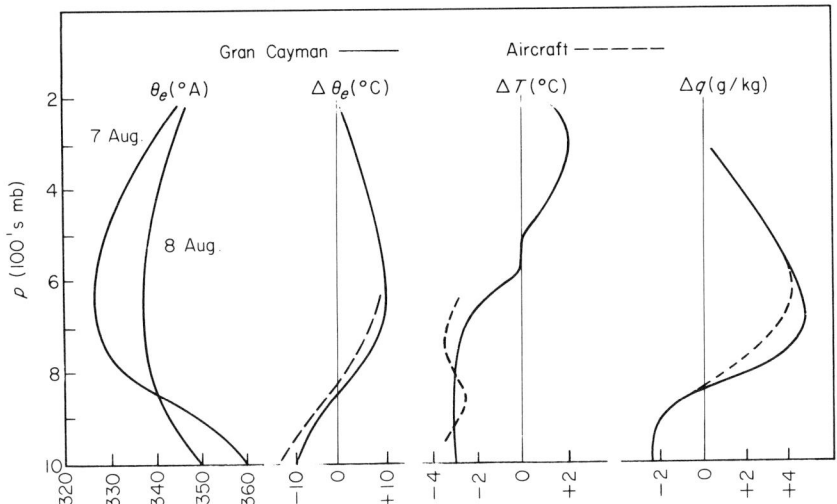

Fig. 4.17. The turbulent overturning of the atmosphere by convection: 24-hour changes across the wave shown in Fig. 4.16 as observed at Gran Cayman (19°N, 81°W), for equivalent potential temperature (left), temperature and specific humidity. Low level cooling and humidity decrease are very marked as are upper warming and humidity increase (68).

axis. Mass ascent from the subcloud layer produces a large energy increase in the middle and upper troposphere; downdrafts carry low energy toward the surface and destroy the subcloud layer on the outflow side of the system. Because of the synoptic-scale convergence and associated net mass ascent, the average θ_e of the whole troposphere is increased, but the profile has become almost uniform, a result of the vertical mixing. Both temperature and humidity are lowered in the low atmosphere, down to the surface where air-sea exchange is greatly enhanced. They are raised at upper levels with broad implications for the maintenance of the whole synoptic system, analysed in Chapter 8.

Turbulent exchange in the cloud layer

The convective clouds must be able to carry out the upward moisture transport, in the case of cumulonimbus, the transport to the high troposphere. At first the action of a very large number of trade wind cumuli will be considered near cloud base.

Cumulus moisture transport

In the general trade wind setting, the broadside vertical motion is downward. We shall neglect this motion for the moment and choose a model with turbulent transport only, i.e. there will be no net mass flow up or down over a square perhaps 100 km long. Inside this square there will be cloudy areas or cloud streets with active ascent and clear spaces with sinking motion. However, not all cloud matter rises. Since the growing stage of cumuli consumes only one-third or less of the total life span of the cumuli, most water particles are either just floating or descending. The specific question to be answered will be: what fraction of the cloud area must consist of actively rising plumes in order to accomplish the total moisture flow? Consider a reference level above cloud base across which all transport takes place (Fig. 4.18). The whole area $A = A_a + A_c + A_d$ where the subscript a refers

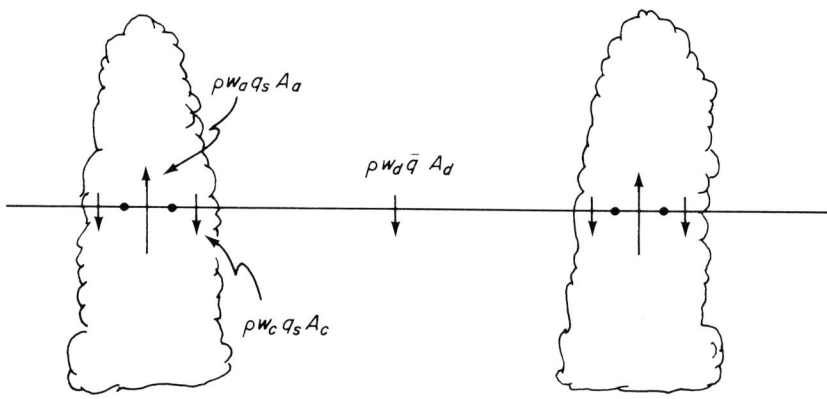

Fig. 4.18. Illustrating vertical moisture transport in cloud layer.

to actively ascending cloud portions, c refers to stationary or descending cloud and d refers to descending clear air. Climatological atlases give the mean low cloudiness in the trades as five-tenths of sky cover, but this has been found to be too high. We shall assume a cloud cover of one-third, still high in parts of the trade wind region. Then

$A_a + A_c = \frac{1}{2} A_d$. Setting A, the whole area, $= 1$, $A_d = \frac{2}{3}$.

The vertical velocities in the three areas will be w_a, w_c and w_d. In the clear, the vertical temperature and moisture distribution may be that of the tropical standard atmosphere (Table 2.1). In the clouds, the temperature will be assumed to be equal to that in the clear to include entrainment effects in the model, and the moisture, of course, must be saturation specific humidity q_s. The net moisture flux F_m is the residual of all upward and downward transports of water vapour through a reference level which will be placed at the top of the subcloud layer, or at cloud base. Then

$$F_m = \varrho \, (w_a q_s A_a + w_c q_s A_c + w_d \bar{q} A_d) \qquad (4.21)$$

where the density has been assumed uniform, a very small approximation, and $\bar{q}$ is the specific humidity of the clear environment. Mass continuity for air, then, demands that

$$w_a A_a + w_c A_c + w_d A_d = 0. \qquad (4.22)$$

Combining Eqs. (4.21) and (4.22),

$$w_d = \frac{F_m}{(\bar{q} - q_s) \varrho \, A_d}. \qquad (4.23)$$

F_m will be assumed as $0 \cdot 30$ cm/cm^2/day, somewhat less than average evaporation to allow for storage in the subcloud layer. Given $\bar{q} = 13 \cdot 5$ g/kg, $q_s = 16 \cdot 5$ g/kg, and other quantities as stated above, $w_d = -1 \cdot 5$ cm/s, a reasonable value.

Solving Eq. (4.22) next for A_a,

$$A_a = -\frac{w_d + \frac{1}{2} w_c}{w_a - w_c} A_d. \qquad (4.24)$$

There are still two unknown quantities on the right side of this equation. A solution requires either more factual knowledge or more modelling. For instance, part of the mass flow compensating against the updrafts occurs through downdrafts when, following the model of Fig. 4.14, water falls into clear air and evaporates. If such downdrafts are omitted for the present computation in non-precipitating cumuli the mean descent is not likely to exceed that of the clear spaces so that $w_c = w_d$. Also, from visual observation and from time-lapse photography, it is known that cumulus clouds rise with vertical velocities on the order of metres per second or they could not grow by 1 km/10 min, for instance. Thus, $w_a \sim 10^2 \, w_d$. Inserting these approximations in Eq. (4.24),

$$\frac{A_a}{A_d} = -\frac{1\cdot 5 \, w_d}{w_a} = 1\cdot 5\% . \qquad (4.25)$$

Actively rising towers need occur in only 1% of the whole area to achieve the required moisture transport. This rather extreme result indicates clearly that cumulus convection is an effective transport mechanism, capable of moving evaporated moisture upward without difficulty. As already seen from Fig. 4.8, the turbulent moisture transport is nearly always upward inside the subcloud layer and almost at the rate of evaporation. Above cloud base, this transport becomes concentrated in narrow ascending cumulus plumes. On the basis of an aircraft experiment near Puerto Rico, LeMone and Pennell (42) conclude that a large fraction of the energy transport out of the subcloud layer is due to motions on the scale of individual cumulus clouds.

The preceding simple model calculation brings out the basic mechanism of vertical energy transport through turbulent motions. It cannot place this transport in proper context to the general circulation of the trade winds; this will be attempted in Chapter 5. Maintenance of the turbulent motions is due at least in part to transport of cooler air over warmer water, something which a "one dimensional" model such as the foregoing cannot show but which must not be overlooked or forgotten. Severe constraints operate otherwise against turbulent exchange. As shown experimentally by Pennell and LeMone (64), initial production of turbulent energy in the subcloud layer is derived mainly from wind shear, but buoyant motion tends to be suppressed toward the top of the subcloud layer in view of Fig. 4.5. When air columns do reach saturation, their further ascent is of course aided by release of latent heat of condensation. Yet entrainment acts quickly to reduce this new energy source in the case of a sounding with a dry layer such as shown in Fig. 4.4(b), in contrast to Fig. 4.4(a) where horizontal mixing will hardly hinder vertical acceleration. Water drops falling out of clouds and evaporating on the way down in the case of non-precipitating cumulus will tend to stabilize the temperature lapse rate and provide another brake on convection (8).

One might think of differential radiation along the vertical as a potential mechanism for sustaining turbulent exchange, in addition to the destabilizing effects over the oceans at night already mentioned. In this one may be encouraged by older profiles of net radiation along the vertical, all of which show net cooling increasing by 1 — 1·5°C/day from the ground to the middle troposphere. Indeed, in the presence of a low overcast or broken cumulus deck, long-wave radiation is held to almost zero below cloud top (Fig. 4.19); there may even be an energy

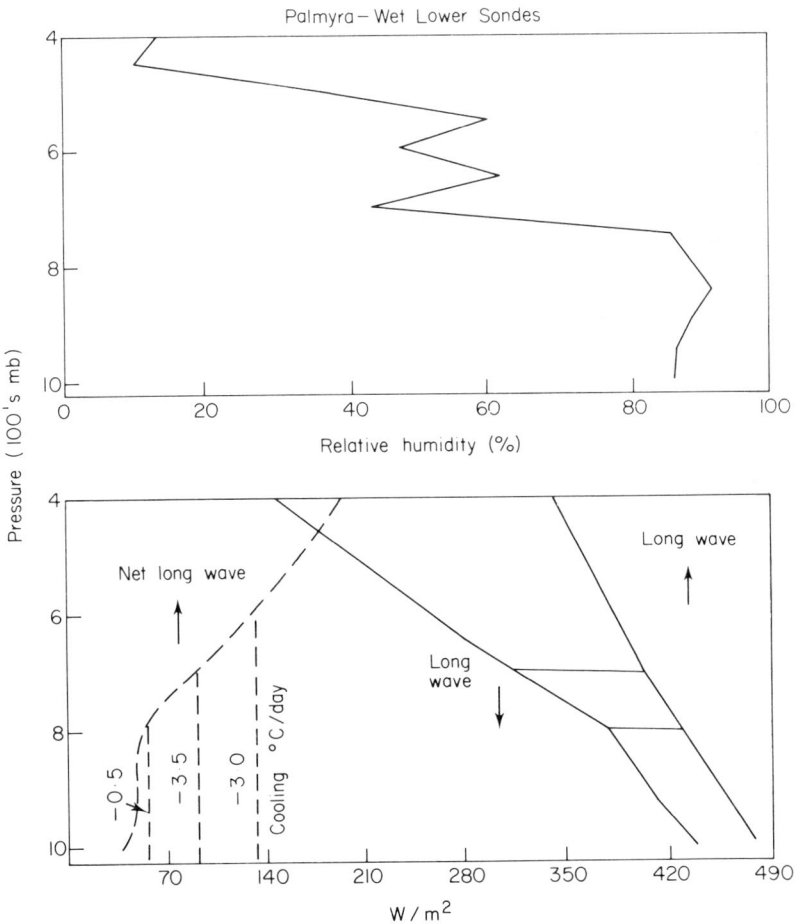

Fig. 4.19. Long-wave radiation composite on days with nearly saturated low levels topped by rapid moisture decrease, for Palmyra (6°N, 162°W) during Line Islands experiment. Upper: relative humidity. Lower right: long-wave radiation up and down. Lower left: net flux and cooling rates for layers below and above top of moist layer (cloud top).

gain. Radiation becomes very large at cloud top and above; net cooling may increase by as much as an order of magnitude or nearly so, seen in the lower part of Fig. 4.19. Here, indeed, is a strong destabilizing factor which may be due to the convection itself. The cloud top cooling, however, will lead to additional condensation which counteracts the cooling (31)! Especially in the low troposphere, where dq_s/dT is large, the net cooling may be reduced to half and even one-third of that computed from the radiative flux divergence. Destabilization of lapse rate above cloud tops is mainly observed at higher levels, where radiation cooling of 1-2°C does not evoke such a large counteracting response through induced condensation.

Even more serious is the fact that long-wave radiometer sondes have shown decreasing cooling rates from the ground upward (20). An example is given in Table 4.2 for Christmas Island during the Line Island experiment, a dry place indeed at that time. Even after allowing for short-wave heating, the net radiative cooling decreases steadily upward (74). Cox (20) computes a very large decrease of long-wave radiation with height, from —3·5°C/day at the surface to —1·5° between 700 and 600 mb, partly with the same data used in Table 4.2. He concludes that the "cooling of the lower layers would tend to stabilize the atmosphere and [this] helps explain why the lower layers in the tropics are not 'boiling' with convection." This point is well taken. The problem is that we have now enumerated an array of

Table 4.2. Radiation cooling at Christmas Island (°C/day) (74)

Pressure (mb)	Long-wave Christmas Island Radiometer[a]	Short-wave London (1957) Computations[b]	Net
100	0	+0·3	+0·3
150	—0·4	+0·3	—0·1
200	—0·8	+0·3	—0·5
250	—1·0	+0·3	—0·7
300	—1·3	+0·3	—1·0
400	—1·6	+0·5	—1·1
500	—1·9	+0·6	—1·3
600	—2·2	+0·8	—1·4
700	—2·5	+0·8	—1·7
800	—2·7	+0·7	—2·0
900	—2·9	+0·5	—2·4
1000	—3·1	+0·4	—2·7

[a]Observations courtesy S. K. Cox.
[b]London, J. (1957). A Study of Atmospheric Heat Balance. Dept. Meteor. New York Univ., Project No. 131. Final Report. (99 pp.)

factors all tending to suppress convection locally. Yet the "boiling" does go on, at least to the degree necessary to effect the sea-air energy exchange. Perhaps we see here one of the reasons why evaporation does not rise to values permitting local energy balance in the ocean and placing the burden of all latitudinal energy exchange in the atmosphere. However, a mechanism is needed to maintain the trade wind convection; failing all else, it must be the transport of trade wind air from higher latitudes over an ocean that manages to retain the isotherm pattern of Fig. 2.12, perhaps because evaporation is kept in certain bounds. This would be a problem for an oceanic energy budget. For the atmosphere, it follows that a two-dimensional model including large-scale advection of cool and dry air is needed in order to retain a viable mechanism for convection to occur and continue. Another factor is evaporation of water drops from cloud tops (end of Chapter 5).

Downdrafts

As a second step, preliminary to computing cumulonimbus transports, we shall examine downdrafts made buoyant through evaporation of rain which can reach the surface (cf. 48). The difference from the preceding calculation is that downdrafts introduce a strong and sometimes concentrated shot of low energy air from the middle troposphere into the lowest levels; thereby they also contribute to upward energy transport.

Consider Fig. 4.20. A downdraft is initiated from a cloudy region in the middle troposphere. At 600 mb its temperature may be the average tropical atmosphere temperature of $+1°C$, its specific humidity 5 g/kg, and, therefore, its relative humidity 70%; $\theta_e =$ 332 K. Without any source of water, the downdraft would take the path marked "no evaporation" and $q =$ constant at 5 g/kg. Clearly, the descending air would immediately become warmer than its surroundings so that the sinking motion would stop quickly. Apart from radiation, which is effective on a much longer time scale, there is no known way for a convective downdraft to penetrate into the low levels unless water is available for evaporation, which will cool the air and thus enable the descending air to remain buoyant. If the cooling takes place at constant pressure its magnitude is given by

$$dT = -\frac{L}{c_p} dq \qquad (4.26)$$

from the first law of thermodynamics for a closed system. The constant $L/c_p = 2\cdot 5$ if q is expressed in g/kg. Then the air will cool

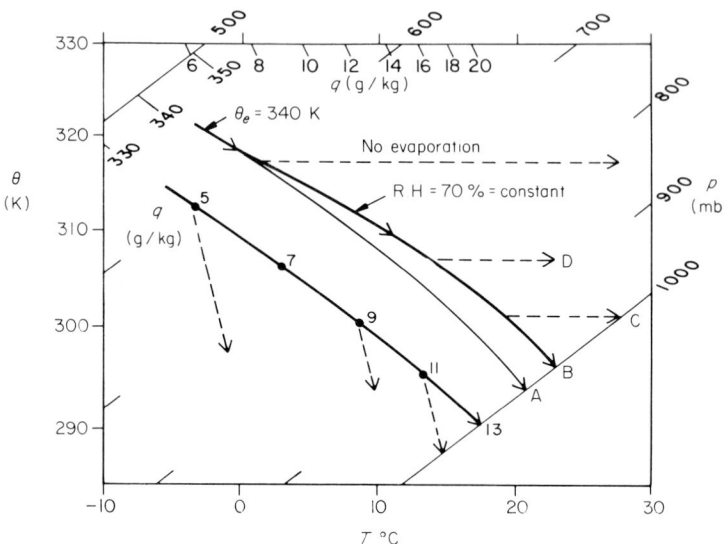

Fig. 4.20. Tephigram showing model of downdraft initiated by evaporation from falling rain. End of downdraft depends on where available water runs out (71). Numbers along q-line indicate downdraft q.

2·5°C for every gram of water evaporated per kilogram of air; during ascent, of course, the reverse applies and the air is heated and kept buoyant by condensation. Depending on the amount of water available, a wide variety of descent paths can be drawn from the 600 mb level. One of these is shown in Fig. 4.20, descent at a constant relative humidity of 70%. This descent path almost exactly follows the mean tropical atmosphere lapse rate of the low troposphere. This and other downdrafts are "unsaturated" and invisible after some downward bulge of the cloud near the origin, in contrast to descent following the moist-adiabatic lapse rate from cloud down which would require a saturated downdraft, indicated in Fig. 4.20 by the curve marked "A". Such downdrafts at times, no doubt, occur, but they are considered relatively rare. The descent at relative humidity of 70% may also reach the surface, indicated by point "B", if eight grams of water are evaporated per kilogram of air. This is a very large requirement for available water, indicated by the "q" or dewpoint line in Fig. 4.20. If 11 g are released during ascent of air from the surface to 600 mb (say from 18 to 7 g/kg) three-quarters of the condensed water are re-evaporated quite close to the reversible state, leaving little for precipitation and net heating of the troposphere. Nevertheless, this type of trajectory occurs frequently; it can, of course, also be generated from lower levels requiring less evaporation. It arrives at the

VERTICAL ENERGY TRANSFER

surface with $T = 23°C$, an almost constant value widely observed in downdrafts, and $q = 13$ g/kg. Given the ocean temperature $T_0 = 28°C$ and the normal air temperature near the surface of 27 to 27·5°C, specific humidity = 18 g/kg, the temperature difference $T_w - T$ jumps by a factor of five to ten, the humidity difference $q_w - q_a$ from about 6-11 g/kg; a factor of two. From Eqs. (4.10) and (4.11) sensible

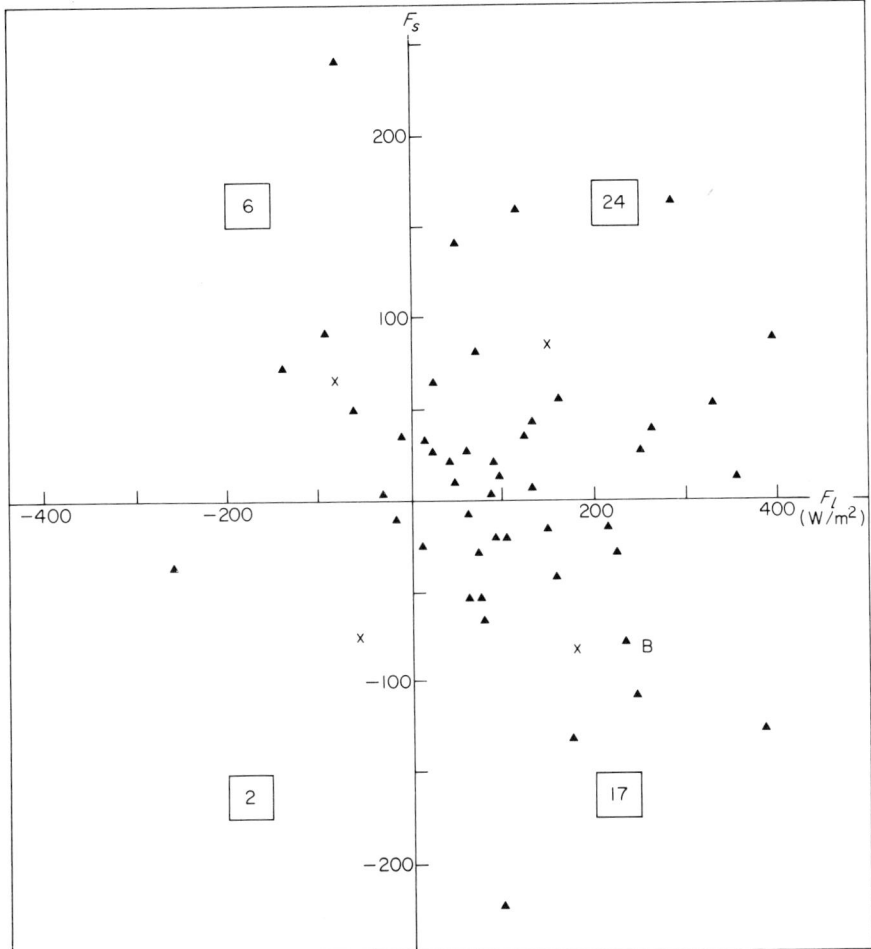

Fig. 4.21. Scatter diagram of F_s and F_l observed in the subcloud layer in active rain areas by DC-6 aircraft during the GATE project. In the experiments marked with solid triangle the aircraft itself was in rain; in the ones marked x it was just outside (71).

heat transport from ocean to atmosphere increases by a factor of ten and latent heat transport by a factor of two, leading to a marked increase in the ratio Q_s/Q_e in convective areas. Such energy increase from the ocean is likely to be followed by buoyant updrafts, shown in the upper right quadrant of Fig. 4.21, in the subcloud layer.

Assuming that only 6 g of water are available, the evaporation will cease at 900 mb from where the further descent will be dry-adiabatic to point C in Fig. 4.20. The downdraft will arrive warm and dry, yielding a point in the lower right-hand quadrant of Fig. 4.21. This type of case, from the turbulent flux measurements, is just as frequent as the cold downdraft, so that the sensible heat flow in the subcloud layer becomes nearly zero while moisture transport is almost always upward. Given a water supply of only 4 g/kg, dry-adiabatic descent will stop at 800 mb and the sinking will continue to point "D", a very warm and dry point injected into a sounding as in Fig. 4.22, where a

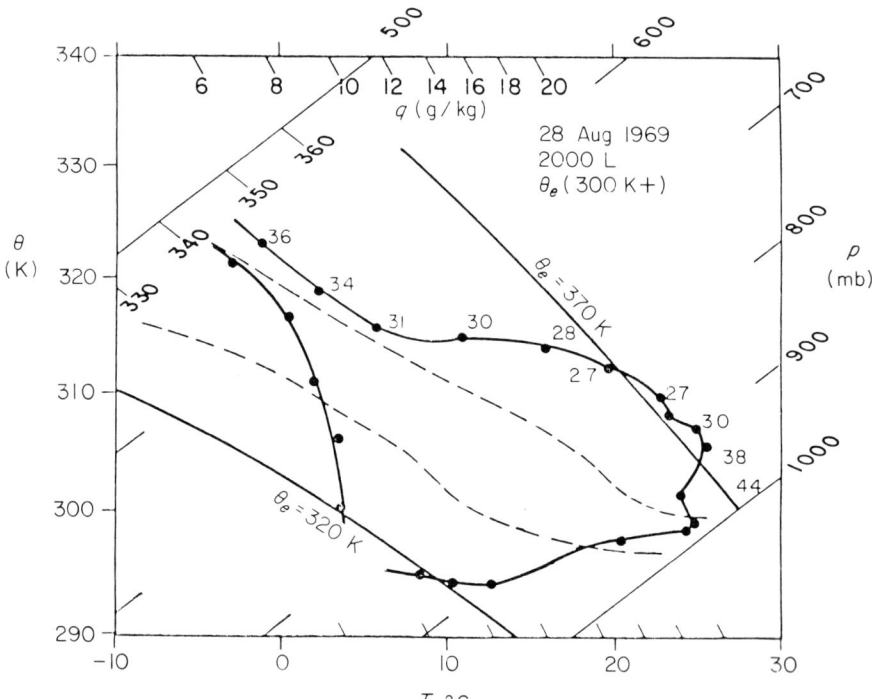

Fig. 4.22. A gross case of downdraft giving out before being able to reach the ground. Numbers along temperature curve denote θ_e (300 K+). Note very low θ_e in downdraft between higher energy values above and below. Observed at Anaco, Venezuela, on 28 August, 1969.

low energy class is found enclosed by higher energy both below and above. Evidently, the mass has run out of buoyancy and is unable to reach the surface. It has also become detached from its layer of origin and now is found wedged in a vertical structure with higher energy above as well as below. Such a structure is often observed at the exit side of deep convection, also the right hand diagram of Fig. 4.14 when applied to tall clouds.

Figures 4.20 and 4.22 illustrate the great variety of downward paths and downward penetrations that may be encountered, hence also the manifold structure of radiosonde profiles of temperature and humidity in the low levels of convective areas. If the downdraft arrives with vigour in the subcloud layer (rate of descent several metres per second) it will acquire strong horizontal kinetic energy from the well-known "thunderstorm high pressure" under the precipitating cloud (Fig. 8.11). As outflow, it will create new lines or zones of convergence at the leading edge. These provide the mechanism for induced new updrafts of the unmodified subcloud layer ahead of the spreading outflow. The latter will be gradually transformed to subcloud layer air through energy exchange from the surface in the course of some hours or longer (over land); it will then be able to become buoyant and ascend as updraft. In this way the "cycle" of vertical turbulent mass exchange is closed.

Energy exchange by cumulonimbus

It is tempting to assume that the whole mass converging into the tropics rises exclusively in cumulonimbi, whereas the heat exchange by cumulus is exclusively turbulent as described earlier. If such undilute ascent indeed provided balance against radiation cooling in the tropics plus heat for export to higher latitudes, a constraint on the observed convergence of mass into the equatorial zone would have been found and this would shed much light on the general circulation.

This pretty picture is not likely to be fully encountered in nature since downdrafts and their recycling are an inevitable by-product of precipitation, so that the precipitation from a cumulonimbus tower will virtually never be equal to the condensate. On the average, the ratio precipitation/condensation *(P/C)* is near 50%; it is clear that the higher this ratio the more efficient the thermodynamic engine, culminating in the hurricane which has the highest efficiency of converting latent heat to kinetic energy. In all other rain systems, the downdrafts disorganize the ascent by destroying the high-energy and nearly-mixed subcloud layer with the intrusion of low-energy and mid-tropospheric air, and no increase of kinetic energy whatever is often encountered (Chapter 8). Because of the role of downdrafts in vertical energy exchange, we should expect that energy transport based on the net convergence of mass into tropics alone should fail to meet balance requirements. A few brief calculations will establish the point.

As was deduced at the end of Chapter 3, the vertical motion in cumulonimbi, based on seeing the growth of towers, is three orders of magnitude larger than the mean vertical motion computed from the equatorial mass convergence, for instance Fig. 1.18. Therefore, the ascending towers will occupy only one per mille of the equatorial area. Given $A = 4\cdot4 \times 10^7$ km^2 for a 10° latitude belt near the equator and the area of one undilute ascent as 25 km^2, large enough for entrainment effect to be excluded, then, if 1‰ of this whole area is covered by such towers, their number $N = 4\cdot4 \times 10^7 \times 10^{-3}/25 = 1800$. Roughly this number should be expected to be found active over the whole belt continuously in the average over 24 hours. For a smaller choice of the area of an updraft, which experimentally is not yet well known, the number of towers would be greater, but this does not affect the calculation.

From Fig. 1.18 the vertical mass ascent in the equatorial zone $M_{z0} \approx 2\cdot4 \times 10^{-4}$ g/cm^2/s or $11\cdot2 \times 10^{13}$ g/s for a latitudinal belt of 10° width. The energy transported upward by the towers from a well-mixed subcloud layer will be $Q_{\text{surface}} = (c_p T + Lq)_s$, the surface "static" energy. On average $T_{as} = 300$ K and $q_{as} = 18$ g/kg, so that $Q_s = 300 + 46 = 346$ J/g ($\theta_e = 350$ K). The energy transport through, say, the 500 mb surface then will be $M_{z0}Q_s = 38\cdot8 \times 10^{12}$ kW. The outflow in the upper troposphere toward higher latitudes is only slightly less, 38·4 such units. In Chapter 12 the energy balance is treated thoroughly. Here our interest lies in examining vertical transport mechanisms. Net radiation cooling of the troposphere above 500 mb will be 1·7 to $2\cdot1 \times 10^{12}$ kW, depending on whether a cooling rate of 1°C/day or a little more is used. Thus, the upward energy transfer fails by no more than $2\text{-}3 \times 10^{12}$ kW to provide energy balance, and this is the amount that must be supplied by turbulence. This is only 6·5% of the upward transport and one should think that it is well within the limits of accuracy of knowing Q_0 and M_{z0} so that turbulence can be neglected. This line of reasoning, however, is grossly misleading and has produced much confusion. The large transport values arise from the fact that they are referred to the zero point of the absolute temperature scale. Thus, in calculations using total energy per unit mass, huge amounts of energy are always carried around which are meaningless and completely obscure the true relative role of transports by mean and turbulent vertical motions where only differences such as $T' - T$ are considered and the huge quantity of heat measured from the absolute reference point drops out.

Clearly, energy transfer with respect to some zero reference level should be introduced; a problem solved by Lorenz (44) for potential energy through his definition of "available potential energy". For the present case, a corresponding approach has not yet been developed. One should consider the domain T, q and choose the reference such that variance about the reference will be maximized. In previous calculations (73) the problem was alleviated, though not solved, by choosing a reference level of 70 cal/g, a value representative of the climate belt between tropics and higher latitudes. The same procedure will be followed here; for convenience, $\overline{\overline{Q}} = 300$ J/g (about 72 cal/g) will be subtracted as a round number. Better choices may be proposed.

With this convention $\overline{M}_{z0} \overline{\overline{Q}} = 33 \cdot 4 \times 10^{12}$ kW is to be subtracted, leaving a residual upward transport of 5·4 units and outward transport in the upper troposphere of 5·0 units. The difference, of course, remains the same, but the radiation cooling achieves the same order of magnitude as the transports and, most important, the estimated turbulent transport is close to 50% instead of 6-7% of the mean transports. It is now evident for what purpose the turbulent transport is mainly needed. If the upper troposphere was in radiation balance, the up-flowing mass would just turn poleward in the high troposphere at constant energy, making things very simple.

Let us consider how the problem of transport against the mean gradient, or the negative viscosity, may be handled. The turbulent mass flows M'_z up and down must cancel by definition as before. Above the level of minimum energy $gz + c_p T + Lq$ increases upward (Fig. 4.23). If we postulate conservation of energy rising and descending columns, we find (Fig. 4.23) that turbulence transports higher energy down and lower energy up—quite the opposite from what is needed, compounding the problem! The classical approach fails to handle the negative viscosity problem.

Next we examine the approach highly popular for computer calculation, whereby one defines any quantity as $x = \bar{x} + x'$. Here $\bar{x}$ may be the climatic value of a latitude belt, for instance as given by the mean tropical atmosphere, and x' any deviation therefrom. The vertical transport of energy in this system is given by $\overline{M_z} \bar{x} + \overline{M'_z x'}$. This formulation disregards the fact that $\overline{M}_z$ in reality does not go up at $\overline{Q}$, the mean energy value in the mid-troposphere, but at the undilute value Q_s. In Fig. 4.23, the midtropospheric mean energy in the vicinity of the trough centre $\overline{Q} = 325$ J/g. Subtracting $\overline{\overline{Q}} = 300$ J/g as before, the upward transport $M_{z0}\overline{Q}' = 3 \cdot 1 \times 10^{12}$ kW, only 57% of the transport with undilute ascent. If this mean transport really existed, the heating deficit of the upper troposphere would indeed become enormous with cooling over 2°C/day. Again we have compounded, rather than solved the problem and all transport must be thrown into a turbulent term $\overline{M_z'Q'}$ fictitiously enlarged to compensate for the deficit arising from the equally fictitious mean transport term $\overline{M_z Q}$. The turbulent transport must be based on some other principle from that shown in Fig. 4.23; for another approach see Squires and Turner (81).

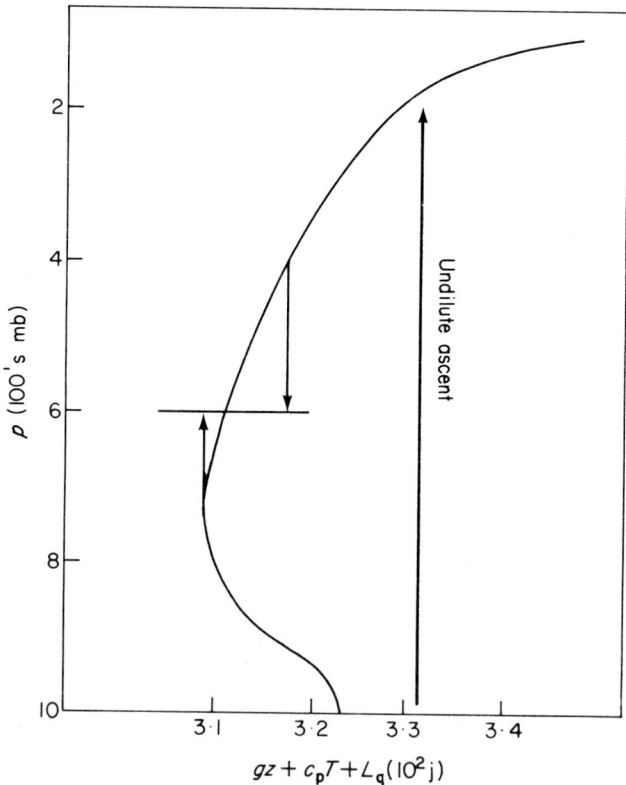

Fig. 4.23. Mean vertical distribution of energy in the equatorial trough zone. Arrows shows vertical mixing according to the classical concept which transfers energy down. Undilute ascent, portrayed on right side of diagram (a moist-adiabat on thermodynamic charts), must be invoked.

The actual physical process is most closely portrayed by writing

$$F_z = M_{z0} Q_s + M_z' (Q_s - \overline{Q}), \tag{4.27}$$

where $\overline{Q}$ is the average heat content of downdrafts. F_z, defined as the total required upward transport, $= 5\cdot 0$ units for outflow and about $2\cdot 0$ units for radiation, total $7\cdot 0$ units. Of this amount, $M_{z0} Q_s$ contributes $5\cdot 4$ units so that the residual is $1\cdot 6$ units. Solving for M_z' $= 1\cdot 6 \times 10^{12}$ kW/$Q_s - \overline{Q})$ with $Q_s - \overline{Q} = 346 - 325 = 21$ J/g, $M_z' = 7\cdot 6 \times 10^{13}$ g/s, or over 66% of the transport due to equatorial mass convergence. Finally $M_z = M_{z0} + M_z' = 18\cdot 8 \times 10^{13}$ g/s, close to a previously calculated value (73). It is evident that turbulent mass exchange needed to supply residual energy transport for balance, which at first seemed so small, is really very large indeed when the calculations are performed to the reference level of 300 J/g.

The vertical transports M_z up and down were separately computed by Betts (7) for cumulonimbi observed in Venezuela (Fig. 4.24); the

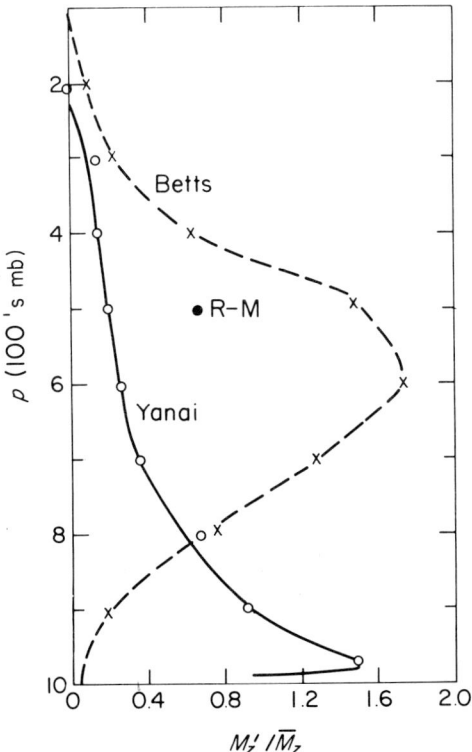

Fig. 4.24. Ratio of M_z' to $\overline{M}_z$: continental (7); oceanic (95).

difference, M_z, is the net transport upward. He finds that $\overline{M}_z$ is only one-third of the total transport in the middle troposphere but that it approaches 100% in the high troposphere. Clearly, the downdrafts originate mainly below 300 mb and attain their maximum strength at 500-600 mb; farther down they diminish toward the surface. Another profile is due to Yanai (95), who adopted Eq. (4.27) for his convective modelling. Here, M_z' remains a small fraction of $\overline{M}_z$ in the high troposphere. Then it increases all the way to the surface, evidently proportional to the amount of water available for evaporating into downdrafts. The large value in the subcloud layer probably portrays low energy outdrafts on one side of the convective cloud mass, and undilute inflow on the other side or, as often observed, above the cold "thunderstorm nose". Our value of 0·67 at 500 mb (marked R-M for Riehl-Malkus) lies between the two profiles, with a somewhat different approach for computing net mass with the assumption $\overline{M}_z = M_{z0} =$ const (p) from the subcloud layer to about 400 mb, where the organized outflow of the high troposphere begins. The net mass inflow in the low troposphere above the subcloud layer is pictured as first descending to the ocean to acquire the mean value Q_s for undilute ascent. Much is to be said for the Yanai presentation over the ocean. On the other hand, from Fig. 4.20, a mechanism for downdrafts decreasing downward is also envisaged through their running out of water for evaporation, which will occur preferably over land.

Convective modelling†

The preceding section has shown a number of processes relevant to cumulus clouds and its various functions which, in the aggregate, are really quite complicated. We have treated undilute ascent, the constraint of entrainment, cumulus clouds in a turbulent field with no net mass movement up and cumulonimbus with net upward mass flow; compensatory motions by large areas surrounding cumuli going through their life-cycle, the mechanism of downdrafts and, finally, net upward motion and turbulence with updrafts and downdrafts together. If all of these processes and their evolution have to be treated explicitly in modelling the large-scale structure of the trades or of the equatorial trough zone, a complex computer operation would be required as well as very detailed observations. However, the subject cannot be ignored, for cumulus clouds have a feedback on synoptic scale weather systems. Therefore, a large field of research has developed with objective to summarize the effects of convection into

†This section was written partly by F. Baer.

VERTICAL ENERGY TRANSFER

tractable computational formulae or, to use the cumbersome word that has been formed, to "parameterize" convection. Although some modelling without parameterization has been reported (Chapter 11), it will be of value to present a survey of some of the researches which, of course, are in a continual state of evolution.

The overwhelming influence of convection on tropical storms and hurricanes led modellers of the tropical vortex to evolve a convective parametric scheme in the early 1960s. At about the same time, general circulation modelling reached the point where proper specification of moisture, especially in the tropics, was found to be necessary and parameterization evolved. These two developments led to separate approaches.

The general circulation approach, although not concerned with the details of convection, allows for latent heat release and transport of heat, moisture and momentum by the large-scale variables (55). However, if lapse rates in the evolving thermodynamic field become unstable, time integration cannot proceed. The parameterization scheme assumes that through convection the lapse rate becomes neutral, moist-adiabatic in a saturated environment and dry-adiabatic in unsaturated layers. This adjustment is achieved under the constraint that total energy is invariant. Efforts with this procedure subsequent to Manabe (55) are concerned with questions of the degree of saturation in unstable regions, speed with which the adjustment takes place, removal of condensed water, etc. Serious doubts have been expressed as to whether convection can be parameterized by lapse rate adjustment alone, and have led to more careful analyses of the alternate scheme.

This alternate parameterization utilizes the flux of moisture into the convective column and the subsequent exchange of moisture, momentum and heat of the convective process with the environment. The method was first evolved by Charney and Eliassen (19) with application to the developing hurricane. They presumed that the predominant effect of convection was instigated by convergence of moist, warm air in the boundary layer. Using a simple formula, which they had developed for the induced vertical velocity at the top of the boundary layer and denoted as "Ekman pumping" (18), they allowed the moisture to be brought out of the boundary layer and the resulting latent heat to be released in proportion to the saturation mixing ratio in the deep vertical column, with the distribution function predetermined from statistical information based on observations. Similar presentations were made by Ooyama (63) and Ogura (62). The technique proved modestly successful when applied to hurricane models. Sophistication in the vertical distribution of this convective heating was introduced in (94) and (85) among others.

A serious deficiency of the boundary layer flux parameterization is the neglect of environmental mixing and lateral flux between the convective cloud and its surroundings. This was recognized by Kuo (38) who developed a scheme whereby the environment mixed with the cloud at all levels and distributed its latent heat according to the difference in the temperature between the cloud and its environment, including the flux of moisture from the surface through the unstable cloud region.

Kuo considered the question: how much water vapour do we have available in relation to that needed for cumulus growth? The setting shall be the typical tropical atmosphere with a conditionally unstable lapse rate. Consider a potential cloud layer between cloud base, determined from the subcloud layer properties to good approximation, and some upper level where a moist-adiabatic path would intersect the mean sounding of the area and, thus, become stable. Enough moisture must be introduced so that, during ascent, the temperature of the rising mass will be increased from the temperature computed from dry-adiabatic ascent to that along the moist-adiabatic temperature curve, expressed by the moisture increment

$$\Delta M_1 = \frac{c_p}{L} \int_{p_{\text{cloud top}}}^{p_{\text{cloud base}}} (T_{\text{sat}} - T_{\text{ad}}) \frac{\mathrm{d}p}{g}. \tag{4.28}$$

where T_{sat} denotes the temperature of the saturated mass and T_{ad} the adiabatic temperature which would have been attained during dry ascent. If an entrainment factor is assumed, T_{sat} will lie close to the environmental temperature T and then additional moisture is needed to saturate the unsaturated admixture from the sides of the cloud. The storage requirement for moisture due to this effect and also that of filling the previously unsaturated volume of the cloud with enough water vapour to give 100% relative humidity is stated by

$$\Delta M_2 = \int_{p_{\text{cloud top}}}^{p_{\text{cloud base}}} [q_{\text{sat}}(T_{\text{sat}}) - \bar{q}] \frac{\mathrm{d}p}{g}, \tag{4.29}$$

where $\bar{q}$ is the specific humidity of the environment.

VERTICAL ENERGY TRANSFER

The moisture increment $\Delta M = \Delta M_1 + \Delta M_2$ is drawn from the subcloud layer and may be denoted by $\Delta \hat{q} \times \Delta p_0/g$, where Δp_0 is the pressure depth of the subcloud layer and $\hat{q}$ its mean specific humidity. Discounting evaporation on the short time scale, the only source of moisture will be flux convergence (Q) in the subcloud layer

$$Q = \hat{q}\, \widehat{\nabla \cdot \mathbf{V}}\, \frac{\delta p_0}{g}$$

for a mixed subcloud layer and uniform convergence without moisture advection. These restrictions are not needed; they are introduced here merely for the sake of a simple picture. From this relation, the time δt needed to establish the cloud can be computed. For instance, let $\Delta M_1 + \Delta M_2 = \Delta M = \Delta q\, \delta p/g$, where Δq is the specific humidity increment of the cloud layer and $\delta p/g$ its pressure depth. Then

$$\delta t = \frac{\Delta q}{\hat{q}} \frac{1}{\widehat{\nabla \cdot \mathbf{V}}} \frac{\delta p}{\delta p_0}. \tag{4.30}$$

Suppose that $\Delta q = 3$ g/kg, $\hat{q} = 18$ k/kg, $\widehat{\nabla \cdot \mathbf{V}}$, the synoptic convergence, $= 1 \times 10^{-5}\mathrm{s}^{-1}$, $p_0 = 100$ mb and $p = 200$ mb determined from an ascent computation using a radiosonde ascent or its equivalent in a multi-layer model. Then $\delta t = 3 \cdot 3 \times 10^4$ s, which is unrealistically large. We now must remember that the cumulus ascent area A_a in the notation used earlier is only about 1% of the total area. Therefore, the synoptic convergence must be multiplied by 10^2 if all convergent mass rises in cumuli. Herewith $\delta t = 330$ s or 5·5 mins, not far from the observed growth rate for trade wind cumuli. The decision on the fractional cloud area A_a/A is made using the temperature lapse rate: whether it is greater or less than moist-adiabatic. Even with $A_a/A = 0 \cdot 01$, a synoptic convergence much less than $10^{-5}\mathrm{s}^{-1}$ would lead to excessive times for cumulus development. The reader may try to estimate the time needed for development of cumulonimbus with the same value of synoptic convergence. He will find that the time required for the cloud to grow to 200 mb is far in excess of one hour. For a realistic time estimate a stronger synoptic convergence than $1 \times 10^{-5}\mathrm{s}^{-1}$ is needed (cf. also 27).

On the whole the foregoing shows in a simple way how fields of cumuli and the degree of their development can be deduced from large-scale parameters of the tropical atmosphere alone. Conversely, the per cent area covered by convective clouds may be estimated from the moisture convergence (33). Kuo's procedure allows for self termination due to saturation as well as statistics on the number of such clouds in the domain represented by each given large-scale value. Recognizing the variety of cloud sizes and the fact that his procedure parameterized only undilute cumuli, Kuo (39) modified his scheme to include convection with entrainment. This scheme has been quite popular and used by numerous modellers for convective parameterization. It not only gives reasonable energy input to the large scales, but is self-regulating and interactive with large-scale variables at all vertical levels. Nevertheless, its effectiveness is strongly dependent on forecast time and may be less than complete in determining cloud population statistics.

This latter limitation has been considered in great detail by Arakawa and Schubert (3). Their primary intent is to determine the vertical distribution of vertical mass flux by ensembles of clouds, the detrainment of mass from the ensemble and the thermodynamic properties of the detraining air. This information, interpreted in terms of the large-scale variables, completed the parameterization. It is evident that this theory requires a distribution of different sizes and characteristics of clouds, denoted as ensembles. Furthermore, the influx of mass and moisture through the boundary layer in the model is considerably more sophisticated than that of Charney and Eliassen. A "cloud work function" is calculated which determines the transport and mixing of the detrained air from the clouds to the environment for spectrum of cloud types and sizes, and effectively establishes the necessary forcing by the parametric convection process into the large scale.

Although parameterizing of convection as it affects the large atmospheric scales has come a long way, and its application has shown some success, there is yet no assurance that when applied over a longer integration time, predictions will show systematic accuracy. This is so because small errors due to parameterization which must occur because the process is at best a statistical one, will grow throughout an integration period, ultimately destroying the accuracy of the computation. As we hope for improved forecasts over a period of several days, we may wish for an alternate forecast scheme and/or parameterization over a longer time period. Despite our successes, there is much still to be learned from the environment concerning the details of convection and its effect on planetary and synoptic scale flow.

A practical application

In view of the foregoing it may be useful to develop rather simple procedures of parameterization. Reed and Recker (67) led the way by

VERTICAL ENERGY TRANSFER

establishing a heat and moisture budget for the West Pacific Islands based on large-scale parameters only. First they took up the heat and moisture balance in the dry area surrounding convective cloud concentrations. Then they considered downdrafts and evaporation and finally the convective core itself. Since the surroundings at large offer no great problem—energy transfer down the gradient—only the intense part of convective cloud masses will be the object of a sample calculation here.

A useful and convenient starting point is provided by a composite vertical profile of divergence for the western Pacific where the centre of a 4° latitude-longitude square was set on the centre of well-marked satellite cloud masses (77). The magnitude of the divergence was $10^{-6} s^{-1}$ for the synoptic scale area (Fig. 4.25), rather small. One observes lower inflow and upper outflow above a high level of non-divergence near 400 mb. The outflow is concentrated near 200 mb; the inflow is spread vertically and not confined to the lowest layers with high

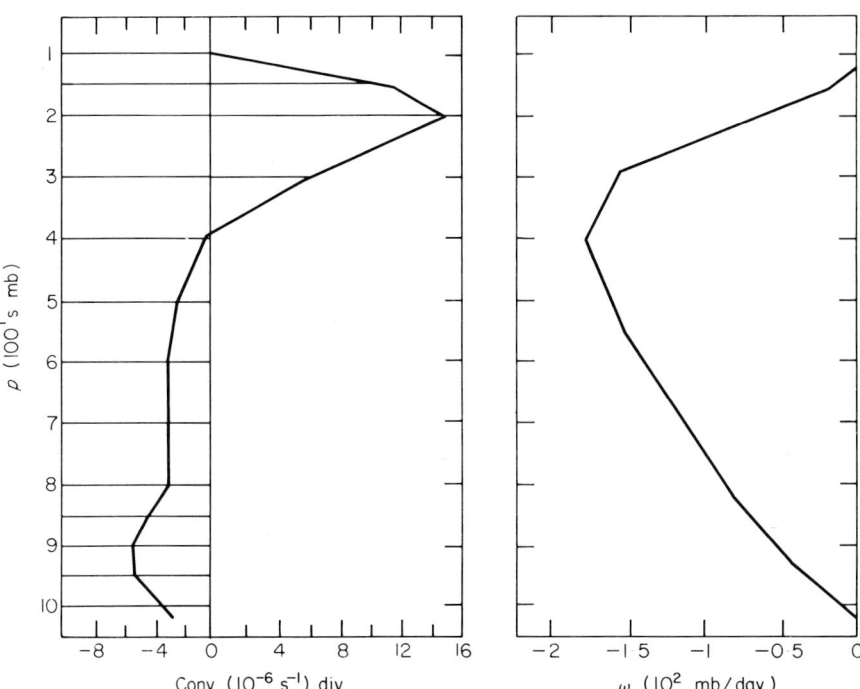

Fig. 4.25. Vertical distribution of horizontal divergence (left) and of vertical pressure-velocity (right) for a large sample of West Pacific satellite cloud masses centred in a 4° latitude-longitude square (adapted from 77, 88).

thermodynamic energy. This distribution is typical of many cases, found by various investigators. The vertical pressure velocity (right side of Fig. 4.25) increases upward to 400 mb; at first sight one may be inclined to say that mass is entrained in an updraft on the way up, but on second thoughts this is quite impossible in view of the vertical energy distribution of Figs 2.23, 2.24 and 4.11.

Entry of mid-tropospheric energy with the low values of θ_e observed would erode any rising tower and render it quite impossible for the mixture to reach the 200 mb outflow layer through convective ascent. Consider Fig. 4.26 which shows energy tubes, i.e. undilute ascent paths for the thermodynamic data given in (77) and similar ones in (88). The diagram portrays the beginning of such tubes from levels at 100 mb intervals, the ascent path to condensation and then the continuing moist-adiabatic path. As is readily seen, trajectories starting at 800 and 700 mb are very cold and should never be able to ascend freely at all. Now it is quite clear that air initially situated

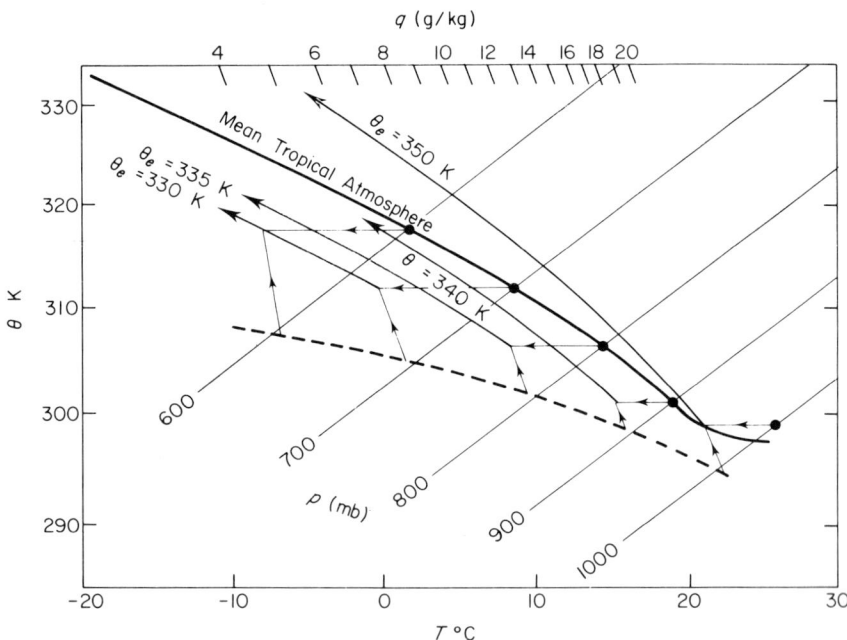

Fig. 4.26. Tephigram showing the mean sounding in large cloud masses (77) and undilute ascents of air from several levels, θ_e marked at top. Below cloud base the ascent is dry-adiabatic at constant specific humidity; above cloud base it is moist-adiabatic. Scales for spacing of isotherms and lines of equal specific humidity indicated.

above 900 mb cannot form cumuli. Therefore, an alternate solution must be sought.

It is not obvious why the layer of net convergence is so deep, but, as demonstrated in Chapter 8, the synoptic systems possess a cold core with pressure relative to the outskirts lowest in the middle troposphere. It follows that a pressure gradient with respect to the outside air so introduced will accelerate the latter toward the convective core; if so, cold core and deep inflow must develop together. Irrespective of this sequence, it is proposed to solve the problem by letting all air converging above 900 mb descend in downdrafts and satisfy mass continuity as before with Eq. (4.27), by letting low-level outside air with high energy enter the system in compensation. Therewith, the ratio of turbulent to net flow will look rather similar to Yanai's profile in Fig. 4.24 (95). The point computed by us earlier lies at a higher ratio, but now multiple constant energy tubes are considered. We have no need for turbulent exchange above 400 mb since an ice adiabat will be used.

Given this approach, the mass flow diagram will be that of Fig. 4.27. The downdraft starts near 400 mb and becomes cumulatively larger to the top of the subcloud layer in which it spreads out horizontally as a "thunderstorm nose." Of course, one could let the

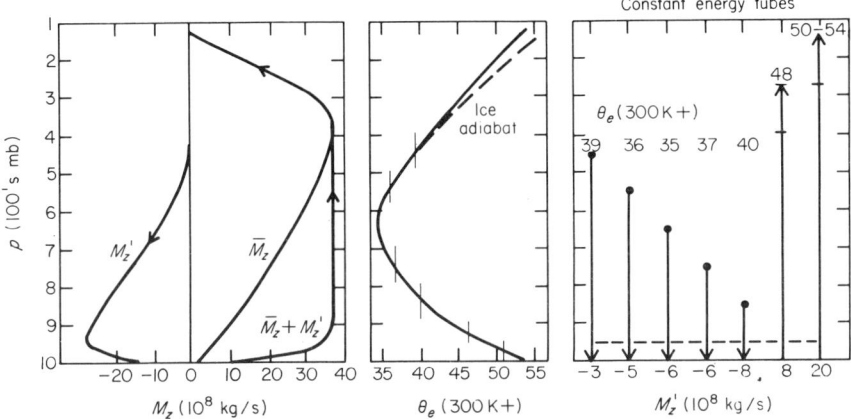

Fig. 4.27. Left: vertical mass flow from Fig. 4.25 assuming that all air converging above 900 mb descends in downdrafts to the subcloud layer and that compensatory inflow of subcloud layer air takes place below 900 mb. Centre: vertical distribution of θ_e for the mean sounding (77) with ice adiabat added for upper part; vertical bars denote layer used for energy tube calculations.
Right: five downdraft and two updraft energy tubes. Turbulent mass flux given at bottom, θ_e at top.

downdraft diminish gradually throughout the lower troposphere with the model of Fig. 4.20 which shows water running out at different levels. However, much more complex modelling than here intended would be needed; a modification certainly would be required over land.

The procedure followed was graphical, broken up into 100 mb steps. In the top layer from 400 to 500 mb, for example, net mass inflow from Fig. 4.25 is computed as 3×10^8 kg/s. Average specific humidity of the inflow is two g/kg, therefore, moisture import 6×10^8 g/s. Descending adiabatically to 500 mb, the temperature of the air there would be $-2°C$ compared with the 500 mb temperature of $-5°C$ on the sounding. Using Eq. (4.26), 1·2 g/kg of water must be evaporated to keep the temperature of the descending air the same as that of the environment. Then the downward moisture flux (F_z') through the 500 mb level will be $9·6 \times 10^8$ g/s. In the next layer, from 500 to 600 mb, the downdraft would reach 600 mb with $+9°C$ compared to $+2·5°C$ as observed. For the difference of 6·5°C, 2·6 g/kg moisture must be evaporated. Total specific humidity is 5·8 g/kg at 600 mb and downward moisture transport $F_z' = 17·3 \times 10^8$ g/s. The procedure is repeated to 950 mb where the outflow into the subcloud layer and the lateral outdraft from there is 40×10^8 g/s. Altogether the transport has increased by $F_z' = 34 \times 10^8$ g/s in this downdraft energy tube at $\theta_e = 339$ K.

Now a second energy tube down is started in the layer 500-600 mb, then a third in the layer 600-700 mb, etc. Table 4.3 gives the total accounting. It is seen that we have here used a model of an unsaturated downdraft, which adopts everywhere the temperature of the rain area. This procedure assumes a minimum downdraft rate compared to a maximum saturated downdraft applied in (73). The latter would be a strong and violent downdraft. For making the choice between these extremes or some intermediate solution, we may note that the total downdraft outflow mass at 950 mb $M_z = -28 \times 10^8$ kg/s. The downdraft, a mesoscale feature, by our previous hierarchy of circulation scales will take place in 0·1 of the synoptic 4° square, or in 2×10^4 km² given the total area of 20×10^4 km². There will be then 10 mesoscale systems with the average area of 2×10^3 km² operating at any time. Noting that $M_{zd} = \varrho w A_d$, the area of a mesoscale downdraft, $w_{950 \text{ mb}} = -12·5$ cm/s. Given zero vertical velocity at 400 mb, the downward acceleration has the exceedingly small value of $1·15 \times 10^{-4}$ cm s^{-2} in order to arrive at 950 mb with 12·5 cm/s. The reader may verify that the buoyancy acceleration is no more than $10^{-7}g$! From Eq. (4.15), $T_v' - \overline{T_v} = -3·3 \times 10^{-5}$ °C and, with

Table 4.3. Calculation of evaporation in downdrafts

θ_e Downdraft (Fig. 4.27)	339		336		335		337		340		$\Sigma F'_z$	$\Delta \Sigma F'_z$	Inflow	Evaporation	Percentage per layer
p (mb)	q (g/kg)	F'_z (10^8 g/s)	q	F'_z	q	F'_z	q	F'_z	q	F'_z					
400											0				
500	3·2	10									10	10	6	4	3
600	5·8	17	5·0	27							44	34	20	14	9
700	8·4	25	8·0	40	7·2	43					108	64	36	28	18
800	10·8	33	10·4	52	9·6	57	10·2	62			204	96	54	42	28
900	12·6	38	12·2	61	11·4	68	12·0	72	13·0	104	343	139	96	43	28
950	13·4	40	13·0	65	12·2	73	12·8	77	13·8	110	365	22	0	22	14
												212	212	153	100
Evap. (g/kg)	12·0		9·0		6·2		3·8		1·8						
Inflow (10^8 g/s)		6		20		36		54		96					
Increment (10^8 g/s)		34		45		37		23		14					
Percentage of evap. per downdraft		22		30		24		15		9					

Evaporation (cm/day) = 0·67.

Eq. (4.14), $q' - \bar{q} = -2 \times 10^{-4}$ g/kg. These values are all well below observational capability. Thus, the minimum assumptions for the downdraft calculation are clearly the most appropriate model for the average synoptic system divergence profile, hence for the largest fraction of synoptic-scale weather systems. It may be noted that the compensatory updraft, lodged in hot towers occupying another order of magnitude less area, still will have a speed of metres per second, ten times larger than the downdraft.

From Table 4.3, the downdraft evaporation averaged over the 4° synoptic area is 0·67 cm/day which may be compared with average ocean evaporation of around 0·4 cm/day. In terms of the five downdraft tubes, portrayed in Fig. 4.27, the tubes starting between 400 and 700 mb use up most water. Hence, for efficiency of precipitation heating, a lowered level of non-divergence will be a most potent mechanism. From another viewpoint, also seen in Table 4.3, evaporation will also be reduced substantially, if the downdrafts run out of water already at 800 mb. In this event, however, there will be a warm and dry lower layer, not conducive to augmenting new cumulonimbi.

Turning now to the ascent, $4 \cdot 5 \times 10^8$ kg/s mass flow are buoyant in the layer 1000-950 mb and 5·5 such units in the layer 950-900 mb. Total mass flow up then will be 32·5 units at 950 mb and 38 units from 900 to 400 mb as undilute towers, satisfying the mass continuity requirement. In this way the problem of having low energy air entering the updrafts has been completely averted. Using an average $q = 18$ g/kg in the layer 1000-950 mb and 16 g/kg in the next layer, upward moisture transport $F_z = (590 + 90) = 680 \times 10^8$ g/s. The specific humidity difference between the outdraft and indraft in the lowest 50 mb is close to 4 g/kg, hence the "eddy" moisture transport F'_z through the boundary of the 4° square is $F'_z = M'_h (q_{\text{in}} - q_{\text{out}}) = 28 \times 10^8$ kg/s $(17 - 13$ g/kg$) = 114 \times 10^8$ g/s, where M_h is horizontal mass flow. Divided by the area of the square, an equivalent evaporation of 0·5 cm/day is thereby added; this is rather more than the, roughly, 50% of evaporation exported by the trade winds showing, as in Chapter 1, that the equatorial trough zone and other weather systems occupy only one-third of the area of the trade winds.

The budget of water vapour and water is presented in Table 4.4. Inflow plus oceanic and atmospheric evaporation produce a large increment of 3·19 cm/day in vapour form, whereas condensation is slightly less. In our case the difference, though within range of error of all calculations, must be assigned to local moistening of the atmosphere which, of course, frequently does occur. One finds—apparently a contradiction—that the air is moister after a rain than before! Of the condensate, 10% are assumed to be advected out of the 4° square by the high-level winds, often seen as long satellite cloud bands mostly formed by cirrus. Of the balance, 0·67 cm is evaporated into downdrafts, leaving almost 2 cm precipitation, a value close to that estimated empirically (77). The ratio precipitation/ condensate is 0·67, somewhat higher than the ratio of 50% found in some studies; the ratio evaporation/precipitation is 0·34. In summary, of the condensate, 10% escapes from the area, 23% is used in the downdraft mechanism and 67% falls out as rain. The latter amount will sustain the heat balance for an area double the size considered against radiation cooling.

Table 4.4. Moisture budget per unit area and per day of a 4° square.

Water Vapour	cm/day +	cm/day −	Water plus Ice	cm/day +	cm/day −
Updraft			Condensate	2·96	
Mass inflow	0·70		Cloud outflow		0·30
Eddy inflow	2·06		Downdraft		
Ocean evaporation	0·90		evaporation		0·67
	3·16		Precipitation (residual)		1·99
Condensation		2·96			
Imbalance ($\partial/\partial t$)	+0·20				
Downdraft					
Inflow	0·92				
Rain evaporation	0·67				
Outflow	1·59	1·59			

We may also note that 10 mesoscale systems with average area of 2000 km^2 and duration of two hours will cover the whole area in 20 hours if their average precipitation is 2 cm each. This value is slightly less but still close to the average yield of 2·5 cm per mesoscale system estimated earlier. Thus, on the whole, a consistent and rather practical method for calculating the mesoscale rain production has been achieved and parameters that will change the rain effectiveness have also become evident through the computation.

One may ask further how long the "eddy" exchange of moist-air-in, versus dry-air-out in the surface layer can be maintained before the whole area becomes covered with a low-energy subcloud layer presaging the end of the life of the system. If the whole depth of the subcloud layer (Δz) is filled with low energy air, $M'_z \delta t = \varrho A \Delta z$. $M'_z = 28 \times 10^8$ kg/s from the foregoing, $\varrho = 1 \cdot 15 \times 10^{-3}$ g/cm³ and A will again be the 4° square. Then $\Delta t = 4 \cdot 1 \times 10^4$ s or very nearly half a day. This, however, is very close to the estimated recovery time needed for transformation of the downdraft outflow back to normal subcloud layer air. Thus, no brake will be applied from the lowest atmosphere, unless outside convective systems interfere, and this is perhaps a reason for the mean divergence field found for the size of persistent synoptic systems. The whole transaction could repeat twice daily and the system could continue to live for days, even if it does not propagate. If it does move, the propagation rate and large-scale moisture advection must be added to the picture, but these data are given by the large-scale flow and thermodynamic fields and can be incorporated readily into computer calculations, as can all of the steps outlined above.

As a final step, we shall determine the contribution made by the turbulent mass exchange in the example to maintenance of the atmosphere's kinetic energy. The contribution made by the net inflow of 10×10^8 kg/s in the layer 1000-900 mb may be computed additionally by the reader. The work done through release of potential energy, widely derived in text books, is given by

$$W = -\frac{R}{g} A \int \widehat{\omega^* T^*} \, d \ln p, \qquad (4.31)$$

where R is the gas constant for air, and $\omega \equiv dp/dt$, the vertical pressure-velocity, a symbol unfortunately in competition with ω denoting the earth's angular rotation. Further, the circumflex denotes area averaging at a given pressure and the star deviations from the area average. Now $M_z = -\omega A/g$. If a simple model of a single updraft and downdraft temperature is introduced in Eq. (4.28), not unjustified because all the downdrafts in Fig. 4.27 descend at the same temperature, we can write

$$W = R \int \widehat{M_z(T_{v_{\text{up}}} - T_{v_{\text{down}}})} \, d \ln p. \qquad (4.32)$$

For the turbulent transport we find that $T_{v_{\text{up}}} - T_{v_{\text{down}}} \approx +2°C$. Then $W = 32 \times 10^{-4}$ j/cm²/s or 32 W/m², 6% of the effective solar constant. This value may be compared with average global dissipation of kinetic energy of about 1 W/m², as also found in regional calculations in the trades and equatorial zone. Of course, we must remember that we are dealing here with a synoptic-scale calculation and reduce the result by an order of magnitude to encompass the whole tropical area. Even then we exceed the average estimated requirement for maintenance of air motion by a factor of three. Evidently, the desire to include the energy release in tropical mesoscale and synoptic systems for general circulation calculations and predictions is very justified.

Only well-established equations have been used in the preceding discussion, which can be programmed for computers. The steps for including the model in practical calculations would be the following:

(1) It is assumed that charts of horizontal divergence and of wind and thermodynamic variables are available from large-scale grids, also information on surface temperature and dewpoint. Divergence profiles with respect to selected satellite cloud masses can be developed from the charts, also a mean sounding. The propagation of the cloud masses and large-scale thermal advection of heat and moisture should also be given.

(2) The sounding is tested for updraft undilute towers, their strength and maximum height or pressure of penetration. Vertical acceleration and motion are computed with the buoyancy formula reduced for carrying a fraction of, say, one-third to one-half of condensed water. An ice adiabat should be used above 400 mb.

(3) Next the slope of the cloud with respect to the vertical axis is computed from $\tan \alpha = w/\Delta U$, where α is the angle between cloud axis and the vertical, w vertical motion at, say, 500 mb and ΔU the shear of the basic current U between cloud base and 500 mb, or the level of maximum wind if lower. If there is a second layer with reverse shear, often encountered in the trades, a second calculation should be made above the level of strongest wind, which then will be more often near 750-600 mb rather than 500 mb. Assuming the cloud tilts by no more than a given angle, say 45° (not tested), the hot tower development will occur. At larger slopes the tower will erode as described earlier in this chapter (Fig. 4.15) and the computation should not be continued. The precise threshold angle for the cloud slope needs to be determined empirically and theoretically.

(4) Assuming that convection tubes to the high troposphere exist and that the cloud inclination is below any critical value, the steps described above are carried out yielding the contribution of the subgrid energy exchanges and precipitation to the large-scale fields. The surface energy value must be watched continually in the calculation to be sure that the subcloud layer is not depleted of high-energy air.

References

(1) Ackerman, B. (1956). Buoyancy and precipitation in tropical cumuli. *J. Meteor.* **13:** 302-310.
(2) Ackerman, B. (1963). Some observations of water content in hurricanes. *J. Atmos. Sci.* **20:** 288-298.
(3) Arakawa, A. and Schubert, W. H. (1974). Interaction of a cumulus ensemble with the large-scale environment; part I. *J. Atmos. Sci.* **31:** 674-701.
(4) Austin, J. and Fleischer, A. (1948). A thermodynamic analysis of cumulus convection. *J. Meteor.* **5:** 240-243.
(5) Battan, L. J. (1958). Influence of the environment on the initiation of precipitation in tropical cumuli over the ocean. *Tellus* **4:** 466-471.
(6) Bénard, H. (1900). Les tourbillons cellulaires dans une nappe liquide. *Rev. Gén. Sci. Pur. Appl.* **II:** 1261-1271, 1309-1328.
(7) Betts, A. K. (1973). A composite cumulonimbus budget. *J. Atmos. Sci.* **30:** 597-610.
(8) Betts, A. K. (1976). The thermodynamic transformation of the tropical subcloud layer by precipitation and downdrafts. *J. Atmos. Sci.* **33:** 1008-1020.
(9) Betts, A. K. (1976). Modeling subcloud layer structure and interaction with a shallow cumulus layer. *J. Atmos. Sci.* **33:** 2364-2382.
(10) Bjerknes, J. (1938). Saturated-adiabatic ascent of air through a dry-adiabatically descending environment. *Quart. J. Roy. Meteor. Soc.* **64:** 325-330.
(11) Braham, R. R., Jr. (1952). The water and energy budgets of the thunderstorm and their relation to thunderstorm development. *J. Meteor.* **9:** 227-242.
(12) Braham, R. R., Jr. (1958). Cumulus cloud precipitation as revealed by radar. *J. Meteor.* **15:** 75-83.
(13) Brocks, J. (1955). Wasserdampfschichtung über dem Meer und "Rauhigkeit" der Meeresoberfläche. *Arch. Meteor. Geoph. Biokl.* (A) **8:** 354-383.
(14) Brook, R. R. (1978). The Influence of Water Vapor Fluctuations on Turbulent Fluxes. *Bound. Layer Meteor.* **15,** 481-489.
(15) Brooks, C. E. P. (1925). Distribution of thunderstorms over the globe. *Geoph. Mem.. Lond.* (3) **24:** 147-164.
(16) Bunker, A. F., Haurwitz, B., Malkus, J. S. and Stommel, H. (1949). Vertical distribution of temperature and humidity over the Caribbean Sea. *Pap. Phys. Oceanog. Meteor.* (Mass. Inst. Tech. and Woods Hole Oceanog. Inst.) **11:** No. 1. (82 pp.)
(17) Byers, H. R. and Braham, R. R. (1949). "The Thunderstorm." US Government Printing Office, Washington DC. (287 pp.)
(18) Charney, J. G. and Eliassen, A. (1949). A numerical method for predicting the perturbation of the middle-latitude westerlies. *Tellus* **1:** 38-54.
(19) Charney, J. G. and Eliassen, A. (1964). On the growth of a hurricane depression. *J. Atmos. Sci.* **21:** 68-75.
(20) Cox, S. K. (1969). Observational evidence of anomalous infrared cooling in a clear tropical atmosphere. *J. Atmos. Sci.* **26:** 1347-1349.
(21) Cressman, G. P. (1946). The influence of the field of horizontal divergence on convective cloudiness. *J. Meteor.* **3:** 85-88.
(22) Deshpande, D. V. (1964). Heights of cumulonimbus clouds over India during the southwest monsoon season. *Indian J. Meteor. Geoph.* **13:** 47-54.
(23) Emmitt, G. D. (1978). Tropical cumulus interaction with and modification of the subcloud region. *J. Atmos. Sci.* **35:** 1485-1502.

(24) Fraedrich, K. (1967). Zur Struktur der Tropischen Konvektionsbewölkung über Südostafrika. *Meteor. Rundsch.* **20**: 168-171.
(25) Fraedrich, K. (1974). Dynamic and thermodynamic aspects of the parameterization of cumulus convection; part II. *J. Atmos. Sci.* **31**: 1838-1849.
(26) Fraedrich, K. (1976). A mass budget of an ensemble of transient cumulus clouds determined from direct cloud observations. *J. Atmos. Sci.* **33**: 262-268.
(27) Gruber, A. (1973). Estimating rainfall in regions of active convection. *J. Appl. Meteor.* **12**: 110-118.
(28) Holle, R. L. (1968). Some aspects of tropical oceanic cloud populations. *J. Appl. Meteor.* **7**: 173-183.
(29) Holle, R. L. (1975). Proposed new code for state of the sky in the tropics. *Bull. Amer. Meteor. Soc.* **56**: 55-58.
(30) Johnson, R. H. (1977). Effects of cumulus convection on the structure and growth of mixed layer over South Florida. *Mon. Wea. Rev.* **105**: 713-724.
(31) Knollenberg, R. G. (1972). On radiational cooling computations in clouds. *J. Atmos. Sci.* **29**: 212-214.
(32) Kraus, E. B. (1966). Aerodynamic roughness of the sea surface. *J. Atmos. Sci.* **23**: 443-445.
(33) Krishnamurti, T. N. (1968). A calculation of percentage area covered by convective clouds from moisture convergence. *J. Appl. Meteor.* **7**: 184-195.
(34) Kuettner, J. P. (1959). The band structure of the atmosphere. *Tellus* **11**: 267-294.
(35) Kuettner, J. P. and Soules, S. D. (1966). Organized convection as seen from space. *Bull. Amer. Meteor. Soc.* **47**: 364-370.
(36) Kuettner, J. P. (1967). Cloudstreets, theory and observations. *Aéro-Revue* **42**: 42-56, 109-112.
(37) Kuettner, J. P. (1971). Cloud bands in the earth's atmosphere. Observations and theory. *Tellus* **23**: 404-426.
(38) Kuo, H.-L. (1965). On formation and intensification of tropical cyclones through latent heat release by cumulus convection. *J. Atmos. Sci.* **22**: 40-62.
(39) Kuo, H.-L. (1974). Further studies of the parameterization of the influence of cumulus convection on large-scale flow. *J. Atmos. Sci.* **31**: 1232-1240.
(40) Langwell, P. A. (1948). Inhomogeneities of turbulence, temperature and moisture in the West Indies trade wind region. *J. Meteor.* **5**: 243-246.
(41) LeMone, M. A. (1973). The structure and dynamics of horizontal roll vortices in the planetary boundary layer. *J. Atmos. Sci.* **30**: 1077-1091.
(42) LeMone, M. A. and Pennell, W. T. (1976). The relationship of trade wind cumulus distribution to subcloud layer fluxes and structure. *Mon. Wea. Rev.* **104**: 524-539.
(43) Lilly, D. K. (1968). Models of cloud-topped mixed layers under a strong inversion. *Quart. J. Roy. Meteor. Soc.* **94**: 292-309.
(44) Lorenz, E. Ref. 21, Chapter 2.
(45) Ludlam, F. H. (1966). Cumulus and cumulonimbus convection. *Tellus* **18**: 687-698.
(46) Madden, R. A. and Zipser, E. J. (1970). Multi-layered structure of the wind over the equatorial Pacific during the Line Islands experiment. *J. Atmos. Sci.* **27**: 336-342.
(47) Malkus, J. S. (1949). Effects of wind shear on some aspects of convection. *Trans. Amer. Geoph. Union* **30**: 19-25.
(48) Malkus, J. S. (1955). On the formation and structure of downdrafts in cumulus clouds. *J. Meteor.* **12**: 350-354.

(49) Malkus, J. S. and Scorer, R. S. (1955). The erosion of cumulus towers. *J. Meteor.* **12:** 43-57.
(50) Malkus, J. S. (1956). On the maintenance of the trade winds. *Tellus* **8:** 335-350.
(51) Malkus, J. S. (1957). Trade cumulus cloud groups: some observations suggesting a mechanism of their origin. *Tellus* **1:** 33-44.
(52) Malkus, J. S. and Riehl, H. (1964). Cloud structure and distributions over the tropical Pacific Ocean. *Tellus* **16;** 275-287.
(53) Malkus, J. S. (1963). Cloud patterns over tropical oceans. *Science* **141:** 767-778.
(54) Malkus, J. S. and Williams, R. T. (1963). On the interaction between severe storms and large cumulus clouds. *Meteor. Monogr.* (5) **27:** 59-64.
(55) Manabe, S., Smagorinsky, J. and Strickler, R. F. (1965). Simulated climatology of a general circulation model with a hydrologic cycle. *Mon. Wea. Rev.* **93:** 769-798.
(56) Miller, M. J. and Betts, A. K. (1977). Traveling convective storms over Venezuela. *Mon. Wea. Rev.* **105:** 833-848.
(57) Moncrieff, M. W. and Miller, M. J. (1976). The dynamics and simulation of tropical squall lines. *Quart. J. Roy. Meteor. Soc.* **102:** 373-394.
(58) Mukherjee, B. K. and Murty, B. V. R. (1978). Features of lower troposphere on occasions of contrasting rainfall at a tropical coastal station. *Tellus* **30:** 110-117.
(59) Neuman, C. J. (1965). Mesoanalysis of a severe South Florida hailstorm. *J. Appl. Meteor.* **4:** 161-171.
(60) Newton, C. W. and Newton, H. R. (1959). Dynamical interactions between convective clouds and environment with vertical shear. *J. Meteor.* **16:** 483-496.
(61) Oliver, V. J. and Oliver, M. B. (1963). Cloud patterns: some aspects of the organization of cloud patterns. *Mon. Wea. Rev.* **91:** 621-632.
(62) Ogura, Y. (1964). Frictionally controlled, thermally driven circulation in a circular vortex with application to tropical cyclones. *J. Atmos. Sci.* **21:** 610-621.
(63) Ooyama, K. (1964). A dynamical model for the study of tropical cyclone development. *Geof. Inter. Mexico* **4:** 187-198.
(64) Pennell, W. T. and LeMone, M. A. (1974). An experimental study of turbulence structure in the fair-weather trade wind boundary layer. *J. Atmos. Sci.* **31:** 1308-1323.
(65) Ramakrishnan, K. P. and Rao, D. S. V. (1955). Distribution of thunderstorms over the world. *Indian J. Meteor. Geoph.* **6:** 171-175.
(66) Ramaswamy, C. (1956). On the subtropical jet stream in the development of large-scale convection. *Tellus* **8:** 26-60.
(67) Reed, R. J. and Recker, E. E. (1971). Structure and properties of synoptic-scale wave disturbances in the equatorial Western Pacific. *J. Atmos. Sci.* **28:** 1117-1133.
(68) Riehl, H. (1965). Varying structure of waves in the easterlies. In "Dynamics of Large-Scale Atmospheric Processes." Proc. Int. Symp. Moscow, June 1965. (A. S. Monin Ed.) pp. 411-416.
(69) Riehl, H. and Augstein, E. (1973). Surface interaction calculations over the Gulf of Tonkin. *Tellus* **25:** 424-434.
(70) Riehl, H. (1977). Vertical distribution of energy transfer and radar echo tops in the equatorial trough zone. *Mon. Wea. Rev.* **105:** 230-231.
(71) Riehl, H., Greenhut, G. and Bean, B. R. (1979). Energy transfer in the tropical subcloud layer measured with DC-6 aircraft during GATE. *Tellus* **30,** 524-536.
(72) Riehl, H. and Miller, A. L. (1978). Differences between morning and evening temperatures of cloud tops over tropical continents and oceans. *Quart. J. Roy. Meteor. Soc.* **104:** 757-764.

(73) Riehl, H. and Malkus, J. S. Ref. 27, Chapter 2.
(74) Riehl, H. Ref. 28, Chapter 2.
(75) Rosenthal, S. L. (1978). Numerical simulation of tropical cyclone development with latent heat release by the resolvable scales. I: Model description and preliminary results. *J. Atmos. Sci.* **35:** 258-271.
(76) Ruprecht, E. (1971). Der Einfluss der seitlichen Vermischung auf die Schauerentwicklung. *Beitr. Phys. Frei. Atmos.* **44:** 1-16.
(77) Ruprecht, E. and Gray, W. M. (1976). Analysis of satellite-observed cloud clusters. I: Wind and dynamic fields. *Tellus* **28:** 391-414. II: Thermal, moisture and precipitation. *Tellus* **28:** 414-427.
(78) Simpson, J. (1969). Models of precipitating cumulus towers. *Mon. Wea. Rev.* **97:** 471-489.
(79) Soane, C. M. and Miles, V. G. (1955). On the space and time distribution of showers in a tropical region. *Quart. J. Roy. Meteor. Soc.* **81:** 440-448.
(80) Squires, E. (1958). Penetrative downdraughts in cumuli. *Tellus* **10:** 381-389.
(81) Squires, P. and Turner, J. S. (1962). An entraining jet model for cumulonimbus updraughts. *Tellus* **14:** 422-434.
(82) Starr, V. P. (1968). "Physics of Negative Viscosity Phenomena." McGraw-Hill Book Co., New York. 256 pp.
(83) Stommel H. (1947). Entrainment of air into a cumulus cloud. *J. Meteor.* **4:** 91-94.
(84) Stommel, H. (1951). Entrainment of air into a cumulus cloud; part II. *J. Meteor.* **8:** 127-129.
(85) Sundqvist, H. (1970). Numerical simulation of the development of tropical cyclones with a 10-level model. *Tellus* **22:** 359-390.
(86) Ulanski, S. L. and Garstang, M. (1978). The role of surface divergence and vorticity in the life-cycle of convective rainfall. *J. Atmos. Sci.* **35:** 1047-1069.
(87) Warner, J. (1955). The water content of cumuliform cloud. *Tellus* **7:** 449-457.
(88) Williams, K. T. and Gray, W. M. (1973). Statistical analysis of satellite-observed trade wind cloud clusters in the Western North Pacific. *Tellus* **25:** 313-336.
(89) Wilson, B. W. (1960). Note on surface wind stress over water at low and high wind speeds. *J. Geoph. Res.* **65:** 3377-3382.
(90) Woodcock, A. H. (1940). Convection and soaring over the open sea. *J. Marine Res.* **3:** 248-253.
(91) Woodcock, A. H. and Wyman, J. (1947). Convective motion in air over the sea. *Ann. N.Y. Acad. Sci.* **48:** 749-777.
(92) Woodcock, A. H. (1953). Salt nuclei in marine air as a function of altitude and wind force. *J. Meteor.* **10:** 362-371.
(93) Woodcock, A. H. (1960). The origin of trade-wind orographic showers. *Tellus* **12:** 315-326.
(94) Yamasaki, M. (1969). Large-scale disturbances in the conditionally unstable atmosphere in low latitudes. *Pap. Meteor. Geoph.* **20:** 289-336.
(95) Yanai, M., Esbensen, S. and Chu, J.-H. (1973). Determination of bulk properties of tropical cloud clusters from large-scale heat and moisture budgets. *J. Atmos. Sci.* **30:** 611-627.
(96) Zobel, R. F. and Cornford, S. G. (1966). Cloud tops over Malaya during the southwest monsoon season. *Meteor. Mag.* **95:** 65-68.

5
The Trade Wind Inversion

The trade wind inversion, an important regulatory valve of the general circulation, has attracted the interest of meteorologist and geographer since its discovery about the middle of the nineteenth century. It acts as a strong lid to oppose vertical cloud development; aridity prevails throughout the inversion regime. The desert-like, cold-water coasts on the western shore of Africa and the Americas lie at the source of the trade current. Islands with high mountain peaks that stand in the path of the trade often have heavy rainfall at low altitudes. As one ascends the mountain slopes, the forests and fertile fields vanish abruptly and give way to a region that is treeless and barren.

During the summer of 1856 an expedition under the direction of C. Piazzi-Smyth visited the island of Teneriffa in the Canary Islands with the purpose of making astronomical measurements for several months near the top of the Peak of Teneriffa (3700 m). On two of the trips up and down the mountain slopes, Piazzi-Smyth (16) carefully measured the temperature, moisture and wind at numerous intervals during the journey and obtained the first detailed observations of the inversion. Figure 5.1 is the plot of temperature and relative humidity during the first of these trips, in August 1856. The trip took all day, so that diurnal variations had to be eliminated as best possible. A clear picture of the temperature inversion with upward rise of over 5°C is evident, but the sharp moisture discontinuity so often found in later soundings is washed out by the mountain circulation.

Piazzi-Smyth observed that the inversion was not located at the top of the northeasterly trade regime, overlain by a layer of hot air coming from Africa, as sometimes thought. On the contrary, it was located in the very middle of the trade current, and thus could not be explained

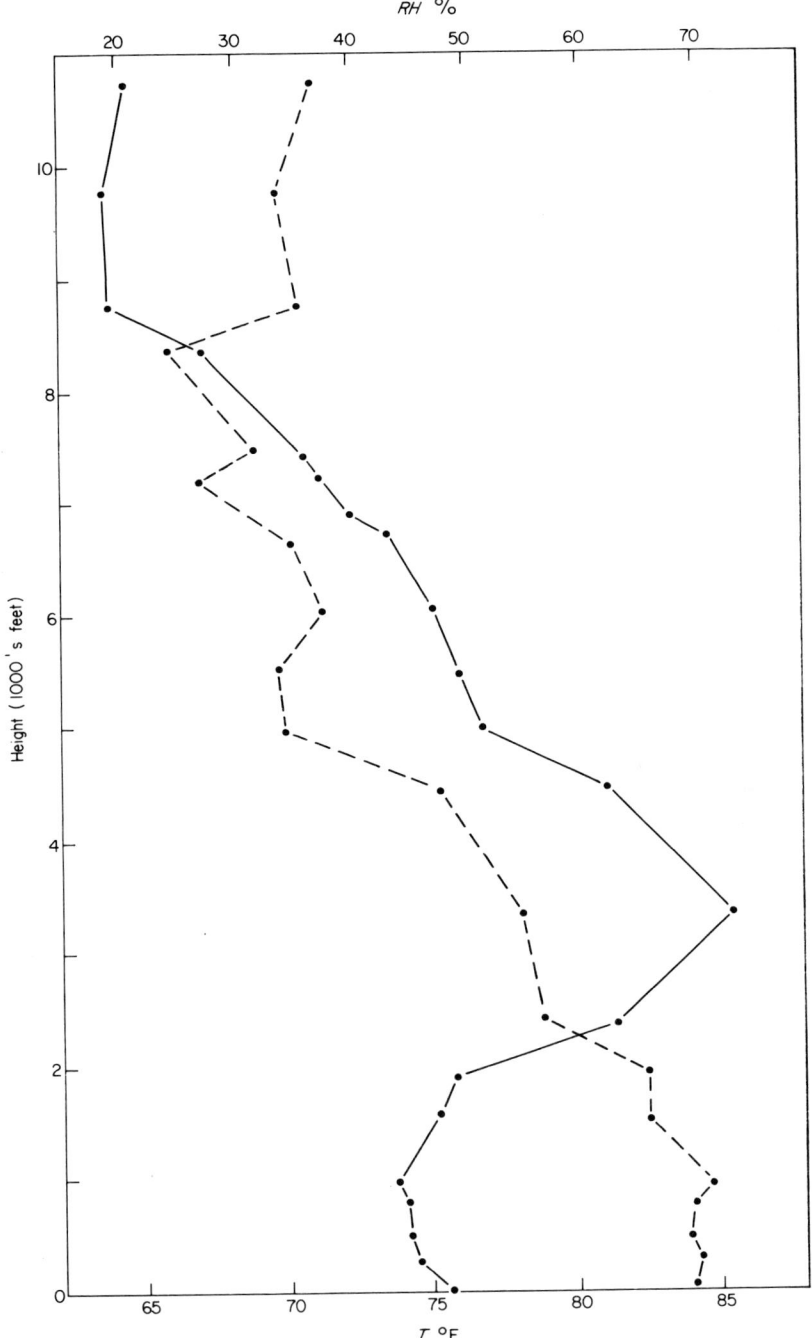

Fig. 5.1. Temperature (solid) and relative humidity (dashed) observed by Piazzi-Smyth (16) during descent from the Peak of Teneriffa.

as a boundary of two streams of air coming from different directions. Uniform wind direction through the inversion has been confirmed at many stations in the trade wind regimes in the century since Piazzi-Smyth's expedition. Nor do wind speed and wind direction vary discontinuously with height at upper or lower inversion boundaries comparable to extratropical fronts. Vertical changes are either small or gradual; this chapter will give some examples.

The trade wind inversion has been the subject of investigations by several research expeditions, listed at the front of this book (p. xii). Subsequent to early exploration by single ships (*Meteor I, Carnegie*), multiple ship (fleet) expeditions formed patterns designed to calculate processes dependent on knowledge of space gradients. These were all undertaken in the Atlantic Ocean and the three major configurations designed in 1969 and 1974 can be seen in Fig. 5.2. Efforts in the Pacific Ocean have been more modest. A line of observations from air-sea rescue vessels near the end of World War II was routinely available

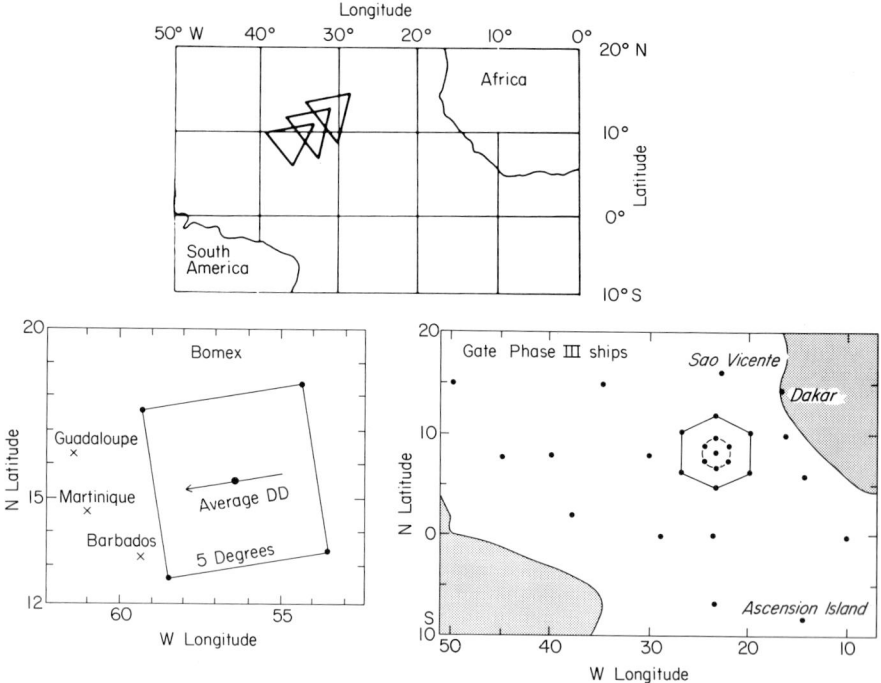

Fig. 5.2. Ship configuration of 3 experiments: Top: ATEX, 3 ships, drifting; lower left: BOMEX, 5 ships, stationary; lower right: GATE, 30 ships, stationary.

on the California-Hawaii great circle route. Several oceanographic cruises in the same region provided more complete inversion data in the 1950s.

Vertical structure

Wind

Profiles of average wind structure through the inversion are seen in Fig. 5.3 for the 1969 Atlantic Trade Expedition (ATEX) (winter) (4, 5) and for the northeast trade of the Pacific Ocean (summer). In the Pacific, the wind was almost uniform through the inversion; direction turned clockwise 6° underneath, speed was 6 m/s already at ship's deck level and increased only slightly from there to the lowest part of the cloud layer (17). Regarding the large temperature inversion found on and off the California coast, the air below, in and above the inversion forms part of the same current (15). In the Atlantic, much closer to the equator and in winter, direction turns gradually clockwise through the inversion. The inflow of mass toward the equator is confined to the layer below the inversion (right side of Fig. 5.3). The zonal wind component already attains its strongest value near 400 m height, below cloud base (6) and then starts to decrease in response to the poleward temperature gradient of winter and attendant "thermal

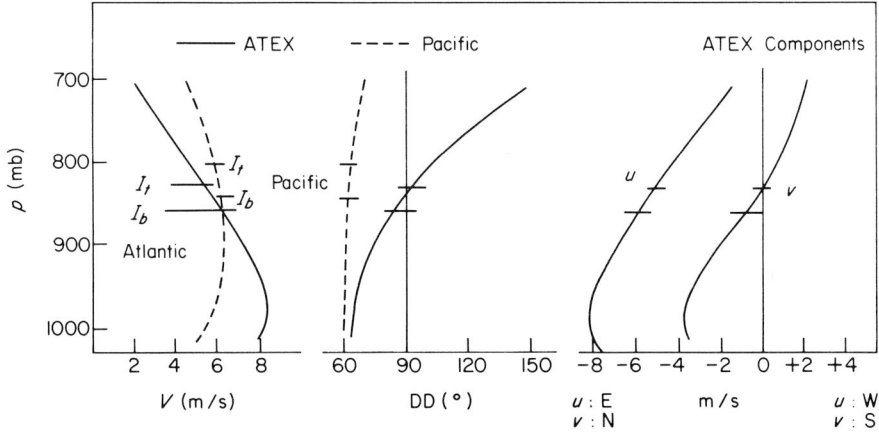

Fig. 5.3. Mean vertical profiles of wind speed, wind direction and zonal and meridional components measured during Atlantic Tropical Experiment (ATEX) of 1969 (2, 3, 4). Also mean wind speed and direction profiles for northeast trade of the Pacific Ocean July-October 1945.

Table 5.1. Comparison of trade wind inversions

	p_b	p_t mb	Δ_p	T_b	T_t °C	ΔT	q_b	q_t g/kg	Δ_q	θe_b	θe_t K	$\Delta\theta_e$
ATEX Cape Verde 1974, June-Sept.	855	825	30	14	19	+5	10	6	−4	330	325	−5
Kurchatov Sept. 1974	920	880	40	17	23	+6	12	5.5	−6.5	333	323	−10
	835	795	40	13	14	+1	8	6.5	−2.5	325	322	+3
Ascension Is.	840	790	50	12	14	+2	8	2.5	−5.5	328	314	−14
Pacific 32°N, 136°W	880	840	40	7	12	+5	7	4	−3	314	314	0
California coast	960	820	140	12	20	+8	9	4	−5	313	322	+9

wind". Flohn (9) has surveyed the wind climatology over the Atlantic Ocean.

During GATE in 1974, soundings made by the Russian ship *Kurchatov* stationed at 0°, 23·5°W could be compared with comparable inversion soundings at the Cape Verde Islands (16°N, 24°W) and Ascension Island (8°S, 14°W). The data are given in Table 5.1 (cf. Fig. 5.11). As in the case of ATEX, the low-level flow converging toward the equatorial convergence zone died out below the inversion with gradual clockwise turning at the Cape Verdes and counterclockwise turning at Ascension. Thereby, the air in and above the inversion does not enter the equatorial trough zone. Rather it moves westward in the two trade wind belts and is modified gradually as described below. On the equator itself, the inversion air is often stagnant or drifts slowly from any direction except south as indicated by the *Kurchatov* record.

Temperature and humidity

We see the mean tephigram of the ATEX ships in Fig. 5.4 for a week of steady trade winds. Here, and in the other inversion presentations, base (I_b) and top (I_t) of the inversion were first located from individual soundings and then averaged following I_b and I_t in order to describe the inversion clearly. Obviously, it is the heat and dryness above the inversion that is the anomalous features of the inversion regime. A combination of heat and dryness always suggests subsidence. We believe that the inversion is formed by broad-scale descent of air from high altitudes near and poleward of the eastern end of the subtropical high pressure cells. These underlie average cold high-tropospheric troughs in summer (Fig. 5.5). From vorticity reasoning air converges in the high troposphere there and, thus, produces the quasi-stationary, high-level, cyclonic vorticity regions. The converging air sinks, and turns anticyclonically toward south and southwest during the descent with low-level divergence. Very dry air in the inversion (Fig. 5.12) may well have come from as high as 300 or 400 mb in the cold upper troughs with allowance for radiation cooling. While descending toward the ocean, the current meets the opposition of low-level marine air flowing equatorward, which is turbulent. The inversion base is situated at the meeting point of these two strata which flow in the same direction. The height of the base is a measure of the depth to which the upper current has been able to penetrate downward. In the mean ATEX sounding the equivalent potential temperature decreased upward through the entire layer, just as the

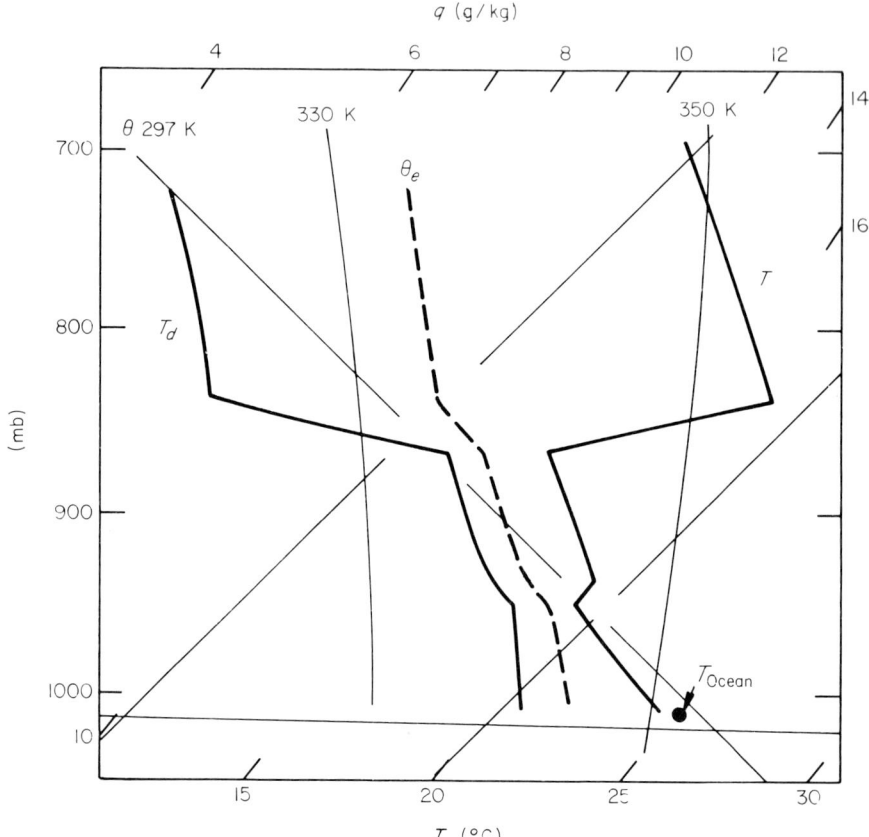

Fig. 5.4. Tephigram of mean ATEX sounding 7-12 February, 1969 (2). Temperature and dewpoint solid, θ_e dashed. Also mean ocean surface temperature.

mean tropical atmosphere, but at lower values. The strongest decreases are observed in the inversion layer.

A partial exception to this explanation of inversion structure occurs in the immediate vicinity of the American and African coasts. Here the surface air is cooled from below as it moves over the narrow strip produced by upwelling along these coasts, discussed earlier in Chapter 2, assuming that average surface streamlines can be taken as steady state trajectories. Turbulent motion distributes this cooling throughout the layer below the inversion, which is thus strengthened, not only by subsidence higher up, but also by cooling from below. The combined action leads to extreme situations where the inversion base descends to very low levels (Fig. 5.6(a)), and occasionally even to the ship's mast (Fig. 5.6(b)) (13)!

THE TRADE WIND INVERSION

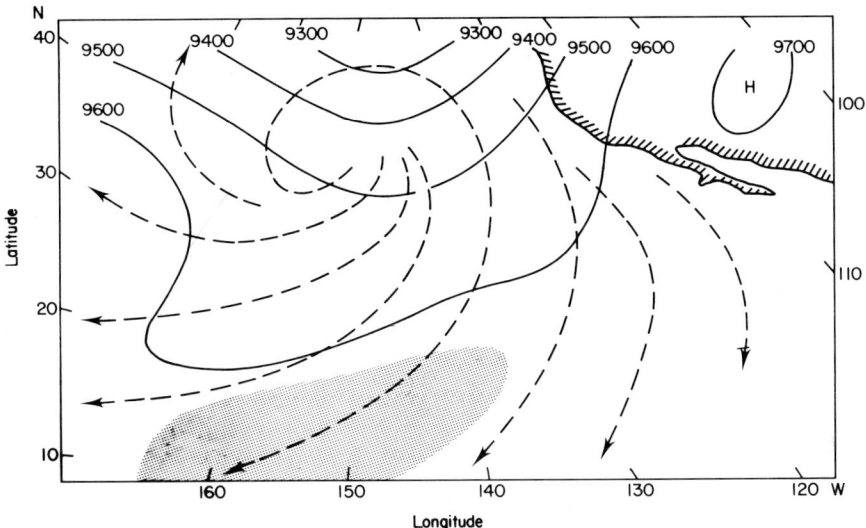

Fig. 5.5. Mean surface streamlines and isotachs for July (speeds above 12 knots shaded, and 300 mb contours (m).

A question arises concerning the mechanism maintaining the lower turbulent motion. Turbulence derived from mean wind or wind shear alone is considered insufficient. Differential radiation from the solid stratus deck, generally situated below the inversion in these situations, is proposed as a mechanism to steepen the lapse rate and contribute to generations of turbulent kinetic energy (14). This matter was discussed in Chapter 4 in connection with sources of thermal instability for vertical overturning (Fig. 4.19). The radiation effect on temperature lapse rate was shown to be very strong but counteracted partly, especially at low levels, by the additional condensation resulting from radiation cooling. Thus, the differential radiation no doubt contributes to maintenance of the inversion when there is an undercast below inversion base; but it is not the decisive factor.

Piazzi-Smyth, from his high vantage point on the Peak of Teneriffa, noted that the top of the cloud layer correlated with the inversion height. He suggested that strong reflection of incident sunlight from the cloud tops or evaporation of water from the upper cloud surface —another mechanism for creating turbulent kinetic energy (7)— accounted for the inversion. This implies again that the inversion is produced by cooling the stratum at and below the base. Even if the differential long wave radiation just discussed is added, the argument is not fully convincing. Firstly, very little of the air forming the inversion passes over cold water areas, and secondly, the undercasts are variable and intermittent as seen, for instance, from satellite photos (Fig. 5.21). Often, there are no clouds or only small cumuli under the inversion, even close to the North African coast as was the case during ATEX. The tephigram (Fig. 5.4), however, makes it very clear that the temperature at the inversion base was in equilibrium with the temperature of the underlying ocean surface, i.e. parcels ascending from the surface will reach the inversion base with a temperature which is not very different from that prevailing there. Thus, the temperature at the base is the "normal" one; it is the heat and dryness of the air in and above the inversion that is the anomalous feature. The height of cloud tops and inversion base are correlated, not because the cloud tops act to produce the inversion, but because the inversion is a formidable lid which the lower cumuli enter but can penetrate only rarely.

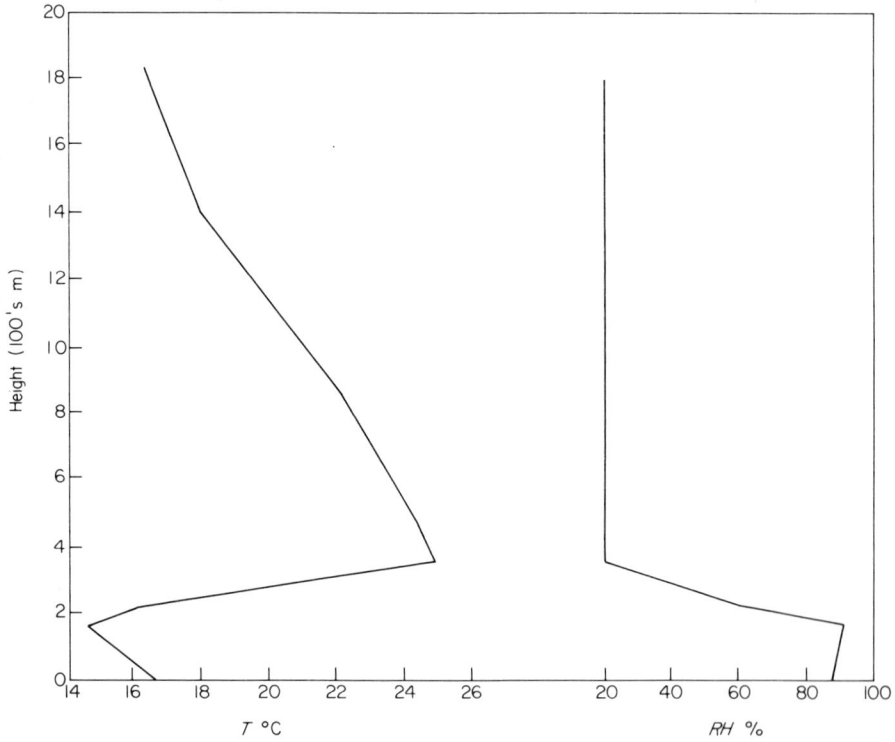

Fig. 5.6. (a) Vertical profiles of temperature and relative humidity, 28 July, 1926, at 22°S, 8°E (from *Meteor* data).

THE TRADE WIND INVERSION

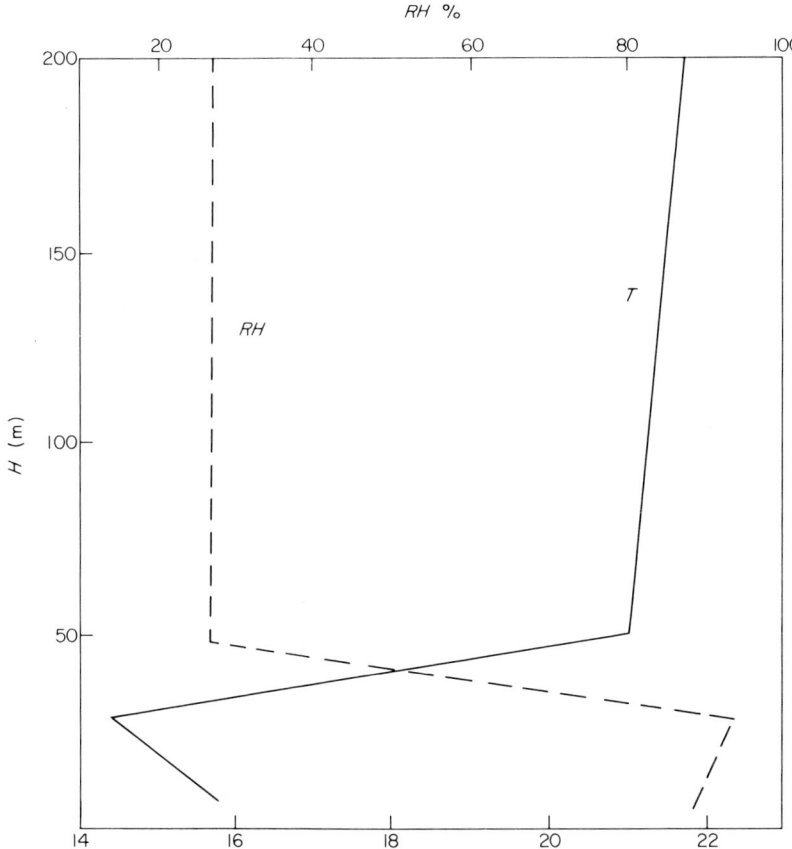

Fig. 5.6. (b) Expanded scale for the lowest 200 m 29 July, 1926, 22°S, 11°E.

Variations of trade inversion structure

Early information on inversion structure over the whole North and South Atlantic Oceans is due to the German *Meteor I* expedition of the 1920s, directed by Professor Erich Kuhlbrodt (13). Its observational programme, its projected route and the scope of its objectives remain unparalleled.

A survey of the papers written on the meteorological part of the expedition leads one to note quickly that, in spite of the great wealth of data collected, these data proved disappointing in some respects. Apparently, the expedition plans were built on the classical hypothesis that conditions were nearly steady in the tropics from day to day but that they varied slowly with the season. Actually the expedition encountered what has become widely accepted as routine: strong synoptic variations of the inversion layer and, presumably, also its large small-scale variability. In order to be able to prepare any chart material, it proved necessary to smooth the observations strongly.

Along the northern and southern coasts of Africa, the height of the Atlantic inversion has a distinct minimum (Fig. 5.7); from there it rises westward (26). Both hemispheres are as symmetrical as can be expected. Poleward termini of the inversion over the eastern parts of the oceans are not found. As already suggested, the "roots" of the inversion extend far into the middle latitudes. Certainly, this is also true in the Pacific Ocean where the beginnings of the inversion lie in the fogbound regions of the northern California and central Chile coasts in the summer season, or even farther poleward.

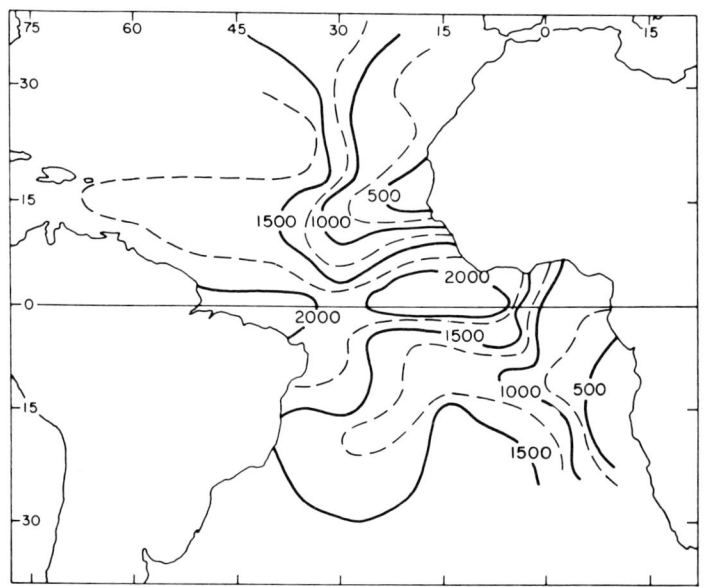

Fig. 5.7. Height of the base of the Atlantic trade-wind inversion (m) (26).

Equatorward from latitudes 15° the inversion base rises both westward and equatorward. The highest contour of 2000 m encircles the equatorial trough zone. As confirmed by the great mass of soundings during GATE, a vertical structure not unlike that of Fig. 2.18 prevails there in the rainy season; a trade inversion intrudes only in rare synoptic situations. Referring to the streamlines of Fig. 1.7, we see that the inversion rises downstream along the streamlines where

THE TRADE WIND INVERSION

they curve clockwise through the whole trade wind belt to the western end of the oceans. Because of high wind steadiness, these streamlines may be largely regarded as surface air trajectories. Where the turning of wind with height below the inversion base is as small as in the Pacific trade (Fig. 5.3), they may be taken as representative of the whole layer underneath the inversion.

The result of four Finnish voyages across the equator to South America in the four seasons of the year during the middle 1950s (22) largely confirms the Meteor findings. Along their ship's trajectory, the ocean north of the equator would be representative of the eastern side, below the equator, toward Brazil, of the western side. From Fig. 5.8 the inversion is strongest where it is lowest; the temperature increase diminishes as the inversion rises. Except on one of the voyages, the Finnish data gave linear correlation coefficients from —0·41 to —0·56 between inversion height and increase of temperature.

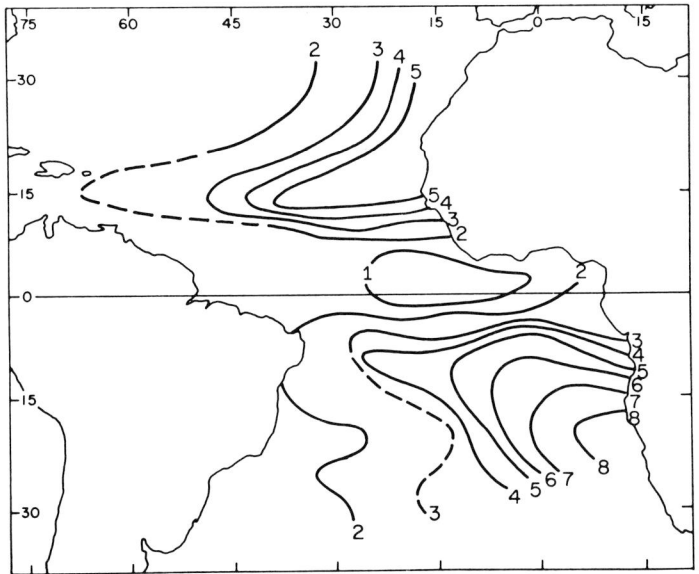

Fig. 5.8. Amount of temperature increase (°C) from bottom to top of inversion (26).

Relative humidity also has the sharpest drop over the eastern portions. A large temperature increase corresponds to a large humidity decrease, while the areas with small temperature increase coincide with the areas of small humidity decrease (Fig. 5.9). Thus, the inversion is strongest in all aspects where its height is lowest, and

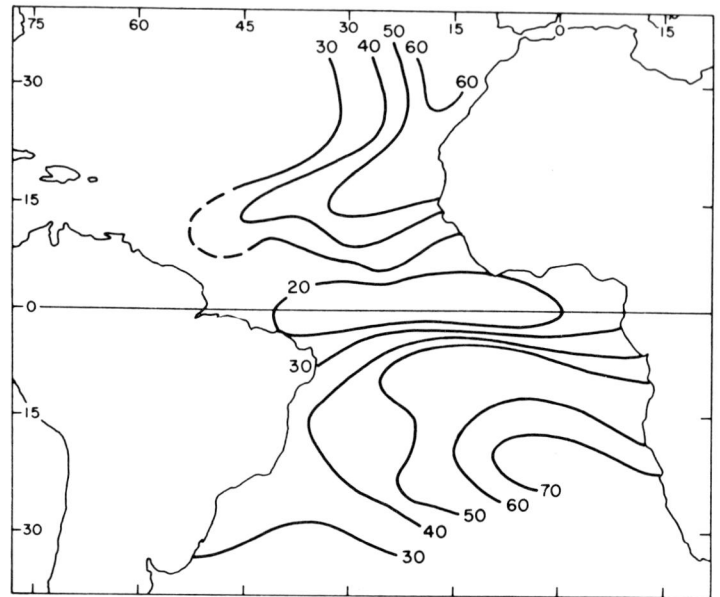

Fig. 5.9. Decrease of relative humidity (per cent) from bottom to top of inversion (26).

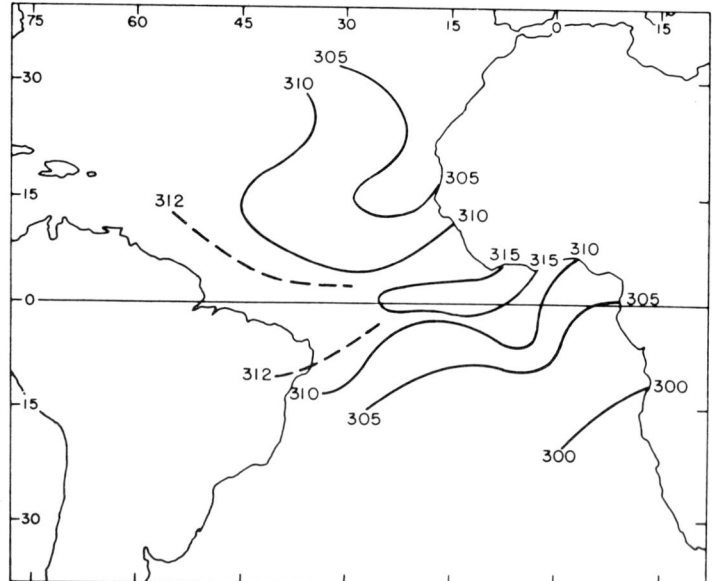

Fig. 5.10. Potential temperature at top of trade inversion (computed from the *Meteor* data).

its intensity weakens as the height rises. A plot of potential temperature along the inversion top (Fig. 5.10) also reveals a pattern that is well correlated with Fig. 5.7. Potential temperature rises downstream along the inversion top; therefore, also the sensible heat $c_p\theta$ of any parcel moving along the inversion surface, in spite of radiation cooling. It follows that the inversion top cannot be a material surface.

As seen from Fig. 2.21, temperature is warmest at the centre of the equatorial trough relative to its outskirts on the average, largest near the surface and decreasing upward slowly over the oceans. The trade inversion areas of the eastern parts of the oceans form an exception. Because of the strong subsidence, the middle troposphere in the inversions regimes is warmer than the equatorial trough zone, seen strikingly by the temperature profile of three stations bordering the GATE area (Fig. 5.11). At Sal and *Kurchatov*, temperature in the inversion rises well above that of the equatorial-type troposphere whereas at Ascension, farther removed from the convergence zone, temperatures above 800 mb equal those of the trough. In all three instances we find turning of wind with height away from the convergence zone, counterclockwise at Ascension and also *Kurchatov*, clockwise at Sal. Thus, the dry, low-energy air never enters the convergence zone; rather, it is sent westward across the whole width of the Atlantic Ocean and, in weakened form, appears occasionally on the western side even in the rainy season, loaded with Sahara dust (Chapter 1).

In some parts of the world, for instance the eastern Pacific (25), the cold core trough may be maintained to high levels; in most areas, however, the sense of the surface-temperature gradient reappears in order for the mean structure of Figs 2.21 and 2.24 to be valid. Striking examples can be found in the records of ATEX during the second week of observations when the southernmost ship, *Meteor II*, was overtaken by an equatorial trough zone with a very marked and narrow satellite cloud band, and the usual ENE-WSW orientation. *Planet* and *Discoverer*, the other ships of the triangle 500 km farther north, remained in the trade wind regime while the *Meteor* soundings became typical of equatorial trough structure. Comparison of *Planet* and *Meteor* soundings at a specific time (Fig. 5.12) clearly reveals the reversal of temperature gradient near 900 mb where *Planet* temperature jumps 6°C in a layer only a few millibars thick. In this case the reverse temperature gradient was not maintained to 700 mb, as the *Meteor* ascent encountered a warmer and drier layer, but at 600 mb the curves crossed again.

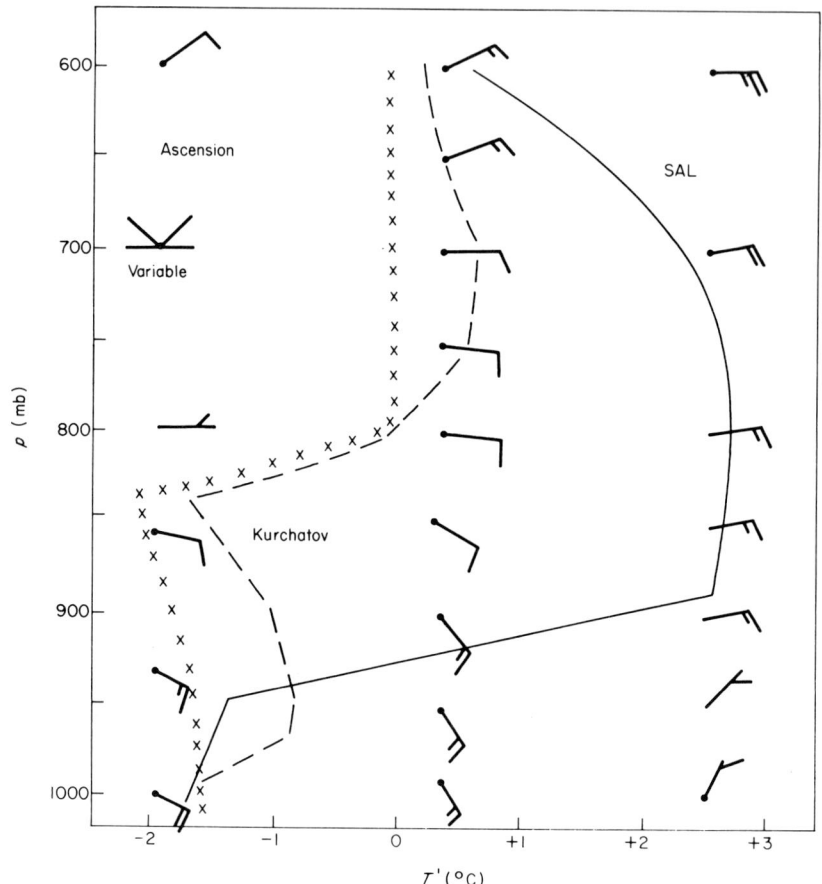

Fig. 5.11. Temperature deviation from mean temperature at equator in the GATE area at Sal, Cape Verdes (16°N, 24°W) (solid), USSR ship *Kurchatov* (0°, 23·5°W) (dashed), and Ascension Island (8°S, 14°W) (crosses). Values averaged with respect to inversion base and top, resultant winds added (see Table 5.1).

The θ_e profile for the *Planet* sounding contains an extreme dry wedge. At 850 mb, where specific humidity is only 1·4 g/kg, θ_e is 309 K. Higher up, q and θ_e rise again. Such a wedge of air with extremely low energy is by no means uncommon. The sounding depicted in Fig. 5.12 should be accepted as correct, but the origin of the dry wedge is likely to be quite different from that of Fig. 4.22 where a convective downdraft mechanism was invoked. One obtains the impression that much of the sinking motion in the trades does not occur uniformly over a deep atmospheric layer, but that it is intensely

THE TRADE WIND INVERSION

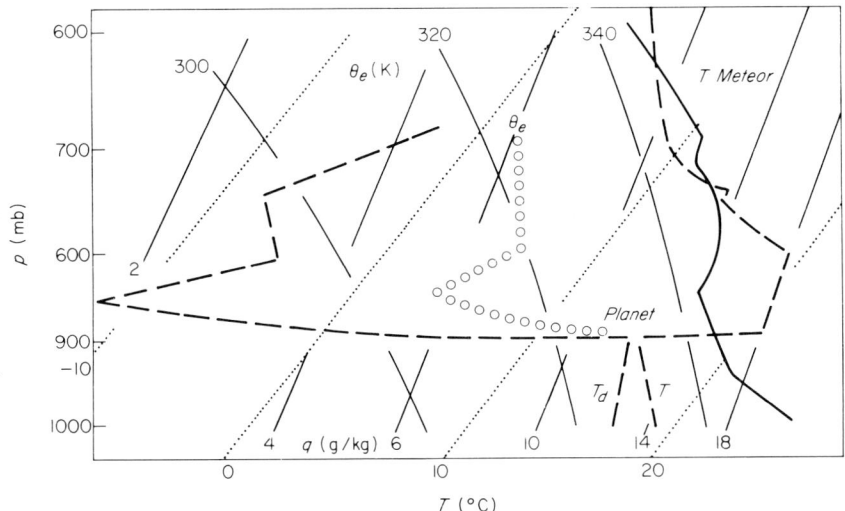

Fig. 5.12. Tephigram showing soundings on 17 February, 1969, 0600 GMT, for the ships *Meteor II* and *Planet* during ATEX. *Meteor:* solid; *Planet:* dashed, also θ_e profile for *Planet*.

concentrated in thin isentropic sheets that slant downward along the air trajectories. Evidence from the Atlantic (1) and from the South China Sea (19) indicates that such shallow layers can be picked up first by soundings in the middle troposphere and that these then sink and intensify, until they finally merge into the top of the main inversion and in this way restore the inversion structure against the effects of turbulence from below.

Sinking in narrow layers without help from evaporation of rain, as depicted in Fig. 5.12, must be spread over very large parts of the trade wind belts. Indeed, one can find evidence of the shallow dry layers all over the tropical oceans intermittently. In Venezuela, the levels of minimum energy and strongest east wind (15-25 m/s) were often observed to coincide closely. Figure 5.13 illustrates an extreme case; θ_e was only 315 K at the level of lowest humidity (21). Six soundings spread over the day assured that the structure shown was representative and not due to instrumental problems of one sonde. Wind in this and other cases was nearly easterly with only slight turning, but a tendency for strongest speed (15-25 m/s) was evident at the base of the layer.

Aircraft flying north-south perpendicular to the basic current even as low as 900 mb have penetrated into such downdrafts, encountering a knife-edge humidity drop of great magnitude, for instance from 11 to 4 g/kg (21) while temperature rose only slightly. The north-south extent of the phenomenon was not very large, only about 100 km; but the northern edge had a distinctly front-like appearance. Upstream sondes at the South American coast and at GATE ships over the Atlantic have encountered similar dry layers. Evidently all these stations are lined up along the same sinking easterly trajectories, again stressing the very large-scale character of the drought-producing phenomenon.

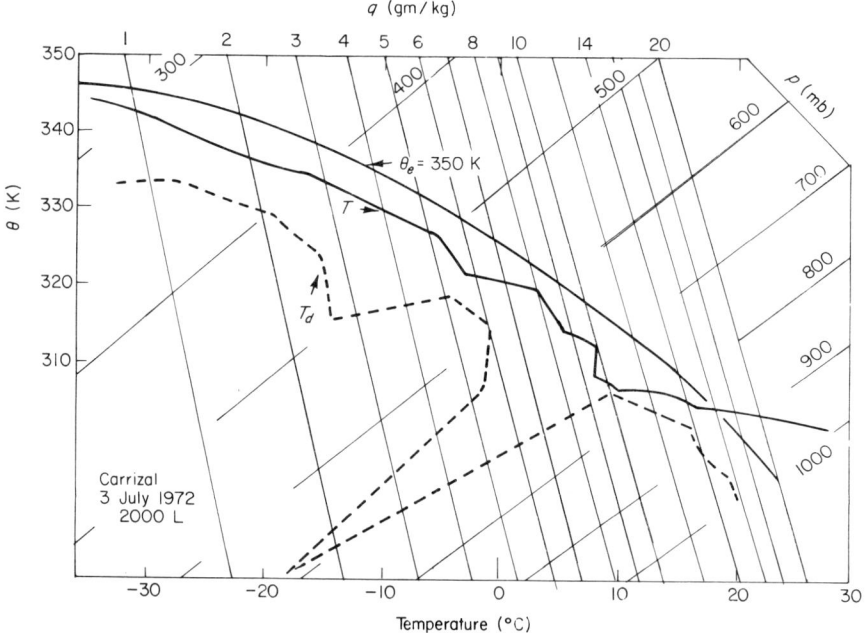

Fig. 5.13. Tephigram of rawinsonde ascent at Carrizal (9·5°N, 67°W), 2000 LT 3 July, 1972; temperature *(T)* and dewpoint *(T_d)* showing shallow layer with extreme dryness, also ascent at $\theta_e = 350$ K (21).

Dissolution of the trade wind inversion

The downstream weakening and eventual dissolution of the trade wind inversion indicated in Figs 5.7 to 5.9 carried the early implication that absorption of heat and moisture from the ocean by the air converging into the equatorial trough zone gradually wipes out the inversion from below and thickens the low trade wind layer until the latter eventually becomes unstable and breaks through into the upper troposphere

THE TRADE WIND INVERSION

(26). In this model the inversion was seen as a material boundary, an assumption already placed into doubt by the increase of potential temperature along the inversion top (Fig. 5.10). Moreover, most of the trade wind trajectories avoid the equatorial trough, as we have seen, and the rise of the inversion starts far from any such trough, as in Fig. 5.7. Clearly, another mechanism for the dissolution of the trade wind inversion must be sought.

It appears that the large-scale dynamics of air on the rotating earth have been overlooked. These are well expressed through Rossby's well-known vorticity theorem (23)

$$\frac{d}{dt}(f + \zeta) = -(f + \zeta)\,\mathrm{div}_2\,V, \tag{5.1}$$

where f is the Coriolis parameter and ζ the relative vorticity about the vertical axis. With the aid of mass continuity Eq. (5.1) may be integrated to yield

$$\frac{f + \zeta}{\Delta p} = \text{constant}, \tag{5.2}$$

known as the theorem of conservation of potential vorticity. The quantity Δp is the pressure difference between top and bottom of a column considered. In the trade winds covering large latitude intervals, the variation of ζ is very small compared to that of f and may often be neglected, so that

$$\Delta p = \frac{f}{f_0} \Delta p_0, \tag{5.3}$$

where the subscript 0 refers to the starting latitude of a trajectory. It is immediately evident that during equatorward motion Δp decreases, for instance by 30% when going from latitude 30° to 20°, and by 50% from latitude 20° to 10°. Thus, the model of Fig. 5.14 arises, which shows shrinking air columns spreading out laterally, yet a strongly rising trade wind inversion along the trajectory.

The July divergence chart (Fig. 1.12) confirms the presence of surface divergence over most of the subtropical Pacific and Atlantic Oceans, and, also, the South Indian Ocean south of 20°S. For the upper air, earlier one-dimensional estimates were placed on a firm footing through ATEX which was designed to determined the two-

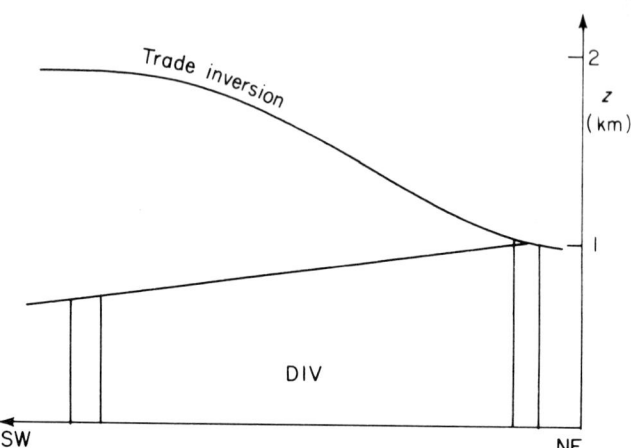

Fig. 5.14. Model of rise of trade wind inversion against air trajectories sinking due to planetary vorticity effect.

dimensional divergence through the triangle formation of the ships (Fig. 5.2). Divergence is very large in the subcloud layer and then decreases to zero at inversion base, an interesting coincidence (Fig. 5.15). Slight convergence prevails in the inversion and higher up. The downward motion is very large, over 600 m/day at inversion base; obviously, the inversion is as solid as a sieve! From Eq. (5.1) another expression for divergence can be obtained assuming it is due solely to what has been termed the "planetary vorticity effect". We neglect variations in relative vorticity again so that the vorticity advection $df/dt \equiv \beta v$ on the left side and

$$\text{div}_2 V = -\frac{\beta v}{f}. \tag{5.4}$$

With equatorward winds divergence is positive, with winds away from the equator the air converges. It is of interest that the v component in Fig. 5.3 changes from northerly to southerly near inversion base where the sign of divergence also changes. Evidently, the planetary vorticity term is important in the Atlantic trade and has been computed for the triangle data, shown dashed in Fig. 5.15. We see that the term accounts for over half, but not all, divergence and downward motion in the limited ATEX region where a strong development of trade winds was then taking place. Around the globe the planetary effect accounts for the whole divergence in winter (Chapter 12).

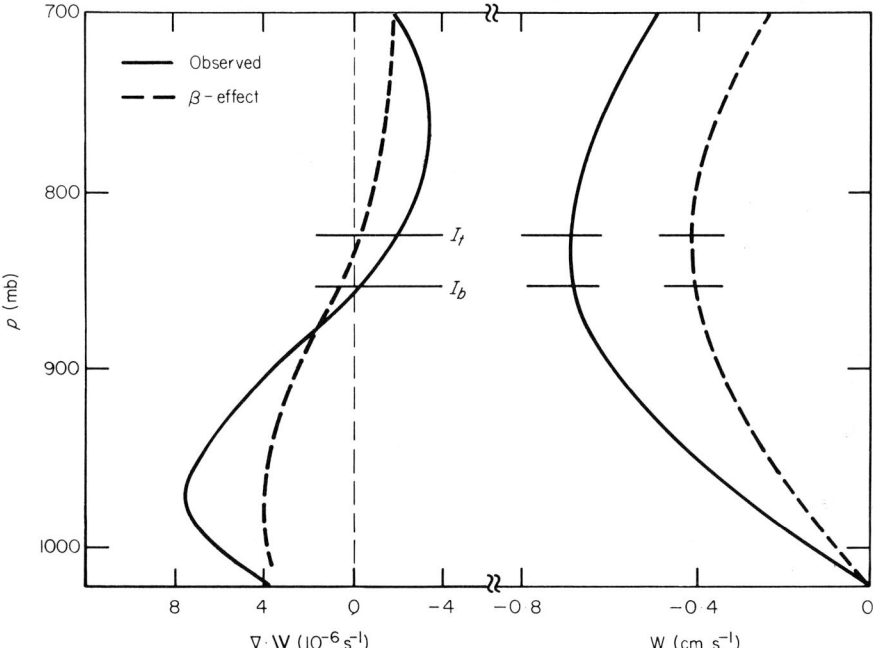

Fig. 5.15. Profiles of average horizontal Velocity divergence ($10^{-6} s^{-1}$) and vertical motion (cm/s) February 7-12, 1969, during ATEX (solid) (2). Profiles computed from planetary vorticity effect dashed. I_b and I_t denote bottom and top of trade wind inversion.

What evidence is there for a highly fragmented inversion such as must exist if the inversion is to remain in steady state as an average feature and yet permit a large, net, downward mass transport as Fig. 5.15 shows? Apart from sonic and laser observations, said to show very rapid fluctuations of the inversion, the best evidence again is provided by ATEX where three-hourly soundings were scheduled at the ships. However, the three-hour time interval is still too large and not all soundings were taken. Nevertheless, a time section of inversion base and top for *Planet* during six days (Fig. 5.16) is very revealing. Obviously the inversion jumped up and down irregularly over considerable pressure intervals. We might, however, be dealing merely with wave-like undulations of the same "airmass" in the inversion. This possibility is removed by the plots of specific humidity and total energy, here taken simply as $Q = c_p \theta + Lq$. Clearly the energy varies substantially, more at the top than at the bottom which, from Fig. 5.4, is controlled from the ocean surface. It is obvious,

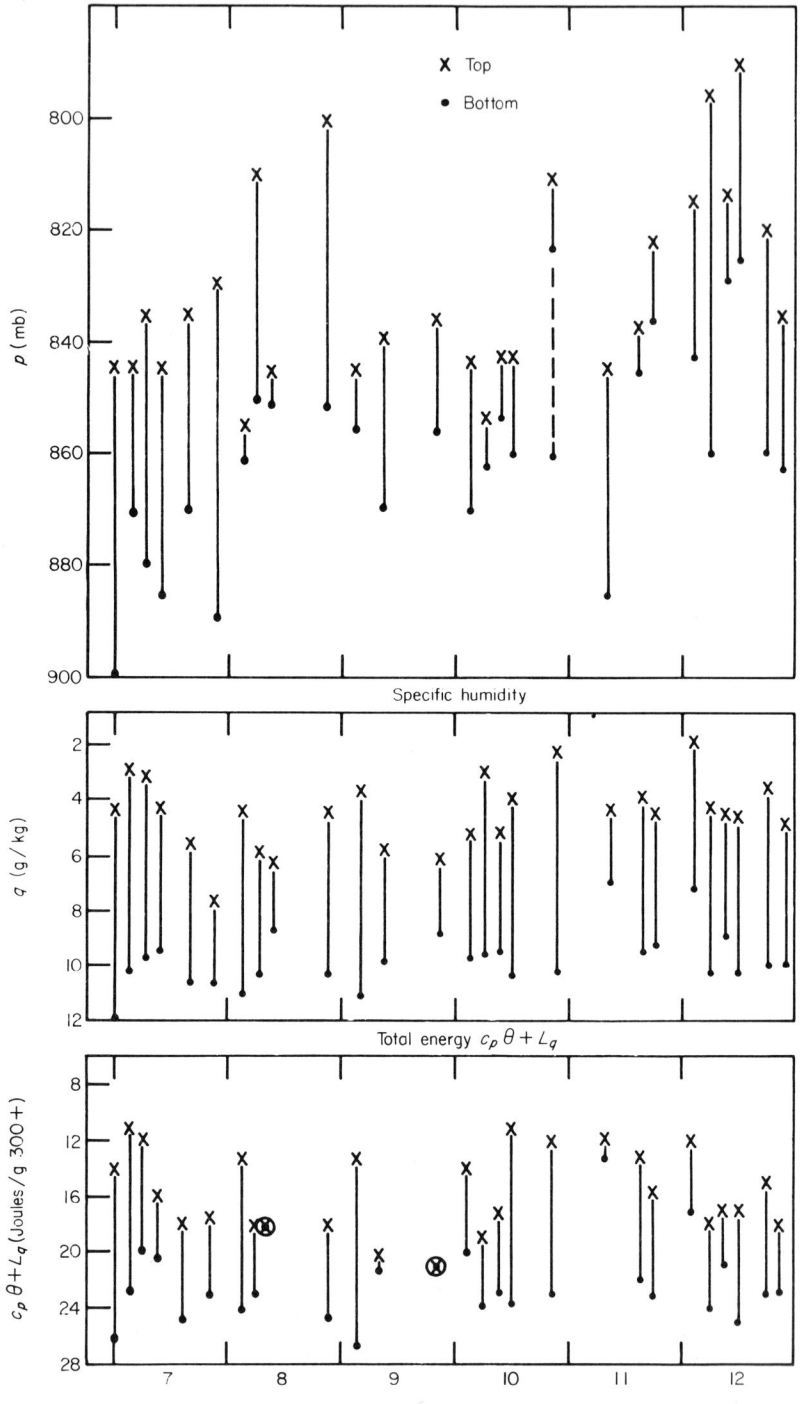

Fig. 5.16. Variation of trade wind inversion on individual observations by the ship *Planet* during ATEX 7-12 February, 1969;
Upper: pressures and pressure differences of I_b and I_t (bottom and top of inversion); middle: differences in specific humidity; bottom: differences in thermodynamic energy.

Fig. 5.17. Upper: Typical trade wind cumulus evaporating in trade inversion. Photo presented by Dr Erich Kuhlbrodt.

Fig. 5.18. Below: NOAA 5 satellite photo for eastern Asia, 9 December, 1975.

however, that energy is transferred through the inversion and that the latter is not a material surface.

We now put forward a model whereby upward transport of heat and moisture from the ocean, through cumulus clouds intruding into the inversion layer, modifies the air present there and incorporates it into the cloud layer through heat exchange and evaporation. Figure 5.17 is a very fine picture showing a typical non-precipitating trade wind cloud evaporating in the inversion, as photographed by the late Professor Erich Kuhlbrodt during the *Meteor I* expedition. Here is the kind of cloud envisaged as the active element in the discussion of turbulent moisture transport in Chapter 4, illustrated by Fig. 4.18. To the several constraints on maintenance of convective turbulence stated there, we should now add another one. The turbulent mass exchange with cumulus updrafts and compensatory downdrafts can be maintained only if θ_e decreases upward (14) or, at least, if θ_e in the inversion is lower than temperature at the inversion base, so that the descending air will remain colder. Figure 5.4 showed that this condition was fulfilled during ATEX.

However, we do observe extensive areas of stratus overcast under the trade inversion in the easternmost parts of the oceans where the

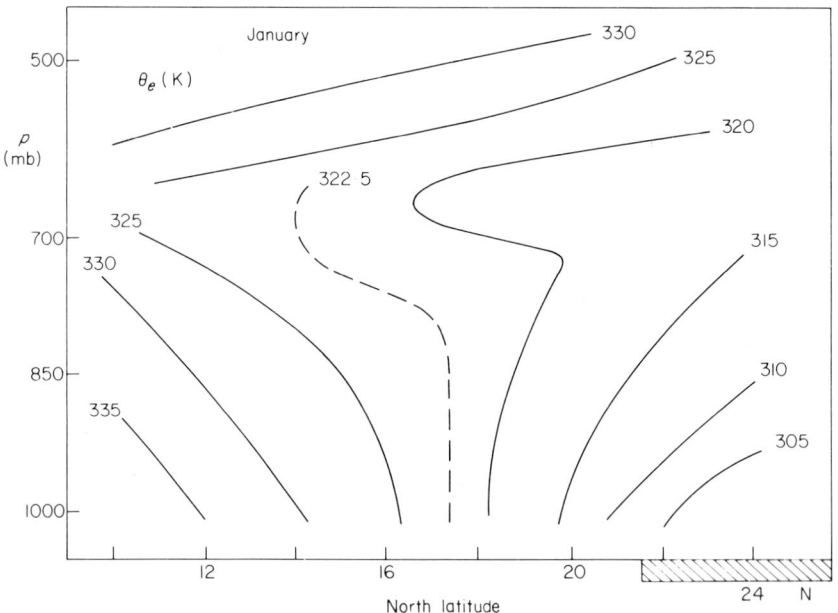

Fig. 5.19. Mean vertical north-south cross-section of θ_e (K) at longitudes 105-110°E for January. Solid shading: Land.

THE TRADE WIND INVERSION

inversion is lowest. In the overcast areas which, from all satellite observations, vary from day to day and are never in steady state, θ_e should increase with height. Where θ_e becomes constant and then begins to decrease upward, the stratus should break up and go over into cumulus along an air trajectory. During the northerly winter monsoon from China into the South China Sea (not a "trade", but closely related), satellite photographs regularly show the low stratus cloud from China held to a stationary limit near 16° to 17°N along the Vietnamese coast and sloping northeastward from there (Fig. 5.18). A mean cross-section for January along 108°E indeed shows the change in sign of the vertical θ_e gradient; the latitude of reversal is 16°N and coincides closely with the edge of the stratus (Fig. 5.19). The Northeast Pacific Ocean furnishes another excellent example.

The northeast trade of the Pacific Ocean

Flying from the west coast of the United States to Hawaii, a rapid transition from coastal stratus to tropical cumulus is often encountered; many travellers have noted it and commented on it. The vertical difference between θ_e at inversion top and bottom reverses near longitude 135°W (Fig. 5.20). The mean longitude of the western edge of the stratus, determined from a series of summer months of

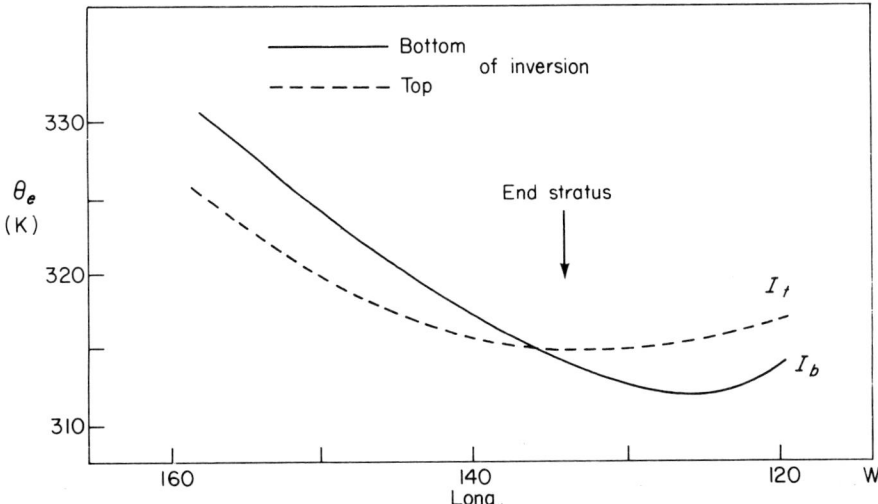

Fig. 5.20. Profiles of θ_e (K) along bottom and top of Northeast Pacific trade wind inversion on the route from San Francisco to Honolulu for the northern summer, data from Neiburger (15).

satellite pictures, turned out to be 133°-134°W, a very close coincidence. Figure 5.21 provides a typical example—the western edge of the stratus is quite sharp, a band of convection cells follows and then cloudiness decays to small tropical cumuli.

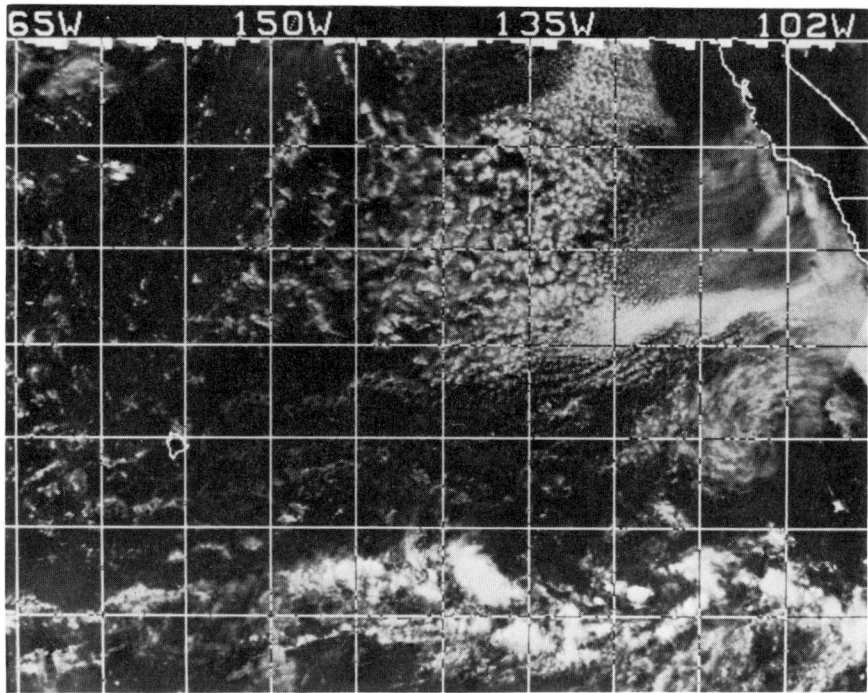

Fig. 5.21. NOAA 5 satellite photo for the Northeast Pacific Ocean, 1 August, 1975.

The northeast Pacific trade illustrates dissolution of the trade wind inversion over a long distance. During 1944 and 1945, the United States Navy maintained stationary air-sea rescue vessels along the California-Hawaii air route (Fig. 5.22). The meteorological observations of these vessels for the period July-October 1945 were used by the author and his collaborators (17) to determine the structure of the trade and compute the processes leading to the dissolution of the inversion quantitatively. Knowledge of the inversion structure was improved substantially by Neiburger (15) who had detailed inversion observation from fixed stations and oceanographic expeditions. Previously only standard-level values were available, which tended to wash out the sharp inversion features. Accordingly the following cross-sections along the trade trajectory have been improved by introducing Neiburger's inversion analysis.

THE TRADE WIND INVERSION

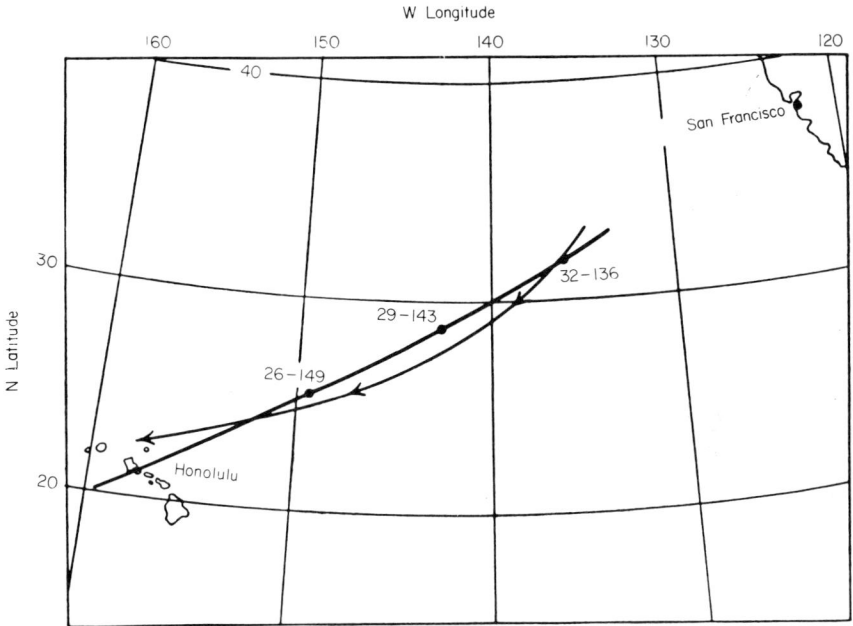

Fig. 5.22. Location of weather ships July-October 1945, and mean surface air trajectory for period (17). The average clockwise turning of wind direction to 700 mb was only six degrees.

Wind structure

Figure 5.3 already showed average profiles of wind direction and speed in comparison with ATEX. Wind speed attains a value close to maximum already at ship's deck level and then increases slowly to 925 mb, within the cloud layer, in contrast to ATEX. Higher up, a slight northward temperature decrease is felt, resulting in a slow decrease in trade wind speed. Direction is remarkably steady, so much so that the northeast Pacific trade can, with justification, be treated as a two-dimensional current, a great advantage for calculations. The hypothesis has been advanced that counterclockwise turning of wind with height, connected with cold air advection, is cancelled by clockwise frictional turning. Uniform wind direction is generally observed behind cold fronts in the United States in the friction layer, whereas during warm front situations the Ekman spiral appears very markedly. Thus, there is good substantiation for the above suggestion; it is not borne out, however, in ATEX, where winds turn markedly clockwise with height (Fig. 5.3) in spite of cold advection. Wind steadiness of the Pacific trade is very high, 80-95% increasing downstream, below the inversion. In and above the inversion it decreases rapidly as the regulatory influence of the connection with the ocean is lost and synoptic wind variations become prominent.

Temperature and humidity

Temperature increases downstream below the inversion. The mean temperature difference through the inversion is 4°C at the upstream end and about 1°C at Honolulu, quite in accord with the Atlantic (Fig. 5.8). The illustration also depicts the presence of four layers: the subcloud layer, extending from the surface to the bases of the cumuli; the cloud layer, which ends at the inversion base; the inversion layer, and the air above the inversion top. The subcloud and cloud layers together form the moist layer.

Specific humidity also increases downstream below the inversion, very little within the inversion and almost not at all above the inversion (Fig. 5.24). Relative humidity, computed from Figs 5.23 and 5.24 is

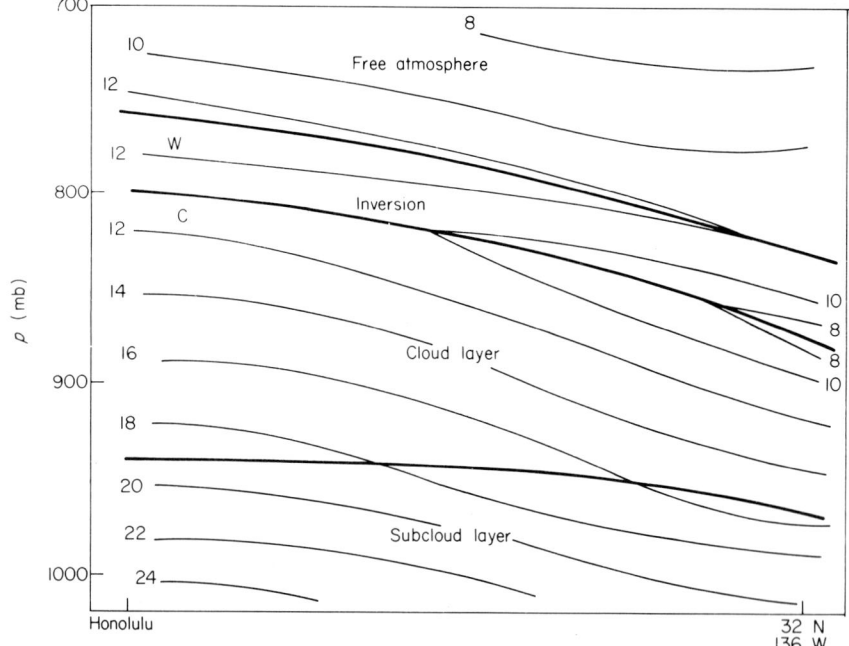

Fig. 5.23. Vertical cross-section of temperature (°C) along the trajectory of Fig. 5.22. Also illustrating the layers of the trade wind zone.

THE TRADE WIND INVERSION

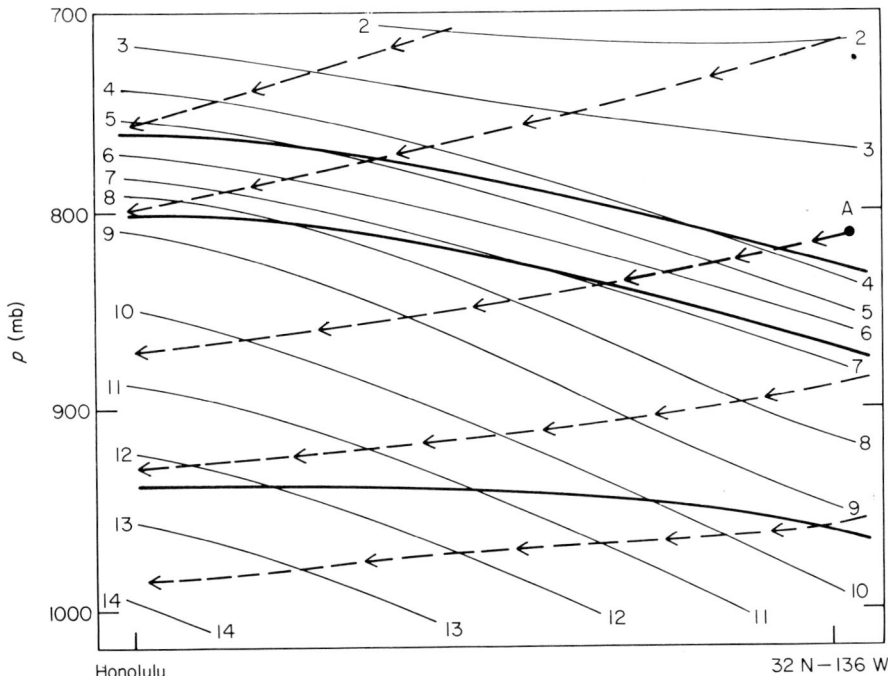

Fig. 5.24. Vertical cross-section of specific humidity (g/kg) and steady state trajectories.

80% or more below the inversion. The decrease through the inversion is over 50% at the upstream end and slightly below 40% at Honolulu; thus, the decrease is a little weaker than in Fig. 5.9, but not substantially.

Divergence and vertical motion

The northeast Pacific trade section first indicated the presence of deep divergence in the trades and the rise of the inversion against the vertical mean motion depicted in Fig. 5.14. However, the divergence was one-dimensional because the vertical section along the ships is two-dimensional only. Thus, downward motion was known to be underestimated. The ATEX divergence profile is not readily applied since the divergence there is very large and changes sign with height together with the meridional wind component, whereas in the Pacific a north component of about 4 m/s is present through the entire layer up to at least 700 mb. Neiburger (15) determined several trajectories with sinking over the whole northeastern Pacific; he found values of divergence up to $3 - 4 \times 10^{-6}$ s^{-1} at 700 mb.

In the following, divergence is computed from the planetary vorticity effect using Eq. (5.4); the resulting downward trajectories may still be underestimated. Values of divergence will be smaller than for ATEX since the latitude and, therefore, the Coriolis parameter f are much larger. Latitude decreases downstream along the cross-section, but so does the v-component from Fig. 5.22 and the ratio v/f is almost constant. Since v is almost constant with height, a single determination of $\text{div}_2 \, V$ from Eq. (5.4) will suffice. It turns out that $\text{div}_2 \, V = 0.96 \times 10^{-6} \, \text{s}^{-1}$. We apply the mass continuity equation using Eq. (5.4), in the integrated form

$$w_H = \frac{\hat{\varrho}}{\varrho_H} \frac{\beta \hat{v}}{f} H \qquad (5.5)$$

where H is distance above the ground and the caret indicates vertical averaging. From this equation $w_{3 \text{ km}} = -250$ m/day compared to -60 m/day for the one-dimensional divergence. Since the divergence is assumed to be uniform along the vertical, $w = dz/dt = \text{const} \, z$ and $\ln z = \text{const} \, t + K$ on integration. Evaluating K with $t = 0$, $z = z_0 = 3 \, \text{km}$,

$$\ln z = \ln z_0 - 0.083 \, t, \qquad (5.6)$$

where t is expressed in days. Since the horizontal distance of the whole trade section examined is 2200 km and the daily wind movement 520 km, the air travels from one end to the other in four days.

With the foregoing assumptions and information, two-dimensional trajectories in the vertical plane along the steady streamlines can be drawn (Fig. 5.25). We see that whereas the inversion ascends downstream, individual columns descend. Again the inversion is shown not to be a material surface separating an upper current, dry and potentially warm, from a lower current, moist and potentially cold. Trajectories cross both inversion top and base. Large masses of air, located above the inversion at 32°N, 136°W, become a part of the cloud layer before they reach Honolulu.

If we consider the previous history of the air passing the Hawaiian Islands, we can make the following divisions in Fig. 5.21:
    (1) The air that has remained below the inversion throughout the journey from 32°N, 136°W,
    (2) The air that has been incorporated into the cloud layer,
    (3) The air that has been incorporated into the inversion layer.
The grossly serrated inversion structure (Fig. 5.16) and the intrusion

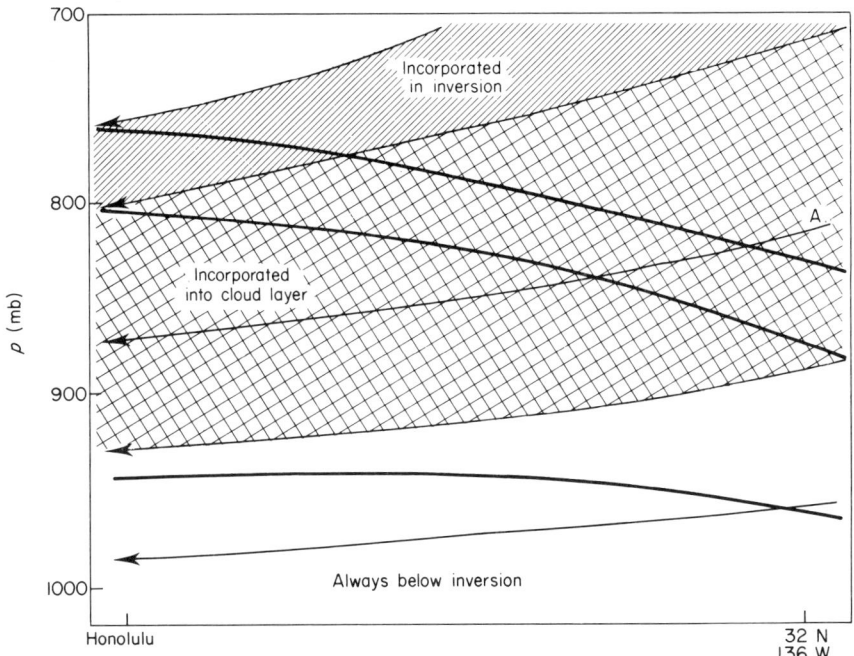

Fig. 5.25. The flow of mass through the trade inversion.

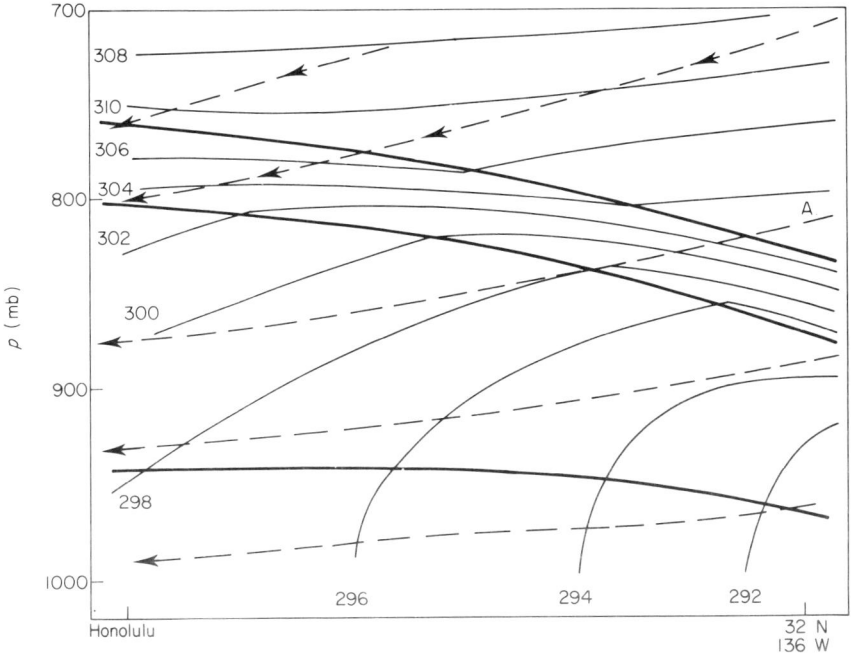

Fig. 5.26. Vertical cross-section of potential temperature (K) and sample mean trajectories from Fig. 5.25.

of cumuli into the inversion layer (Fig. 5.17) have already indicated the mechanism by which the dry upper air is transformed along its trajectory downward. We have also seen that sinking and intensification of shallow dry wedges from the middle troposphere is a principle mechanism to restore the inversion top so that in the end a mean stable inversion structure is attained.

In the subcloud layer the air moves steadily to higher temperature and potential temperature (Fig. 5.26). Higher up, the potential temperature decreases strongly during descent through the inversion (Fig. 5.26). After passing the inversion base, potential temperature again increases. Let us follow the single trajectory marked "A" in the energy diagram of $c_p\theta$ vs Lq (Fig. 5.27). From the start in the

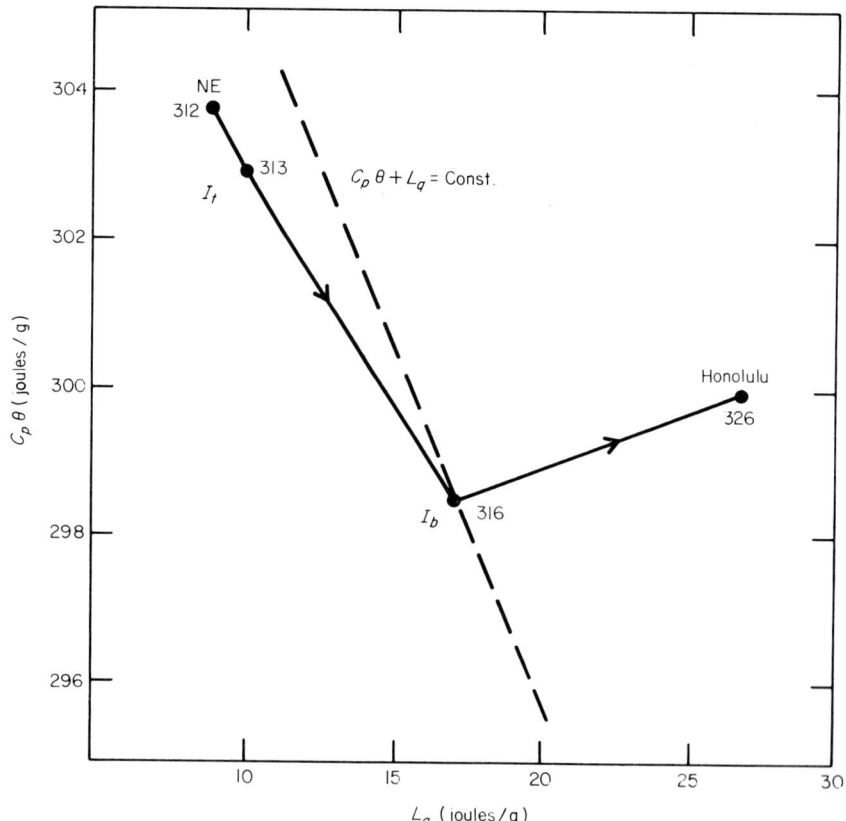

Fig. 5.27. Variation of $c_p\theta$ and Lq (J/g) along the mean trajectory marked "A"-in Fig. 5.25-26. I_b and I_t denote bottom and top of trade inversion. Dashed line gives slope for constant energy $c_p\theta + Lq$.

THE TRADE WIND INVERSION

northeast down to inversion base, energy is almost constant, i.e. parallel to the sample line for constant energy given by the dashed line. Descent through the inversion is accomplished partly by giving up heat to evaporate droplets from intruding cumuli. Hereby the descending air is assimilated into the cloud layer. After crossing the inversion base, both $c_p\theta$ and Lq increase, leading to a rapid rise in total energy. It is evident that, as is common, the variations in Lq greatly exceed those of $c_p\theta$. Nevertheless, even the sensible heat increases against the radiation cooling.

What we have found on one trajectory is generally true. A cross-section of θ_e is proportional to a section of total energy (Fig. 5.28). All trajectories cross toward higher θ_e, mostly in the cloud and subcloud layers. θ_e decreases everywhere with height, most rapidly in the inversion. The slope of the isotherms decreases toward the eastern end, but the longitude of slope reversal of Figure 5.20 is not quite reached. Trajectories also cross toward increasing θ_e above the inversion top indicating that there were several interruptions of the inversion regime due to synoptic systems which induced higher cloud systems. About 20% of the days in the area are without inversion (15).

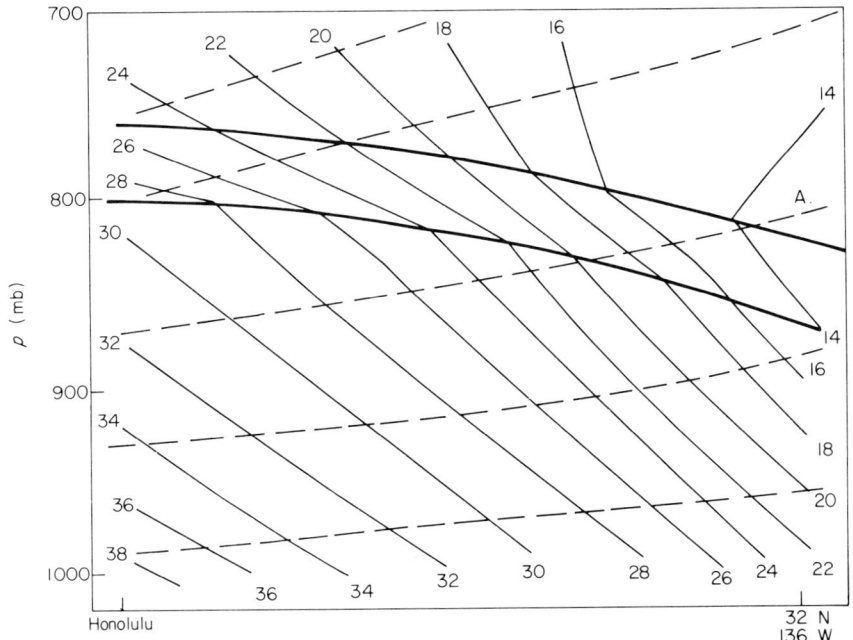

Fig. 5.28. Vertical cross-section of equivalent potential temperature (K, 300° +).

Trade wind energy budgets

In Chapter 3 we saw how the atmospheric general circulation, through its ordered belts of ascending and descending motion, lays down zones where evaporation is high and precipitation low, mainly in the trade winds regions, and that the equatorial trough zone has a great excess of precipitation over evaporation. Here, we shall be concerned with the question how the trade wind belts manage to exist while, at the same time, exporting a large fraction of the energy absorbed by them, mainly in the form of water vapour. Apart from such budgetary questions, we must inquire how the laws of motion are satisfied so that the trade winds can continue to blow; it will be seen that sensible heat addition to the low atmosphere is a key factor in this calculation.

Energy sources and sinks

The atmosphere is an "open" system. It exchanges heat through radiation with space, through radiation, conduction, and the water cycle with the ground. The total energy source for the atmosphere

$$H = Q_s + Q_e + R_a, \tag{5.7}$$

where R_a denotes net radiative heating or cooling of the atmosphere. As is well known, the atmosphere as a whole is a cold source. The energy source H may be subdivided into heat and moisture sources separately:

$$H_s = Q_s + LP + R_a, \tag{5.8}$$

$$H_l = L(E_0 - P). \tag{5.9}$$

Precipitation now appears explicitly, as a source in the heat budget and as a corresponding loss in the moisture budget.

Over the earth as a whole and over the year, Eqs (5.7)-(5.9) will all be zero for a steady climate. Precipitation heating and sensible heat transfer from the ground balance the radiation cooling; all evaporated water is returned as precipitation and run off in rivers to the ground for steady ocean levels. In smaller areas and time intervals things are more complex. For instance, in the trade wind belts evaporation exceeds precipitation, yet the moisture does not accumulate there. There must be a divergence of latent heat flow (Fig. 3.2) so that

THE TRADE WIND INVERSION

$$\text{div}(Lq) = L(E_{\text{ocean}} - P) \tag{5.10}$$

or, in terms of mass of moisture,

$$\text{div}(q) = E_{\text{ocean}} - P. \tag{5.11}$$

For heat, the corresponding form is

$$\text{div}(h_s) = Q_s + LP + R_a. \tag{5.12}$$

Eqs. (5.10) and (5.12) may be added to yield div (H), the divergence of total energy omitting only the energy of motion (18).

The northeast trade of the Pacific Ocean

As our first case, we shall compute the energy balance in the Pacific trade. This circulation is nearly steady in the summer season depicted, and the wind turns very little with height. Thus, a steady and nearly two-dimensional model may be assumed for a first approximation. The computation will be made for the two layers indicated in Fig. 5.29, following the mean stream tubes. There is the great advantage

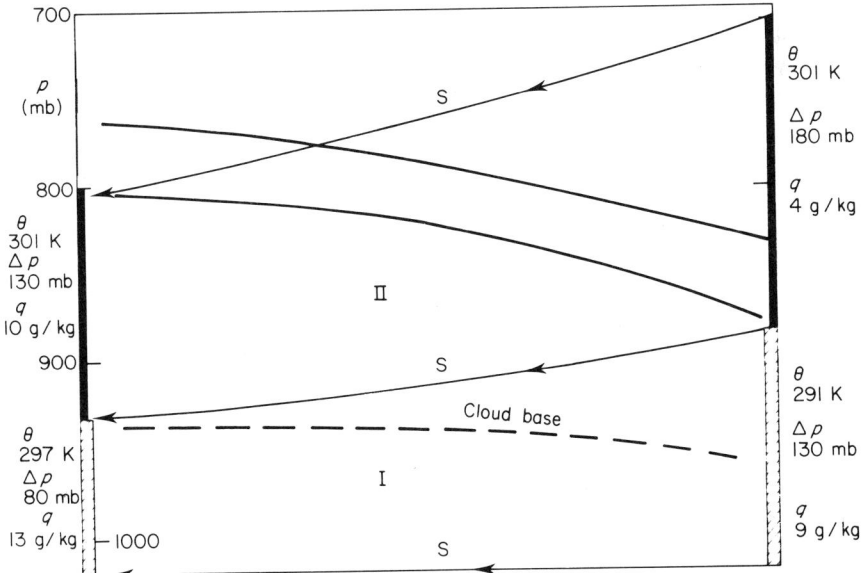

Fig. 5.29. Layer and trajectory arrangement for Pacific trade energy budget.

that only one calculation of transport by the mean winds need be made since, by definition, the mean motion cannot cross its own trajectories! The lower layer (I) in Fig. 5.29 is entirely below the inversion; the upper layer (II) contains the air that is incorporated into the cloud layer from above during its four-day journey over the distance of 2200 km. The upper layer has a pressure thickness of 150 mb in the middle of the traverse, the lower 100 mb. Wind speed averages 5 m/s in the upper layer and 6 m/s in the lower one (Fig. 5.3).

The divergence of moisture

$$\text{div}(q) = \int V \frac{\partial q}{\partial s} \frac{\delta p}{g} \delta s \, \delta n, \qquad (5.13)$$

where the integration extends over the distance δs in the cross-section, n is perpendicular to s and $\delta p/g$ is the pressure thickness. Both s and V are slanting in the cross section, i.e. they follow the stream tubes. The ratios of vertical to horizontal distance and of vertical to horizontal wind speed, however, are so small that, for practical purposes, the horizontal distance and the horizontal wind speed may be used. Evaluation then can be made directly from Eq. (5.13) by taking the moisture difference between downstream and upstream ends and using the mean wind speed just given to obtain $V \frac{\partial q}{\partial s}$. The mass transport of air along a tube

$$M_{air} = \int V \frac{\delta p}{g} \delta s \, \delta n \qquad (5.14)$$

will be determined in the middle using $\delta p/g$ there and setting $\delta n = 1$ cm for convenience. All data have now been given. As the reader can check, the divergence of moisture flow is 38×10^6 g/day in the upper layer and 21 such units in the lower layer; total 59×10^6 g/day or 150×10^6 kJ/day in energy units. In order to acquire some feeling for the meaning of the numbers, it is helpful to ask what evaporation from the ocean would be needed to supply the same amount of moisture. Dividing by the horizontal distance, $E_{equiv} = 0.25$ cm/day. The actual evaporation E_{ocean} is estimated as 0.37 cm/day. Thus, the residual available for precipitation, from Eq. (5.11), is 0.12 cm/day or 3.6 cm/month.

The calculation can now be "closed", as one says, given a good network of rain gauges. Over many land areas, these exist and over the ocean we expect radar and satellite to furnish such data. For 1945 data, however, the only possible verification is by means of a heat budget. We shall see how much rain is needed in order to keep the temperature field steady. The divergence of heat flux is obtained from the first law of thermodynamics in the form

$$\frac{dh_s}{dt} = c_p \frac{T}{\theta} \frac{d\theta}{dt} \, [\text{J/g/s}], \tag{5.15}$$

where dh_s is sensible heat addition. Analogous to Eq. (5.13) the horizontal divergence

$$\text{div}(h)_s = c_p \int \frac{T}{\theta} V \frac{\partial \theta}{\partial s} \frac{\delta p}{g} \delta s \, \delta n. \tag{5.16}$$

The factor T/θ is almost unity in the low atmosphere, but it is carried along since it is readily computed from preceding cross-sections. Proceeding as for moisture, the difference in θ between upstream and downstream ends is so small and uncertain in the upper layer that it is best to call it zero. In the lower layer, however, θ increases from 291 to 297 K, i.e. by 6°. Evaluating Eq. (5.16) as before, div (h) = 30 × 10⁶ kJ/day. We can make this number meaningful by converting it to equivalent radiation cooling. At constant pressure Eq. (5.15) becomes, now for arbitrary mass M,

$$dh_s = c_p \, dT \, M. \tag{5.17}$$

Evaluating with $c_p = 1$ J/g/D and $M = 100$ g/cm², dT (equivalent) $= -1.3$°C per day after division by the distance ds.

The radiation cooling is not well known. Estimates of net radiation under a cloud layer plus inversion indicate no more than $R_a = -1$°C/day. Applying this rate to both layers, $dT = -1.0 + (-1.3) = -2.3$°C/day in the lower layer and -1.0°C in the upper layer. Then, for both layers together,

$$c_p \frac{dT}{dt} M = -380 \, \text{J/day}.$$

The sensible heat flux from the sea, Q_s, is known to be small in the trades. Formerly, it was assessed at about 5% of Q_e. In view of Eq. (4.5), this ratio must now be raised to 15%. For evaporation of

0·37 cm/day, the equivalent latent heat flux is 925 J/cm^2/day, of which 15% is 140 J/cm^2/day. This, then, is the sensible heat source, which compensates partly for the loss computed above.

Another internal transformation between sensible and latent heats must be considered, since the sinking air evaporates cloud drops as suggested in Fig. 5.27. For an estimate, of the evaporation of 0·37 cm/day only 0·30 cm/day passes through cloud base, discussed under turbulent transport by cumuli in Chapter 4. The residual of 0·3 cm should condense between the condensation level in the middle of the section of Fig. 5.29 and the inversion base of 800 mb, if undilute ascent is assumed. In reality, entrainment occurs reducing the water vapour in the updraft. If half the adiabatic water content or 0·15 cm actually condenses then, after deducting 0·12 cm for precipitation as obtained from the moisture budget, a residual of 0·03 cm cloud drop water is present per cm^2 of surface which must be evaporated by the air sinking through the inversion of Fig. 5.27. The heat required for such evaporation within the atmosphere is 20×10^6 kJ/day as the reader may verify. Given an inversion thickness of 30 mb, the decrease of potential temperature during descent will be 3°C for evaporating the cloud water compared to an actual decrease of 4°C, which includes radiation cooling.

The whole drop evaporation has been placed into the inversion in this sample calculation, not entirely correct, of course. It is emphasized, however, how the transfer of mass through an inversion, usually considered an impenetrable boundary, is accomplished with evaporation and the attendant turbulent up and down motions of the inversion in Fig. 5.16. In this case the cumuli provide a thermally destabilizing effect—condensation heating lower down and re-evaporation higher up. In our two-layer model, both precipitation and re-evaporation have been placed in the upper layer; in a more complex three-layer model the vertical arrangement with lower condensation heating and upper evaporation cooling would stand out more clearly. This model contrasts with that presented at the end of Chapter 4 and in various published analyses of downdraft evaporation. When the evaporation occurs low down from evaporating raindrops, the evaporation may occur at lower altitudes than the condensation and, hence, contribute to thermal stability. Such stabilization will appear again in a marked way in Chapter 8 when we consider the freezing of upper cloud portions and later melting of the ice products much farther down.

Figure 5.30 summarizes the budget for both latent and sensible heat. At the interface between the two layers, turbulent moisture

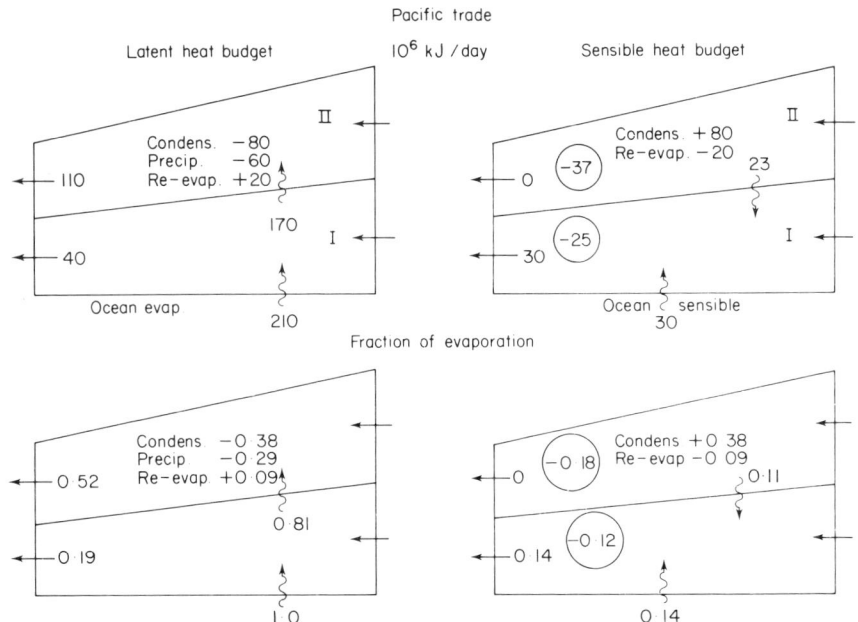

Fig. 5.30. Upper: Energy balance for Pacific trade (10^6 kJ/day); Left: latent heat; right: sensible heat. Transport by mean motions, solid lines, by turbulence, wavy lines. Lower: In per cent of evaporation. Precipitation, of course, falls out.

transport is upward, that of sensible heat downward, providing partial cancellation as in the lower right quadrant of Fig. 4.21. Further, two-thirds of the evaporated water is exported to other regions of the globe, notably the equatorial trough zone from Fig. 3.2. Nevertheless energy balance has been closely achieved without any requirement for "assistance" from other regions of the globe. Hence, these trades are driving members of the general circulation, a fact that will be related shortly to the large ratio of sensible to latent heat export, which is 0·20. The latent heat energy does not immediately affect the circulation of the trades, but the sensible heat export is very effective in maintaining it. The ratio of precipitation heating to condensate is high (75%), but the ratio of precipitation to ocean evaporation is low (29%).

In the lower part of Fig. 5.30 the budgets are restated in non-dimensional form, in the percentage of the principal key process which is operative, the ocean evaporation. Such presentation immediately gives a perspective for evaluation of the relative magnitude of all transformations that take place, and it makes the area treated comparable with other areas of different size.

In the BOMEX experiment (11) (see also (10) for the Caribbean trade inversions), during a five day period, divergence was found to be $5 \times 10^{-6} s^{-1}$ close to the ground, stronger than in the Pacific calculation. Maximum subsidence occurred near the base of the trade wind inversion which, in that region, had a magnitude of only about 1°C temperature increase. As in Fig. 5.30, turbulent fluxes transport moisture upward, but downward flux by the mean motion predominated. This is quite in contrast to areas with strong inversion regimes, where the upper layer is extremely dry. The Bowen ratio was about 0·1; Eq. (4.5) was not yet available, else it would have risen to 0·2, a sizeable value. It is interesting to read that "condensation and evaporation processes associated with the development and dissipation of trade wind cumulus can make a significant contribution to the heat balance of the cloud and trade inversion layers" just as in Fig. 5.30.

ATEX energy budget

Energy balances in ATEX which have been widely computed and utilized (2, 7, 27), suffer from the defect noted in Chapter 4 in connection with cumulonimbus energy balance: a very large "base" amount of energy is carried around swamping all the budget components of interest. As in the last chapter, we shall adopt the method of eliminating a base value of energy which, in the different area and season, must be lower than for the equatorial trough zone, if the important fluxes are to appear in correct perspective. The base value removed from sensible heat flux transports is 250 J/g. It may be noted that with the method employed in the Pacific trade we were able to circumvent this problem.

Because of the ATEX ship configuration (Fig. 5.2), so useful for definite divergence and vertical motion determinations, we do not have a long trajectory as in the Pacific Ocean and cannot work with a coordinate system following slanting trajectories, a simple and powerful technique. Instead, the evaluations must be made in a coordinate system locally fixed in the centre of the ATEX triangle. The gradual drift of the triangle need not be included.

The most interesting feature of ATEX is, that while inversion and pressure fields are similar to the other experiments, there was no precipitation, especially in the first week of the experiment when the triangle found itself in the subsiding trade wind stream. Hence, a set of balance diagrams must result which is rather different from Fig. 2.30 and the interpretation of the maintenance of the whole trade wind inversion layer must be different. Because of the locally fixed coordinates the novelty is that transports, vertical and horizontal, through the slanting inversion surface are presented, a very useful and illuminating variant. The area enclosed by the triangle was 25 × $10^4 km^2$, very much larger than that used for the Pacific trade. Hence, the budgets will be given in percentage of evaporation (0·55 cm/day) only, a much clearer mehod of presentation in any event (Fig. 5.31). One evaporation will be called a "unit".

The moisture budget of the layer above I_t is established by the mean motions alone. There can only be very little inflow from the top (Fig. 5.4) and the lateral convergence arises mainly from the convergence of mass above the inversion with sinking motion throughout the triangle (Fig. 5.3).

In the inversion layer there is a distinct deficit from the mean motions and this must be filled by the intrusion of trade wind cumuli into the inversion base, as exemplified for just this part of the world by the spectacular Fig. 5.17 from the *Meteor I* expedition. Both water vapour and water are involved.

In the cloud layer, the mean motion exports 0·53 units through the sides against the inflow down through the inversion. If the upward turbulent transport of 0·58 units at the inversion boundary is added, the requirement for such upward transport at cloud base is 0·86 units, very similar to the Pacific. In the subcloud layer, curiously, net outflow and evaporation are nearly identical. Unless we approach the budget from the top down, one might think that all evaporation is caught up in the subcloud layer and transported away, a simple picture but grossly erroneous. The total budget reaffirms what the direct turbulence measurements and many other investigations have shown: latent heat flux passes through cloud base only slightly diminished from Q_e, whereas sensible heat flow must stop there, caught in the stable layer with strongly increased slope of the θ profile (Fig. 4.5).

In summary, the budget exercise has brought out one very cogent fact: the vertical turbulent moisture flux and its disposition. As to inversion slope, and maintenance of the whole trade wind regime, the budget cannot give a clue. We now turn to the budget of $c_p T + gz$, closely abbreviated by $c_p θ$ in the low levels. We recognize that we must succeed without any precipitation heating. Radiation cooling, as best estimated, is shown in circles in Fig. 5.31 (right).

Starting from the top as before, the mean motions above inversion top produce a convergence of 0·32 units or, after radiation, of 0·23 units which, in view of Fig. 5.16, should be assigned to turbulent transport down.

In the inversion, the mean vertical and horizontal motions import 0·11 units. Of the vapour influx from below we assume, as before, that one-half, or 0·22 units (see moisture budget), mix with the surroundings through entrainment and exit as part of the water vapour outflow. The balance of the vapour first condenses and then re-evaporates into drier air from above without any net impact on the heat budget. We therefore have a surplus of 0·31 units after radiation, composed of convergence of moisture by the mean motion plus turbulent import of water vapour plus water by the intruding clouds. Before deciding how to assign this surplus, let us look at the lower layers.

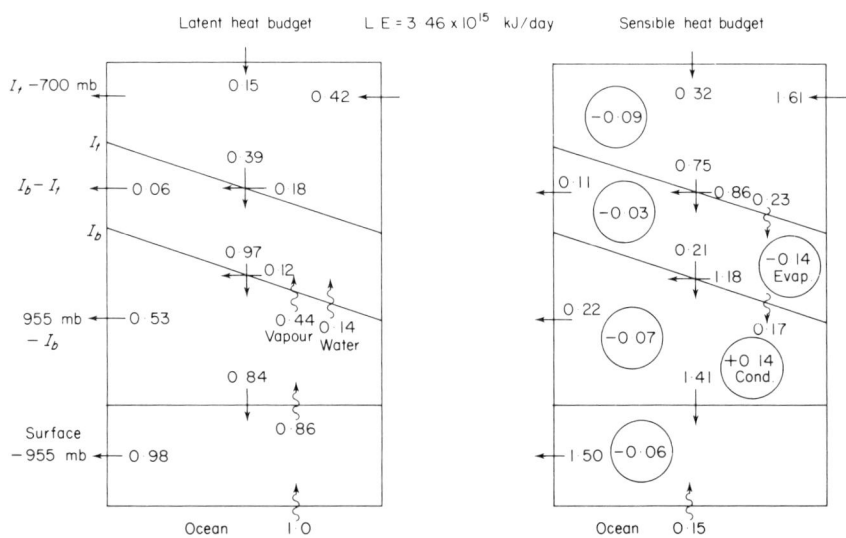

Fig. 5.31. Energy balance for ATEX in per cent of evaporation (after 2, 27). Vertical coordinate arbitrary. Other notation as in Fig. 5.30.

In the subcloud layer, there is virtually balance, a net export of 0·09 units plus radiation cooling being balanced by Q_s. The specific heat of water vapour is included in surface source and heat export. No sensible heat transport upward takes place through cloud base as expected.

In the cloud layer the mean motions produce a net export of 0·24

THE TRADE WIND INVERSION

units. Therefore, there is a deficit of 0·31 units after radiation. Water vapour convergence is 0·28 units from the moisture budget. If half of the converging vapour again mixes with drier air and is transported out, 0·14 units condense. Now, the condensed drops should be of cloud drop size only in view of the shallow depth of the cloud layer (90 mb) compared to about double that depth in the middle of the Pacific trade trajectory. From several studies (Chapter 4), a cloud thickness of 90 mb is quite insufficient to produce raindrop-size particles that can acquire appreciable fall velocities and reach, for instance, the subcloud layer. More likely, all these small drops are swept upward into the inversion by the updraft due to the cumuli. Then there will be no heat of re-evaporation required in the cloud layer. The condensation heat of 0·14 units accrues to the layer and reduces the energy deficit there, whereas the heat of re-evaporation is drawn from the surplus of the inversion layer. This distribution of heating and cooling has been entered in the heat budget.

Returning now to the inversion layer, the surplus of 0·31 units is diminished by the 0·14 units for re-evaporation of drops; the residual of 0·17 units is a turbulent transport down the gradient of potential temperature from inversion to cloud layer, a major part of the maintenance of the whole ATEX trade wind region. Herewith, balance has been achieved in all layers as close as is needed.

The main result of the ATEX budget analysis is that we have managed to find a reasonable and thermodynamically consistent path for the air sinking from above, to pass downward through the inversion, and to acquire the characteristics of the cloud layer through exchange of sensible to latent heat.

Since the moisture budget only complicates the picture but does not have a net impact on the heat budget, we can determine the latter in summary from the integral over the boundary in Fig. 5.31. We find inflow of 1·93 units into the top layer, outflow of 1·83 units from the lower layers and a surface source of 0·15 units. Net import is 0·25 units to balance radiation cooling, which turns out to be 1·1°C per day, a reasonable value. Most important is that the heat inflow is at the top and moves down the potential temperature gradient with the mean motion. Through the sinking the potential energy is lowered, and this sinking provides the near-balance against radiation cooling. The potential energy release is effected through pressure-boundary work by outside portions of the general circulation. We must consider the Atlantic trade as "driven" dynamically from the outside. From the energy viewpoint, however, the region is still a major source inasmuch as the total evaporation from the sea is exported to the tropics at large.

Maintenance of the trade winds

We ask the question whether the Pacific trade section, though part of the general circulation, can be regarded as independently dynamically maintained on average. We are encouraged in doing so by Figs 5.23 and 5.24 which show temperature and moisture increasing downstream in the subcloud and cloud layers. The trade wind inversion is the top of the convective turbulent layer. Above it, the atmosphere is subject to the large-scale features of the general circulation and its synoptic systems. These undoubtedly interfere from time to time with any internally maintained circulation underneath, but if the lower circulation should be capable of internal maintenance, there will always be the tendency toward a "return to normal" from any deviation pattern; such a strong tendency toward steady state is indeed observed climatically, as is well known.

The wind stream of the trades has been discussed in (8, 24), among others. Here we choose a coordinate system s,n, where s is fixed along the surface flow (Fig. 5.22) as a straight coordinate. Velocity components are v_s ($= V$ at the surface, where V is total speed) and v_n, positive to the right of v_s. The horizontal equations of motion in this coordinate system are

$$\frac{\varrho d v_s}{dt} + \varrho f v_n = -\frac{\partial p}{\partial s} + \frac{\partial \tau_{s,z}}{\partial z}, \tag{5.18}$$

$$\frac{\varrho d v_n}{dt} + \varrho f v_s = -\frac{\partial p}{\partial n} + \frac{\partial \tau_{n,z}}{\partial z} \tag{5.19}$$

where $\tau_{s,z}$ and $\tau_{n,z}$ are the shearing stresses along s and n and the frictional forces are given by their vertical derivatives. Lateral stresses may be neglected; the acceleration terms are small and will be omitted. It was hoped that $\partial \tau_{n,z}/\partial z$ would be zero, i.e. that the flow could be treated as fully two-dimensional and that the v_s-component would be geostrophically balanced, but this did not prove to be the case (6). The frictional force along n makes about a one-third contribution to the balance of forces in the layer below the inversion; the wind turns with height. We shall adopt the convention that the top of the boundary layer is situated at the level where streamlines and isobars first become parallel, which is slightly below (20) but possibly at the trade inversion, above which the magnitude of the turbulent motions becomes small in the mean, therewith also $\partial \tau/\partial z$ but not necessarily τ itself.

From observation, the surface pressure difference is 4·5 mb from upstream to downstream end of the section. Using Figs 5.23 and 5.24 temperature increased by slightly over 3°C over the distance of 2200 km for the layer 1000 to 800 mb and specific humidity increases by somewhat over five g/kg, hence $\delta T_v \approx +4°C$. For convenience we shall use the layer 1020 to 820 mb, even though the inversion is situated in the top portion of this layer at the upstream end. Given the hydrostatic equation differentiated between constant pressure surfaces

$$\frac{\delta D}{\hat{D}} = \frac{\delta T_v}{\hat{T}_v}, \tag{5.20}$$

where D is thickness between isobaric surfaces. For $\delta T_v = 4°C$, $T_v = 290$ K, and $D = 2000$ m, $\delta D = 28$ m, so that $\delta p_0 \approx -3$ mb if there is no pressure-height gradient at 820 mb. The surface heat source thus accounts for a major part, but not all of the surface pressure difference. Evaluating the Coriolis term in Eq. (5.18), $v_n = +1\cdot0$ m/s at 820 mb, for force balance without friction at the top of cloud layers. Give $V = 6$ m/s, the turning from the surface wind direction is 10°, compared to 6° found from the observations.

However, due to the increase of T_v downstream, the surface isobars with higher pressure to the north must rotate counterclockwise with height, a feature found in Atlantic, Pacific and BOMEX (11). The result is a compromise (Fig. 5.32), neither invariant isobar direction as postulated in Ekman spiral computations, nor the purely two-dimensional flow where the isobars must rotate fully into the direction of the streamlines for geostrophic balance. At the surface the flow is 18° to the left of the surface isobar direction, toward lower pressure. After isobaric turning of 10° to the left and of wind turning of 6° to the right, the two curves first intersect but do not become parallel. For a discussion of pressure-wind relations higher up see (7). The level of intersection is 1600 m, slightly less than the 2000 postulated above; thus, the first estimate was not a bad one.

It would be interesting to compute the pressure gradient at the top of the turbulent layer if there were no heat source. If the hydrostatic equation is differentiated at constant temperature (A-6)

$$\delta p_H = \frac{p_H}{p_0} \delta p_0, \tag{5.21}$$

where H is the top of the layer considered. For $p_0 = 1000$ mb, $p_H = $

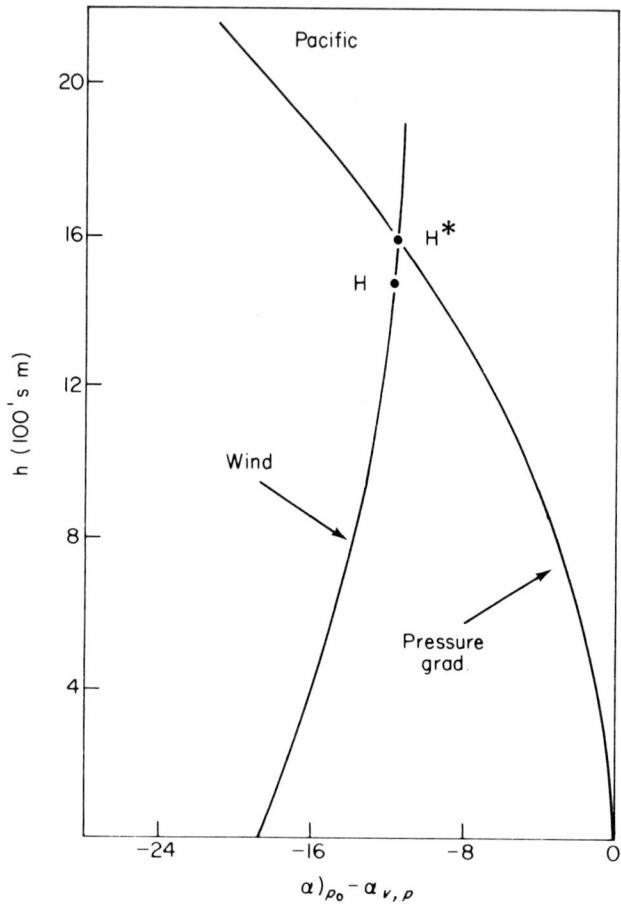

Fig. 5.32. Vertical variation of wind direction and isobar direction in the Northeast Trade of the Pacific Ocean, relative to the direction of the surface isobars (20). Negative angles denote counterclockwise deviations. H and H* are computed values of boundary layer height.

800 mb, and $\delta p_0 = -5$ mb as before, $\delta p_H = -4$ mb. A large pressure gradient would still exist at the top of the turbulent layer. This pressure drop may be interpreted as the means by which distant circulation would maintain the trades, which become an energy consuming branch of the general circulation when there is no sensible heat source. This is the situation of ATEX. The Pacific pressure difference, after taking the turning of the isobars into account, is 1 mb, only 0·2 of the total pressure difference, which herewith appears

THE TRADE WIND INVERSION

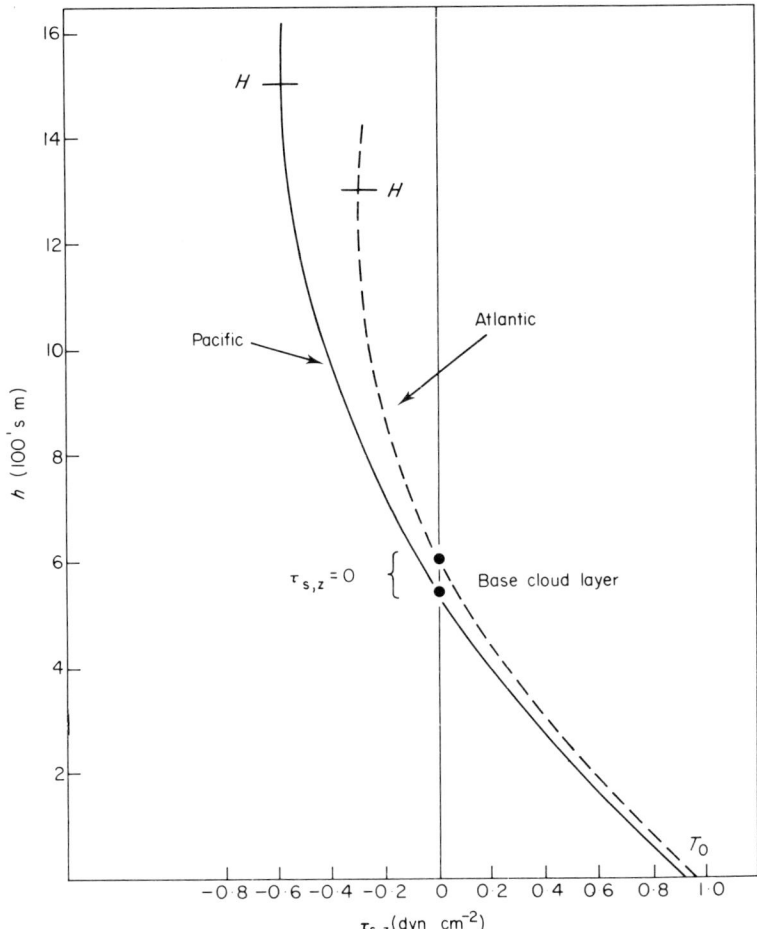

Fig. 5.33. Vertical distribution of shearing stress in s—direction for Pacific trade and ATEX (20).

as the major component establishing that internal energy sources largely provide the pressure field for maintaining the motion of the trade winds against friction.

The vertical distribution of shearing stress is obtained by integrating Eq. (5.18) with the assumptions made. Thus

$$\tau_{s,h} - \tau_0 = \int_0^h \frac{\partial p}{\partial s} \delta z + \int_0^h \varrho f v_n \, \delta z. \tag{5.22}$$

At the surface the total shearing stress appears, thought to lie along the wind direction very near to the ground. With nearly uniform temperature gradient over the section of Fig. 5.23, the pressure gradient may be assumed to decrease linearly upward over any short vertical distance, i.e.

$$\int_0^h \frac{\partial p}{\partial s} \delta_z = \frac{1}{2} \frac{\partial}{\partial s} (p_0 + p_h) h . \tag{5.23}$$

The last term of Eq. (5.22) may be integrated numerically or with modelling (20). Resulting vertical stress profiles were computed both for the Pacific trade and ATEX (Fig. 5.33); they are very similar. Surface stress was taken from the *Meteor* observations for ATEX and computed with Eq. (4.9) for the Pacific, and in both cases the stress becomes zero very close to cloud base. Lower momentum is imported from above and below to maintain the trades against downstream acceleration by the pressure force. The levels of zero stress and maximum east wind do not coincide. In ATEX the strongest wind was located at 300 to 400 m, in the Pacific and also BOMEX well up in the cloud layer. The definition of shearing stress given in Eq. (4.8) appears not to be valid in turbulent flow.

From the foregoing, large portions of the trade wind zone, such as in the Pacific, may be regarded as self-sustaining and internally closed-off circulations. Against this background of "return to normalcy" one can then discuss enforced deviations from the mean (12, 14a) which can, among other solutions, produce stable oscillations of the trade wind belt about the mean, and thereby influence the general circulation.

References

(1) Augstein, E. (1972). Untersuchungen zur Struktur und zum Energiehaushalt der Passatgrundschicht. Ber. Inst. Radiometeor. und Maritime Meteor., Univ. Hamburg. No. 19.

(2) Augstein, E., Riehl, H., Ostapoff, F. and Wagner, V. (1973). Mass and energy transports in an undisturbed Atlantic trade-wind flow. *Mon. Wea. Rev.* **101:** 101-111.

(3) Augstein, E., Schmidt, H. and Ostapoff, F. (1974). The vertical structure of the atmospheric planetary boundary layer in undisturbed trade winds over the Atlantic Ocean. *Bound. Layer Meteor.* **6:** 129-149.

(4) Brocks, K. (1971). *Trop. Meteor.* Jahrbuch der Fraunhofer Gesellschaft, München 1970/1971, Beispiele angewandter Forschung. 10 pp.

(5) Brocks, K. (1972). Die Atlantische Expedition 1969 mit dem Atlantischen Passatexperiment. *"Meteor" Forschungserg.* A, **10:** 30 pp.
(6) Brümmer, B., Augstein, E. and Riehl, H. (1974). On the low-level wind structure in the Atlantic trade. *Quart. J. Roy. Meteor. Soc.* **100:** 109-121.
(7) Brümmer, B. (1976). The kinematics, dynamics, and kinetic energy budget of the trade wind flow over the Atlantic Ocean. *"Meteor" Forschungserg.* B, **11:** 1-24.
(8) Charnock, H., Francis, J. and Sheppard, P. A. (1956). An investigation of the wind structure in the trades: Anegada 1953. *Phil. Trans. Roy. Soc. London.* (A) **249:** 179-234.
(9) Flohn, H. (1957). Studien zur Dynamik der Atmosphäre. I. Horizontale und vertikale Windkomponenten auf dem Atlantik. *Beitr. Phys. Atmos.* **30:** 18-46.
(10) Gutnick, M. (1958). Climatology of the trade wind inversion in the Caribbean. *Bull. Amer. Meteor. Soc.* **39:** 410-420.
(11) Holland, J. Z. and Rasmusson, E. M. (1973). Measurements of the atmospheric mass, energy and momentum budgets over a 500-kilometer square of tropical ocean. *Mon. Wea. Rev.* **101:** 44-55.
(12) Kraus, E. B. (1959). The evaporation-precipitation cycle of the trades. *Tellus* **11:** 147-158.
(13) Kuhlbrodt, E. (1928). *Z. Geoph.* **4:** 385.
(14) Lilly, D. K. Ref. 43, Chapter 4.
(14a) Malkus, J. S. (1956). Ref. 50, Chapter 4.
(15) Neiburger, M., Johnson, D. S. and Chien, C. W. (1961). "Studies of the Structure of the Atmosphere over the Eastern Pacific Ocean. I: The Inversion over the Eastern North Pacific Ocean." Univ. Calif. Press. 94 pp.
(16) Piazzi-Smyth, C. (1858). Astronomical expedition on the Peak of Teneriffa. *Trans. Roy. Soc. London* **148:** 465-533.
(17) Riehl, H., Yeh, T. C., Malkus, J. S. and LaSeur, N. E. (1951). The northeast trade of the Pacific Ocean. *Quart. J. Roy. Meteor. Soc.* **77:** 598-626.
(18) Riehl, H. and Malkus, J. S. (1957). On the heat balance and maintenance of circulation in the trades. *Quart. J. Roy. Meteor. Soc.* **83:** 21-29.
(19) Riehl, H. and Augstein, E. Ref. 69, Chapter 4.
(20) Riehl, H. and Soltwich, D. (1974). On the depth of the friction layer and the vertical transfer of momentum in the trades. *Beitr. Phys. Atmos.* **47:** 56-66.
(21) Riehl, H. (1977). Venezuelan rain systems and the general circulation of the summer tropics. I: Rain systems. *Mon. Wea. Rev.* **105:** 1402-1420.
(22) Riissanen, J. (1966). The trade wind inversion according to observations of the Finnish Atlantic Expeditions 1957-1958. *Geophysica* (Helsinki) **8:** 325-329.
(23) Rossby, C.-G. and collaborators. (1939). Relation between the intensity of the zonal circulation of the atmosphere and displacements of the semi-permanent centers of action. *J. Mar. Res.* **2:** 38-55.
(24) Sheppard, P. A. and Omar, M. H. (1952). The wind stress over the ocean from observations in the trades. *Quart. J. Roy. Meteor. Soc.* **78:** 583-589.
(25) Simpson, R. H. Ref. 32, Chapter 2.
(26) VonFicker, H. (1936). Die Passatinversion. *Veröff. Meteor. Inst. Berl.* (1) **4:** 1-33.
(27) Wagner, V. (1975). Zusammenhänge zwischen der troposphärischen Zirkulation und den Energetischen Prozessen im Bereich der Hadleyzirkulation über dem Atlantik. Ber. Inst. Radiometeor. und Maritime Meteor., Univ. Hamburg. No. 26.

6
Diurnal and Local Controls

The broad-scale aspects of wind, temperature and rainfall discussed in earlier chapters can be interpreted to furnish the traveller to the tropics with a good, general indication of what he is likely to encounter. Such an interpretation is indeed needed for there is a widespread notion that "tropics = steamy jungle", and that simply is not true. The jungle areas cover only 10% of the world's land, often concentrated around mountain chains having relatively high precipitation. Elsewhere in the tropics there are open fields and forests, savannah, and even deserts on the equator; a very varied picture in the global aspects.

Apart from seasonal changes (dry season; rainy season), the diurnal weather pattern tends to govern life strongly in the tropics. Absent are the large day-to-day temperature changes of higher latitudes. Climate can change literally by the mile, especially near mountainous coasts. The traveller arriving at the Mombasa airport (4°S, 40°E) on the Indian Ocean shore of equatorial Africa may find great heat and humidity; upon going westward up the slope to the highlands near 1600 m altitude, where Nairobi lies, he will encounter a most pleasant temperature climate and little oppressive humidity if he chooses his season correctly. Arriving at Honolulu on the other side of the world, the traveller may go to the beach district around Diamond Head, where sunshine is profuse, rain is rare and the days are often hot in summer. Only a few miles away, on the hill slopes, temperature will be much lower during the day, the sky mostly overcast and the relative humidity so high that care is necessary to ensure that clothing and other equipment do not mould. In touring the island of Oahu, the traveller can drive in one hour from dense tropical jungle across

DIURNAL AND LOCAL CONTROLS

fertile, pineapple and irrigated sugar plantations to country so dry that only thorny cactus can survive. Going west from Recife on the Brasilian Coast (8°S, 35°W) the same spectacle is encountered: at first dense jungle and farms on the upslope from the oceans, then, past the crest, a sudden change to near-desert conditions. However, starting up from the ocean on the other side of the continent, say at Trujillo in Peru, also at 8°S but 79°W, the desert lies along the ocean shore and as the road rises, first through stark boulders and organ pipe cactus, a very beautiful and fertile stretch of land is encountered near 1500 to 2000 m altitude on the slope of the Andes, so aptly named the "tierra templada" by the South Americans.

Travellers should be prepared for unexpected cold weather. The author remembers needing thick sweaters for a boat ride near Hong Kong in March, and feeling cold in a hotel in the interior valley of Puerto Rico with temperature not much above 10°C and solid morning fog. He also noted the cold at the Aswan Dam in Egypt in December, though the location of 23·45°N is termed "subtropical". Lastly, the traveller should be prepared nowadays for near-refrigeration in the indoor "climate" which has sprung up in increasing numbers of establishments throughout the tropics.

These few examples of extreme variety within short distances illustrate the important place local climate occupies in the tropics. Textbooks containing regional climatology offer a wide array of local conditions. Here we may note that local and diurnal climates are thought important for large-scale processes and that important diurnal variations exist not only over land but also over sea, especially cloudiness and rainfall. It will be well, therefore, to start our examination of diurnal and local controls by contrasting air-ground exchange processes in the course of the day over land and over sea.

Ground-air interactions through heating

Radiation

Two similar surface radiation stations were established at Palmyra (6°N, 162°W) during March-April 1967 and at Anaco, Venezuela (9°N, 64°W) during August-September 1969 by Dr Stephen K. Cox (University of Wisconsin, later Colorado State University). Both locations had a mild rainy season during the experimental periods and are quite comparable. The instruments at Palmyra were located on a pier piling approximately 15 m from the shore in Palmyra lagoon (2) and looked at the water, which had a mean surface temperature of 28°C. Therefore, the installation should give a reasonable first approximation to conditions over water and can be compared with the solid land station, Anaco.

The course of incoming radiation at about 3 m height is quite similar (Fig. 6.1) at both stations, except that Palmyra had somewhat

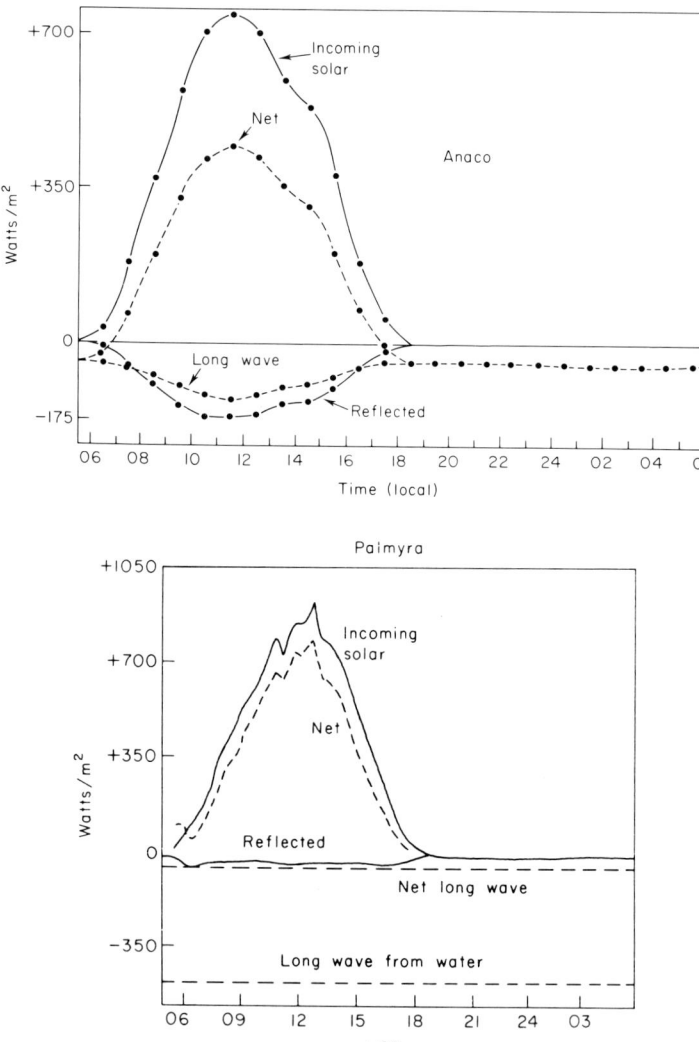

Fig. 6.1. Comparison of diurnal course of radiation over land and on the ocean shore: Anaco (9°N, 64°W) (28) and Palmyra (6°N, 162°W) (2).

larger values which presumably are related to lower cloudiness. For comparability, the Anaco radiation was adjusted to that at Palmyra (Table 6.1). Reflected radiation was much less at Palmyra, though still 8%, compared to over 20% in the area of low grass and sandy soil at Anaco. Net long-wave radiation reaches a distinct maximum there during daytime hours due to the heated soil (Fig. 6.1). Presumably, outgoing radiation was constant at Palmyra, but this was not fully established for all hours of the day since long-wave radiative flux downward was not measured in that experiment.

Table 6.1. Comparison of ground-air interaction over land and sea (W/m^2)

	Anaco, Venezuela (9°N, 64°W) (28)	Palmyra (2) (6°N, 162°W)
Short-wave radiation reaching ground	250 (adjusted to Palmyra)	250
Reflected radiation	−50	−23
Solar radiation absorbed in ground	200	227
Net long-wave radiation from ground	−65	−23
Net radiation absorbed in ground	135	204
Q_s/Q_e	0·60	0·15
Q_e	85	150
Q_s	50	22
Evaporation (cm/day)	0·28	0·50
Balance	0	32 (into ocean)

Net radiation absorbed in the earth turns out to be 50% larger over sea than over land. Since land storage may be assumed negligible, the net radiation must be returned to the atmosphere through sensible and latent heat transfer (Q_s, Q_e). From the change in temperature measured by balloon during the morning hours (Fig. 6.2), Q_s at Anaco was estimated as 50 W/m^2, thus $Q_s/Q_e = 0·6$ and $Q_e = 85$ W/m^2 equivalent to evaporation of 0·28 cm/day for the assumed enhanced solar radiation (actual 0·22 cm/day). Balance is zero. Over water we do not have the requirement of zero balance. Q_s/Q_e was taken as 0·15 corresponding to measurements made in many areas and using Eq. (4.5). Q_e for the area and season is given as 150 W/m^2 corresponding to an evaporation of 0·5 cm/day. This was confirmed in November-December 1978 by overflights with gust probe measurement at 148°W (pers. comm. B. K. Bean, NOAA). Total heat transfer from the ocean then is 172 W/m² and a residual of 32 W/m² remains in the water to be carried off to other areas of the globe, unless the ocean is warming locally. This is a large value, 13% of the short-wave radiation reaching the ground.

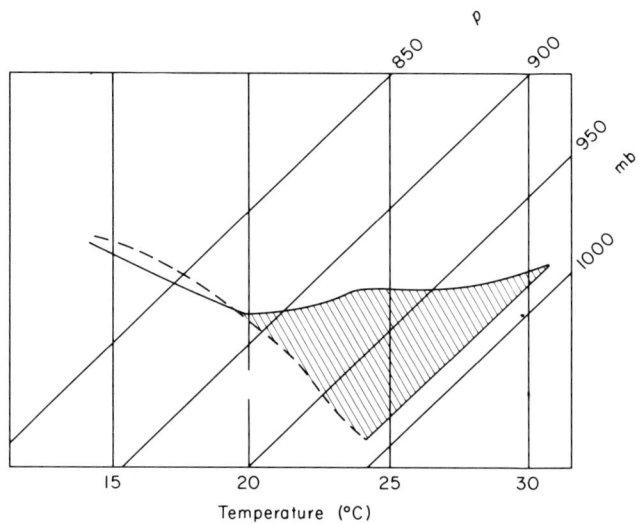

Fig. 6.2. The morning heating of the low atmosphere at Anaco. Curves meet near the convective condensation level (28).

Incoming solar radiation of 250 W/m² at the surface has been observed at other locations even in the equatorial trough zone (27). However, on cloudy or rainy days, the radiation may be reduced to as little as 100 or even 50 W/m². On such days the ground-air energy exchange becomes virtually zero over land, whereas over the ocean some of the stored energy is given off, often even at an enhanced rate, on windy days.

Daily course of surface temperature and humidity

Based on the differences in the disposal of the incident solar radiation at the surface, a very different course of surface temperature and energy will be expected over land and sea, or at a coastal station with ventilation from the ocean. We shall compare the Venezuelan research station, Carrizal (9·5°N, 67°W) and San Juan, Puerto Rico (18·5°N, 66°W) in the rainy season (July to September). At San Juan

(Fig. 6.3) the daily temperature range is 5°C, large compared to 0·5—1·0°C over the ocean in spite of the proximity of the instrumentation to the coast; at Carrizal, the temperature range is 8°C, not large for continental interior locations. Both night-time minima and daytime maxima exceed those of San Juan. On clear days the temperature range has attained 13-14°C, which is also the range encountered at mountain stations 3-4 km high with little water vapour above them. The early maximum at San Juan must be ascribed to the mitigating effect of the sea breeze which normally sets in during the late forenoon.

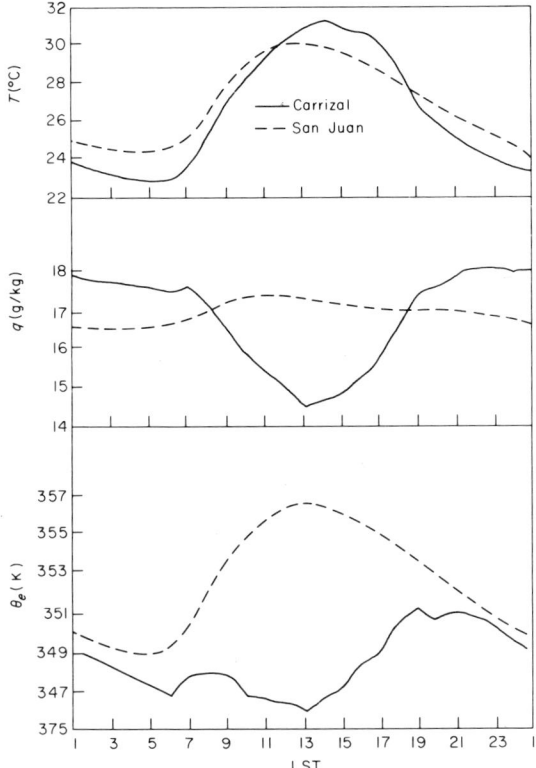

Fig. 6.3. Comparison of the diurnal course of surface parameters at San Juan, Puerto Rico (18·5°N, 66°W) and at Carrizal in Venezuela (9·5°N, 67°W) during the rainy season.

The sea breeze should also be responsible for the remarkably constant specific humidity there, with even a slight increase in the noon hours, whereas Carrizal typifies a widely encountered continental regime including such distant locations as Berlin, Germany. During the night, moisture from evaporation accumulates near the ground. As daytime turbulence commences, the high moisture is carried up and replaced by drier air from above, so that surface specific humidity falls against the evaporation cycle, which has a maximum during the warmest hours of the day (Fig. 6.4)! As a result, equivalent potential temperature (θ_e) follows the course of temperature at San Juan, whereas the downward mixing of dry air at Carrizal actually leads to a morning decrease and an early afternoon minimum. Evidently the diurnal course of cloudiness and precipitation there is strongly influenced by the turbulent mixing. On days without synoptic weather systems, clouds only build up in the later afternoon when, with higher moisture aloft plus evaporation from the ground, total moisture in the tropospheric air column increases, surface moisture values begin to recover and clouds can form near the convective condensation level. The maximum build-up

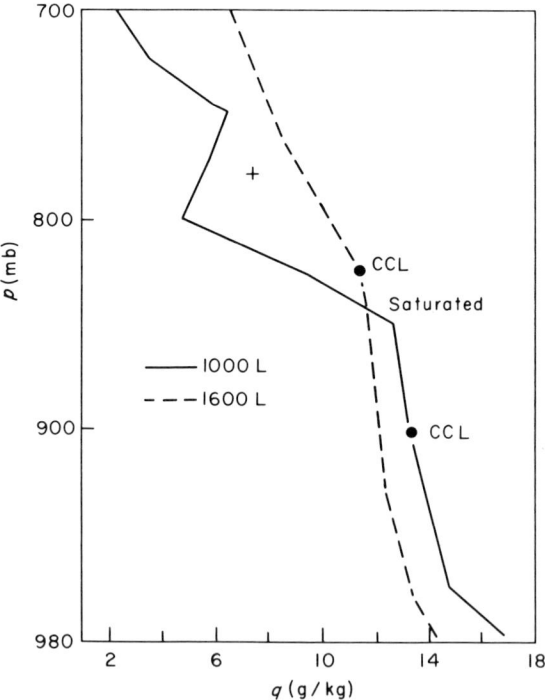

Fig. 6.4. Illustrating the diurnal redistribution of moisture along the vertical by turbulence: Carrizal, Venezuela, 16 August, 1972.

of large clouds and radar echoes is attained at the end of the diurnal heating period near sunset (Fig. 6.5).

In considering the daily variation, care must be taken with thermometer shelter data on days with more than just traces of precipitation. The wood of the shelter accumulates moisture and acts as a large wetbulb thereafter, so that values in the shelter become quite unrealistic and unrepresentative of outside values (31, Chapter 7).

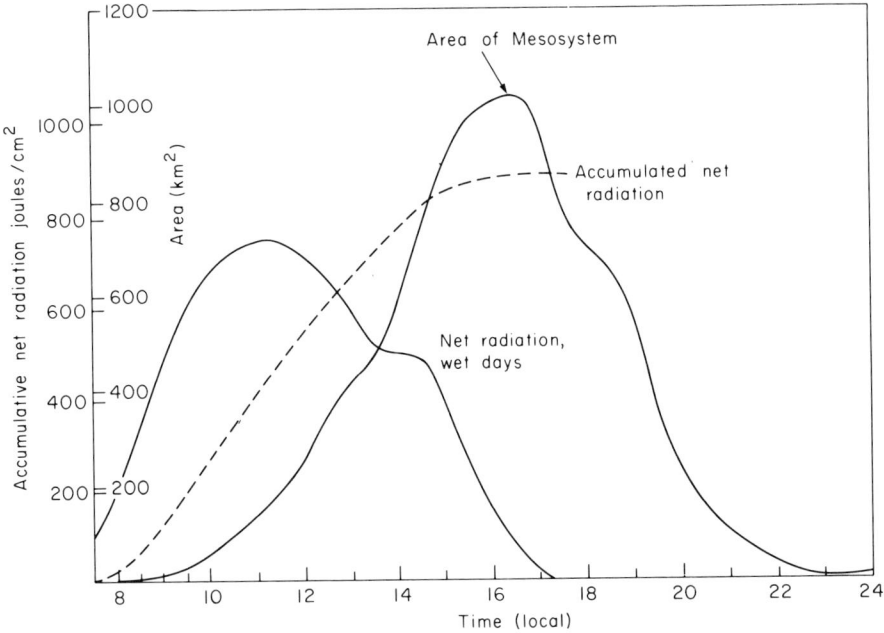

Fig. 6.5. The daily course of the area of radar echoes (mesoscale size only) compared to the net solar radiation absorbed in the soil and to the accumulated net radiation, during an experiment at Anaco, Venezuela (9°N, 64°W) in 1969 (3).

Heating over islands

Islands, large and small alike, will act as an isolated heat source within the surrounding body of ocean, following the scheme of Table 6.1. They become, then, literally "heat islands"! While potential temperature is virtually constant in the subcloud layer over the sea (Fig. 6.6), the inward-moving air in daytime will heat up and increase

258　　　　　　　　　　　　　　CLIMATE AND WEATHER IN THE TROPICS

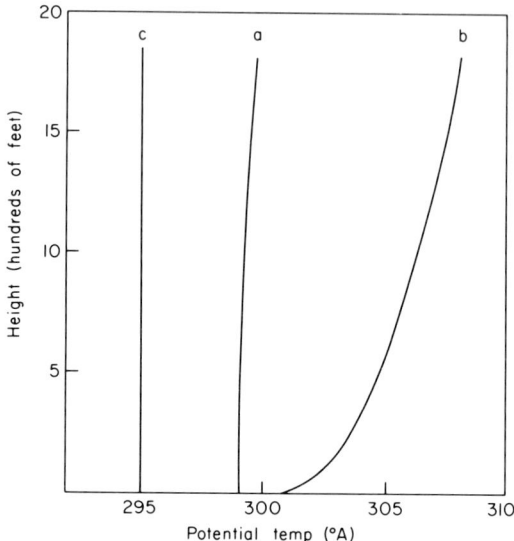

Fig. 6.6. Vertical gradient of potential temperature over Puerto Rico in the rainy season: (a) over sea; (b) on north slope of mountains at time of highest temperature; and (c) on the slope at time of lowest temperature.

its potential temperature along the mountain slopes during travel of one to two hours as shown by curve 'b' of Fig. 6.6. During night, in contrast, downward drainage tends to establish an adiabatic temperature profile at a value lower than that over the ocean.

With heating and little change in specific humidity along the daytime trajectories, the height of the convective condensation level inevitably must also rise so that one observes a dome-shaped array of cloud bases over islands; the mountains, when low, are unlikely to be in cloud until after the onset of rain. Malkus (23) has given beautiful demonstrations of the heated island effect on Puerto Rico (Fig. 6.7).

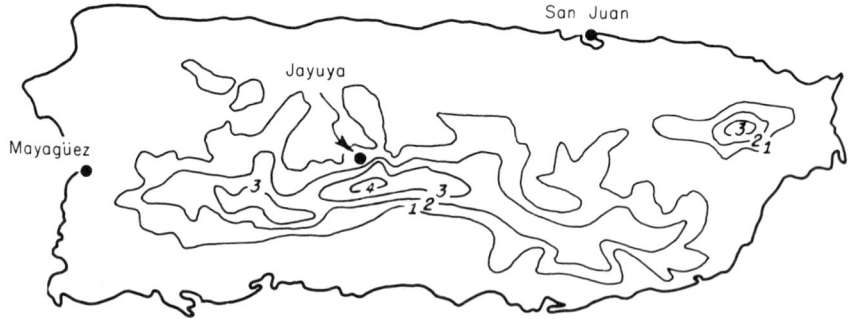

Fig. 6.7. Outline of the island of Puerto Rico, with contours at 300 m intervals.

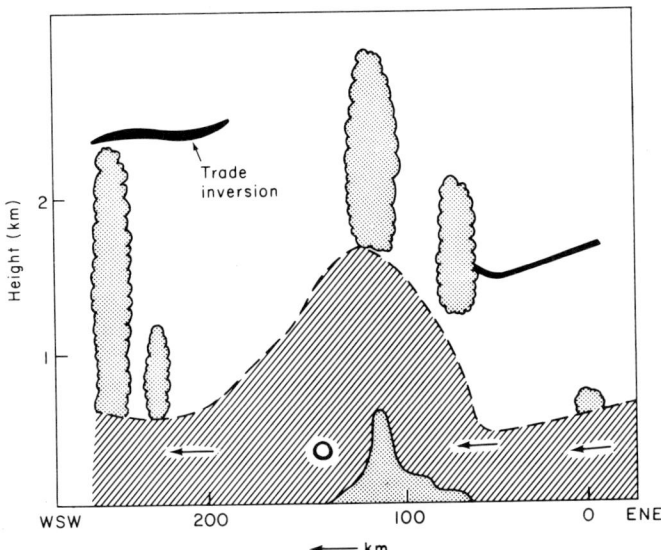

Fig. 6.8. Cross-section of cloud base and of orographic cumulus clouds across Puerto Rico from ENE to WSW (Fig. 6.7) during undisturbed trade wind regime (23). Heavy solid shading denotes island, heavy lines the trade wind inversion interrupted over land, and the arrows indicate wind of about 5 m/s along the ENE trade wind. Observe circle indicating calm just to lee of mountains.

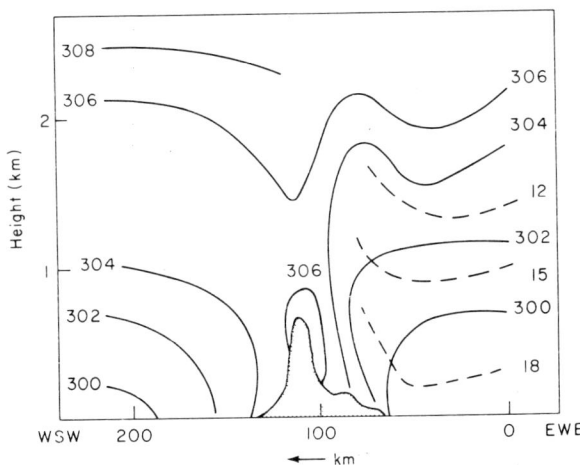

Fig. 6.9. Cross-section of potential temperature (solid) and of specific humidity (dashed) for the section of Fig. 6.8. The island heating effect, the upward transport of moisture and the descending motion off-shore are indicated. Downwind side distribution estimated (23).

The cloud base rises on a traverse across the mountains (Fig. 6.8) and potential temperature becomes distributed as in Fig. 6.9. Under these circumstances there is, of course, thermal instability giving rise to the cloud pattern. The heated island has been likened to an "equivalent mountain" (22). Temperature differences between the free atmosphere and actual mountain ranges increase with the height of the mountains. They are quoted as 4-9°C, comparing the Tibetan plateau to nearby radiosonde-derived climatic means (7). The elevated heat source has a marked effect on the general circulation in the area, supporting advance and strength of the summer monsoon.

Seasonal effects on islands

The combination of radiation differences between sea and air, topography, nature of the underlying surface and relation of stations with respect to the prevailing winds leads to a great variety of diurnal and seasonal temperature variations. We shall consider Puerto Rico again, an island located within the trade regime of the Atlantic. Wind direction on average varies only slightly through the year; the large seasonal changes of the diurnal regime, which characterize the monsoon regimes with their complete wind reversal between seasons, are missing.

The geographic structure of Puerto Rico is simple. The coasts form a rectangle (Fig. 6.7), with the long side oriented east-west. Its length is about 180 km, and its width 70 km. A mountain range averaging 800 m in height, with highest elevation above 1300 m, extends from east to west across the whole length of the island. We shall compare three stations shown on the map (Fig. 6.10): San Juan, situated on the windward northern coast, has good sea breezes during the daytime. At night a weak land breeze reaches the city only in periods of weak trade. Jayuya lies at an altitude of 500 m in a valley of the gently sloping northern foothills. Here we can expect cold-air drainage at night; indeed, the whole valley is often covered with dense radiation fog in the morning. This situation is by no means unusual, for pockets of radiation fog, even fog blankets over jungles, are far more common than once was believed. Mayagüez, finally, is situated on the leeward western shore, where at night drainage from the mountains reinforces the trade so that wind speed is often high and gusty. The sea breeze tends to move inland from the west during the day, but to do so it must overcome the trade. Therefore, the time of arrival of the sea breeze is quite variable; it is often delayed till 2-4 p.m., which permits the temperature to reach a high maximum.

DIURNAL AND LOCAL CONTROLS

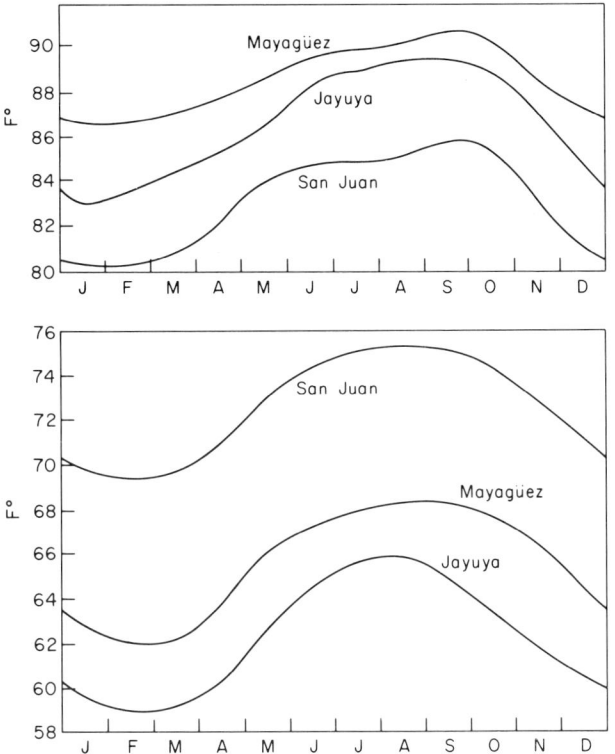

Fig. 6.10. Seasonal variation of mean daily maximum and minimum temperature (°F) at three Puerto Rican stations shown in Fig. 6.7.

This description suggests that we should expect an "oceanic" climate at San Juan and more "continental" climates at Jayuya and Mayagüez, though the latter is located on the coast, as is San Juan. All lee areas have relatively continental conditions, and the present case is no exception. The average diurnal temperature range is 12°C at Mayagüez and Jayuya, and only 5°C at San Juan. Remarkably enough, the range is almost invariant at all three stations throughout the year. As the weather changes from the dry season (January-March) to the wet season in summer, with a large increase in the total moisture content of the air, a reduction of the diurnal temperature range may be expected, but this is not observed. Nor is the range less than over middle-latitude continents where the moisture content is appreciably lower than over Puerto Rico.

Figure 6.10 also shows the annual course of mean daily maximum and minimum temperatures. Perhaps most striking is the parallel course of all lines. Throughout the year, Mayagüez has the highest maximum temperature with a peak in September-October at the height of the rainy season. Jayuya temperatures are a close second, especially in summer, when daytime showers occur almost daily and when we might look for relatively low maximum temperatures. San Juan is considerably cooler, and this is easily explained by the sea breeze. The slight hump in July-August results from particularly strong trade winds in these months.

At night the inverse picture holds. San Juan is warmest, in accord with the local wind circulation. The small diurnal range at San Juan, compared with the other stations, comes both from lower daytime and higher night-time temperatures—an illustration of the predominant marine influence. From the viewpoint of human comfort, Jayuya and Mayagüez are to be preferred. The night-time temperatures, 3-5°C lower than at San Juan, permit better sleeping, and one needs a light blanket toward morning. The daytime temperatures, though higher, are coupled with strong wind and low relative humidity so that the cooling power is much higher than at San Juan.

Table 6.2. Temperature extremes at three Puerto Rican stations

	Maximum	Minimum
San Juan:		
Winter	31	17
Summer	34	21
Jayuya:		
Winter	33	8
Summer	37	14
Mayagüez:		
Winter	33	11
Summer	37	16

Table 6.2 gives the temperature extremes on record. Considering the heat waves which the city dweller in the United States must endure, one might go in summer to Puerto Rico to keep cool. It is highly unlikely that the temperature there will reach 40°C. Even though the mountains are only of moderate height, rather cool temperatures can be found, including 10°C in valleys. Similar temperature records can be found at thousands of stations in the tropics.

Diurnal wind variation

The diurnal wind variation over land and along coasts in the tropics takes a course similar to that of higher latitudes in summer. A close connection exists between gradients of heating and local wind systems. At night, wind speed approaches zero as vertical momentum exchange is curtailed by increased thermal stability, which also permits moisture accumulation near the ground from the slight night-time evaporation. When the stable layer is wiped out by radiation, during the morning hours, wind speed increases and often becomes gusty when air with high momentum is mixed down from upper levels. In rarer situations, when the air aloft is nearly calm, irregular local wind systems develop during daytime.

Land and sea breeze†

The most famous local circulation along coastlines in and out of the tropics is the land and sea breeze wind system, initiated by differential heating and cooling over land and sea. The classical description prevails. Land heating, illustrated in Figs 6.6 and 6.9, will inevitably lead to expansion of air columns and higher pressure over land than over sea above some level. Through this pressure gradient, air is accelerated along from land to sea, raising surface pressure over the water and lowering it over land. The reverse surface pressure gradient, so initiated, provides for the first acceleration of the sea breeze from water to land. The vertical circulation is in the sense that converging warm air ascends over land and cooler air sinks over the water creating a widely observed ring of cloudless air around islands (Fig. 6.11), evident from satellite photography.

The sea breeze, generated by the ascent of warm and descent of cool air, will continue to accelerate until ground friction offers enough resistance to compensate for the pressure gradient, or until the land-sea temperature gradient is reduced by the movement of cooler air inland and by subsidence over the sea. Then the sea breeze becomes steady or goes through several cycles of increase and decrease. When radiation lessens in the afternoon as the sun goes down or as clouds, stimulated by the sea breeze (Fig. 6.8), shield the ground from the sun's rays, the sea breeze weakens and dies. While the beaches often are clear and sunny in the middle of the day, when huge clouds are balled inland, these clouds with attendant precipitation tend to move into the coastal plain near sunset.

†A compilation of 450 references on sea breezes has been published by Jehn (A-8).

At night, the foregoing tends to apply in reverse; the energy for developing the land breeze is given by the temperature difference between curves (a) and (c) in Fig. 6.6. Because of thermal stability the land breeze normally is confined to the lowest levels where ground retardation usually holds speeds to 1-2 m/s compared to 5-10 m/s for the sea breeze.

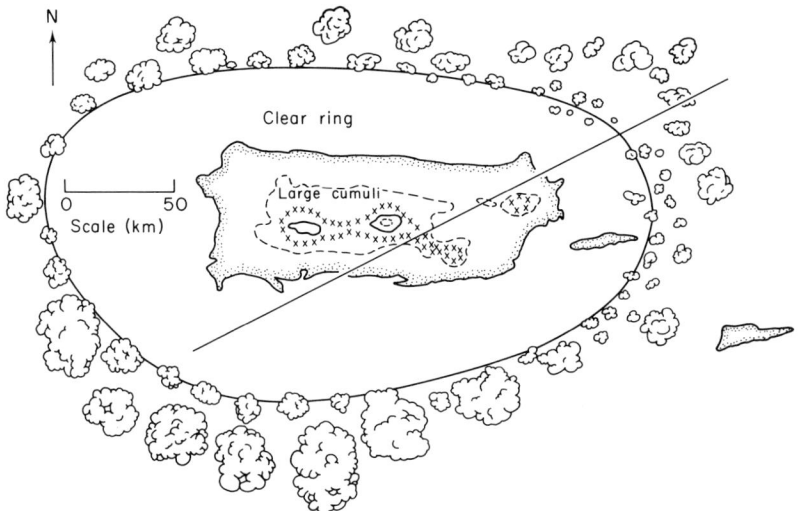

Fig. 6.11. Sketch of clear ring from subsidence around Puerto Rico (23).

Factors in land and sea breeze development

The foregoing describes the principle of a "simple heat engine" upon which, however, a number of other factors are superimposed. A dynamic influence, dependent on latitude, is the Coriolis parameter. In equatorial regions, this parameter is small and goes to zero; the land breeze moves in on a straight path and may have time to penetrate 100 km and more. In middle latitudes, the winds are gradually deflected, clockwise in the Northern Hemisphere, so that after some time the sea breeze blows parallel to the coast. Penetration of the sea breeze is limited to a coastal strip roughly 20 to 50 km wide.

A second factor, noted in connection with the diurnal temperature cycle, is the prevailing wind. At San Juan and vicinity (Fig. 6.7), the sea breeze normally reinforces the trade; one merely notes a gradual strengthening of wind speed during the morning. At Mayagüez, however, the sea breeze arrives from the west and so must overcome the trade. This requires a stronger surface pressure gradient. As the morning progresses, land heating is not interrupted and the cooler air over the sea, trying to enter, piles up in depth, held off-shore by the trade. In the end, the landward-directed surface pressure gradient may become so strong that the sea air begins to accelerate and comes inland as a miniature cold front, with gusty winds and showers or even a squall line at its forward edge. Temperature drops rapidly—in extreme cases on the African coast from 35° to 20°C—while relative humidity may jump from 20 to 90%. When the general wind opposing the sea breeze exceeds 7 m/s, development of a sea breeze is unlikely.

Time of arrival of the sea breeze depends, of course, on the distance from the coast. At Ahmedabad, near the Arabian Sea (23°N, 73°E), the sea breeze comes from the Gulf of Cambay, over 80 km to the south-southwest (32). Although the summer monsoon reinforces the sea breeze from this direction, time of onset is near or even after sunset. Highest frequency is in May, before the onset of monsoon rains, when the strongest, diurnal land-sea temperature contrasts develop.

We have already seen the effect of a relatively large island with mountains in stimulating the sea breeze, but even a small, flat island in the trades can give rise to long cloud streets (Fig. 6.12). Satellites have shown that some of these, for instance from Curaçao, can become 50 to 100 km long and remain quite straight and persistent.

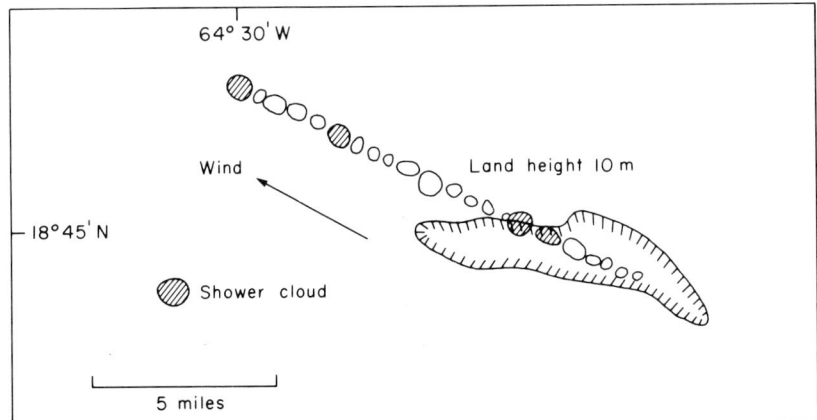

Fig. 6.12. Cloud street forming over the small, low island of Anegada in the Virgin Islands near noon parallel to wind and extending downstream far beyond island. Measured in dry season on 26 March, 1953 (24).

The depth of the sea breeze may be quite variable. Famous is van Bemmelen's (35) diagram on sea breeze structure at Djakarta (Fig. 6.13). The sea breeze reaches its greatest strength there in the afternoon at a height below 300 m. The return current centres at 2000 m, and it is clear that it cannot descend from that height back to sea level within a few hours. Often, the sea breeze is reported to be better developed with clear rather than with cloudy skies, and thus is strongest in the dry season. At Bombay, however, the sea breeze depth is lowest in winter and rises toward the monsoon rainy season (Fig. 6.14) suggesting that the precipitation process and release of latent heat of condensation may well be another strong stimulus in extending the sea breeze depth (4).

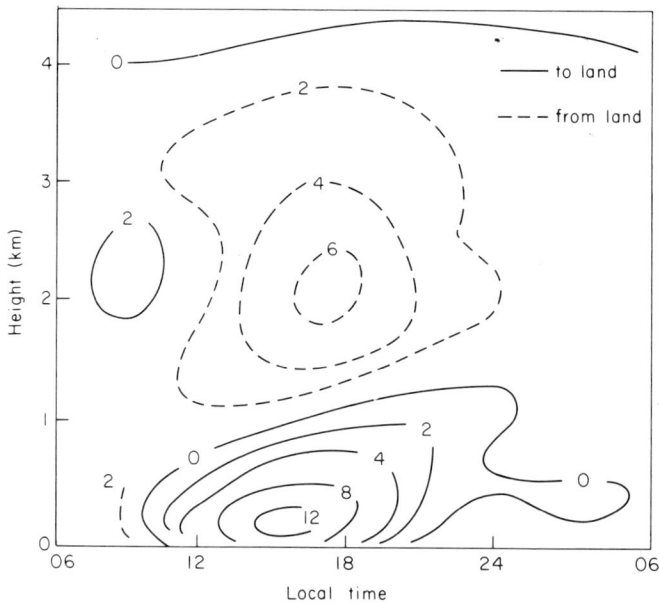

Fig. 6.13. Diurnal variation of wind speed and reversal of wind direction at Djakarta (6°S, 107°E) (35).

Sea breeze depth depends also on vertical stability of the air. Under a strong trade wind inversion, as on the slopes of the main island of Hawaii (Fig. 6.15), the upward moving air may be stopped entirely and forced to go around the great volcanoes (16). Very strong winds, apparently downslope, develop in the saddle between the volcanoes,

DIURNAL AND LOCAL CONTROLS

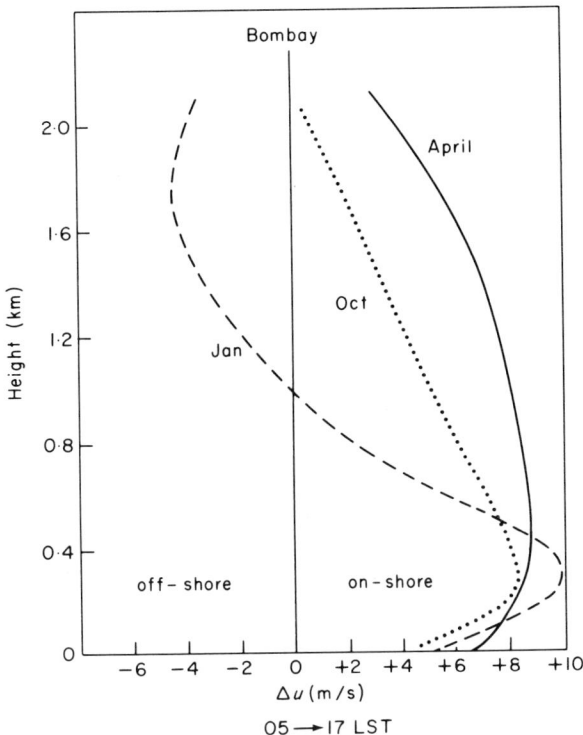

Fig. 6.14. Difference in onshore and offshore wind component from 0500 to 1700 L for three months at Bombay (19°N, 73°E) (4).

suggestive of a Venturi effect. Strong, low-level wind maxima in the early morning hours also have been observed on the island of Barbados in the West Indies (Fig. 6.16) (5), and at the research station, Anaco, in Venezuela in the rainy season of 1969. Another type of cloud development with sea breeze, in this case on the Pacific side of northern Peru, is sketched in Fig. 6.17. In these cases either the predominant easterly flow itself may become unstable, or the surface air enters from the shore as a sea breeze against the general current leaving the oceanic trade wind inversion completely behind. It then acquires much heat on the upward trajectory and eventually stimulates large precipitating cumulonimbi, possible on the Andes slope (17) because the setting is a very grand and high one, with the mountains, and the upward trajectories reaching to 4000 m and beyond along the slopes.

Fig. 6.15. Termination of sea-breeze trade wind clouds along east slope of the volcano Mauna Lea on the island of Hawaii. Courtesy Dieter Henning (16).

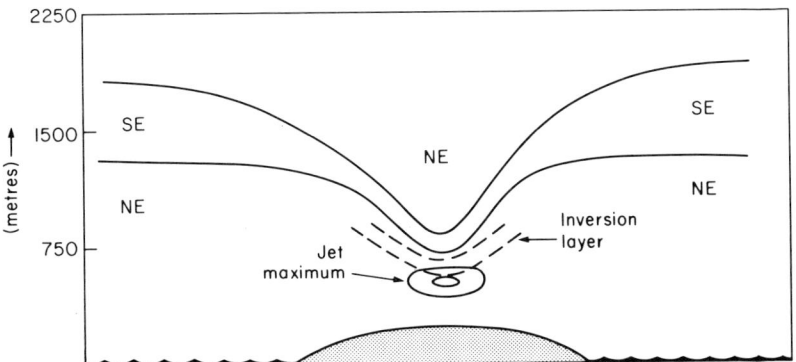

Fig. 6.16. Sketch of drop in height of trade wind inversion during the morning hours over the mountains on northern Barbados (West Indies) and attendant strengthening of wind speed under inversion (5).

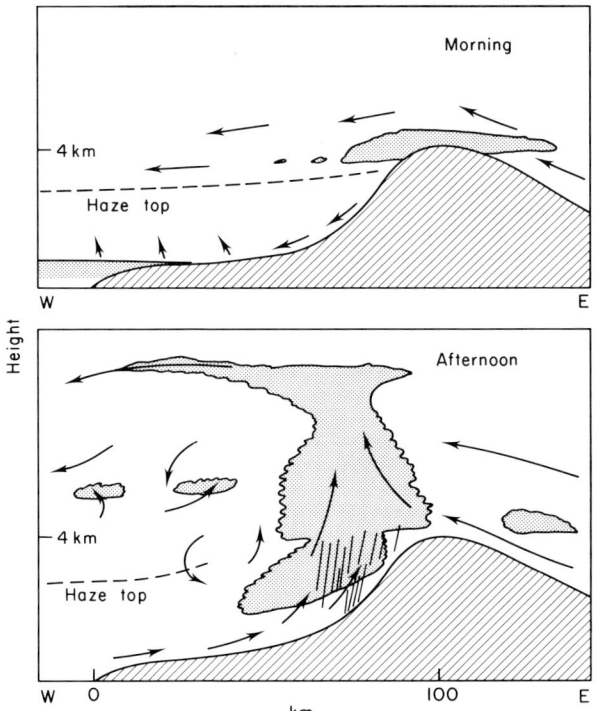

Fig. 6.17. Sketches of diurnal variation after of wind circulation and clouds on the western slope of the Andes near 8°S (after 17). The "haze top" indicates upper limit of off-shore trade wind inversion regime.

Mountain and valley breezes

Many of the foregoing illustrations can be equally termed examples of mountain-valley or land-sea breezes, since these topographic features tend to become coupled near coast lines. Interior mountain and valley circulations show the sea breeze pattern: uphill flow during day and downhill flow during night, but the variety of actual wind systems superimposed on the simple heat engine scheme is very great and the reader is referred to Jehn (A-8) for a listing of the very extensive literature. Much appears to depend on height and slope of the mountains, breadth and extent of the valleys. Only one valley-breeze scheme is shown here (Fig. 6.18), after the well-known late German geographer, Troll (34, cf also 9). On many expeditions through the world he noted that vegetation was very sparse and often desert-like in the middle of broad valleys on the south (rainy) slope of the Himalayas (33), and on the eastern slopes of the Andes. The suggestion is that the up-slope currents induce compensatory downward motion in daytime over the valley, resulting in dissolution of all potential clouds. At night, with reverse circulation, clouds may

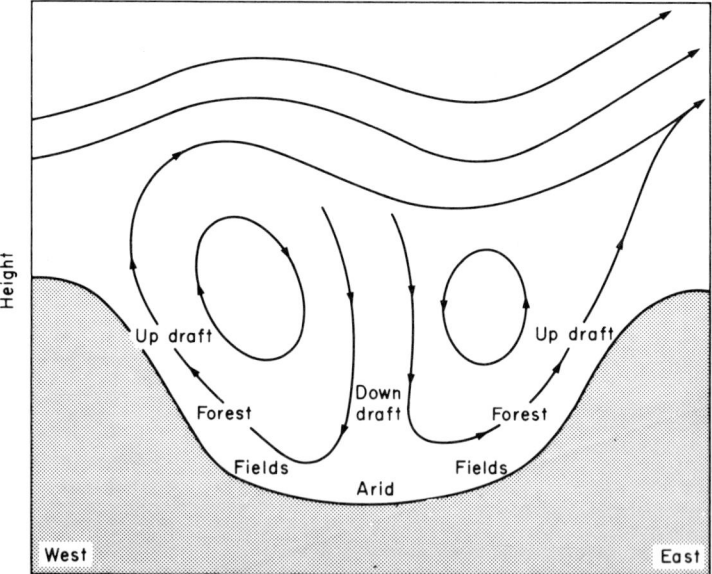

Fig. 6.18. Daytime valley-mountain circulation found in broad Himalayan and South American valleys, adapted from Carl Troll (34).

develop in the centre of the valley, but this happens only rarely due to thermal stability at night. Of course currents perpendicular to our illustration along the valley can, and sometimes do, add convergences and divergences, and in this way the whole precipitation pattern can become very complex (29). It should be pointed out that the pattern of heavy precipitation over mountains coupled with much sunshine over the valleys is an ideal setting for irrigation agriculture; such developments can be found in various mountain chains where agricultural engineers have taken advantage of natural conditions.

Sea breezes on large scales

The diurnal temperature variations over sea are small and rather uniform over large areas. Nevertheless, in the Northeast Pacific trade regime, described in Chapter 5, wind speed increases by 1-2 m/s in the whole subcloud and cloud layers from noon to midnight, July-October. At the same time the trade inversion lowers several hundred metres (21) so that continuity of mass is satisfied with $VD = $ const, where V is layer-mean wind speed and D depth of the layer under the inversion. How does this curious phenomenon come about? As meteorologists from the University of California have pointed out, an exceedingly strong west wind develops along the whole shore of Southern and Central California in the noon hours above the height of the mountains in the summer season, and this strong wind toward the interior deserts persists until far into the night. The extent of this circulation is such that it should be labelled "mini-monsoon" rather than sea breeze. It is entirely conceivable that this monsoon is felt far out over the ocean; then the sinking of the trade inversion would be related to strong wind divergence along the shore of over 1000 km length. Similar suggestions about vast continental-oceanic, diurnal mass exchanges have been made for other oceans.

Diurnal variation of rainfall

Variation over sea

According to a simple classical scheme, diurnal shower activity increases over land in the afternoon (Fig. 6.5) and diminishes in the evening, while clear skies predominate in the late night and morning hours. Over sea, the reverse has been observed by oceanographic expeditions, that is, rainfall increases during the night to attain a maximum at sunrise; it decreases during the daytime hours. The Atlantic weather ships, though at higher latitudes, confirm the same cycle (19). Some statistical analyses of satellite radiation temperatures (Table 6.3) also suggest that the largest build-ups over land and sea follow this scheme (30). However, concentration of rain in the noon and afternoon hours has been found off the coast of Africa along the equatorial trough (15); afternoon cirrus cloudiness increases in hurricanes (1). Thus, a generally valid statement cannot be made even over the oceans although the nocturnal-early morning rain type does appear to predominate according to an extensive summary by Gray (14).

Table 6.3. Morning-evening comparison of satellite radiation temperatures (30) in percentage of 0900 and 2100 LT observations, June-September 1974

Area	Temperature (°C)	Morning	Evening
Venezuela	−30 to −39	38	21
	< −50	10	35
West Africa	−30 to −39	28	14
	< −50	29	55
Arabian Sea	−30 to −39	27	39
	< −50	32	25

Variation over land

On the shore and over land, rainfall types vary very widely and often in a way not yet understood. Certainly, powerful synoptic rain systems will often precipitate at any hour of day or night. The large fraction of precipitation often due to such systems (Chapter 3) will distort diurnal precipitation curves, unless the "undisturbed" days are separated out. One can readily understand how sea breeze convergence during onset can lay down mid-morning maxima on the shore, as observed along the eastern coast of Texas; with inland movement the coast clears up and one sees the large afternoon thunderstorms well inland over cities such as Houston. As already mentioned, convergence tends to return to the shore area in the late afternoon along many coasts, especially where the normal trades blow parallel to the coast, and lines of evening showers then may form around 2100 LT over the beaches and reappear intermittently during the night.

DIURNAL AND LOCAL CONTROLS 273

Beyond this point, however, an enormous variety of diurnal rainfall types is encountered, especially in mountainous areas (7). For instance, during the summer monsoon season in Nepal, precipitation will occur at any hour during day or night and even the morning minimum is weak, though still in evidence at stations such as

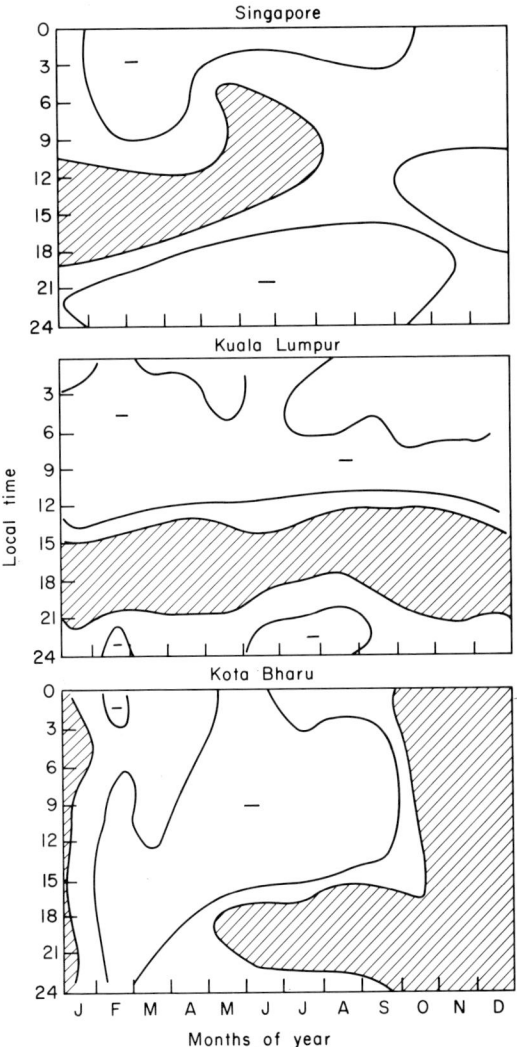

Fig. 6.19. Annual course of diurnal variation of rainfall at three stations in Southeast Asia. Shaded portions of diagrams: median hourly rainfall over 5 mm; area marked with minus sign; less than 1·3 mm (adapted from 25).

Katmandu (28°N, 85°E) (6). In Malaya, relations have been established between rainfall types and coastal configurations, mountain chains and also the season of the year—monsoon or dry (25). Even there, a large variety of "irregularities" remains. As an example, Singapore (see Fig. 3.13 for precipitation) at the southern tip of Malaya (1°N, 104°E) is quite mixed, though a minimum around midnight persists throughout the year (Fig. 6.19). At Kuala Lumpur (3°N, 101·5°E), on the western side of the peninsula, the late afternoon maximum is amazingly regular in dry and rainy seasons. In contrast, Kota Bharu (6°N, 102°E) on the east shore and backed by mountains, has no diurnal peak at all during the main rainy season, October-December, while an early evening maximum appears in most other months with distinctly clear skies from midnight to 1500 LT.

Preferred convection points and long convection trails from such points (Fig. 6.12) compound the confusion of diurnal rainfall types. Over most land areas the ground is at least slightly hilly, and vegetation and soil types vary, so that the absorption of the sun's rays is not uniform. Clouds will tend to form where heating is strongest and where the general air stream is forced upward in mountainous country. We refer to such areas as "convection points". Not infrequently do we see here a stationary cloud persisting for hours even

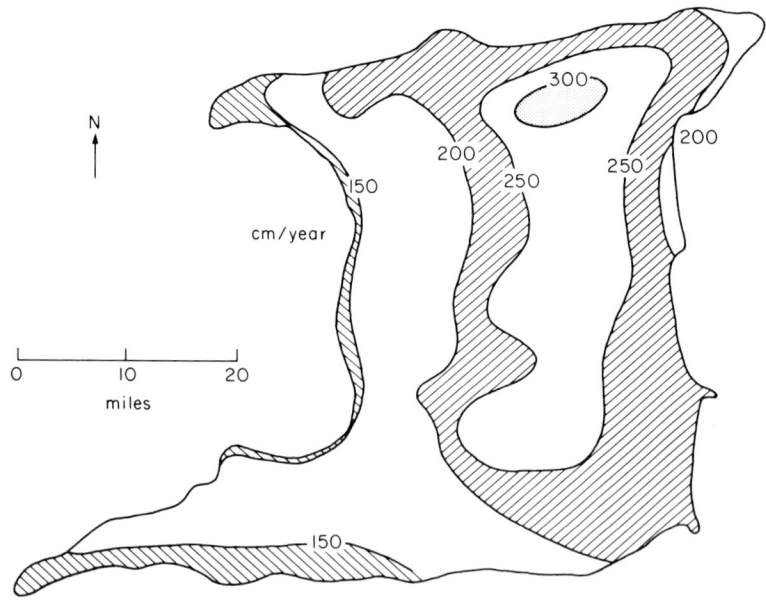

Fig. 6.20. Annual precipitation (cm) on the island of Trinidad (West Indies).

DIURNAL AND LOCAL CONTROLS

though the wind may be blowing at a high rate. Trains of quasi-stationary clouds exending downstream from such convection points may complicate diurnal rainfall curves as far as 100 km distant, as is observed, for instance, at San Juan, Puerto Rico.

For a very well documented case, let us glance at the island of Trinidad in the Lesser Antilles just north of Venezuela. A mountain chain with peaks near 1000 m lies east-west along the northern coast; only smaller hills and much marshland is found on the bulk of the island. Annual precipitation (Fig. 6.20) is largest in the northern mountains (13) and from there a belt of heavy precipitation extends southward, paralleling the east shore over which the trade winds move onto the island. Synoptic weather systems control the annual march of precipitation (Fig. 3.9), but the island obviously exerts a strong influence. It is most curious that the eastern coast has its peak rainfall at night and the west coast during the afternoon, both during rainy and dry seasons (Fig. 6.21). Convergence is enhanced in the east at night, when the land breeze meets the easterly trades. In the

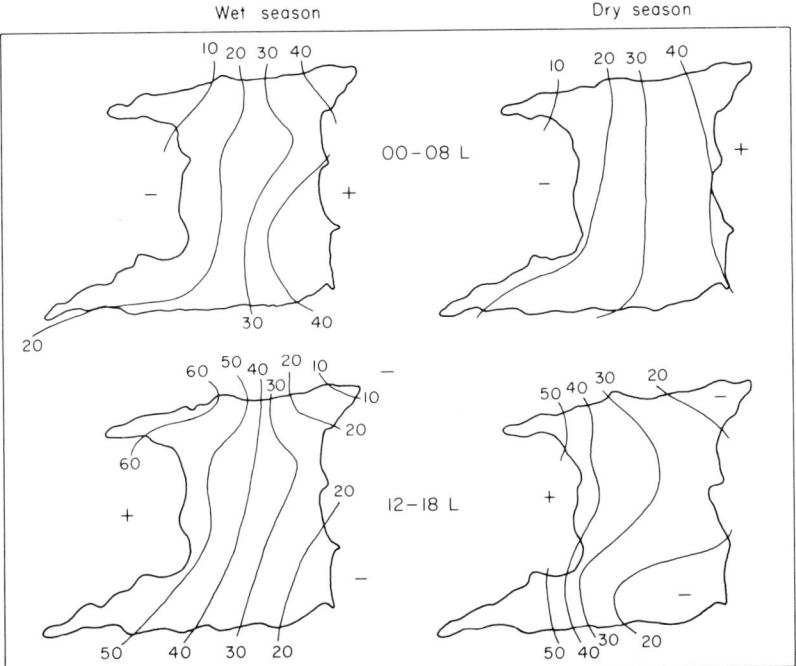

Fig. 6.21. Per cent of precipitation falling on different parts of Trinidad during the indicated hours during dry and wet seasons (adapted from 13).

afternoon, the convergence moves to the western shore when the sea breeze tries to enter the island from that direction. One has the impression of a maritime rain type in the east, giving way to a continental type during the traverse of the island. The north-south orientation of the rain maximum centred in the eastern part of the island of course suggests general trade wind convergence upon striking land, mostly nearly at right angles. On the whole, the Trinidad precipitation, well analysed by Garstang (13), is complex and yet quite understandable.

The control of climate by orography

Although this topic has already arisen several times in connection with Puerto Rico, Trinidad and Malaysia, it is worthwhile to consider it further, since the presence of mountains exercises a most profound influence on total available water and on the whole annual course of precipitation. A striking illustration is furnished by the small island of St Helena (16°S, 6°W), located in the severe southeast trade inversion region of the South Atlantic Ocean (20). The island, not much more than 16 km long, has a central mountain range over 600 m opposing the trade wind flow at right angles (Fig. 6.22). In spite of the trade inversion, the low mountains are able to assert themselves and produce a very strong, local, annual rainfall pattern which has been sketched, as best possible with the few stations, in the illustration. The rain enhancement by the low mountain chain is enormous but by no means uncommon. If the northern coastal station (Jamestown) is representative of the oceanic precipitation with its leeside position, the mountain effect raises island-averaged precipitation by a factor of three.

The published rain data for the one year shown in the illustration had an irregular seasonal course, with a late summer and mid-winter maximum at most stations. It is of interest to read that cumulus and cumulonimbus clouds were seen with greater frequency than may be expected in the strong trade inversion regime. Most precipitation could fall during interruptions of the inversion with passage of weather systems classified as "cold fronts" but which, upon closer inspection, may reveal more tropical features as encountered elsewhere (see Chapter 8 for Hawaii), even though July and August 1955 were exceptionally cold months.

A debate of long-standing concerns the variation of precipitation with height along mountain slopes. At one time the opinion prevailed

DIURNAL AND LOCAL CONTROLS

that precipitation is heaviest near one km height and diminishes upward from that level since the saturation specific humidity, and therewith the maximum water content in a column above the surface, decreases. Undoubtedly, precipitation diminishes for this reason near the summit of very high ranges, but rainfall is a function not only of water content of the air, but also of its rate of ascent. Depending on local conditions, this rate may increase to unknown heights. It is not surprising that, in the course of the years, measurements of runoff have proved that in many high mountain ranges the precipitation far exceeds early estimates. The "1000 year flood" has occurred not just once, but twice, shortly after the completion of certain reservoirs and spillways. Explorers climbing the high Himalayas have encountered severe snow-storms, and many expeditions have perished because of them.

Nevertheless, Flohn (9) maintains that the foregoing applies only to extratropical regions and that in the tropics the highest precipitation occurs in the altitude range 1000 to 1400 m, according to his

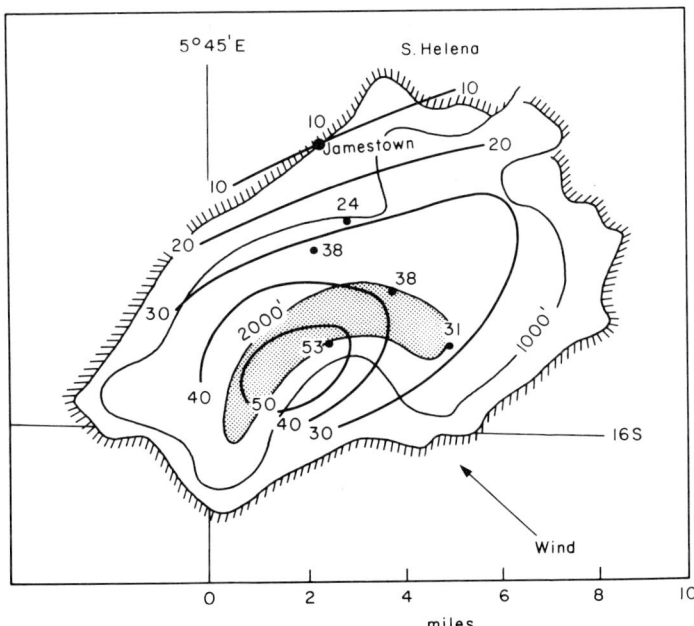

Fig. 6.22. Sketch of island of St. Helena (16°S, 6°W) and annual precipitation during 1954-1955 (20). Analysis sketched (inches).

references which are world-wide. He suggests that the difference is due to the fact that wind speed, therewith upward motion and moisture transport on the windward side of mountains, increases to the height of the mountains in the westerlies, whereas in the tropics highest speeds are found in the low troposphere. This is often true in the trades and also the equatorial westerlies. Flohn's example from Kilimanjaro, the highest mountain of Africa, is impressive (Fig. 6.23). However, precipitation on Ruwenzori (0°, 30°E) is much larger from upsurges of moist air from the Congo basin. A large glacial system is found on the upper 500 metres of the mountain, maintained in part through protection from solar radiation by persistent heavy cloud cover. The decrease of precipitation with height starts there only at 3000 m (36).

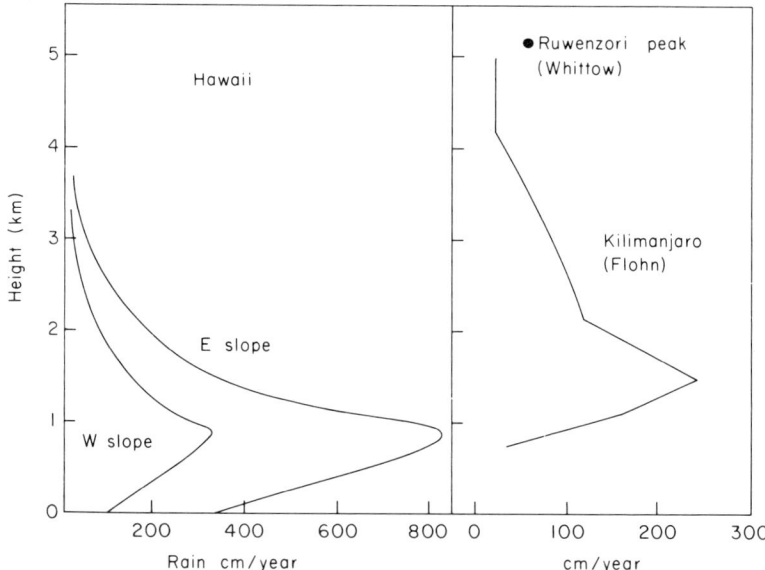

Fig. 6.23. Variation of annual precipitation with height (cm) on east and west slope of the main island of Hawaii (left) and on Kilimanjaro (right) (9). Also annual precipitation estimated at the peak of the Ruwenzori range (0°, 30°E) (36).

Flohn's diagram is matched by the rain profiles on the slopes of the main island of Hawaii helped in large measure by the trade wind inversion. The eastern slope faces the trade winds and thereby has much heavier precipitation than the west in a shallow layer below the inversion, one of the world's largest annual values. The decrease of precipitation downward below cloud base is sometimes ascribed to

DIURNAL AND LOCAL CONTROLS

evaporation of falling drops. This interpretation is very realistic, especially in arid regions where cloud base is high and relative humidity low underneath. Evaporation into downdrafts of deep convective systems as discussed at the end of Chapter 4 would act in the same direction. In the extreme situations on the slopes of Hawaii, of course, these mechanisms do not apply; the increase of rainfall from sea level to inversion base results from increased convergence, upslope motion and, possibly, "horizontal precipitation", i.e. masses of clouds with small particle sizes driven by the trade winds into the jungles on the slopes.

Like St Helena, precipitation over the Hawaiian Islands is estimated to exceed that over the surrounding ocean by a factor possibly as high as three through the extreme mountain effects. Except for the islands with the high volcanoes, mountain height does not exceed trade inversion base height. There, precipitation is raised to extreme values over large portions of the island interiors, notably on

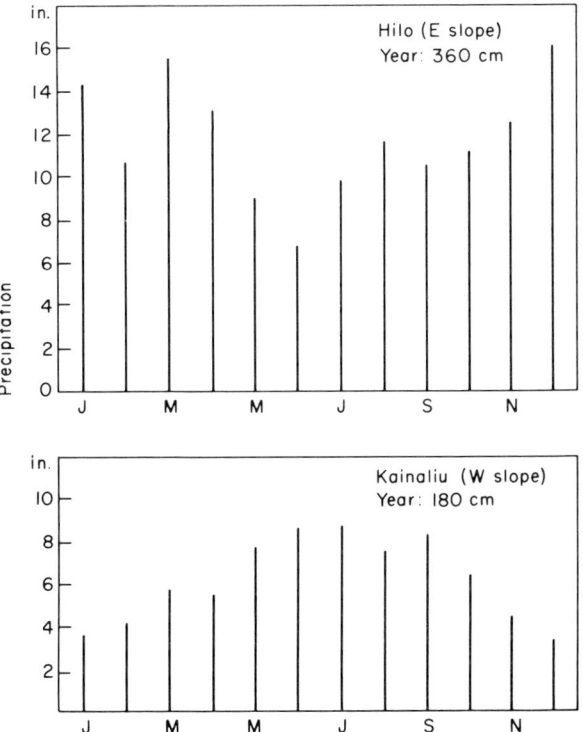

Fig. 6.24. Seasonal course of rainfall at Hilo and Kainaliu on east and west sides of the main island of Hawaii.

Kauai where the world's highest annual precipitation is reported (Chapter 3), slightly more even than on the east slope of the main island, not far from 1000 cm/year.

The islands all exhibit the windward-leeward difference in annual precipitation by factors ranging from two to four. Most spectacular is the climate divide of the main island seen in Fig. 6.23. There, an actual inversion of the annual course of precipitation from the average winter maximum (Fig. 3.12) is encountered. While Hilo (20°N, 155°W) follows the average march of precipitation, Kainaliu, 100 km to the west on the Kona coast, has a distinct summer maximum (Fig. 6.24). The reverse sea breeze against the trade winds toward the pass—a picture similar to that of the Andes (Fig. 6.17)—is thought to be responsible for the summer rain peak. If so, the Venturi effect between the high volcanoes, somewhat similar to that over the hills of Barbados (Fig. 6.16), would assume great importance climatologically.

Inversion of annual rainfall regimes across mountain chains are quite common in the tropics, for instance along the northern coast of Venezuela where a finger of the Andes extends eastward. The main rainy season occurs with southeast trades, which are downslope along the north coast from the mountains; clouds tend to end abruptly near the mountain ridge. Cold-water upwelling along the coast contributes to the drying effect so that the shore area, and also islands such as Curaçao, are nearly deserts. In winter, during surges of the trade from the subtropical anticyclone of the Atlantic Ocean, on-shore northeast winds can become very strong (over 25 m/s) and then the north slope of the mountain range receives its precipitation when the south is dry or has only overhanging clouds. Very complex local rainfall regimes are found in the Philippines and Indonesia.

Possible increases of precipitation through reservoir construction

We have now seen, amply, that orography and coasts of large water bodies profoundly influence the daily rhythm and also the annual course of precipitation, superimposed on the annual course of synoptic weather systems. Islands increase the precipitation over that of the surrounding ocean by sizeable amounts—factors of two to three, possibly larger, through heating and mountain effects. We now touch on a point of importance in the modern world, namely, whether large inland water bodies tend to raise precipitation around them as often claimed. Many engineering projects for creating large inland lakes have been proposed to alleviate deserts, notably in the Sahara.

At first sight, the evidence is all negative. The world rainfall chart (Fig. 3.4) shows that the vast trade wind regimes of the globe have very low precipitation due to sinking of the general circulation, and may be termed "oceanic extensions of the continental deserts". For higher latitudes it has been amply shown that evaporation from large water bodies such as the Great Lakes is near zero during the rain periods, coupled with warm fronts on the forward side of cyclones. The evaporation largely takes place into the dry air to the rear of the cyclones; this water vapour, in the subsiding cold air, may be swept thousands of kilometres before it enters a zone of convergence far away.

One may also question the occurrence of very large evaporation from lakes with a roundish shape, rather like Lake Victoria, where surface air will move over the lake for a long time, at least some hours. Normal evaporation from computations in Table 6.1 will have the order of perhaps 0·6 cm/day or about 2 m/year. Sizeable enhancement is possible only when much warmer air moves over the water—a difference of 10 to 15°C between water and air temperature is by no means uncommon. Then, through the advection of such warmer air, sensible heat is transferred from air to water, the net heat source of the water is increased, and the evaporation at first sight can rise almost without limit—very poor for the function of a reservoir as a storage basin. However, the sensible heat transfer very quickly induces vertical stability in the air, a low temperature inversion can form and the temperature of the air approaches the temperature of a lake. Turbulence and evaporation then die out. Due to formation of the inversion over tropical lakes the enhancement of evaporation may be quite small (12), a vital factor for reservoir construction. The potential effect on local precipitation similarly would be small and, unless other mechanisms enter the picture, little is gained that cannot be achieved by irrigation.

An elongated, but large reservoir, such as Lake Nasser along the Nile, with reservoir length of 300 km will gain little protection from the effect described, unless wind direction follows the river exactly. In the case of the Nile this happens to be partly, but not sufficiently, true. Annual evaporation has been closely estimated at 280 cm/year (26); the climate around the lake is said to have changed very little.

A somewhat different outcome is reported from Lake Kariba, which was formed in 1959-1960 on the Zambesi River between latitudes 16-17°S and longitudes 27 and 29°E. The length of the lake is given as

256 km, its average width is 32 km, not the best possible shape from what has just been said, but no doubt the only possible one. The area of 8000 km^2 is about 15% that of Lake Victoria. Now Hutchinson (18) has carried out what has always been suggested for occasions of new reservoir construction; he measured the climate—as best as possible—before and after, and determined potential climate changes due to the new reservoir. After a very careful and circumspect analysis of the necessarily fragmentary climatic material and possible natural climate changes, Hutchinson did reach the conclusion that precipitation increased sizeably on the shore and adjacent land with percentages ranging from very little to nearly 50%. He noted that the water vapour picked up by the air crossing the lake cannot provide for the increased rainfall. Consequently, he proposed a *dynamical mechanism* for the increased rain; the formation of a *lake breeze circulation* interacting with nearby mountains, giving rise to convergence and therewith to the formation of cumulonimbi on the shore in the absence of marked stable layers or dry inversion in the air. Through this means, water vapour advected from distant regions is brought to condensation and precipitation in the vicinity of the reservoir.

Hutchinson herewith offers a plausible mechanism, and his analysis may be considered a distinct advance capable of wide utilization. In principle, of course, the mechanisms of land and sea breezes have been well known, but net enhancement of precipitation—not from the local evaporation but from the *dynamical effect*—has not been proved in earlier days. One wonders what stations on the new reservoir would show, even though this reservoir is rather long and narrow. The following has been noted about night precipitation (A-18):

> In summer, a lake the size of Lake Michigan or Lake Victoria in East Africa will give rise to outbreaks of night-time thunderstorms over the lake itself. At night, air tends to flow from colder land to the warmer water and converge there, the opposite of the sea breeze. With heating from the lake water the decrease of temperature with height remains large over the lake during the night, facilitating overturning of the air and formation of cumulonimbus clouds. Of course, lake evaporation is not involved here, but the lake does possess a mechanism for generating local storms which, once started, may move off over land and persist for several hours.

Some of these remarks were prompted by the studies of Flohn and Fraedrich (8, 10, 11) on Lake Victoria (area 5×10^4 km^2) where annual precipitation is highest over the lake itself with displacement of the highest precipitation (over 225 cm/year) toward the western side (Fig. 6.25). From surface winds and pilot balloon observations, divergence and net flow of air toward or from the lake was measured

DIURNAL AND LOCAL CONTROLS

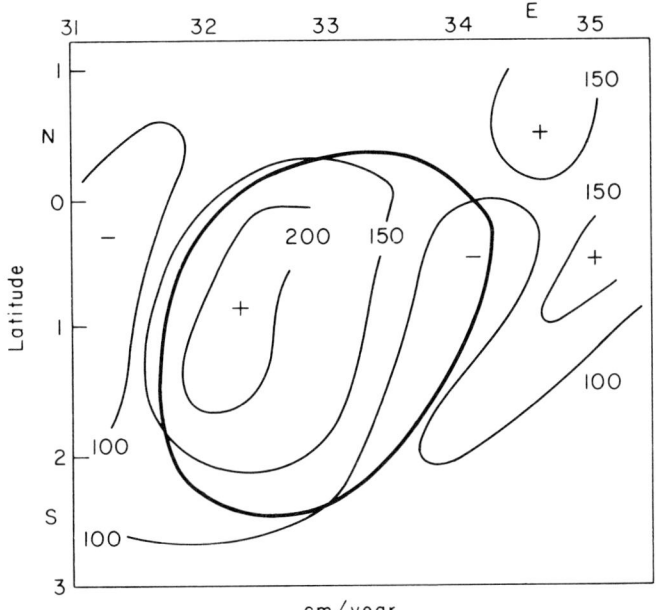

Fig. 6.25. Annual precipitation (cm) over and around Lake Victoria (8).

for two seasons shortly after sunrise and in the early afternoon (Fig. 6.26). Clearly, the integrated flow is toward the lake in the early morning and away from it in the afternoon, confirming the typical land-sea breeze reversal. Divergence is $-2 \times 10^{-5} s^{-1}$ at 0800 LT and over $+3 \times 10^{-5} s^{-1}$ at 1400 LT at the surface. While the varying thermal

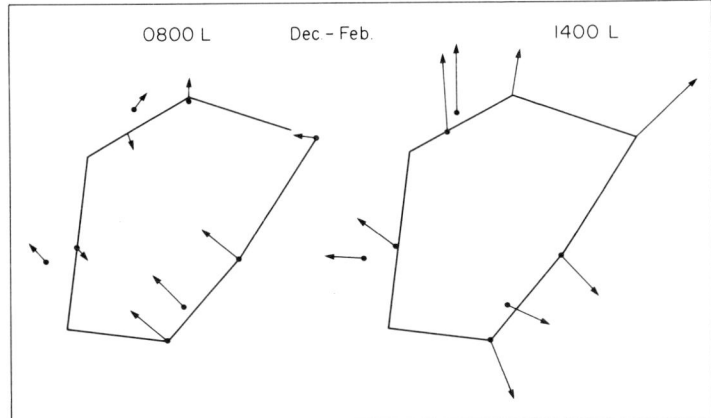

Fig. 6.26. Surface wind directions and speeds measured around Lake Victoria at 0800 and 1400 LT December-February (10).

stratification and moisture content of the atmosphere have their effect, we may suppose that the rainfall maximum over the lake represents an enhancement due to the lake breeze mechanism and not just water drawn from suppression of precipitation over the surrounding land. If this is true, the "reservoir effect" of the large lake is considerable, with a factor averaging about 1·25 from Fig. 6.25, similar to that found by Hutchinson.

Thus, a real possibility may exist for mitigation of desert or at least steppe climates through the introduction of large reservoirs, not for the evaporation of their water but for enhancement of area precipitation through generating organized upward current from the introduction of thermal discontinuities at the surface. The main limit of course is the vertical structure of the atmosphere. In the presence of strong and dry temperature inversions, similar to the trade inversion and coupled with actual subsidence, little can be achieved. However, there is plenty of territory where conditions are moderately favourable to favourable, as the Zambezi has shown.

Air pollution

The principal source of air pollution in the tropics are the extensive, and often violent, dust and sand storms that originate in the deserts, mainly in connection with remnants of extratropical cold fronts moving rapidly and with strong winds equatorward (Harmattan). However, more limited dust storms can originate at any time of year. As seen in Chapter 1, the dust from the Sahara carries all the way across the Atlantic ocean in the northern summer. Satellite photographs show the dust trails clearly. Many measurements were made during GATE through aircraft intercept, mainly with the intention of observing the dust's effect on radiation. At times the dust moves on a counter-clockwise trajectory and so arrives at Dakar (15°N, 17°W) from the northwest, at first a surprising experience.

Man-made pollution for the largest part originates from burning in the broad agricultural band of Africa north of the equator. Again, these veritable dust clouds are shown clearly by satellite pictures covering very large areas. The agricultural burning may be related to the marked retreat of glaciers on the high volcanoes in East Africa through increased ash deposit and lowering of the ice albedo. Otherwise, the tropics experience the typical pollution from city sources, though not to the extent of cities in higher latitudes where atmospheric stability may be very high for prolonged periods.

However, where cities have become large, air pollution has become visible and attains the distinct early morning maximum so well known from higher latitudes. In Caracas, the whole of this large city can be seen submerged completely in smog (called "calina") on mornings of the dry season from hotels located on high ground; only the suburbs on the higher hills remain visible. The smog usually disappears between 9 and 11 a.m. with vertical instability produced by the sun's warming.

Land and sea breezes can produce strange results. At San Juan (Fig. 6.7), the gases from the electric power plant and particulates from burning used to drift out to sea at night with the land breeze, only to return with the northeast trade(!) in the morning hours to the shore residences and hotels. This situation has reportedly now been remedied.

Reduction of tropical forests, especially the jungles with their huge trees, appears to have a serious impact on world climate. Carbon dioxide has been observed to rise gradually over the last 100 years by at least 10%. Since the 1960s, as is evident from measurements made at the high observatory on the peak of Mauna Loa on the Island of Hawaii, far removed from any nearby pollution sources, the rate of increase has accelerated, a development generally credited to fossil fuel burning. However, the reduction of tropical forests may play as great or an even greater role (37). The large trees have been found to retain a very high quantity of carbon dioxide, which is liberated to the atmosphere when the trees are felled and then decay or are burned. The plants on agricultural land replacing the forests retain very little carbon dioxide. Now, next to water vapour, carbon dioxide is the most important atmospheric gas absorbing long wave radiation. A continuing increase in the amount of this gas in the atmosphere (easy to achieve because its absolute magnitude is minuscule) will tend to raise world temperatures with unforeseen and potentially detrimental effects on climate everywhere. Conservation of the tropical jungles appears to be an obvious necessity for mankind.

Fog

By and large the tropics are thought to be free of fog, and it is indeed true that fog problems there are minor compared to higher latitudes. Yet radiation fog does occur and has been responsible for some severe aircraft accidents. When surface relative humidity is high, radiation fog should be able to form readily on quiet nights if water vapour is unusually low in the atmospheric column. Normally, a deep moist column prevents the fall of surface temperatures through strong back radiation. However, we have seen examples of very dry air at low levels (Fig. 5.13) with trade wind inversion. Radiation fog is reported from the high humidity and swampy areas around the mouth of the Orinoco River on the South American coast.

Fog responsible for some of the aircraft accidents has been reported from India and Pakistan. Dense fogs in mountain valleys during morning occurs widely, and cold air drainage reinforces the tendency toward for formation at all latitudes. Advection fogs are most prevalent along the west coasts of the continents where the equatorward-moving trade wind air causes upwelling of cold water along the coast and, hence, the largest negative anomalies of surface temperature found in the tropics (Fig. 2.13). Here, fog is indeed the way of human and plant life. Cities such as Lima, Peru, must cope with fog problems perennially.

Semi-diurnal pressure variation at the surface

Brief mention may be made of the semi-diurnal pressure variation, which is a regular and monotonous feature of barograph traces in low latitudes throughout the year. At times, this variation has been linked to the diurnal course of clouds and rain, especially the predominant suppressed convection of the forenoon when the main semi-diurnal high pressure occurs. Since the divergence associated with the semi-diurnal wave only has the magnitude of $10^{-7}s^{-1}$, it is difficult to accept this hypothesis. There is, however, a distinct difference in the composition of the wave over sea and land. Over the oceans the variation is symmetrical and the amplitude is generally found to be one millibar, still observed at San Juan, Puerto Rico (Fig. 6.27). Over land, however, the wave is produced by a thermal diurnal in addition to the semi-diurnal component. The afternoon thermal and semi-diurnal minima of pressure coincide, resulting in a large afternoon amplitude of the pressure variation. The late-night minimum is correspondingly damped. Due to its thermal character, the diurnal part of the pressure wave varies strongly on clear, compared to cloudy days. On average, an afternoon amplitude of 2·5 mb may be taken as typical; but it rises to as much as 3·5 mb (i.e. pressure falls 7 mb between 1000 and 1600 LT) on days with suppressed convection and highest temperatures in the range 33 to 35°C. On cloudy days the diurnal curve may be completely upset, especially the afternoon minimum when a heavy

DIURNAL AND LOCAL CONTROLS

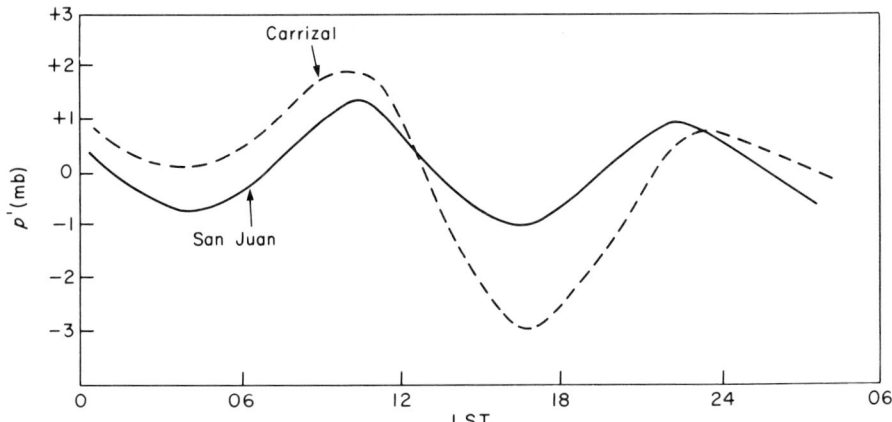

Fig. 6.27. Mean semi-diurnal pressure variation at San Juan, Puerto Rico (18·5°N, 66°W) and at Carrizal, Venezuela (9·5°N, 67°W).

rain shower with its pressure rise of 1-2 mb arrives about the time of maximum radar echo development (Fig. 6.5). Since these pressure rises tend to maintain themselves after rainstorm passage, an element of uncontrollable turbulence is thrown into 24-hour pressure change charts, discussed in the next chapter.

References

(1) Browner, S. P., Woodley, W. L. and Griffith, C. G. (1977). Diurnal oscillation of the area of cloudiness associated with tropical storms. *Mon. Wea. Rev.* **105**: 856-864.
(2) Cox, S. K. and Hastenrath, S. L. (1970). Radiation measurements over the equatorial central Pacific. *Mon. Wea. Rev.* **98**: 823-832.
(3) Cruz, L. A. Reference 8, Chapter 3.
(4) Dekate, M. V. (1968). Climatological study of sea and land breezes over Bombay. *Indian J. Meteor. Geoph.* **19**: 421-426.
(5) DeSouza, R. L., Aspliden, C. I., Garstang, M. and LaSeur, N. E. (1971). A low-level jet in the tropics. *Mon. Wea. Rev.* **99**: 559-562.
(6) Dhar, O. N. (1960). The diurnal variation of rainfall at Barahkshetra and Kathmandu during monsoon months. *Indian J. Meteor. Geoph.* **11**: 153-156.
(7) Flohn, H. (1958). Beiträge zur Klimakunde von Hochasien. *Erdkunde* **12**: 294-308.
(8) Flohn, H. and Fraedrich, K. (1966). Tagesperiodische Zirkulation und Niederschlagsverteilung am Viktoria-See. *Meteor. Rundsch.* **19**: 157-165.
(9) Flohn, H. (1971). Beiträge zur vergleichenden Meteorologie der Hochbebirge. *Ann. Meteor.* **5**: 9-16.
(10) Fraedrich, K. (1968). Das Land-und Seewindsystem des Viktoria-Sees nach aerologischen Daten. *Arch. Geoph. Biokl.* Ser. A. **17**: 186-206.

(11) Fraedrich, K. (1971). Modell einer lokalen atmosphärischen Zirkulation mit Anwendung auf den Viktoria-See. *Beitr. Phys. Atmos.* **44:** 95-114.
(12) Fraedrich, K. (1972). On the evaporation from a lake in warm and dry environment. *Tellus* **24:** 116-121.
(13) Garstang, M. (1959). A study of the rainfall distribution of Trinidad, West Indies. Woods Hole Oceanogr. Inst., Woods Hole, Mass. Contribution No. 185.
(14) Gray, W. M. and Jacobson, R. W. (1977). Diurnal variation of deep cumulus convection. *Mon. Wea. Rev.* **105:** 1171-1188.
(15) Gruber, A. (1976). An estimate of the daily variation of cloudiness over the GATE A/B area. *Mon. Wea. Rev.* **104:** 1036-1039.
(16) Henning, I. and Henning, D. (1967). Abbildung lokaler Zirkulationen durch Wolkenfelder auf Hawaii. *Meteor. Rundsch.* **20:** 109-114.
(17) Howell, W. E. (1952). Local weather of the Chicama Valley (Peru). *Arch. Meteor. Geoph. Biokl.* Ser. B. **5:** 41-53.
(18) Hutchinson, P. (1973). Increase in rainfall due to Lake Kariba. *Weather* **28:** 499-504.
(19) Kraus, E. B. (1963). The diurnal precipitation change over sea. *J. Atmos. Sci.* **20:** 551-556.
(20) Lamb, H. H. (1957). Some special features of the climate of St. Helena and the trade wind zone in the South Atlantic. *Meteor. Mag.* **86:** 73-76.
(21) Leopold, L. B. (1948). Diurnal weather patterns on Oahu and Lanai, Hawaii. *Pacif. Sci.* **2:** 81-95.
(22) Malkus, J. S. and Stern, M. E. (1953). The flow of a stable atmosphere over a heated island. *J. Meteor.* **10:** 30-41.
(23) Malkus, J. S. (1955). The effects of a large island upon the trade wind air stream. *Quart. J. Roy. Meteor. Soc.* **81:** 528-550.
(24) Malkus, J. S. (1963). Tropical rain induced by a small natural heat source. *J. Appl. Meteor.* **2:** 547-556.
(25) Nieuwolt, S. (1968). Diurnal rainfall variation in Malaya. *Meteor. Rundsch.* **6:** 157-165.
(26) Omar, M. H. and El-Bakry, M. M. (1970). Estimation of evaporation from Lake Nasser. Meteor. Dept. Cairo Research Bull. 2, no. 1. 27 pp.
(27) Quinn, W. H. and Burt, W. V. (1967). Weather and solar radiation reception in the equatorial trough. *J. Appl. Meteor.* **6:** 988-993.
(28) Renné, D. (1970). Surface-air energy exchange over Eastern Venezuela as related to Streamflow and Cumulonimbus Systems. Paper no. 166. Dept. Atmos. Sci., Colorado State Univ., Fort Collins. 32 pp.
(29) Riehl, H. and Byers, H. R. Ref. 32, Chapter 3.
(30) Riehl, H. and Miller, A. L. Ref. 72, Chapter 4.
(31) Riehl, H. (1972). Surface temperature and humidity observations in the tropics. *World Meteor. Organis. Bull.* **21:** 227-229.
(32) Sajnani, P. P. (1956). Sea-breeze at Ahmedabad. *Indian J. Meteor. Geoph.* **7:** 49-54.
(33) Schweinfurth, U. (1956). Über klimatische Trockentäler im Himalaya. *Erdkunde* **10:** 297-302.
(34) Troll, C. (1952). Die Lokalwinde der Tropengebirge und ihr Einfluss auf Niederschlag und Vegetation. *Bonn. Geogr. Abh.* **9:** 124-182.
(35) Van Bemmelen, W. (1922). *Beitr. Phys. Frei. Atmos.* **10:** 169.
(36) Whittow, J. B. (1960). Some observations on the snowfall of Ruwenzori. *J. Glaciol.* **3:** 765-772.
(37) Woodwell, G. M. (1978). The carbon dioxide question. *Scient. Amer.* **238:** 34-44.

7

Weather Observations and Analyses

The operations of any weather service are geared to the demands that it must meet. In former days a forecaster in the tropics was required only to give hurricane warnings and in some countries, notably India, to make seasonal rainfall predictions. Many forecast offices were closed entirely during the dry season.

All this has changed with the advent of air routes across the equator, development of hydroelectric power, advances in agronomy, offshore oil drilling and many problems relating to drought on short and long time scales. Weather prediction of rain, wind squalls and ocean waves is in demand. Forecasts are wanted, ranging from a few hours ahead to seasons and much longer periods, and forecasting schedules of weather bureaux and other advisory services extend over the entire year. A successful forecaster's education must transcend the meteorological core knowledge. In order to service special interest groups adequately, such as farmers, water resources engineers and many others, he must become intimately acquainted with the end-users' needs and demands, as has been clearly and forcefully explained (6).

Observations

No part of tropical meteorology has undergone as marked and rapid a development as observation and analysis techniques. The reason is the evolution of radar, satellite and high-speed computer. Forecasters have always been schooled to work with networks of observations, i.e. data far beyond their local horizon. To the "conventional data", satellite photographs, radiation charts, radar photographs and other advanced products have been added, as well as machine-produced analyses and forecasts. Reliance on distant centres of weather analysis and forecasting makes highly reliable communications networks paramount. Only the ground-based radar appears destined to remain a locally controlled tool, albeit with range of up to 300 km. Facilities for interrogating satellites can also be acquired locally. Maintenance of small computers for translating large-area predictions into local weather is advisable. In the preceding chapter we have stressed the enormous variety of local weather patterns that can easily overshadow large-scale weather changes, or modify these in an untold number of ways.

Surface observations

Once the foundation and mainstay of all weather services, the importance of surface observations in the tropics has experienced a continuing eclipse, so much so that the number of stations has been reduced in some instances, and there is also a more general threat mainly due to increasing satellite information. Elimination of the networks would be a gross exaggeration. While the role of surface observations for daily forecasting may have greatly diminished and while the content of synoptic reports may be in need of drastic change, the records of such quantities as temperature and humidity are essential for monitoring climate and the climate trends considered so vital. Further, surface temperatures still do serve an important function occasionally, as was exemplified by the disastrous fall of temperatures in Brazil in the autumn of 1975 (11), when much of the whole coffee crop was lost for years with worldwide economic implications. It is the large deviations from the mean state that should receive much greater recognition than is usually accorded to them in climatic description. Average conditions auspicious for human endeavour matter little if the catastrophes recur too frequently and destroy or nullify what man's effort has built.

Pressure

Although the mercurial barometer is perhaps still the most accurate meteorological instrument for general purposes, and although surface pressure should be the main underpinning of all weather analyses, sea-level isobars with rare exceptions (hurricanes) have proved to be a practically useless tool in the tropics. The best use of pressures has been in the construction of 24-hour pressure change charts. In mountain terrain there are well-known dynamic effects: the pressure drop across a mountainous island can amount to as much as three millibars and in channels between large islands one finds pressure reductions of one millibar or more, but even over large, rather flat and low-lying areas, such as most of northern South America, little is gained by drawing sea-level isobars. Station heights have been found to be in error by 10-20 m, which affects the sea-level pressure field far more than an error of several degrees Celsius in temperature for a station height of 100-200 m; the error should only be of the order of 1/10-2/10 millibars.

In an attempt to arrive at a method for reliable quantitative utilization of surface pressure, an experiment was made with the 21 stations of the Venezuelan Meteorological Service in the rainy season of 1972 (18). In the first place, only station pressures were used. At each station, the semidiurnal pressure tendency was next found and subtracted from each observed pressure value. The semidiurnal variation can be a function of season, and this can be allowed for in modern high-speed machine calculations. Charts of deviation so corrected gave, in general, reasonable patterns, but often all computed deviations were either negative or positive for the whole country, suggesting that a very large-scale pressure wave was present which had little or no relation to actual weather. Such waves have been widely observed and they are thought to be single waves with wavelength of the earth's circumference travelling around the globe once in about five days, i.e. at a speed near 93 m/s. The wave has an amplitude of about one millibar on average, sometimes two; it travels zonally and has no tilt with latitude. Burpee's analysis (2) contains a literature summary, Rudloff (19) suggests ionospheric origin.

The average pressure deviation for a large country like Venezuela may result from this fast travelling wave alone, in which case the average deviation should go through a five-day cycle. However, there may also be longer-term deviations persisting for one or more months, which must be related to real climatic anomalies. In Venezuela, for instance, pressure remained below the long-term averages throughout the 1972 season, during a prolonged drought. This long-term tendency should be tabulated and followed separately in any country just because of its potential significance. For the production of a pressure chart that clearly depicts the synoptic pattern, the entire average deviation must be subtracted from each station value. Figures 7.1 and 7.2 show examples. On 29 August, 1972, pressure was below average over the whole interior of Venezuela except on the north coast.

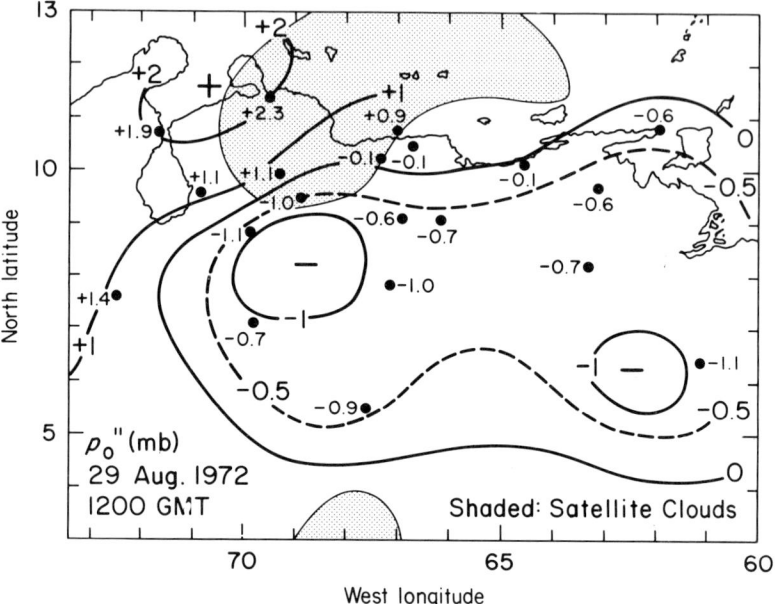

Fig. 7.1. Surface pressure anomaly chart (mb) for Venezuela, 29 August, 1972, 1200 GMT. Shaded areas have satellite cloud cover (18).

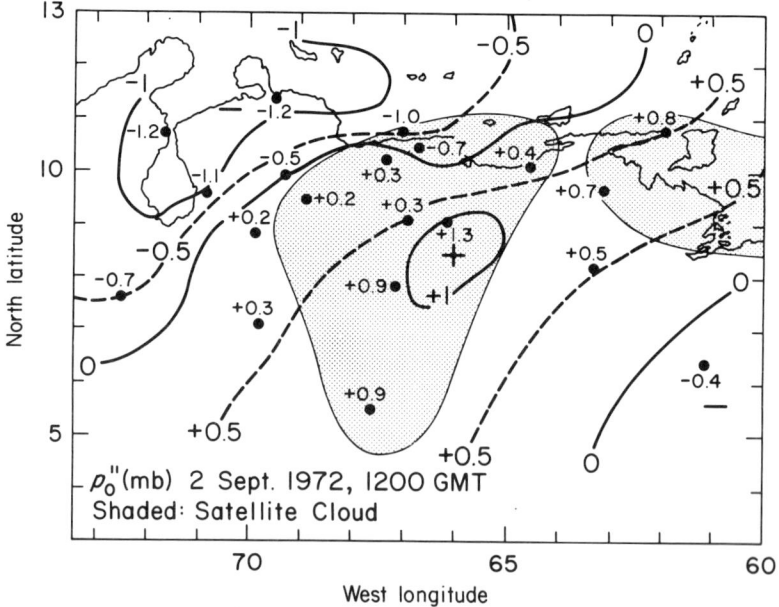

Fig. 7.2. Same for 2 September, 1972, 1200 GMT (18).

Cloudiness and rain were suppressed throughout the low pressure area; a satellite cloud only appeared with the positive pressure anomalies on the north coast. In contrast, on 2 September, a large rain area had entered Venzuela from the south with pressure rises and positive pressure anomalies. The correlations here shown will not be the same everywhere in the tropics, but local forecasters can form a quantitative picture for their own area through the use of such anomaly or anomaly-change charts. These can also be used as bottom boundary charts for the construction of upper-level contour charts, provided the mean monthly 1000 mb height field is known adequately.

Temperature and dewpoint

A troublesome problem arises during the rainy seasons with instrumentation placed in unventilated or poorly ventilated thermometer shelters. These shelters absorb solar radiation; the resulting heat storage is used to evaporate water during and after rain showers (15). An energy source, which is entirely non-existent for the atmosphere at large, is thus made available within the instrument shelters, leading to unreasonably high moisture values and falsifying the vapour pressure used, for instance, for computation of evaporation. Satisfactory results can be obtained by placing an aspirated drybulb and wetbulb hygrometer on a large, well-ventilated, covered verandah. It has been suggested that well-ventilated, drybulb and wetbulb instruments should be installed everywhere in the humid tropics on covered spaces similar to the verandah instead of instrument shelters. It may be added that instrumentation for moisture determination continues to be in need of international standardization. In some areas, discontinuities in dewpoint or specific humidity outline national boundaries quite well. A sudden change of just 1°C in dewpoint, which corresponds to a specific humidity jump of 1 g/kg at 1000 mb, introduces considerable analysis difficulties.

Another problem arises for temperature measured in shelters. At high wind speeds, notably hurricanes, the air at a thermometer is subject to dynamic compression and a correction must be applied, as it is to temperature measured in aircraft. Some aircraft thermometers appear to be constructed so that they permit full dynamic retardation of the air without friction. For that case the first law of thermodynamics may be integrated. Assuming constant density

$$c_p T - \frac{p}{\varrho} = \text{constant} \tag{7.1}$$

Also the Bernoulli theorem applies:

$$\frac{C^2}{2} + \frac{p}{\varrho} = \text{constant}. \qquad (7.2)$$

where C is here the aircraft true air speed. Adding,

$$c_p T + \frac{C^2}{2} = \text{constant}. \qquad (7.3)$$

The constant is evaluated on the outside where $T = T_{\text{air}}$ and the air has the full relative velocity C with respect to the aircraft. Bringing the air to rest at the thermometer,

$$c_p T + 0 = c_p T_{\text{air}} + \frac{C^2}{2}, \text{ or } c_p (T\text{-}T_{\text{air}}) = \frac{C^2}{2}. \qquad (7.4)$$

The difference $(C^2/2c_p)$ to be subtracted from T as correction, is negligible for very slow aircraft; it becomes 0·45°C at $C = 30$ m/s, 1·25° at 50 m/s and 5·0° for a 100 m/s aircraft. The relation can be drawn as a straight line on log-log paper (Fig. 7.3). When the instrument does not permit perfect expansion without a boundary layer, some energy is undoubtedly given off by friction. The correction then will be smaller and is usually determined by flying the aircraft past an instrumented tower for calibration purposes.

Similarly, for air encroaching on a perfect thermometer shelter,

$$c_p (T\text{-}T_{\text{air}}) = \frac{V_{\text{air}}^2}{2}, \qquad (7.5)$$

where V_{air} now is the ground speed of the air (wind) on the outside. The same corrections apply. Unfortunately, a thermometer shelter no doubt is never so engineered as to permit perfect expansion, so an unknown amount of energy is lost in transit of the air through the lattices. At hurricane speeds the thermometer will read only somewhat higher than true air temperature. At velocities of over 50 m/s where this is likely to matter, thermometer shelters usually become inoperative, so the course of air temperature through the region of hurricane winds is best determined from aircraft observations. A second correction in the thermometer shelter would be one due to the measuring instrument itself, also unknown for a thermograph, the only instrument that might keep going through an extreme wind period.

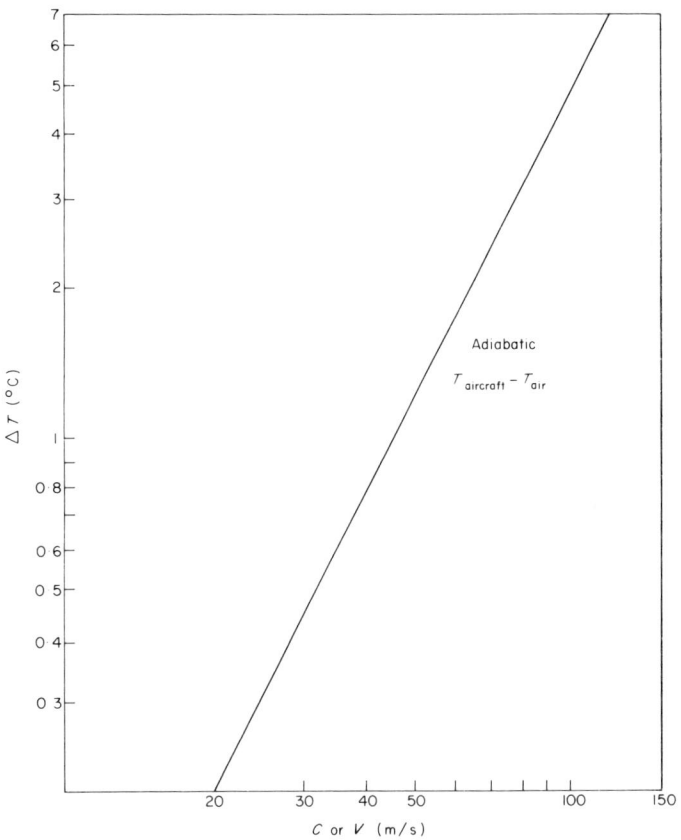

Fig. 7.3. Adiabatic correction for aircraft-measured temperature at true air speed C.

Wind

Surface winds over land are generally unrepresentative except for their immediate environment; over sea, they are generally very useful for synoptic analysis. Nevertheless, surface wind at land stations remains important for local weather, especially land and sea breezes and all other local circulations. Further, anemometer wind speed is required for evaporation measurements and computations of dry and wet cooling power.

Clouds, hydrometers and precipitation

Deficiencies of the international cloud atlas and introduction of a new code designed by Holle (5) with regard to the tropics has been discussed in Chapter 4 illustrated with Figs 4.9 and 4.10. Satellite cloud photography and temperature information have become an obviously indispensable tool. As noted in Chapter 4, hail is rare in the tropics, but it does occur in certain seasons and areas. Thus, reports of hail must be well maintained and, of course, those of thunderstorms, fog, dust storms and other weather variants contained in the international WW tables.

Determination of precipitation by radar is likely to be superior to single station rainfall, which often depends on whether or not a particular heavy rain shower passes over or bypasses the rain gauge. Radar precipitation can be computed for areas of varying size and, thus, will achieve what otherwise can be accomplished only by a local network of rain gauges (Fig. 3.21). However, recording rain gauges will retain full value since they indicate precipitation rates and give information on questions such as what percentage of precipitation from a particular rain system falls in what percentage of the time—information used variously in this text, and also of immediate importance for local flood forecasting and protection. Rain intensities are needed for many engineering purposes, from determining the size of culverts under roads and city drainage systems to the construction of dams and spillways.

Visibility

Because of dust storms, air pollution and fog, observations of visibility remain a requirement for airport observations.

Physical parameters

As shown at the beginning of Chapter 4, observation of physical parameters, in this instance radiative exchange between ground and air as well as latent and sensible heat transfer, dominate understanding of local climates and their potential manipulation to an ever increasing extent. Hence, it becomes important that these factors be transmitted in synoptic reports of first-order stations. This means not only that transmitted observations must be correct (for instance, the dewpoints discussed above), but also that the relevant observations be taken. For radiation, a net radiometer appears to suffice for most purposes. All the details contained in Table 4.1 are not needed for synoptic purposes.

Much study has gone into the problem of evaporation computations. No method of direct measurement, apart from lysimeters, has proved satisfactory, though new instrumental development, of course, is always possible. As explained in speciality texts, evaporation can be computed with a heat balance method such as employed in Table 4.1. Usually, evaporation is determined by modelling turbulent transports. For the method of Thornthwaite and Holzman (20) a mast is needed, perhaps the anemometer mast, for observation of wind speed and vapour pressure at two levels; vapour pressure is derived from wetbulb measurements. This installation is not complicated or expensive. Hourly values may be used as a good first approximation and temperature should be added for stability correction. The heat balance method is more difficult, since it requires knowledge of the ratio of sensible to latent heat exchange. Various empirical approximations have been worked out. For instance, Penman (12) combined heat and turbulence methods, and his formula has since been refined (A-3) for use over water surfaces. It would seem that a single formula generally satisfactory for land evaporation (not potential evaporation) should be worked out, since evaporation enters strongly for many applied purposes. Lacking such a formula for computation from mean data, it would become necessary to consider expensive installations that measure $\varrho_v' w'$ directly.

Upper-air observations

Wind measurements

The mainstay of tropical weather analysis in the upper air have been wind measurements. Except in very light wind situations they have generally been proved reliable; they give good continuity in time at one station and make a sensible, cohesive picture when analysed in network form, given sufficient density of observations such as one wind every 500 km. Near vortex centres, sharp shearlines and waves a higher wind density is of course desired for clear definition and computation of derivatives, especially vorticity and divergence.

The principal tools of observation have been rawinsondes made with direction-finding equipment at the ground. They give the wind distribution of the whole troposphere and sometimes are continued to high levels of the stratosphere. Under the impact of World Weather Watch, the rawinsonde network has been extended to almost all corners of the world where a ground station can be established, and, in addition, there are some stationary weather ships, mainly outside the tropics. The old-fashioned pilot balloon observation remains useful for determining low-level winds; their cost is low, so that dense networks can be maintained. Such networks should be amplified with low-cloud motion vectors from nephoscopes in areas with intense weather. Such observations were regarded as highly useful in the early days of synoptic weather analysis. They have come into disuse, but they are missed in areas where the need for low-level wind data is greatest.

An enormous boon to wind analysis are winds derived from cloud displacement, as seen by geostationary satellites every half hour (10). The assumption must be made that wind and cloud motion are identical, which is probably largely true for middle and upper clouds. For cumuli, the direction has usually been found reliable, especially in the trades with nearly two-dimensional motion. On account of the relative motions introduced by cumulus rotation in shearing wind fields (Fig. 4.14), satellite-derived speeds have been found to be as much as 30% out. Computation is usually confined to low and high clouds, i.e. one obtains only two levels with high wind density (Fig. 7.4). The advantage of such intense global wind coverage is quite overwhelming and leads to rapid advances in understanding weather processes and interaction between widely distant regions, and in analysis as a basis for forecasting purposes. With the advent of Doppler radar and inertial navigation, enormous quantities of reliable winds are also furnished by commercial airlines crossing the world, usually between 300 and 200 mb. Thus, the modern high-tropospheric chart often carries more observations than surface charts and far more data than intermediate levels. Figure 7.5 exemplifies wind coverage in the layer 300-200 mb, a highly dramatic change indeed from the old days!

Upper winds given by the path of a rising balloon can theoretically be measured without local observing equipment from triangulation with the aid of very low frequency radio impulses sent out by stations that may be thousands of kilometres apart. With this technique, the

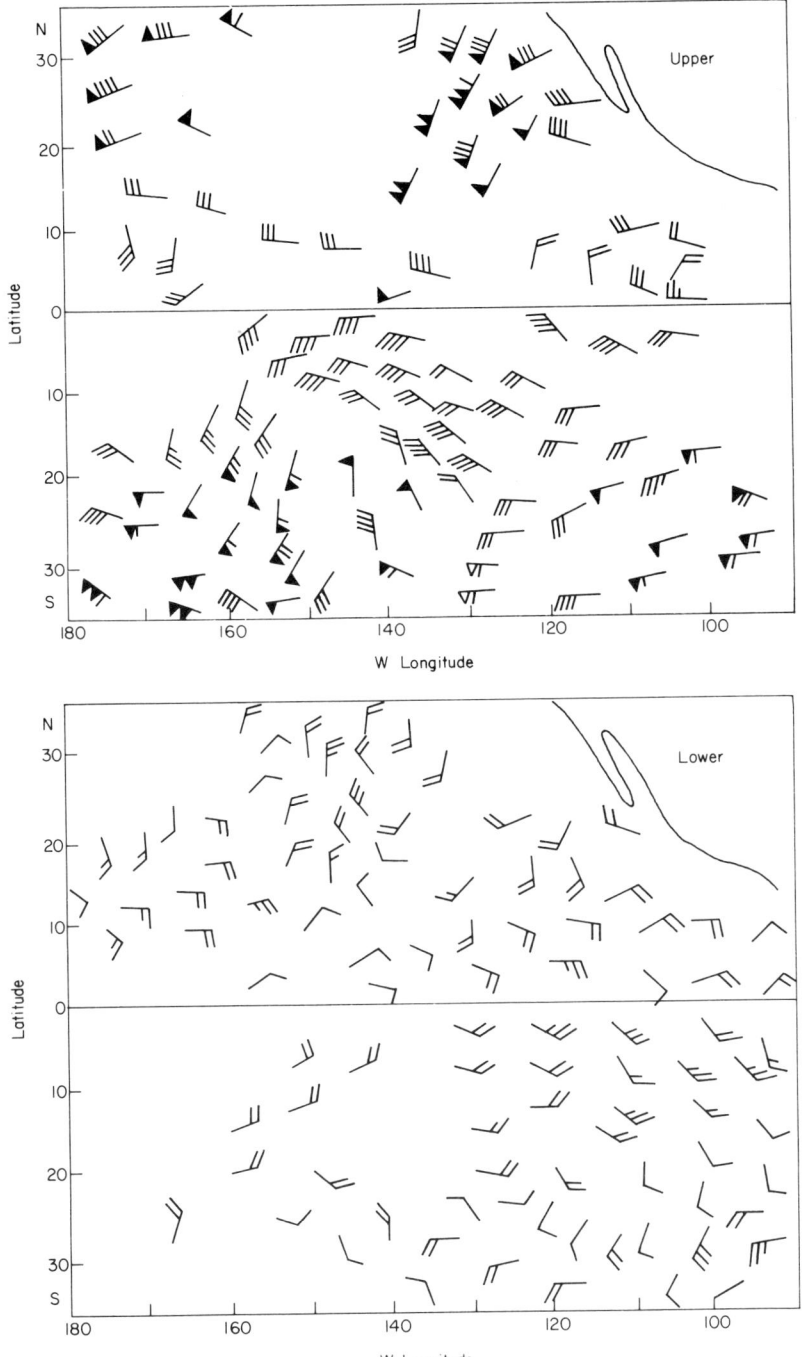

Fig. 7.4. Upper: Satellite-generated winds in eastern Pacific, 25 December, 1977—layer 300-200 mb;
Lower: near 850 mb. Courtesy National Satellite Service, Washington, D.C.

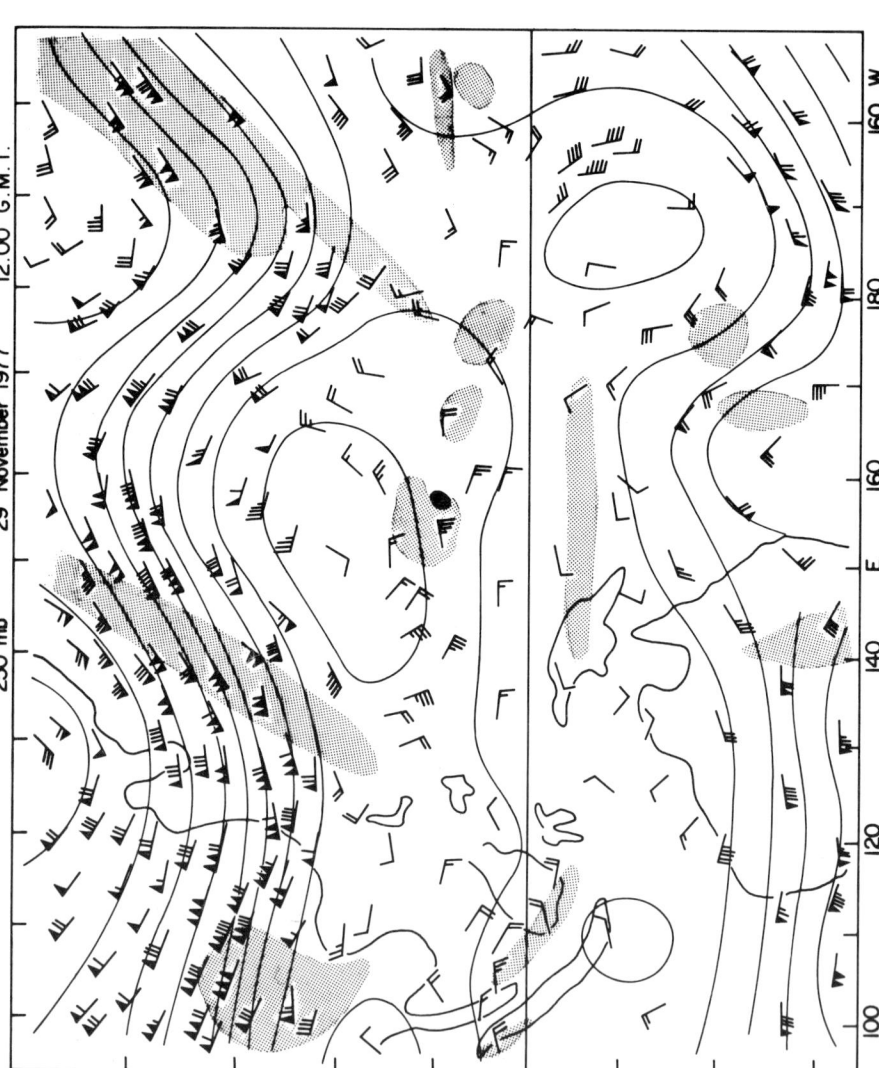

Fig. 7.5. 250 mb chart (winds layer 300–200 mb) in the Pacific, 29 November, 1977, 1200 GMT. Winds from balloon soundings, aircraft measurements and satellite computations. Contours of the 250 mb surface outside the tropics, connected by stream function between latitudes 20°N and S. Shaded: satellite cloud areas. Dot: tropical storm. Courtesy National Weather Service, Honolulu.

whole atmosphere may eventually be filled with winds from balloons sent from the ground or ship, or dropped from high-flying aircraft. Constant pressure balloons encircling the globe at various latitudes (see Fig. 1.22) have proved a highly effective research tool, but they are not considered practical for operation because of potential danger of collision with aircraft if present in great numbers. Chaff released from high-flying rockets and followed by ground radar is a very effective technique for determining winds in the range of 30-70 km and perhaps higher.

Temperature, humidity and heights of Isobaric surfaces

In contrast to the high quality of wind observations, those of temperature and humidity, plus the derived height of isobaric surfaces, have proved disappointing. It is just these quantities, especially that of humidity, which in the tropics are of paramount importance to the understanding and predicting of weather and climate. Yet the data have been so discouraging that meteorologists with good reason have thrown almost all of their weight on wind analysis and many doubt the value of maintaining the radiosonde network. Theoreticians have been at work to derive heights of isobaric surfaces, and thence virtual temperatures, from the wind fields, a remarkable inversion from older days in higher latitudes when the upper pressure or height fields were reasonably well-established and winds, later seen to be rather good ones, were computed routinely with the geostrophic approximation.

Various types of radiosonde instrumentation used around the globe differ from one another by 1°C or more in temperature; this is also the order of range of accuracy for instruments of the same type. From Eq. (5.20) the error in thickness variation between isobaric surfaces, therefore, will be 0·33 to 0·4% depending on the layer to be computed. For 1000-200 mb, $D = 12$ km, $\hat{T}_v = 250$ K so that, for a one degree error, $\delta D = 48$ m. This will be the error of the 200 mb height for integration upward. For reverse integration, the error in surface pressure will be 4·5 mb, given a ratio of 9 m/mb. These errors of 200 mb height or surface pressure often equal the total difference across large portions of the tropics; in Chapter 12 the reason for the small gradients is explored. Even on individual synoptic charts for an area with rather uniform instrumentation and dense sounding network, such as the Caribbean, it is rather hopeless to draw contours without the aid of winds implying quasi-geostrophic modelling. Therewith, of course, the value of independent analysis is lost.

In spite of these strong statements, the reader will see in the next chapter that highly important facts about the structure of weather systems can be elicited from careful analysis of the time sequence of soundings at one station. However, we are then dealing with peak events occurring in about one per cent of the atmosphere at any one time. When a rare sounding inside an active rain area appears isolated in an open network, it is apt to be regarded in error! Small wonder that demand has arisen for "parameterization" of convective events (Chapter 4). In addition to the temperature problems cited, there is still a residual problem, in spite of all engineering efforts, concerning the diurnal temperature variation in the upper air and absorption of solar energy by the radiosonde instruments, or an effect of the radiation from the balloon surface on the instruments. Different results are obtained with different sounding equipment.

Efforts to deduce the temperature structure from satellite radiation measurements in several channels (21) have about the same range of variability as radiosondes, rather good for extratropical regions; improvements, of course, may be expected for the tropics.

Instrumental problems are even worse when the highly important humidity variable is considered, even though that problem has been largely solved for aircraft with the dewpoint hygrometer and other devices. As an example, weather analysts became aware during the 1960s that humidity given by the United States radiosonde always dropped sharply after sunrise and increased again just as quickly after sunset! This remarkable "fact" was subjected to investigation. It turned out that poor ventilation of the thermometer housing and radiation errors in sunlight were responsible (8). A new sonde, introduced in 1972, is free from this error (16), though some residual problems have been noted (1). This is a boon for current analysis, but the error of many years remains ingrained in climatological records, causing no end of problems related to statistical approaches. Considerable difficulty was experienced as a result of this in selecting examples of mean soundings for Chapter 2; fortunately Jordan (7) remained suspicious of the daytime sondes and computed his mean atmosphere from nocturnal ascents alone.

In conclusion, a large question mark remains about the value of maintaining the radiosonde network in the tropics. Low-level sondes have been developed, good to about 500 mb, meant to supplement the temperature structure determined from very high levels downward by satellites, noted above. The low-level sondes retain their value as inexpensive instruments to measure features such as ground and trade wind inversions, and very dry layers as described in Chapter 5. These features are often omitted in general analyses and they are not reported in monthly tabulations (9). Nevertheless, their real regional and local impacts, as well as their influence on the general circulation, does not diminish thereby.

One might advance the suggestion that the World Weather Watch network should take one sonde daily during night hours, even though the release times will no longer be synoptic; the semi-diurnal surface pressure variation must be subtracted. When the global network is used to produce long-term forecasts, such as five days, the irregularity in release time appears minor compared to the errors avoided in mixing day and night soundings and overlooking the semi-diurnal pressure wave. The recommended restriction, of course, does not apply to wind sondes which can be taken as often as desired and at standard times.

Methods of analysis

A distinction must be made between analysis tools for large-area and long-period prognostic charts, and those for the often all-important local forecast.

Large-scale analysis

Charts such as Fig. 7.5 cover very large areas, and they may be extended over the whole globe. They are prepared for constant pressure surfaces and also for deep-layer flow, such as 1000 to 100 mb. Poleward of about latitudes 20°, height values of isobaric surfaces are generated at a regular grid spacing from the observations through numerical analysis techniques (4, 3) using a mixture of pressure heights and winds. Between latitudes 20°N and S, grid point values are generated from the winds alone with weighting factors; then a *streamfunction field* is computed (Fig. 7.5). For the streamfunction the divergent part of the wind is excluded, so that mass transport between two lines is equal just as in the case of isobars of contours of pressure surfaces. For cross-equator analysis around the globe some problems with calculation in small areas with fixed boundaries disappear. The lowest contour drawn in the westerlies on either side of the equator serves as one boundary, i.e. it is considered a part of the streamfunction field, and the condition of continuity with longitude is that any line comes back to its starting place, just like an isobar.

A field of *velocity potential* for the divergent part of the wind and, after integration, the vertical motion, can also be generated. Such a field is found, for instance, in van de Boogaard's tropical atlas for July (A-23).

Figures 7.6 and 7.7 portray another type of machine product, low-level and 200 mb streamline charts, in this instance depicting a hurricane in the Gulf of Mexico with satellite view in Fig. 7.8. The synoptic complications are seen to be quite considerable. The winds are partly from land stations, but a large fraction over sea is satellite-

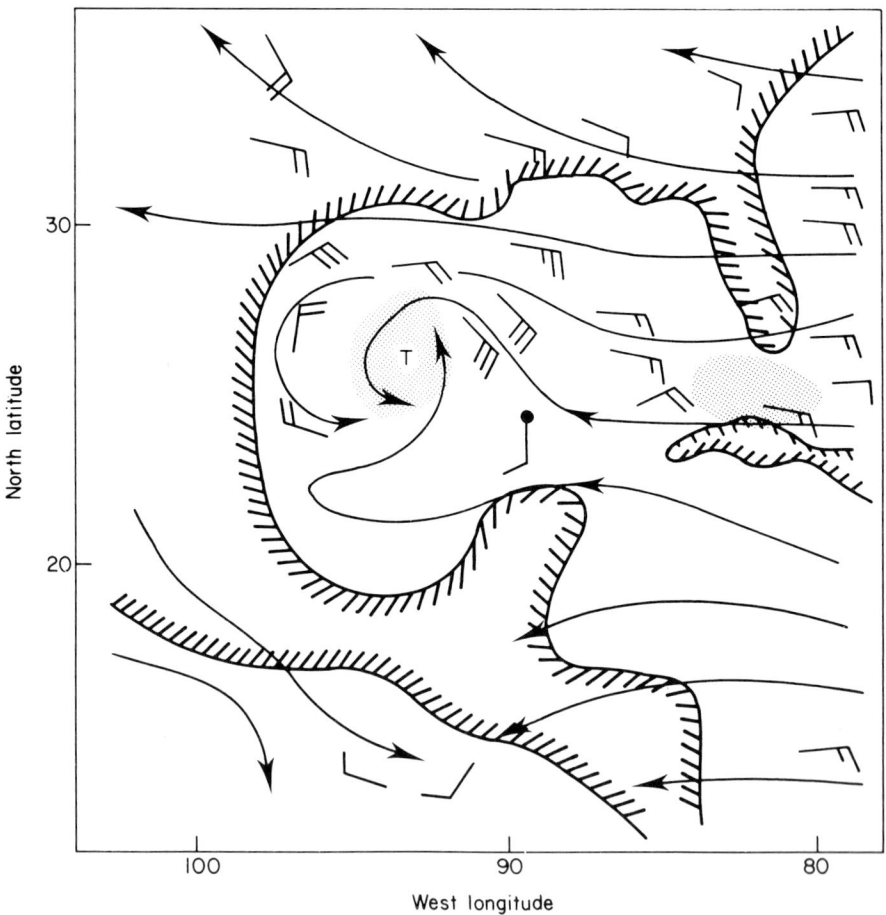

Fig. 7.6. Machine analysed streamlines for hurricane Anita in the Gulf of Mexico and surroundings, 1 September, 1977, 00 GMT near 850 mb; Shaded: satellite cloud areas. Courtesy National Hurricane Center, Miami.

WEATHER OBSERVATIONS AND ANALYSES

computed. Streamlines by definition are parallel to the wind vector everywhere, while the streamfunction lines will cross streamlines, though usually at a small angle. The streamlines are computed so that equal mass transport takes place between them, thus, when wind speed increases or decreases, streamlines must begin or end; where they do so, mass must either leave or enter the level analysed through vertical displacement. Thus, these streamlines differ markedly from those drawn, so that streamlines can either come together or separate along lines or at singular points, such as anticyclonic or cyclonic centres. Such streamlines, of which Figs 1.7 and 1.8 are examples, are very pictorial, but they carry the implication, not meant seriously of course, that infinite convergence of divergence take place where the lines actually join or separate. Analysis of streamlines can be carried out by a computer developed at the University of Wisconsin (22).

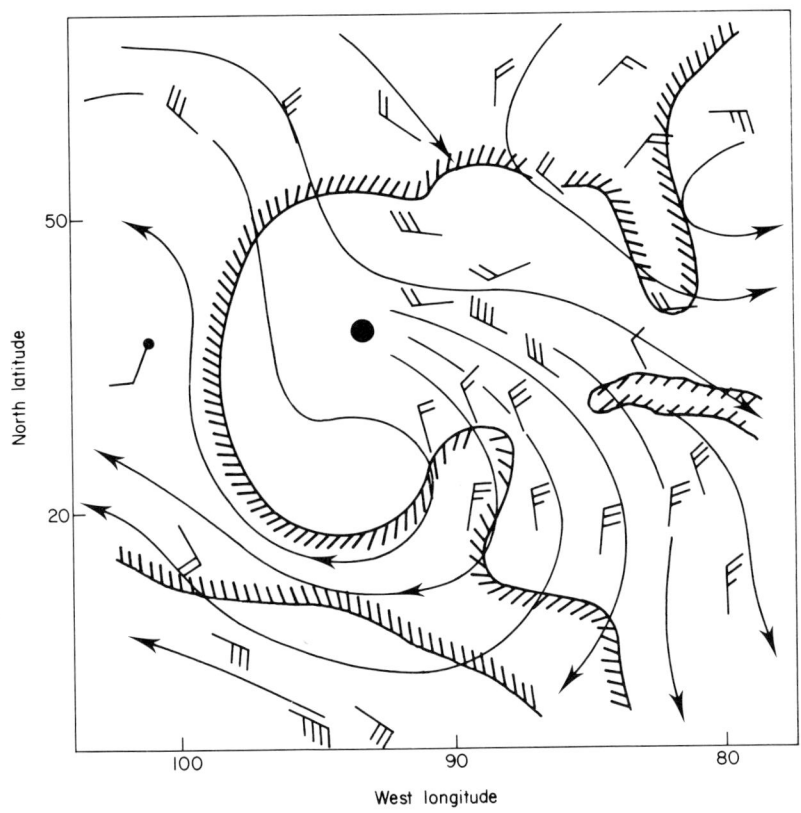

Fig. 7.7. Same for 200 m, 1 September, 1977, 00 GMT.

Local weather analysis

The widespread dissemination of computer products for large-area analysis has made a considerable impact on forecasting offices, which are now relieved from the former burden of drawing these charts by hand once or twice daily. In tight situations, especially during the near approach of a hurricane, the old analysis skills are and will remain of critical importance; otherwise the forecasting staff, to the extent that positions are retained, can concentrate on the local weather and the application of large-scale computer analyses and predictions for their immediate responsibility of producing forecasts, whether for the public, agriculture, air traffic or other purposes. The importance of making prediction, such as of rain showers at particular places and times of day for public functions of many kinds, is well known.

The first requirement is for apparatus that permits interrogation of geostationary satellites every half hour, or arrangements for rapid transmission of such charts from any single interrogation facility that a country may maintain. Of course, the satellite photo, possibly amplified by equivalent radiation temperatures indicating cloud top heights, is only a tool and does not predict the weather by itself. Another tool vital for the interpretation of the satellite information is the complete time section, especially at centres with a rawinsonde station which can give detailed information on how synoptic and mesoscale weather systems affect the immediate area of concern. The time section, very widely used at tropical forecast centres, should contain (a) all upper winds (b) surface pressure anomalies as discussed above (c) 12 and 24-hour changes of these pressures (d) 24-hour changes of the height of upper isobaric surfaces using night-time balloon ascents and (e) a plot of temperature and specific humidity departure of the night-time data from the mean atmosphere of Table 2.1 or a locally determined mean atmosphere. As emphasized above, the radiosonde data do become useful when used in this particular way. Plots should not be confined to standard isobaric surfaces, but should contain all significant points since they are still obtainable locally, especially trade wind inversions, ground inversion and layers with very low and high humidity. Total moisture of the air column should also be carried, mainly for precipitation and fog forecasting. An advanced technique is computation of $gz + c_pT + Lq$ or of θ_e as a function of height, also kept in time section form, and "energy tube" calculations on thermodynamic charts similar to Figs 4.26 and 4.27. Some of this information, especially when computed once a day, may come too late for an effective forecast. At many locations an upstream station exists from which the weather systems come and one can help oneself by analysing this upstream station. It may be noted that 12 and 24-hour time changes should be plotted in the middle of the interval over which the change is measured, not at the beginning and end. In this way the instantaneous tendency is represented as well as possible—perfectly in the case of a linear tendency, but poorly when a trough or other significant weather feature happens to pass in the middle of the time interval.

Local maps, for instance of an island such as Puerto Rico, will complement the analysis. Of particular importance are pressure deviations, surface wind for sea breezes etc., dewpoint for a general indication of the low-level moisture content (which has an impact on energy tube calculations) and a plot of pilot balloon winds and of radar echoes. In general, 5 cm or 10 cm radar sets are recommended for the tropics. The radar information should be photographed for at least two ranges, one with 50 km at the outer limit of the set and another at 200 km, depending on the capability of the set. A time

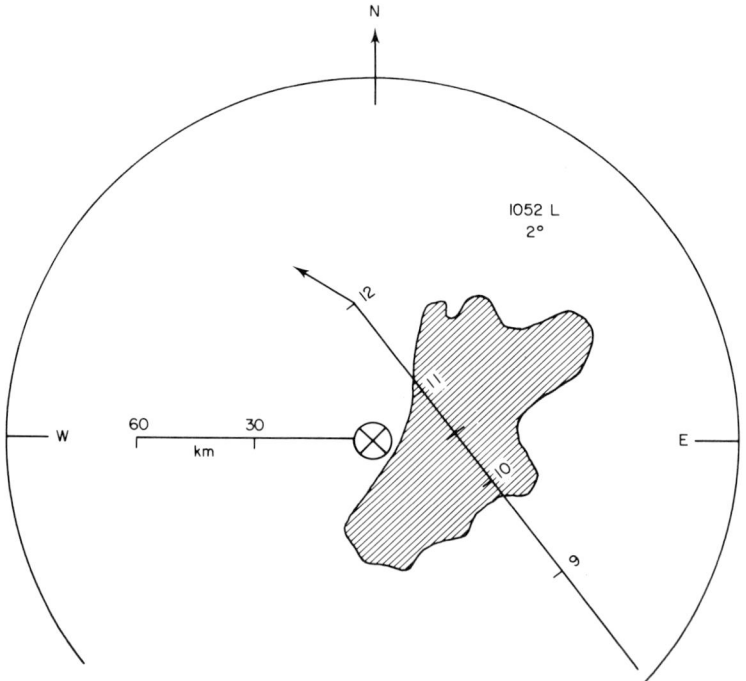

Fig. 7.9. Outline and track of radar echo at 2° antenna tilt on 22 July, 1972 observed with 10 cm radar at Carrizal, Venezuela (9·5°N, 67°W).

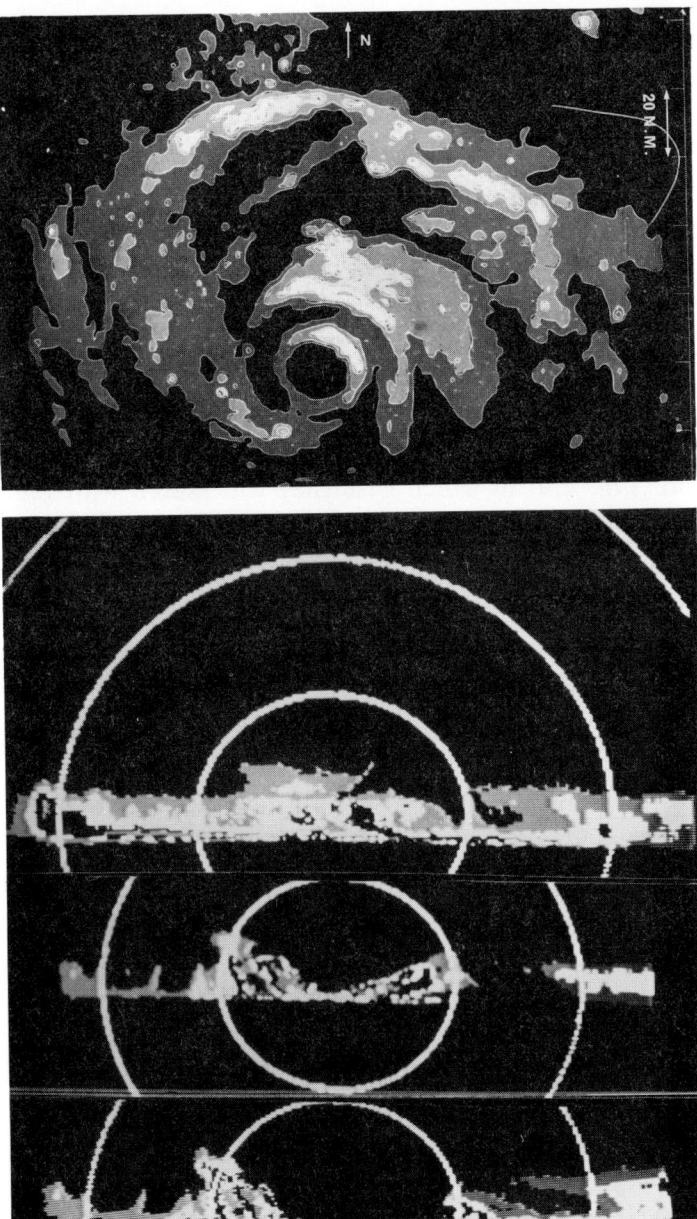

Fig. 7.10. Radar photograph of hurricane Anita, 2 September, 1977, about 0600 GMT. Top: PPI scope; the bright areas inside the dark echo outlines indicate the heaviest water concentrations. Bottom: three RHI sections through eye. Upper: Centre of eye wall; middle: inner rim of eye wall; lower: inside eye. Range markers are for 18 km; highest echo in the upper section reaches to 9-10 km above aircraft. Courtesy Dr R. C. Sheets, National Hurricane Laboratory, Miami.

sequence of the pictures can then be kept. The minimum elevation angle should be two degrees which will carry the beam of a free standing instrument above the surface obstructions. The elevation angle should be varied in order to be able to discriminate between masses of low clouds and higher clouds. Plots of cloud coverage at different horizontal levels can be developed by changing the antenna tilt; some modern radar sets can produce such charts automatically.

Plan Position Indicator (PPI) maps reveal many features of interest, especially whether small echoes are combining to become a mesoscale system. The latter, when formed, can be arranged perpendicular to the wind (Fig. 7.9) in which case they will march with the weather system in which they are embedded. Echoes elongated along the wind and asymptotes of convergence, even very large ones, may remain nearly stationary for many hours, though the synoptic envelope moves along. Where range height indicator (RHI) capability exists, a cross-section in the direction or directions of the most intense clouds present should be photographed every 30, or 15 minutes when the cloud mass is close.

Advanced radar sets carry devices for attenuating radar beam return intensity. By this means, the density of water in particular clouds or parts of clouds can be seen, often a valuable aid in local prediction when a "normal" looking cloud develops a core of great intensity. A fine example is offered by hurricane Anita of 1977 (Fig. 7.10), in the western Gulf of Mexico with central pressure of 925 mb and strongest winds of 80 m/s. Very advanced sets permit computation of motion in clouds, and a combination of several sets leads to a determination of vertical motion as well. Radar technology will undoubtedly continue to present more opportunities for ingenious application by local forecasters—notably with regard to precipitation and its intensity.

Synoptic climatology

So far, we have been concerned with the instantaneous analysis of weather maps and locally important data. There is, in addition, a wide field of statistical analysis which has been tapped here and there, and which, with computer availability, can be widely exploited.

In the first place, it is of great value to ascertain correlations between precipitation and local parameters such as pressure, pressure changes, dewpoint, upper winds and synoptic systems predicted to come into the area of interest. Because of local conditions, different parts of a forecast area may react quite differently. Synoptic climatology can provide for a transition from qualitative to quantitative forecasting and, given sufficiently good correlations, should lead to the development of computer programme predicting an area distribution of precipitation from large-scale analysis and forecast parameters.

Another field for which statistical analysis is important, is the long range outlook on various time scales. There are problems that can readily be solved, and the solution provides considerably aids to the meteorologist. The composition of precipitation and its variability was discussed in detail in Chapter 3. Suppose, now, that we are concerned with a monsoon climate with a single rainy season of three to four months' duration. The first half may have passed with 0, 25, 50 ...% of the average precipitation for this first half. What are the chances that the second half will produce enough rain for the total seasonal rainfall to be 50, 75, 100, 200 ...% of the long period average? This question can obviously be answered and it is often asked by the users who quite clearly have an interest in knowing at least the probable outlook for their survival in the next dry season. Naturally, they would prefer a definite and correct prediction, but since such a prediction is seldom attainable, the probability analysis is much better than nothing. A particular case in point is reservoir management. Both electricity, as well as irrigation and drinking water, may be supplied by a reservoir. If the water level at the start of the dry season is only 50% of the average, all demands until next year's rainy season cannot be met. If the probability of a low reservoir is great, should not conservation measures be started already after the first half of the current rainy season? Usually, the water outflow is largest and most liberal during the rainy season. Conservation measures applied early may go a long way toward alleviating the problems of the next year.

This is a very practical and important problem. Another one which touches on seasonal prediction is the following: in the Caribbean, experience has shown that rain-bearing weather systems coming from the east tend to be weakened and suppressed when the upper westerlies of the main summer rainy period are stronger than average and vice versa (17). Figure 7.11 shows two types of vertical wind profiles for northern Venezuela. In 1969 the easterlies were deep and rain systems with copious precipitation occurred frequently in the northern part of the country. In 1972 very strong westerlies prevailed aloft, and a great drought ensued which included the Caribbean islands and even wider regions.

Now it has often been said, without sound proof, that the general circulation anomalies at the start or even before the onset of the rainy season determine the character of the whole season. This is not likely to be true always and everywhere or many statistical investigations of the past would have come upon such a reliable tool, but the success apparently obtained for West Africa between 10° and 20°N, from the Atlantic to Lake Chad, should be noted (23). A high correlation exists between June precipitation and June-September precipitation. Thus, on the first of July, decisions such as those concerning reservoir storage can be made with considerable, though not absolute, confidence, and this statistical experiment can be run for any area. Since June precipitation must be some function of the general circulation such as shown in Fig. 7.11, a statistical test can include upper wind parameters; enough wind data have been taken in most areas since the 1940s for this purpose. In the Caribbean, with its large upper-air network, area maps of deviations of the general circulation are the most logical way to proceed. In other parts of the world, where only few sounding stations with adequate statistical samples may exist,

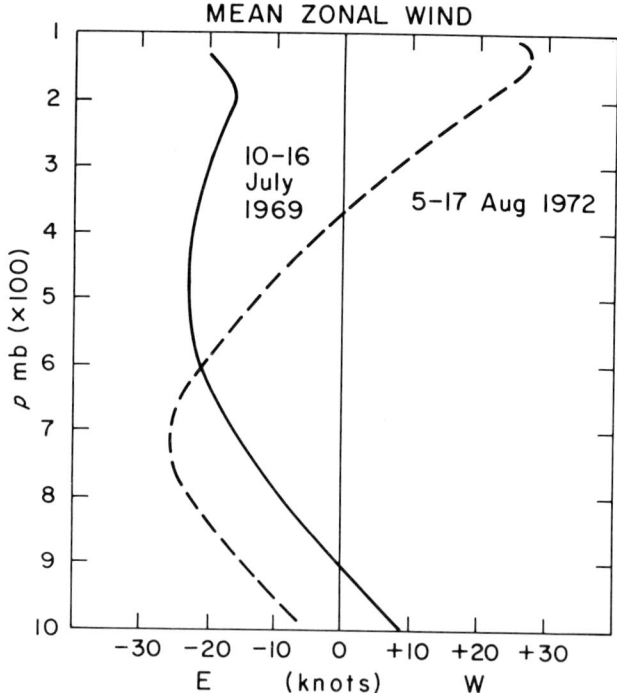

Fig. 7.11. Typical vertical wind profiles of the zonal wind component in northern Venezuela during periods with heavy precipitation (July 1969) and with drought (August 1972) (17).

single station analysis may be the only route to take. At any rate, a good try can be made at evaluating the probability of persistence, exemplified by Priestley (13) for surface pressure and rainfall at Darwin.

Asking such precise questions, with a probability distribution as answer, appears to be a strong technique compared to the many efforts over the last century to produce a deterministic forecast of what the seasonal precipitation actually will be. India is the most notable example where, with high population and marginal food supply, attempts to predict monsoon rainfall date from the 19th century. Success has apparently been limited. Regression equations based on past data, even when divided into several independent samples, appear to fail when applied to the actual future. This is also true for extensive computations of zonal indices in different latitude and longitude belts, their correlation and extrapolation. It is not expected that general circulation models on computers will yield such predictions in the near future, although, of course, attempts at direct prediction are continuing. For the Indian monsoon, the Himalayan snowpack continues to be mentioned as a potential predictor, but this is a very speculative subject, and the purpose of the last few pages has been to show that problems can be selected so that there will be a definite and useful answer on the probability basis for the seasonal outlook. It depends on the ingenuity of regional forecast staff to select the right problems for their areas.

References

(1) Betts, A. K., Dugan, F. J. and Grover, R. W. (1974). Residual errors of the VIZ radiosonde hygristor as deduced from observations of subcloud layer structure. *Bull. Amer. Meteor. Soc.* **55**: 1123-1125.
(2) Burpee, R. W. (1976). Some features of global-scale 4-5 day waves. *J. Atmos. Sci.* **33**: 2294-2299.
(3) Cressman, G. P. (1966). Data requirements for medium-period weather prediction. *Ann. N.Y. Acad. Sci.* **140**: 61-68.
(4) Döös, B. R. (1957). Automation of 500 mb forecasts through successive numerical map analyses. *Tellus* **8**: 76-81. Also: Upper-air analyses over ocean areas. *Tellus* **9**: 184-194, 1958.
(5) Holle, R. L. Reference 29, Chapter 4.
(6) Jacobs, W. C. (1947). Wartime developments in applied climatology. *Amer. Meteor Soc. Monogr.* **1**: 52 pp.
(7) Jordan, C. L. Ref. 16, Chapter 2.
(8) Morrissey, J. F. and Brousaides, F. J. (1970). Temperature induced errors in the ML.476 humidity data. *J. Appl. Meteor.* **9**: 805-808.

(9) National Oceanic and Atmospheric Administration: Monthly Climatic Data for the World. National Climatic Center, Asheville, N.C.
(10) Oliver, V. J. and Anderson, R. K. (1969). Circulation in the tropics as revealed by satellite data. *Bull. Amer. Meteor. Soc.* **50:** 702-707.
(11) Parmenter, F. C. (1976). A Southern Hemisphere cold front passage at the equator. *Bull. Amer. Meteor. Soc.* **57:** 1435-1440.
(12) Penman, H. L. (1948). Natural evaporation from open water, bare soil, and grass. *Proc. Royal Soc.* (London) A, **193:** 120-145.
(13) Priestley, C. H. B. (1962). Some lag associations in Darwin pressure and rainfall. *Austral. Meteor. Mag.* No. **38:** 32-42.
(14) Rao, M. S. V. and Theon, J. S. Ref. 30, Chapter 3.
(15) Riehl, H. Ref. 31, Chapter 6.
(16) Riehl, H. and Betts, A. K. (1972). Humidity observations with the 1972 U.S. radiosonde instrument. *Bull. Amer. Meteor. Soc.* **53:** 887-888.
(17) Riehl, H. (1973). Controls of the Venezuela rainy season. *Bull. Amer. Meteor. Soc.* **54:** 9-12.
(18) Riehl, H. (1978). Objective surface pressure analysis in tropical countries. *World Meteor. Organis. Bull.* **27:** 175-179.
(19) Rudloff, W. (1958). Bemerkenswerte Luftdruckschwankungen in den Tropen. *Ann. Meteor.* **8:** 363-372.
(20) Thornthwaite, C. W. and Holzman, B. (1942). Measurement of evaporation from land and water surfaces. US Dept. Agriculture Tech. Bull. No. 817.
(21) Wark, D. Q. and Fleming, H. E. (1966). Indirect measurements of atmospheric temperature profiles from satellites. *Mon. Wea. Rev.* **94:** 351-362.
(22) Whittacker, T. M. (1977). Automated streamline analysis. *Mon. Wea. Rev.* **105:** 786-788.
(23) Winstanley, D. (1974). Seasonal rainfall forecasting in West Africa. *Nature, London* **248:** 464.

8
Synoptic-scale Weather Systems

Introduction

The unsteadiness of the atmosphere, described in Chapters 1, 3 and 4, may be classified under the summary title of "turbulence", but this turbulence can be subjected to techniques of analysis other than those usually employed in physics for such problems. There one visualizes a large number of random motions within a certain medium; individual motions cannot be discussed. Instead, physical and statistical laws are developed concerning the net effect of the turbulence on the medium as a whole. The concept of "temperature" is an illustration.

To the extent that the sum of all atmospheric disturbances produces the general circulation, we can speak of turbulence, but not all the turbulent elements need be viewed as random fluctuations, accessible to study only by examining their aggregate. It is true that their magnitude in space and time covers a wide range; this "spectrum", however, shows concentrations of frequency and energy of the disturbances in a few "spectral lines". We speak of "centres of action", "synoptic scale weather systems", "convection cells" and so on. Each of these scales of turbulence can be investigated separately for the laws governing its particular appearance and behaviour in time and space.

Weather prediction makes sense only when this procedure is feasible and sound. It therefore has become a principal function of the meteorologist to discover and set down, in the form of models, basic characteristics of turbulence elements, which can be viewed as discrete entities. In attempting to do this, the science of meteorology has confronted itself with a problem of a high order of difficulty. Its concern with this problem, of course, stems from the demand for forecasts. In asking what the weather will be tomorrow or whether the monsoon will bring ample rain, people testify to their own experience of discrete scales of turbulence within the atmosphere. They are also demanding a great deal. While meteorology has travelled far toward an understanding of its medium, much time will pass before all the demands of the public can be satisfied.

In this chapter we shall consider the "daily" or "synoptic scale" weather systems with life-time of days and scale of 1000-3000 km. The "synoptic" weather systems produce about 75% of total precipitation in most tropical areas (Chapter 3). The rain episodes or "rainstorms" number about 20 in the mean for a rainy season of 100 days; average frequency widely has been found to be once in five days. Two or three rain episodes can easily account for half of the precipitation of a rainy season, so that failure to materialize of just one or two of these events can lead to drought conditions; the rarer excess, conversely, can bring floods.

When one notes this dependence of precipitation in the tropics on discrete weather systems and compares their frequency and duration with parallel events in higher latitudes, much similarity is evident. For the higher latitudes, however, it has been shown that the basic westerly current can become unstable in certain ranges of wave length (the synoptic scale) and break down into clockwise and counterclockwise revolving vortices in the low atmosphere. There, energy transformations become concentrated. In the tropics, no general basic current prevails comparable to the westerlies; a general criterion of baroclinic instability is not likely to exist.

Scales of turbulence

The question arises why, nevertheless, a set of tropical weather systems is observed which matches those of higher latitudes in size and duration. At first sight it is not obvious why the tropical branch of the general circulation does not work as a steady and stationary engine—as it was originally conceived. A clear model exists for this branch, dating in parts back to the days of Hadley, after whom it is often named, and amplified later up to the 1940s. In Chapters 1-3 we have successfully identified many features of tropical climates with this model, beginning with the division of the world's surface wind systems into a tropical half with east winds and a higher latitude half with west winds. Nevertheless, the synoptic weather systems occur and concentrate most of the energy release within them. In the global setting of the equatorial trough zone, a hierarchy of the fractional area occupied by various scales of motion has been proposed (71), given in Table 8.1. Average sizes and numbers for the several subdivisions in a circumpolar belt 10° latitude wide in the equatorial trough zone have been added (74). There is, of course, a sizeable frequency distribution attached to all of the average areas, but they are reasonably supported by satellite and radar data for synoptic and mesoscale (Chapter 3, also (27)). For the undilute towers an area of 25 km^2 assures reasonably well that entrainment will not destroy them except in very strong wind shear (see end of Chapter 4). A total of 1600 to 2400 active towers at any time compares well with an estimate of the number of active thunderstorms (3); however, not all undilute towers are thunderstorms.

Table 8.1. Hierarchy of tropical scales

	A	km^2	Average size km^2	$n/10°$ belt
Equator, 10° belt	1	4×10^7		
Synoptic	0.1	4×10^6	3×10^5 $n = 13$	13 (20)
Mesoscale	0.01	4×10^5	2×10^3 $n = 15$	200 (300)
Undilute cb	0.001	4×10^4	25 $n = 8$	1600 (2400)

Synoptic scale — $L = 3000$ km; frequency = 1:5 days; $C = 600$ km/day.

Uncoupling in the troposphere

In intense, well-organized synoptic systems, especially tropical storms, the troposphere acts as a well-coupled fluid. Most of the time, however, the synoptic systems are sufficiently weak, or the primary circulation is sufficiently strong, notably during the Indian summer monsoon, for these systems to be prominent in only half of the troposphere while they are only weak in the wind field (but not the weather) of the other half. Thus, the concept of two or three layer tropical tropospheres has arisen (25).

In the trades we find large fluctuations of high-tropospheric winds above easterlies showing only small changes. In the ten days of Fig. 8.1, a trough in the easterlies passed on 7 July at 850 mb accompanied

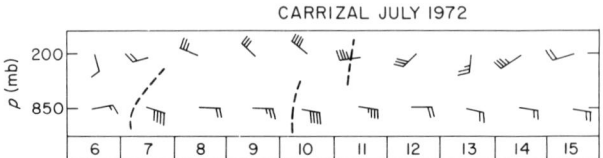

Fig. 8.1. Winds at 850 and 200 mb at Carrizal, Venezuela (9·5°N, 67°W) 6-15 July, 1972. Wind speed in knots. Dashed lines: passage of axes of synoptic systems.

by a small cyclonic wind shift and extension into an upper outflow ridge. On 10 July a surge of the trades arrived almost without change of wind direction but with a doubling of wind speed. Clearly, the amplitude of the 200 mb flow greatly exceeded that at 850 mb. The classical viewpoint of steady antitrades above the lower, equatorward flowing trade currents is contradicted by high-level daily synoptic charts. Synoptic-scale systems exist everywhere; in the trades, their energy is weak at low levels compared to that of the basic easterlies and they merely appear as minor deformations of the streamlines. At high levels, they control the wind field.

Along large portions of the equatorial trough zone, especially over the "continental" half, the picture is reversed. Perturbations are most prominent in the lower equatorial westerlies and give way to an easterly wind regime in the upper troposphere (Fig. 8.2). Although lower and upper troposphere are not fully "uncoupled", just as in the trades, it is quite remarkable how relatively steady the 200 mb winds remain, even when large-scale intrusion of mass from low levels takes place by means of cumulonimbi. It is always the layer with easterly winds which is relatively steady; the large fluctuations occur in westerly basic currents and also in layers without definite zonal basic current.

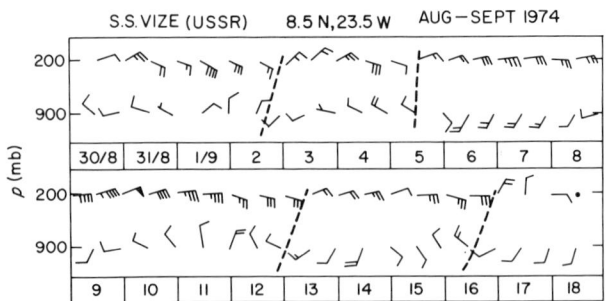

Fig. 8.2 Winds at 900 and 200 mb at the USSR ship Vize (8·5°N, 23·5°W), 30 August-18 September, 1974. Other notations as in Fig. 8.1

Influence of radiation

Through satellite technology vast amounts of information about many aspects of tropospheric radiation are becoming available on the "on-time" basis. An input of such information into analysis and prediction of synoptic weather events becomes feasible. Further, radiometer sondes (35) measuring long-wave fluxes up and down through the troposphere have been developed, and they could become a regular part of rawinsonde observations sent up at night. Upper-air temperatures in the tropics are remarkably steady, even though heat release through precipitation occurs, roughly, in only one per cent of the tropical area. Since net radiation cooling of the troposphere is estimated as 0·8-1·2°C/day, the tropical atmosphere by and large must compensate partly by sinking, so that the radiation cooling is counteracted by compression heating (24).

Quite a different situation prevails in and near the edges of synoptic weather systems. Upper and middle clouds may well cover the whole weather-active 5° square, again evident from satellite photographs. Clouds are known to be strong absorbers and emitters of long-wave radiation, hence the atmosphere below cloud base experiences very

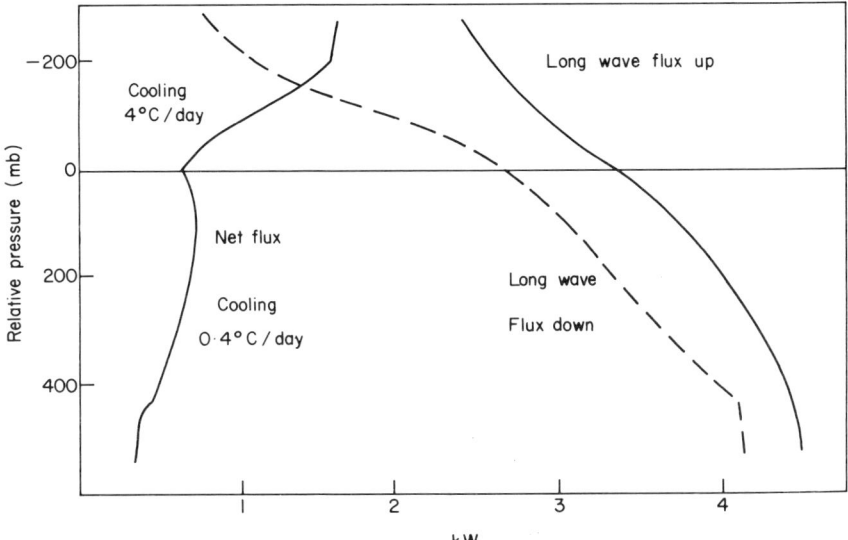

Fig. 8.3. Long-wave flux components, net flux and cooling for nights with deep moist layer at Palmyra, Line Islands (6°N, 163°W) averaged with respect to zero setting of pressure at top of moist layer. (Data courtesy of S. K. Cox.)

little cooling, or even warming, through long-wave radiation (Fig. 4.19). At the top, where downward radiative flux may become very small, large cooling occurs (1). Soundings in a wet upper troposphere have been composited with respect to the cloud top given by the relative humidity, marked as zero-pressure in Fig. 8.3. Radiative cooling is about 0·4°C/day in the whole layer below the clouds and even slight warming prevails in the upper part of this layer, with an abrupt change to cooling of 4°C/day near and above cloud top. From there the downward radiation flux decreases very sharply upward while the upward flux decreases more slowly than lower down, resulting in a jump of the net flux upward. With the passage of cloud covers, very large changes in long-wave radiation may be measured at one station during two successive nights. Thus, at Miami, Florida, radiation cooling on 18 October, 1960 was only 0·3°C in the troposphere (Fig. 8.4); during the following night, it was 2·0°C (72). If this difference in cooling persisted at two neighbouring stations for 24 hours, a surface pressure difference in the range of 5-7 mb would develop, given that hydrostatic balance, height of the 100 mb surface and other conditions remained the same. These values are very large, well in excess of observed tropical pressure gradients and 24-hour changes.

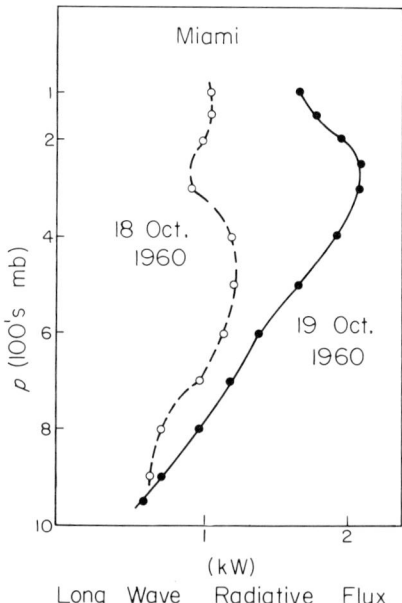

Fig. 8.4. Net long-wave emission on two consecutive nights at Miami, Florida (72).

The tropical rainstorm as a synoptic system

Tropical weather events exhibit a wide and seemingly bewildering variety of shapes and envelopes around the globe. Since, apart from the subtropical margins, they all exist to a large extent on conversion of latent heat energy to potential and then to kinetic energy, it is unlikely that there are as many physical mechanisms as superficial aspects of these storms. With the steady increase in tropical observing networks, plus special research observing programmes, it has become possible to identify features shared by most synoptic entities, apart from tropical storms (see Chapters 9-11). They may be summarized under the name of "tropical rainstorm" (28).

Description

Mesoscale rain areas

One of the striking features indicated by the rainfall observations, reinforced by radar photography, is the very marked concentration of precipitation in small areas—the "mesoscale" rain areas with radar return—and the tendency for individual stations to receive only one strong shower in the majority of cases during passage of one disturbance (Chapter 3). Mesoscale rain areas become efficient precipitation producers when their maximum area, as seen on a PPI screen with at least 2° antenna elevation, attains about 1000 km^2 or more. The average mesoscale area will be taken as 2000 km^2; but the range is wide, from 10^3 to 10^4 km^2, in rare cases even more (Fig. 3.28).

Precipitation varies correspondingly. On average, for the 2000 km^2 radar echo, the mesoscale area yields about 2·5 cm of rain in coordinates moving with the cloud mass in two hours. At individual locations, of course, rainfall may vary widely. Almost always there are some stations in a network which are completely missed by rain, especially during passage of weak rain areas. A main concern in analyzing the mechanisms of synoptic weather systems, therefore, lies in observing the changes in atmospheric structure associated with the mesoscale events.

The general setting

As one striking example of the setting, Fig. 8.5 shows the vertical time section of wind, relative humidity, precipitation and pressure-height

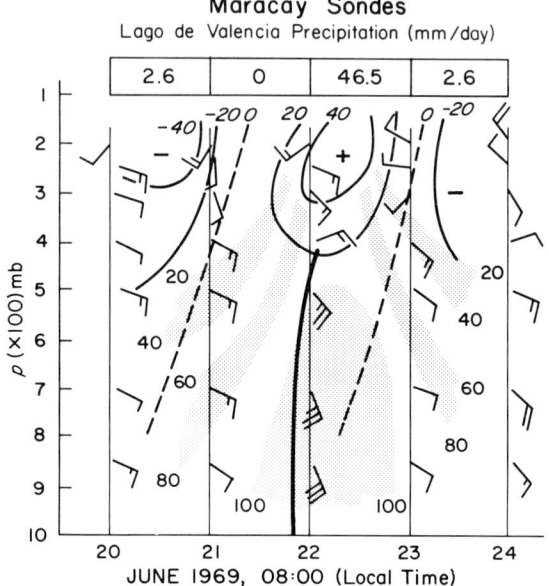

Fig. 8.5. Time section of upper winds (knots, relative humidity (per cent, shaded)) and 24-hour pressure-height changes (m) at Maracay (10·2°N, 67·4°W) for the heavy precipitation event. Daily rainfall from Fig. 3.21 at top. Heavy solid line: axis of weather system from South America at approximate time of passage (A-19).

changes during passage of a strong synoptic system which passed north-northwestward across central Venezuela. The "roots" of this system were far south in Brazil (Fig. 12.19), which illustrates interhemispheric connections. The precipitation record for the Lago de Valencia area of northern Venezuela was shown in Fig. 3.22 and the detailed rain distribution for the largest synoptic event, 22 June, 1969, in Fig. 3.21. A completely rainless day with suppressed convection preceded the event. June 22, 1969 was a rather "ideal" case, a model or prototype event with high cyclonic vorticity at low levels after a sharp wind shift coupled with rainfall well above average for a synoptic passage and overlain by high-tropospheric outflow with upper pressure-height rises and warming temperatures, contrasted with low and middle-troposphere cooling.

From another part of the world we see the passage of a wave on 2-3 September, 1974 through the centre of the GATE network (Fig. 8.6). This wave travelled westward along the equatorial trough in deep easterlies with an understructure of equatorial westerlies (Fig. 8.8).

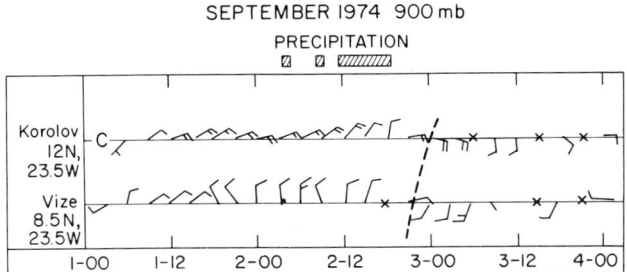

Fig. 8.6. Three-hourly 900 mb winds (knots) at the USSR ships *Korolov* and *Vize*, 1-3 September, 1974. Dashed line marks trough passage moving from east to west. Precipitation periods at *Vize*: shaded boxes.

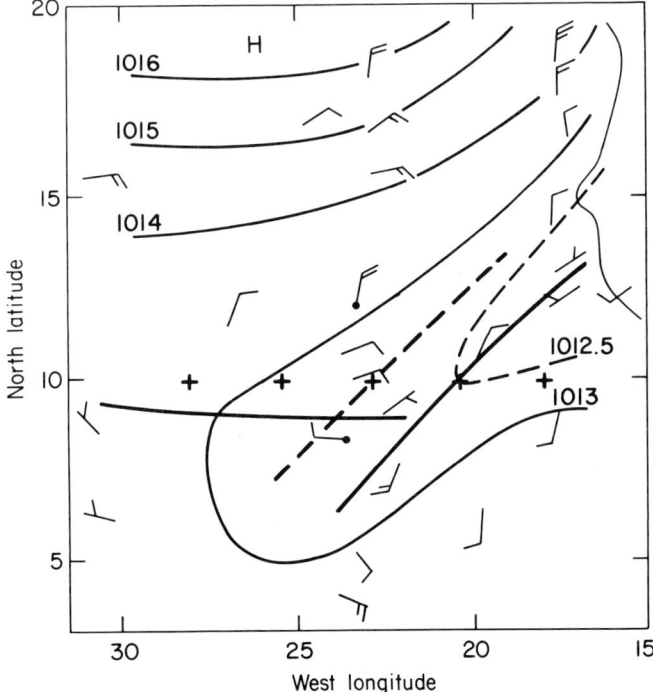

Fig. 8.7(a). Surface map for the GATE area, 2 September, 1974, 1200 GMT. Winds in knots, pressures and isobars in millibars. Heavy solid lines: moving wave in easterlies oriented northeast-southwest, equatorial trough oriented east-west. Dashed line: position of trough 12 hours later. Crosses: 12-hourly trough positions at latitude 10°, 2-4 September, 00 GMT.

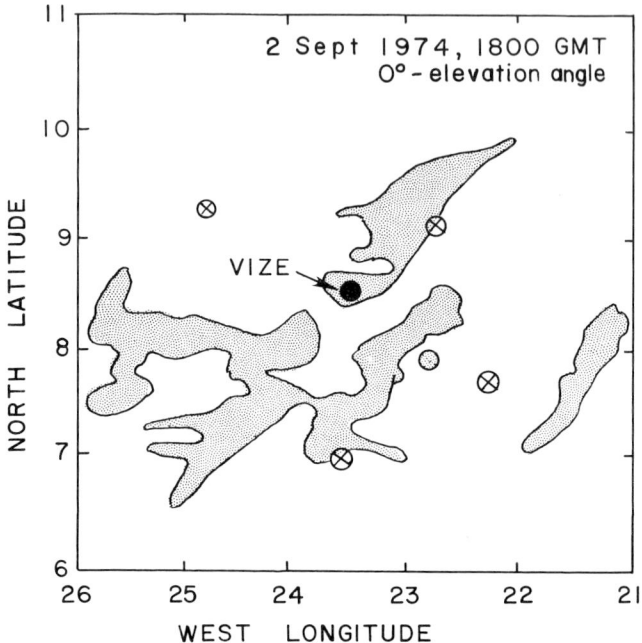

Fig. 8.7(b). Principal radar echoes 2 September, 1974, 1800 GMT.

Three-hourly observations produce much detail. The time sequence of 900 mb winds, at the level of strongest equatorial or "monsoon" westerlies, is very "regular" both at the ship *Korolov* on the north side of the equatorial trough (Fig. 5.2) as well as at *Vize* right in the trough. The surface map at 1200 GMT, 2 September (Fig. 8.7a), shows surface wave position and orientation very clearly. In spite of the low latitude, pressure-wind relations look very similar to those found on middle latitude oceanic charts. Crosses mark the steady 12-hourly progression of the wave through the GATE network for five periods. In spite of various mesoscale features which attended the wave, the broad outline could be followed and it held together very well.

Although vertical wind structure at *Vize* differs from Maracay in that the layers with east and west winds are reversed, the sequence in the temperature and moisture field is rather similar (Fig. 8.9). Of course, there is far more detail. Precipitation over sea was measured both by ship rain gauges and also by radar from several ships. *Vize*

SYNOPTIC-SCALE WEATHER SYSTEMS 325

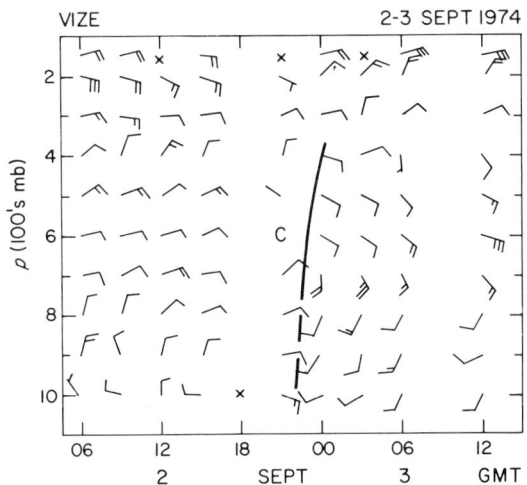

Fig. 8.8. Time section of upper winds at *Vize*, 2-3 September, 1974.

had 17 mm, *Korolov* 10 mm, but several ships measured rain up to 50 mm; the distribution was similar to that of Fig. 3.21. Precipitation preceded trough passage. Figure 8.7(b) shows the lineup of principal radar echoes parallel to wave axis and equatorial trough.

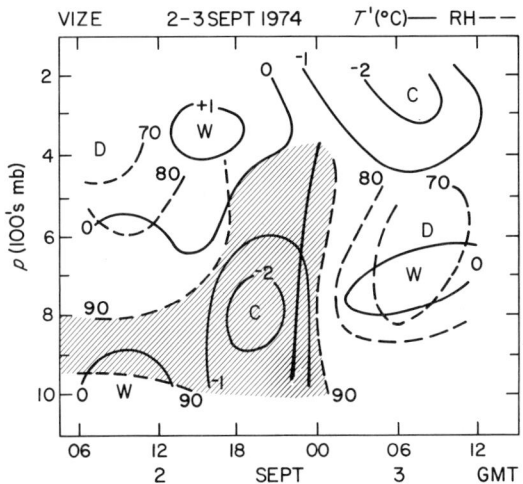

Fig. 8.9. Time section of temperature deviation from the mean (°C, solid) and relative humidity (per cent, dashed) at *Vize*, 2-3 September, 1974, Shaded: >90%.

Temperature and humidity

The outstanding thermal feature of the rain area is the low and mid-tropospheric cold core, overlain by a warm or at least neutral layer in the upper troposphere. This type of anomaly structure, shown as vertical profile for Maracay in Fig. 8.10, has been encountered very widely and may be taken as representative of a large fraction of tropical rain situations. The height of zero anomaly lies close to the top of the convergent layer found in statistical summaries (Fig. 4.25). Over the ocean, the maximum cold anomaly is not found at the surface itself due to immediate restoring action of heat flux from the sea, whereas there is no such ready heat source over land. The dotted line in Fig. 8.10 and the profile of Fig. 4.17 bring out this fact. At higher levels, warm temperatures precede, rather than follow, the *Vize* axis. (Deviations were computed from means for each station individually, in order to avoid contamination resulting from different types of instruments used.)

Specific humidity anomaly for Maracay remained nearly zero at the ground and rose above $+ 2$ g/kg in a deep tropospheric layer with nearly 100% relative humidity. This profile is typical of what has been widely found, except that sometimes surface specific humidities decrease as in Fig. 4.17. At *Vize*, from the relative humidity cross-

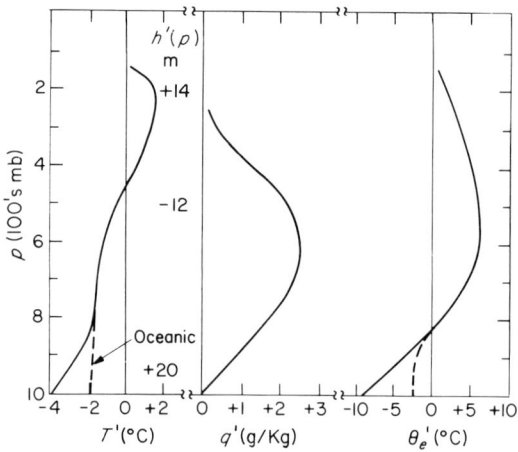

Fig. 8.10. Vertical profile of temperature, specific humidity and equivalent potential temperature deviations from the local mean at Maracay, 22 June, 1969. Dashed: Low-level modifications at *Vize*. Also pressure-height deviations at 1000, 500 and 200 mb.

section, moisture also changed little at the surface and increased throughout the troposphere in the rain area, though less than at Maracay because of the equatorial trough position. There, synoptic weather systems tend to follow each other without strong intermediate anticyclones, resulting in high surface convergence values on the mean seasonal chart (Fig. 1.12). Away from the trough, alternation of cyclonic and anticyclonic systems increases, as does the amplitude of change from deep dry to moist layers and reverse.

The latent heat energy change dominates the variation of total thermodynamic energy expressed by θ_e except close to the ground. Equivalent potential temperature increases markedly above 850 mb both in the cold and warm layers, with largest increase in the layer where normally the θ_e-minimum is situated. Convective activity acts to make θ_e uniform in the troposphere. Near the ground, the decrease over the ocean is less than over land because of the compensating energy flow from the ocean.

Wind profiles

At the arrival of the rain event at *Vize*, winds turned sharply from light southeast to strong east-southeast then southwest, i.e. cyclonic vorticity was associated with the axis and the rain area. The cyclonic vorticity is found concentrated in a very narrow zone over considerable vertical depth at the eastern end of the rain area. In the high troposphere the flow becomes anticyclonic associated with the warm high-level temperature anomalies. Large 200 mb troughs are located one to two days forward and backward from the rain area.

Wind direction turns markedly with height, so that one is tempted to apply thermal wind considerations. In spite of extensive efforts, no uniformly satisfactory result is achieved in contrast to middle latitudes. At *Vize*, for instance, the wind turned clockwise with height at least up to 400 mb ahead and counterclockwise behind the rain area; no positive correlation of thermal advection and temperature anomaly field is evident. In the high troposphere the relation is better. Clockwise turning of wind with height preceded the upper ridge and counterclockwise turning followed it. In the Caribbean and Venezuela the same difficulty has been encountered in the convective layer below 400 mb. In Fig. 8.5 the thermal wind relation succeeded, at least by sign, in the high troposphere. Clockwise turning coincided with warming; many other cases can be cited to demonstrate that this correlation is typical.

Pressure-height changes

Just as in middle latitude convective storms, surface pressure tends to fall for some hours (after removal of the semi-diurnal tendency) ahead of tropical mesoscale events, and then it rises rapidly by about two millibars as the cold tropospheric mass of Figs 8.9 and 8.10 arrives and surface temperature drops to a rather universal value of 22°C. We see a composite pressure profile in Fig. 8.11 where time has been converted into distance with respect to a travelling mesoscale rain area. Just this kind of profile was observed at Maracay on 22 June; the situation is not included in the composite. The pressure rise often starts rather weakly 30-60 minutes ahead of the main precipitation with arrival of the first scud clouds. Given the pressure rise which corresponds to an increase of 20 m in 100 mb pressure-height, hydrostatic integration indicates a drop of 12 m at 500 mb, and an

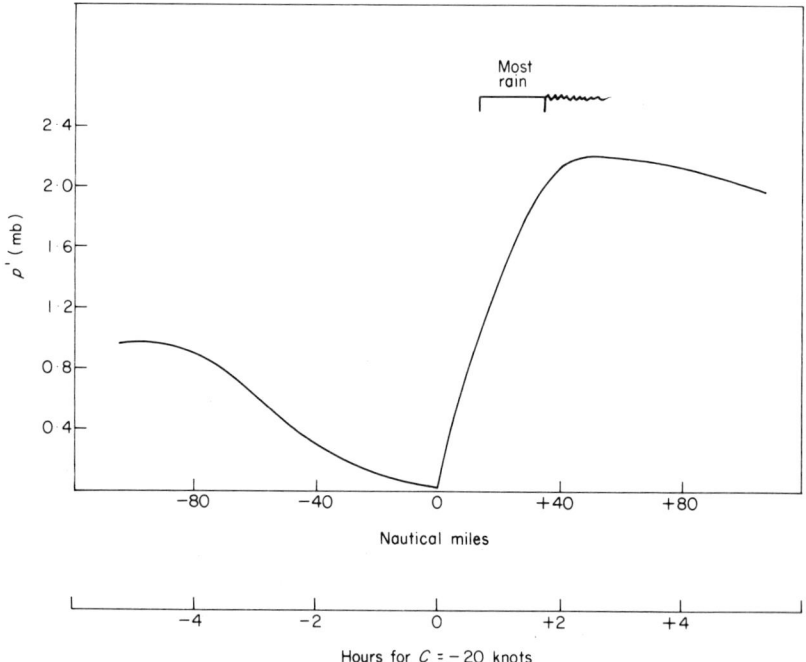

Fig. 8.11. Composite surface pressure profile (deviations from lowest pressure) during passages of mesoscale systems at Carrizal, Venezuela. Semi-diurnal pressure variation removed. Scale in hours and distance with respect to minimum pressure assuming propagation rate of 10 m/s westward (74).

increase of 14 m at 200 mb (Fig. 8.10). These typical values suggest maximum cyclonic vorticity at 500 mb at the top of the cold layer and anticyclonic relative vorticity at 200 mb for quasi-balanced flow. Such vorticity distributions have been widely computed from wind data in many types of synoptic envelopes (79).

Analysis

Now that many of the most important facts about a large portion of synoptic disturbances have been enumerated here, while radar observations were discussed in Chapter 3, we shall examine what they mean for the tropical rainstorm and its mechanisms.

Concentration of mass flow

Foremost is the question whether the mass entering the border of synoptic-scale convergence areas all converges into the mesoscale radar echo areas or whether a substantial fraction goes into mass ascent over the whole synoptic area. The regularly recurring observation that tropical rain systems have low and mid-tropospheric cold cores (70) has been a hindrance in understanding how these systems could be "driving members" of the general circulation. Any hypothesis of sudden vertical instability, as may be produced by the disappearance of a trade wind inversion, fails; the air in the rain areas is very stable compared to the surrounding mean atmosphere. Now, if the whole mass in the convergence zone ascended, temperature would certainly drop. But all θ_e-surfaces would be lifted above the subcloud layer, resulting in the dotted profile marked "no overturning" in Fig. 8.12. It is now a principal fact that, excepting hurricanes, this profile is not observed, the one marked "final" is. Vertical overturning, and a tendency to establish uniform θ_e in the convective layer is indicated, foreseen by Rossby (78). The level of no change quite often is 850 mb. Higher up, θ_e increases while lower down it decreases. Thus, another model must be substituted for that of general mass ascent over a 500 km square with mass convergence.

Other observations have indicated that a model of area-wide updraft could not hold. In a general convergence area, temperatures in the low and middle troposphere should cool long before the arrival of the actual cloud mass, due to dry-adiabatic lifting. Yet, in the series of Venezuela experiments, and also in some cases during GATE, the opposite was encountered (73). Temperatures in the middle atmosphere rose and moisture decreased near the outer edges of the convergence zone suggesting subsidence. The subsidence appeared to terminate about an hour before arrival of the main cloud mass. At times, some towers grew ahead of the main system and evaporated. In order for such moistening to assist later ascent of undilute towers, middle-level winds must approximate the propagation rate (C) of the system fairly well; it is indeed often observed that $U - C$ is a minimum in the middle troposphere (Figs 8.13-14).

Based on these pieces of evidence, the model of general mass ascent can be replaced with that of "energy tube" ascent of Figs 4.26 and 4.27. The 900 to 850 mb level usually marks the upper limit for starting buoyant paths. On 22 June, 1969 air from this level could barely reach 500 mb. The 850 mb level, therefore, becomes the level of no energy change in diagrams such as Figs 4.17 and 8.12. All air converging above this level is stable with respect to moist ascent; it cannot rise except through outside forcing. From this fact we concluded in Chapter 4 that the whole middle-level inflow must become the downdraft to the surface layers where, from contact with

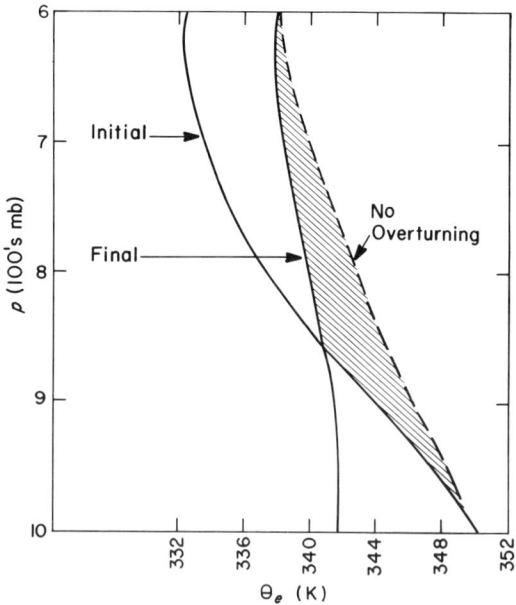

Fig. 8.12. Vertical profiles of equivalent potential temperature at Maracay: "initial" on 21 June, 1969, "final" on 22 June and dashed profile showing expected final profile, if whole convergence area was lifted with mean vertical velocity.

ground or ocean, the energy would increase over 12 hours until this air became capable of entering an undilute tower ascent at a later time.

The middle-level inflow, when first observed, was a distinct surprise. It was difficult to understand it and to fit it into the convective storm picture. However, downdraft cooling and attendant depression of mid-tropospheric pressure surface provides a mechanism for the upper inflow. One must postulate that downdraft and inflow develop together until the temperature profile of Fig. 8.10 is established.

Life-cycle

The "energy tube model" of undilute ascent permits the tropical synoptic systems to do work maintaining themselves and the general circulation. In the computation at the end of Chapter 4 the release of kinetic energy was estimated as six per cent of the effective solar constant (theoretical radiation received at a given latitude) per synoptic system. This amount exceeded by a factor of three the required work of 1 W/m^2 to overcome frictional slowing down of the winds. No doubt there are great differences in the efficiency of individual systems. It appears to be most advantageous for the upper warm area with anticyclonic circulation and outflow to be generated and maintained. In publications showing tropical rain areas associated with upper cold lows or troughs, one finds the heavy rain area (if any) situated asymmetrically in the outskirts of the upper cyclonic centre, with change to clockwise curving flow at 200 mb (Fig. 8.27). In the core of the upper cyclone, pressure-height gradients are weak; rainfall, if any, is light and comes from altostratus(!).

The upper warm core of Fig. 8.10 will disappear when the energy of surface inflow diminishes even slightly, or when, with very strong vertical shear of the horizontal wind, the undilute towers are forced to lean excessively and erode, i.e. they no longer remain undilute. Many synoptic events have a life-cycle of little more than a day (93), yet on satellite time series (Fig. 8.18) westward-moving cloud masses can at times be followed around the globe! These cloud masses, however, do not remain in steady state but go through cycles of intensification and weakening. The tropical synoptic system as a steady state proposition for prolonged periods is no more realistic than a corresponding model of a steady extratropical cyclone. Nevertheless, it is at times convenient to use the assumption of steady state to bring out some important mechanisms; the unsteady state often is a small, though important, difference between large opposing influences.

The decline of mesoscale rain areas within the synoptic envelope may be viewed similarly to the decline of individual cumuli discussed in connection with Fig. 4.14. With positive shear (say east wind increasing upward) the air enters undilute towers of a mesoscale system from west if the line is oriented north-south and travelling westward. Because of the vertical shear, the middle and perhaps upper cloud portions lean toward west. Then the ascending mass drops its water into the air above the inflow; the resulting low energy downdraft enters the inflow and thereby cuts off the supply of high energy air to the cloud. This sequence would take perhaps half an hour or a little more after the start of rain while the clouds are still growing. When the high-energy inflow stops, the peak of the growth has by necessity been reached; further expansion and heavy rain end together. On the right side of Fig. 4.14 the wind shear is such that the downdraft arrives at the ground at the other side of the inflow.

On average, the life of mesoscale systems lasts four hours; the active rain period is two hours and the concentration of precipitation occurs, more often than not, in the half hour before the radar echoes attain their greatest area. One can see clearly how the development of downdrafts puts a brake on the whole mesoscale cloud concentration in most instances.

Location of rain areas

This subject will be discussed in terms of the vertical profile of the wind component U positive in the direction of the propagation vector C. Three principal types of relative wind profiles $U - C$ occur (Fig. 8.13). On the left side the case of 22 June, 1969 is depicted. The system moves with the celerity of the cloud mass which has crossed the equator from south (Fig. 12.19). In the low troposphere winds are very strong and almost exactly in the direction of C (Fig. 8.5). Air moves toward the cold core with high cyclonic vorticity (Fig. 8.14). Considering the vorticity theorem convergence and ascent are taking place in the inflow to the rear of the axis of wind shift. High-tropospheric outflow mostly is to the right of C but also develops a small rearward component in the course of the day. Thus, the warmest air and outflow anticyclone trail the low troposphere axis, very pronounced in Fig. 8.1 where the rain area was moving westward while the 200 mb wind was blowing from west.

Quite a different situation prevailed during 2 September, 1974 at ship *Vize* (right side of Fig. 8.13). Crosses mark the 12-hourly progression of this wave toward west at latitude 10°N (Fig. 8.7).

333

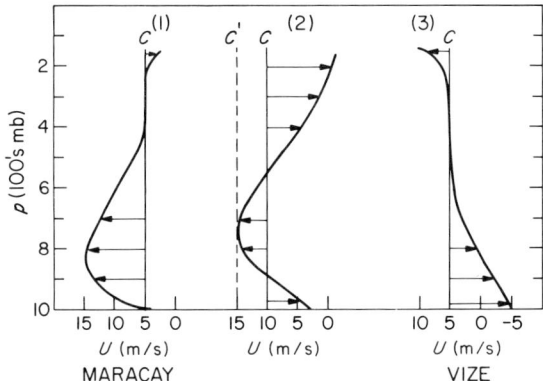

Fig. 8.13. Three models of $U\text{-}C$ relations along the vertical. Positive wind speed along C (direction arbitrary): left; 22 June, 1969 at Maracay; right, 2 September, 1974 at Vize; centre, frequent profile in waves in the trade winds (C' for squall lines).

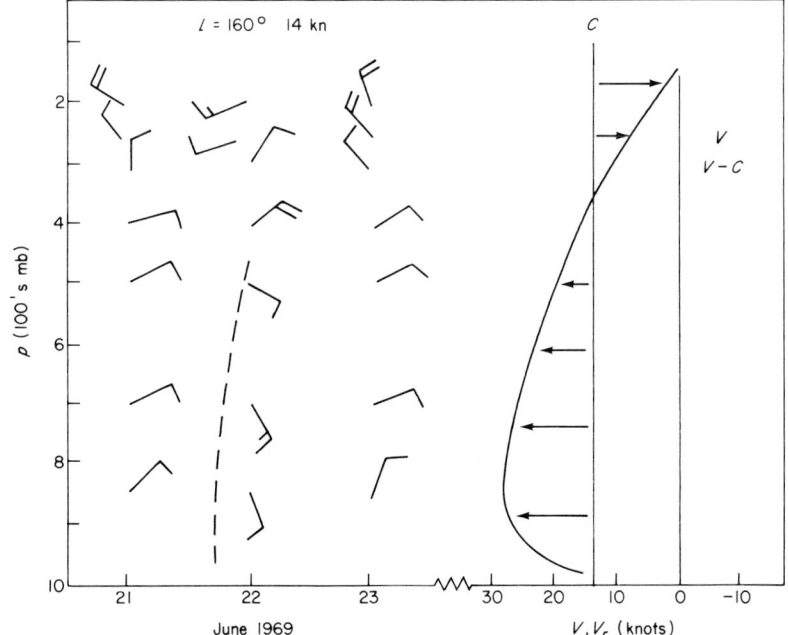

Fig. 8.14. Time section of relative winds at Maracay, 21-23 June, 1969. Right: Vertical profile of V or $V\text{-}C$; direction of C is 160°.

Approximately, $C = 5$ m/s in that direction. In the lower troposphere the relative wind is from west, especially in the lowest 100-150 mb where the equatorial westerlies move in the opposite direction from the wave (Fig. 8.15). Hereby, the strong inflow occurs on the forward side of the axis into the vorticity maximum, and this is where the rain area has been observed to be situated in such equatorial trough cases. At high levels, much outflow is crosswise, but a forward component is also present with easterlies generally increasing upward. Therewith the warmest air aloft and the outflow ridge are located forward (west) of the axis, which explains the large difference in structure of the Caribbean and East Atlantic.

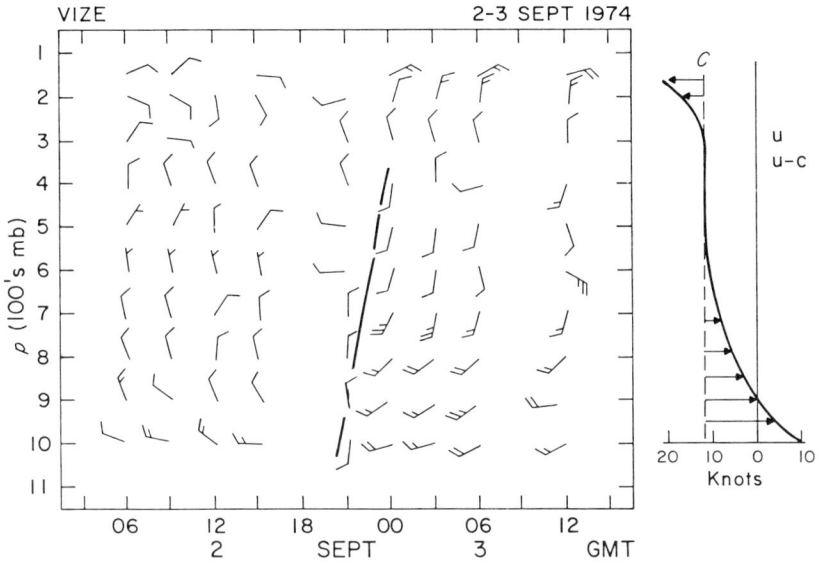

Fig. 8.15. Vertical time section of relative winds at *Vize,* 2-3 September, 1972. Right: Vertical profile of U or U-C; direction of C is 90°.

The central diagram portrays a faster propagation speed with the same wind profile as on the left. Now the relative inflow is shifted to the forward side and air overtakes the system only in a midtropospheric layer; the upper outflow is strong and directed backward. In these cases strongest precipitation occurs very close to the axis itself, but light rains from backward sloping anvils may continue for one or more hours. In extreme situations of squall lines, which may move with 15-20 m/s *(C'),* the line will outrun the air at all levels (44, 46). The cold core is still present, but analysis is more complex than in the wave situations.

SYNOPTIC-SCALE WEATHER SYSTEMS

In the easterly trades, when no equatorial westerlies are present, the reasoning based on Fig. 8.13 remains valid. When waves move slowly and trade wind speeds at low levels exceed C, the left profile of Fig. 8.13 applies. The air enters the cyclonic vorticity area from east and converges and ascends there. The amplitude of the trajectories exceeds that of the streamlines as the air gradually overtakes the wave axis.

When the wave speeds exceeds basic current speed, Fig. 8.16 becomes relevant. Troughs and ridges overtake air initially situated to their west. Air columns then move southward and eastward ahead of troughs in relative motion; the inverse occurs to the rear. Streamlines and trajectories are out of phase. On the path from P to P' in Fig. 8.16 the air acquires relative cyclonic vorticity, as the vorticity centre overtakes the air. Here, then, are convergence, ascent and precipitation. After crossing the trough line toward east in relative coordinates, the high vorticity area "leaves the air behind" as the latter turns anticyclonically and its vorticity decreases. Weather and cloudiness end quickly.

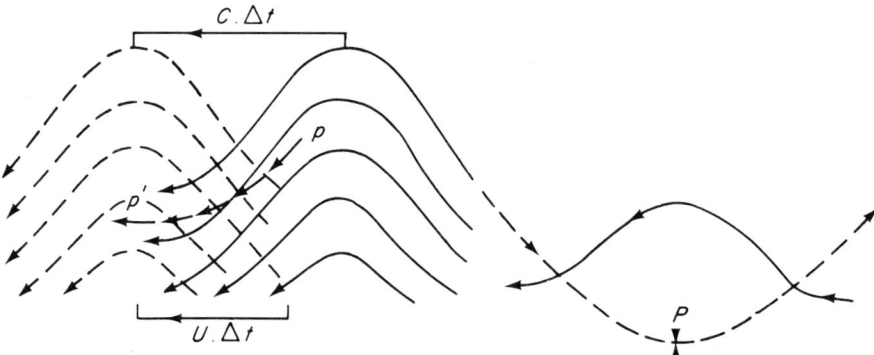

Fig. 8.16. Left: model showing air east of trough being overtaken by trough line. Right: streamline (solid) and trajectory relative to wave (dashed) for wave moving westward faster than air ("Tropical Meteorology" (1954), H. Riehl).

In the case of the slow-moving wave, the change in vorticity east of the axis is due to motion toward higher relative vorticity and also to motion to a higher latitude (β-effect). The associated convergence and rainfall are likely to be large. For the fast-moving wave, the two effects are opposed as in middle latitude cyclones. Normally, the change due to relative vorticity predominates, but in case of a flat wave the two effects may become equal, so that $f + \zeta_{\text{rel.}} = $ constant. Conservation of absolute vorticity is the mechanism of the Rossby wave which travels without convergence or divergence. Accordingly, the weather may remain indifferent during flat wave passages. While a "null" case, hard to detect on weather maps because of small amplitude, these waves nevertheless can be observed occasionally. Further, many waves that move faster than the basic current are attended by only a little rain, even though there may be an extensive middle and high cloud cover.

Holton (29) has discussed the relation between vertical shear of the mean zonal wind and the location of the precipitation maximum from more general aspects of wave mechanics. He does find that Pacific waves studied can be interpreted as Rossby waves driven by release of latent heat of condensation. The structure of the wave disturbance is thought to be very sensitive to the vertical shear of the mean zonal wind just as in the above discussion. With westerly shear in the lower troposphere the precipitation occurs to the east of the surface trough, but with easterly shear the precipitation zone is west of the trough. In principle, this result agrees with that derived above.

While "weather" may be situated variably east or west of an axis in the trades, it will always be located in the equatorial westerlies as long as synoptic weather events follow the predominant westward path (17, 75). Therewith, precipitation tends to be displaced to the south of the equatorial trough axis in the Northern Hemisphere, and this was noted in Chapter 3 as a mean climatic property for the continental half of the equatorial trough zone. Even in the trades, especially near their equatorward margin, a shallow layer with equatorial westerlies temporarily appears in many synoptic situations as air is pulled northward and turns clockwise with increasing latitude. Due to their short duration these westerlies do not appear on mean climatic wind charts, but they are associated with high-precipitation events.

Point convergence

Much effort has been expended in attempting to gain some understanding of the mechanics of synoptic disturbances through calculations of vorticity balance (64, 79). Since wind charts usually locate the cyclonic vorticity centre near 500 mb, where it should be in case of balanced wind, some progress has been achieved with this approach. One obstacle has been the vertical vorticity transport. On the synoptic scale, production of vorticity with convergence in the low troposphere, and corresponding extinction in the high troposphere, far exceed what can be supplied by the balancing link, the vertical vorticity transport. The conclusion has been that cloud-scale vorticity transport must be taken into account.

One can follow this trend of thought further by noting that nearly all net mass flux to the upper troposphere goes up in undilute cumulonimbi occupying no more than 0·1% of the tropical surface area. In essence, the description of synoptic convergent areas approaches the classical case of revolving fluid with a very small sink in the middle where fluid is withdrawn. While it seemed satisfactory to deduce the general location of rain areas from large-scale vorticity distributions and relative winds, it becomes necessary to examine whether the wind observations are compatible with the concept of a contracting mass of fluid which is removed vertically in a very small central area. The general problem for a circular disc with uniform rotation was solved by Rayleigh (62) and later extended to include uniform earth rotation (4). No net convergence takes place during contraction and fluid is evicted only at the centre. A detailed account is given in Brunt's text (A-2).

Consider the polar coordinate system of Fig. 8.17, where v_θ and v_r are tangential and radial wind components. At any radius r the relative circulation $C_{\text{rel}} = 2\pi r v_\theta$. The circulation due to the earth's rotation about the centre of the coordinate system is $\omega \sin \phi r$. Therefore, the absolute circulation $C_{\text{abs}} = 2\pi r (v_\theta + \omega \sin \phi r^2)$. The absolute vorticity averaged over the area (A) enclosed by the circle around which the circulation is determined, $\zeta_a = C_{\text{abs}}/A$, or $\zeta_a = 2 v_\theta/r + f$. If C_{abs} and the absolute angular momentum around a circle with radius r is conserved, i.e. when no forces such as friction are acting,

$$v_\theta r + \omega \sin \phi r^2 = v_{\theta 0} r_0 + \omega \sin \phi_0 r_0^2, \tag{8.1}$$

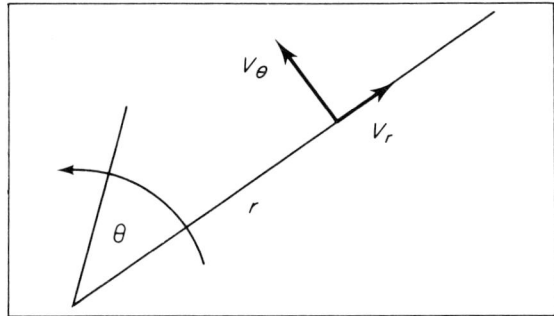

Fig. 8.17. Polar coordinate system for contracting fluid calculations.

where r_0 is some initial radius such as $r_0 = 310$ km for a converging 5° square (area 30×10^4 km^2). This theorem is considered further in Chapter 9. For a clear presentation of the calculations one can also write, if the latitude change may be neglected over short distances within a synoptic envelope, that

$$v_{\theta r} = \frac{r_0}{r} v_{\theta 0} + r\omega \sin \phi \left[\frac{r_0^2}{r^2} - 1 \right]. \qquad (8.2)$$

Consider an area contraction by an order of magnitude going from synoptic to mesoscale as described in the scale hierarchy at the beginning of this chapter. At the outer radius r_0 the relative circulation may be zero, so that only the second term on the right side of Eq. (8.2) need be considered. The factor $r_0^2/r^2 = 10$, therefore, $r_0 = 3 \cdot 2 \, r$. For $r_0 = 310$ km, $r = 100$ km very nearly. Inserting in expression (8.2) and computing at latitude 10° where $\sin \phi = 0 \cdot 17$, $v_{\theta r} = 11$ m/s if the contraction takes place into a single mesoscale area.

One can see at once how important such contraction is for potential hurricane formation, even if the full 11 m/s are not realized because of friction. In the normal synoptic event, such massive concentration does not occur; rather we have a model concentration into 15 mesoscale systems with area of 2000 km^2 each. Because of this scattering, 15 areas of 2×10^4 km^2 contract into 15 areas of 2×10^3 km^2. For this case $r = 25$ km, the equivalent radius of a mesoscale radar area, and $v_{\theta r} = 2 \cdot 8$ m/s. The same result could be obtained directly by setting the absolute vorticity $\zeta_a = 2 v_\theta/r + f = 10 f_0$.

Of course, computation of v_θ/r is merely a convenience; the relative vorticity in actuality may also appear as shear or some combination of shear and curvature. However, we see clearly that the equivalent tangential motion is only 25% of that obtained in the contraction to a single mesoscale system. Hence it is most important for consolidation to occur for starting a cyclone. The *scattering* of the convergence into 15 mesoscale concentrations makes the whole system very inefficient, but this is a distinct property of the "normal" synoptic scale system; the more scattered the average convergence becomes, the less efficient will be the generation of organized circulation, even though there will be generation of kinetic energy through ascent in undilute towers with a warm outflow anticyclone in the high troposphere. It is not known why the overall convergence breaks down into separate mini-convergences with frequency on the order of 10^1.

The circulation of $2 \cdot 8$ m/s around a mesoscale system is realistic.

Considering that v_θ has been assumed zero at r_0, frictional influences are negligible at first. No doubt they will decrease the circulation, especially if higher latitudes are considered where $v_{\theta r}$ would be greater, for instance double our value at latitude 20°. No measurements of line integrals flown around mesoscale rain areas have been published. In view of the short life-cycle of most such areas, it would be difficult to complete a useful circumnavigation with the steady state assumption. In any event, it must now be taken into account that even the mesoscale system does not act as a unit but breaks down further into a series of undilute towers. Their number has been estimated earlier as eight for towers with area of 25 km²; again, this number must be considered as devised for modelling purposes, subject to large variation in individual situations.

It may be remembered that the number eight is derived from the postulate of area contraction by another order of magnitude before the converging fluid is evicted. Equation (8.2) may now be used to calculate the circulation around the towers. Alternately we can take $\zeta_a = 2 \cdot 5 \times 10^{-3} \text{s}^{-1}$ for a total area contraction by a factor of 10^2. To a high degree of approximation $\zeta_a = 2 v_\theta/r$ in this case. Given the area of 25 km² per tower, hence $r_{tower} = 2 \cdot 8$ km, $v_{\theta tower} = 3 \cdot 5$ m/s. Even without friction this velocity around each of the 120 model towers of the synoptic system is vastly less than the 37 m/s which would result through contraction of the whole area convergence to a single area with mass ascent.

The tremendous degradation of the synoptic convergence through the scattering process is evident, and this remains the point providing most insight into the mechanics on the synoptic scale. However, the circulation of 3·5 m/s or rotation rate of once in 90 minutes is large enough so that it should be observed by aircraft or even visually. From general experience with cumulonimbi such a rotation rate is considered highly unlikely. Friction—ground or lateral—no doubt acts to reduce the mean vorticity over the area of a tower. However, the extremely high rotation rates are not requisite for the vorticity balance mentioned above. In general the absolute vorticity at 500 mb measured on the synoptic scale is $2f$, and this is the vorticity considered for vertical transport using large-scale modelling. From the preceding calculations it is obviously possible to increase the transport by almost any amount needed by invoking the capability of the towers in transporting vorticity as well as energy. An increase by a factor of 2-5 is readily achieved. Thus, the undilute tower turns out to be an efficient mechanism for the dynamics of tropical rainstorms as well as for their thermodynamics and energetics.

Frequency of synoptic rain events

If synoptic envelopes travel with a mean speed of 5° latitude per day or 12-13 knots, fast in some areas and slow in others, then 15 such systems would deliver precipitation on 15 5° squares in a day. Since there are 72 such squares in a circumpolar belt 10° latitude wide, the entire belt would receive such precipitation in 72/15 or roughly five days, assuming even distribution. Interestingly enough, this is precisely the time interval that experience and many spectral analyses of rain frequency (86) have shown to be average. Monthly rainfall charts of the equatorial trough zone indicate a mean precipitation of 17 cm/month (Chapter 3). Given six rain occurrences per month, the average rain per synoptic envelope (coming from a single mesoscale cloud per station for the most part) would be close to the 2·5 cm estimated earlier as average mesoscale precipitation.

Synoptic envelopes

The preferred direction of motion of synoptic disturbances in the rainy season is from east to west as has been long known from forecasting experiences and affirmed powerfully by longitude-time composite strips for discrete latitude intervals (11) (Fig. 8.18) (see also 84a for other detailed illustrations). Such "strips" have been used for, say, the height of the 500 mb surface in middle latitudes where this form of representation has become known as "Hovmoeller diagram". From Fig. 8.18 one gains the impression that particular synoptic cloud masses travel all the way around the globe, a matter of long-standing speculation. Of course, they do not remain in steady state, but increase and decrease in intensity. Some disappear completely, while others fade out but reappear again after some days. The longitudes where the latitude interval 10-15°N crosses continental coasts, are most obvious. In the Atlantic, for instance, there is a general trend toward decreasing intensity from east to west across the ocean, until the band strikes Central America near longitude 85°W. The frequency of cloud bands is not easily read, but approximately 20 per three months, not far from the five-day interval discussed above, appears to be a reasonable estimate. In contrast, the rate of propagation, which is fairly uniform, is easy to determine. The slope of the cloud lines indicates one transit around the world in two months, or 6° longitude per day, not far from the average observed celerity of synoptic envelopes. This type of continuity chart should be an ideal

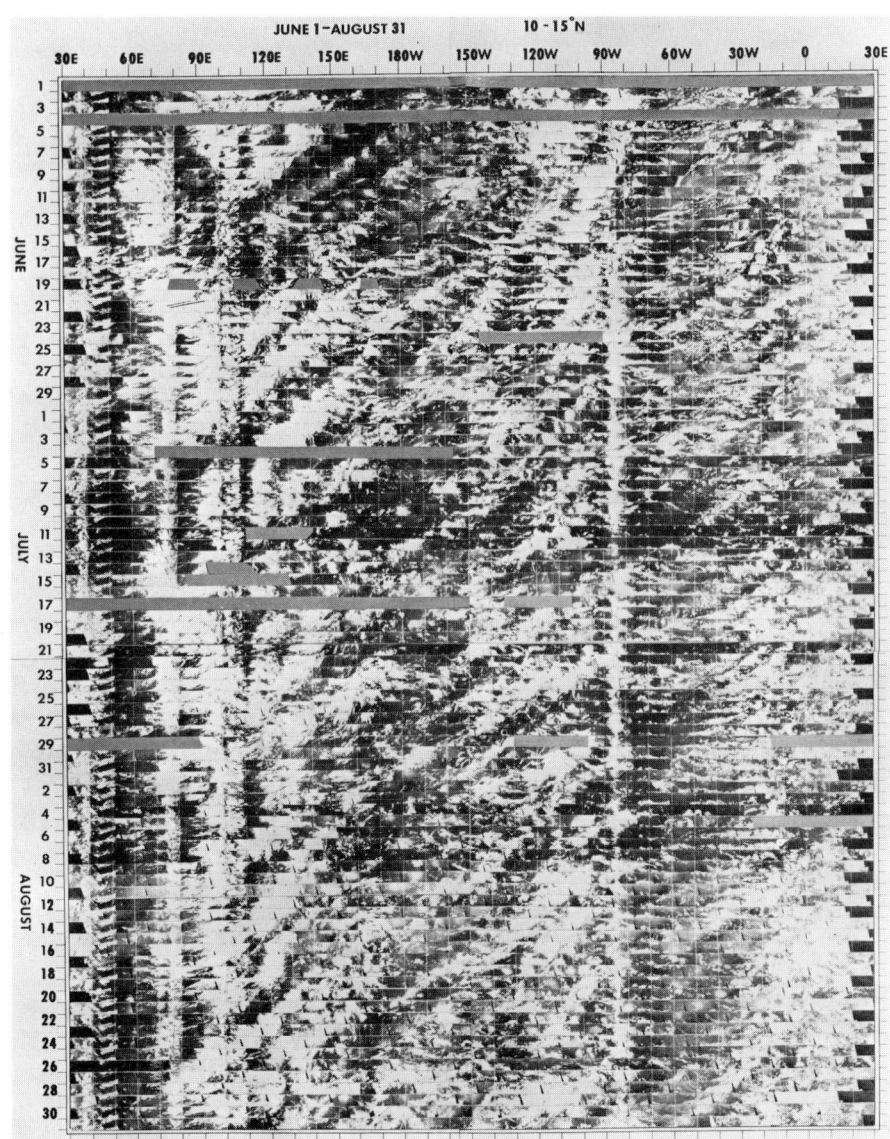

Fig. 8.18. Time-longitude diagram of satellite cloudiness between latitudes 10-15°N, June-August 1967 (11). Courtesy Dr. J. M. Wallace, University of Washington, Seattle.

tool to be kept on a current basis in forecast offices for, say, the five day outlook.

The synoptic envelopes containing most active weather have been identified, for the most part, as waves, vortices and shearlines between opposing currents. Complex interactions take place between lower and upper troposphere and between tropical and extratropical weather systems. On some days the flow, especially in the upper troposphere, is so turbulent in the Caribbean region that it cannot be classified at all in terms of particular shapes. In situations such as the extreme Manila rain episode of 1972, described later, a circulation anomaly but no complete synoptic entity can be identified.

In the face of this complex situation investigators have naturally tended to bend their efforts on recognizable patterns in terms of classical fluid mechanics. Of these, the analysis of waves is the most advanced.

Waves in the easterlies

From an early beginning made in the Caribbean from 1943-1946 (69), description and analysis of waves in easterly currents has grown into a body of literature, the size of which is far larger than for any other tropical weather system except hurricanes. This literature, usually based on series of observations in one or another part of the globe, does not tie together well enough to be presented in a small and compact form. External controls differ greatly, and any resultant wave structure is very sensitive to these. Basic current profiles may differ substantially from year to year in one region (Fig. 7.11). A deep easterly current from low levels to the high troposphere is a requirement for wave development, similar to waves in westerly currents. Where strong upper westerlies intrude into the tropics, the waves should be expected to be damped or non-existent. On account of the general asymmetry of the general circulation, with polar westerlies closer to the equator in the Southern than in the Northern Hemisphere, the waves should be predominantly a Northern Hemisphere phenomenon. This must be true, since their existence has been variously denied by Southern Hemisphere sources. In the Caribbean and the western Pacific (86) the strength of the easterlies varies, so that there will be many in some years, as in the Caribbean in 1969, and very few in other years with strong westerlies, such as 1972 (from Fig. 7.11). Over western Africa, where the easterlies always appear to be deep, such variability of occurrence has not been reported.

SYNOPTIC-SCALE WEATHER SYSTEMS   343

Not every cyclonic curvature in the trade wind streamlines denotes a "wave" (69); it can arise, for instance, from superposition of trough or vortex in the upper troposphere. Figures 8.19 and 8.20 are particularly fine examples from a Caribbean analysis of Yanai (89). Two areas of cyclonic flow are evident at 700 mb, one in the eastern, the other in the western Caribbean. The western area underlies a very sharp trough,

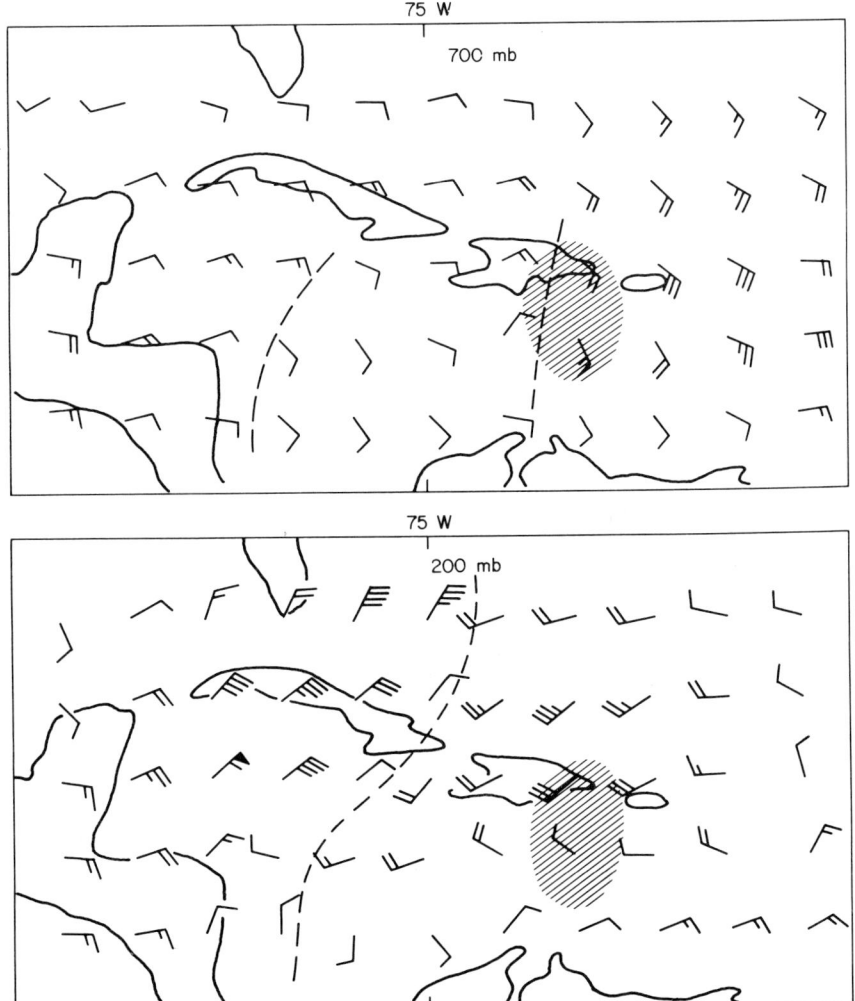

Fig. 8.19. Computer-analysed wind charts (knots) at 700 and 200 mb for the Caribbean area, 23 August, 1962, 1200 GMT (89). Dashed curves mark cyclonic vorticity maxima. Figures 8.19 and 8.20 courtesy of Dr. Michio Yanai, University of California, Los Angeles.

virtually a shearline at 200 mb, it is a downward reflection of this feature. The heaviest concentration of cyclonic vorticity is attached to the eastern area. There, the large satellite cloud is located and upper streamlines curve clockwise, while the sky is almost cloudless along the 200 mb trough with its high vorticity area farther west. Hence, the area around the lower cyclonic curvature in the western Caribbean is also cloudless; the "wave trough" is, of course, the eastern one.

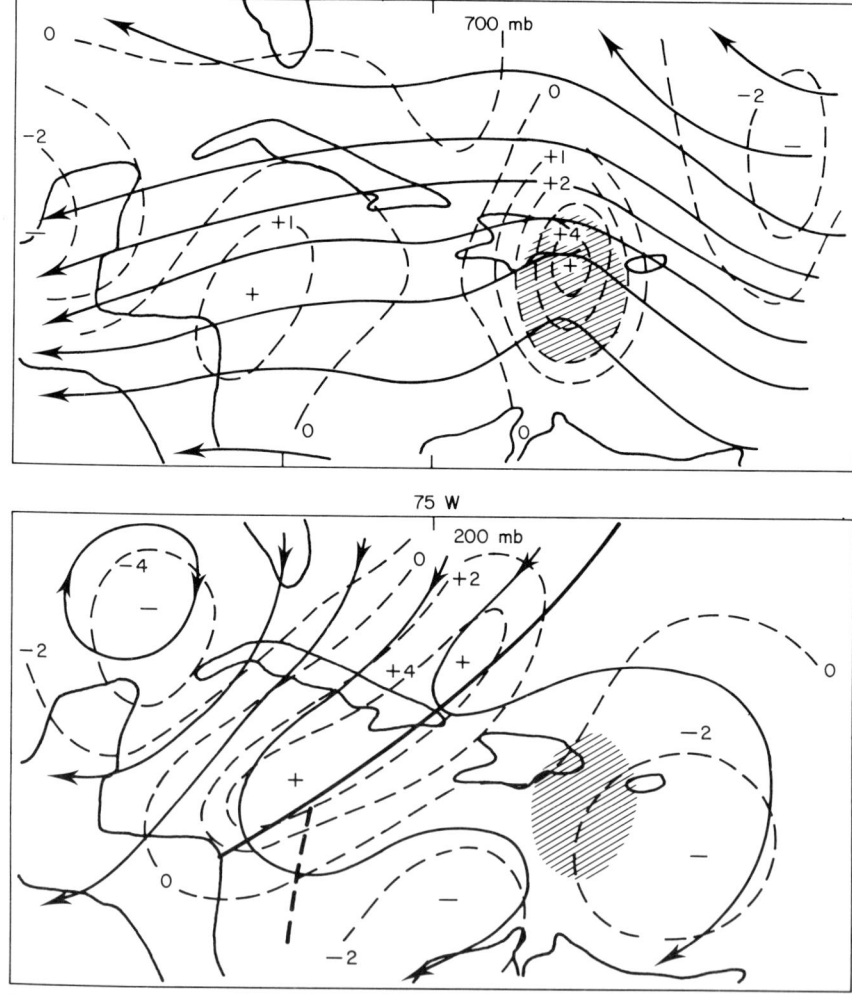

Fig. 8.20. Streamfunction for the winds of Fig. 8.19 and relative vorticity (dashed, $10^{-5}s^{-1}$) (89).

Figures 8.19 and 8.20 and their further analysis by Yanai demonstrate many of the points made in the preceding modelling section; they include such features as convergence and sinking motion in the upper cold trough, rising and diverging motion in the upper warm ridge with resultant conversion of potential to kinetic energy. A clear account of the different kinds of wave structure and of cyclonic flow not related to waves is due to Merritt (43).

Origin

By far the best source of trains of waves, not just solitary wave formations, is Africa. Already in the 1930s (57, 67) the existence of squall lines in West Africa attended by sharp pressure rises has been pointed out. These move rapidly westward into the Atlantic Ocean, and have often been traced all the way to the Caribbean, greatly aided by the arrival of the satellite era (10). Of course, such tracks had been known for many years to hurricane forecasters who coined the term "Cape Verde hurricane" and always kept their eye out to the east for suspicious ship reports and arrival of pressure falls over the Lesser Antilles and on ships farther east.

The waves do not track back to the Arabian Sea, but form downstream from the Abyssinian mountains where the lower easterly current across Africa first becomes organized (5, 9). There, a sharp north-south temperature contrast exists over a distance of several hundred kilometres. With the warmer air in the north, the low equatorial westerlies change quickly to easterlies with height; there is an indication of an easterly wind maximum near 600 mb where this temperature contrast vanishes. At least qualitatively thermal and actual wind shear agree. From vertical north-south cross-sections it has been found that the gradient of potential vorticity reverses or becomes zero between $1°$ and $15°N$ over the eastern part of Africa (5); such reversal is regarded as a criterion for instability (12, 50). The shear of the horizontal wind, both with latitude and with height, contributes to the energy of the waves.

Another viewpoint has been offered by introducing the concept of a hydraulic jump under an inversion as a potential mechanism (19, 20). One-dimensional waves, i.e. weak east wind followed by strong east wind, would appear first. They may later become two-dimensional with the appearance of north and south-wind components. Freeman (19) believes he has found such waves in the Southwest Pacific. Some waves arriving in the Caribbean and northern South America from the Atlantic appear to be very nearly one dimensional from sequences of upper winds and westward travel of satellite clouds elongated east-west with small latitudinal width. An inversion sometimes, but by no means always, precedes these "waves".

Structure

Figures 8.6 and 8.9 have already given a glimpse of wave structure in the eastern Atlantic. An overview of the whole last period of GATE is contained in Fig. 8.2 with sequence of precipitation in Fig. 3.25. Four passages with cyclonic turning are distinctly evident at the surface. At 200 mb, the easterlies are fully dominant and north-south components are out of phase with the low levels. All four wave passages are associated with rain amounts in the class of rainstorms (Fig. 3.25); they contribute to the upper 50% of precipitation.

Analyses of the waves have been undertaken in the African sector (6, 10) and the western Pacific (63). Reed (65) proceeded to evaluate the period shown in Fig. 8.2 in a quantitative and three-dimensional manner. He found eight waves over a 20° latitude interval. In general, the intense part of one wave is confined to 5°, at most 10°, latitude. The waves were normalized for wave length by dividing them into a number of sectors following previously developed methods (26, 23, 6). Reed uses the division into eight compartments (6) where number four is placed at the trough and number eight at the ridge, and the intermediate distance is split up evenly. By definition, the north component of the low-tropospheric wind changes to southerly at the trough, reverse at the ridge. In the north-south direction averaging is performed with respect to the centre of cyclonic vorticity at 500 mb.

The East Atlantic-African composite wave is based on numerically analyzed charts such as Fig. 8.19. Small-scale features of the flow are smoothed out so that the synoptic scale is brought out clearly; average synoptic wave length was 24° longitude. Many short-lived features of smaller area come and go within such an envelope, and these were suppressed by longitudinal smoothing. The eight values in each longitudinal row were analyzed harmonically and only the first two terms were retained.

The composite wind field, presented at four levels (Fig. 8.21) has a very smooth configuration. At the surface, one recognizes the equatorial trough, at 700 mb, the well-pronounced wave and at 200 mb, the wavy easterly flow with large phase shift from the surface. Reversal of the sense of the meridional component with height is well brought out in Fig. 8.22, an east-west cross-section at the central

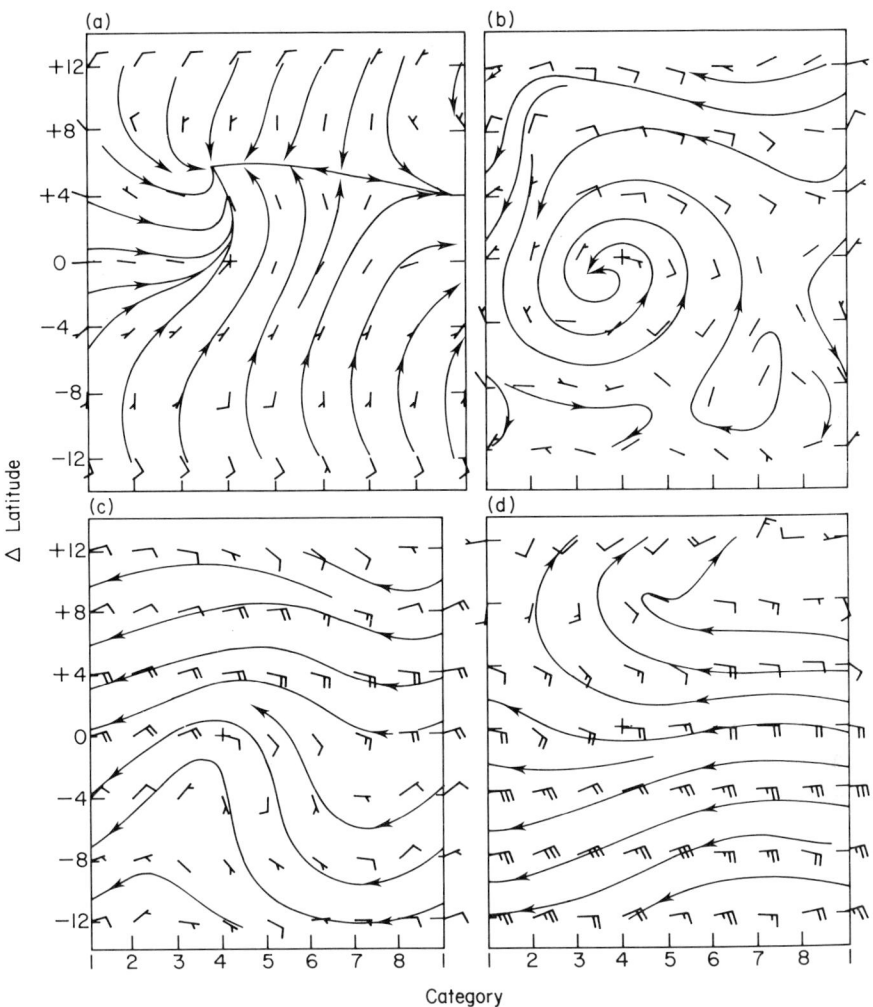

Fig. 8.21. Composite wave in easterlies in the East Atlantic—West African region. Winds (knots) at four levels for the categories described in text (65). Figures 8.21-8.24 courtesy Dr. R. J. Reed, University of Washington, Seattle.

latitude. Along the same section, relative vorticity (Fig. 8.23(a)) has a pronounced maximum at the trough line between 700 and 600 mb, with maximum value as large as the Coriolis parameter. It is most interesting that cyclonic vorticity and convergence (Fig. 8.23(b)) extend almost across the whole wave at the surface. In Chapters 1 and 3 we postulated that the equatorial trough zone must be a cyclonic and convergent vortex street in order for climatic charts such as Fig. 1.12

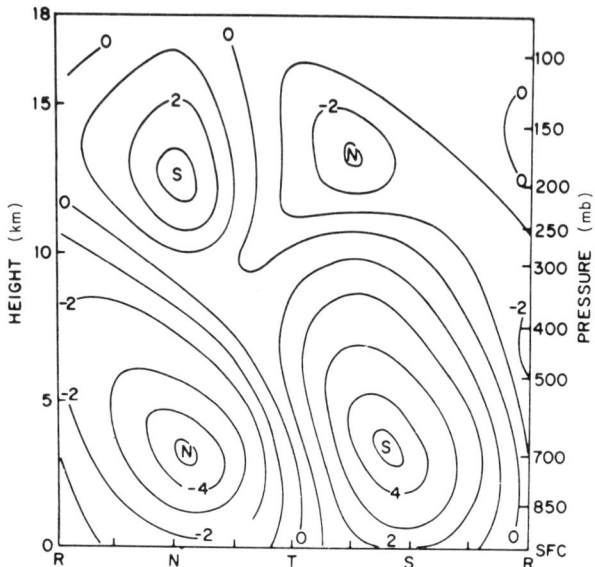

Fig. 8.22. Meridional wind component at the central wave latitude (65).

to exist. For the area under discussion, this postulate has been verified. It also follows that the mean vertical motion is almost everywhere upward (Fig. 8.24(a)), largest west of trough; the precipitation centre (2 cm/day) is associated with the upward velocity maximum.

Since the waves travel westward at an average rate of 8° longitude per day, Fig. 8.13 (right side) and 8.16 apply. The low-level air travels rapidly from west relative to the trough and enters the vorticity maximum from that direction. The general monsoon pressure gradient from the Atlantic Ocean toward Africa accelerates this low-level relative motion. Convergence and ascent occur west of the wave crest as on 2 September, 1974 and in the Caribbean example of Fig. 4.16. Interestingly enough, relative humidity turns out to be lowest in the zone of highest precipitation (Fig. 8.24(b)). This fact would be quite incomprehensible, were it not for the earlier discussion of the concentration of ascent into undilute towers. The low humidity results from downdrafts, these occupy at least ten times the area of the updrafts and, thus, are more frequently represented in the soundings that make up a statistical summary (end of Chapter 4). Similar results on the weather distribution were found for the equatorial trough in the western Pacific (45).

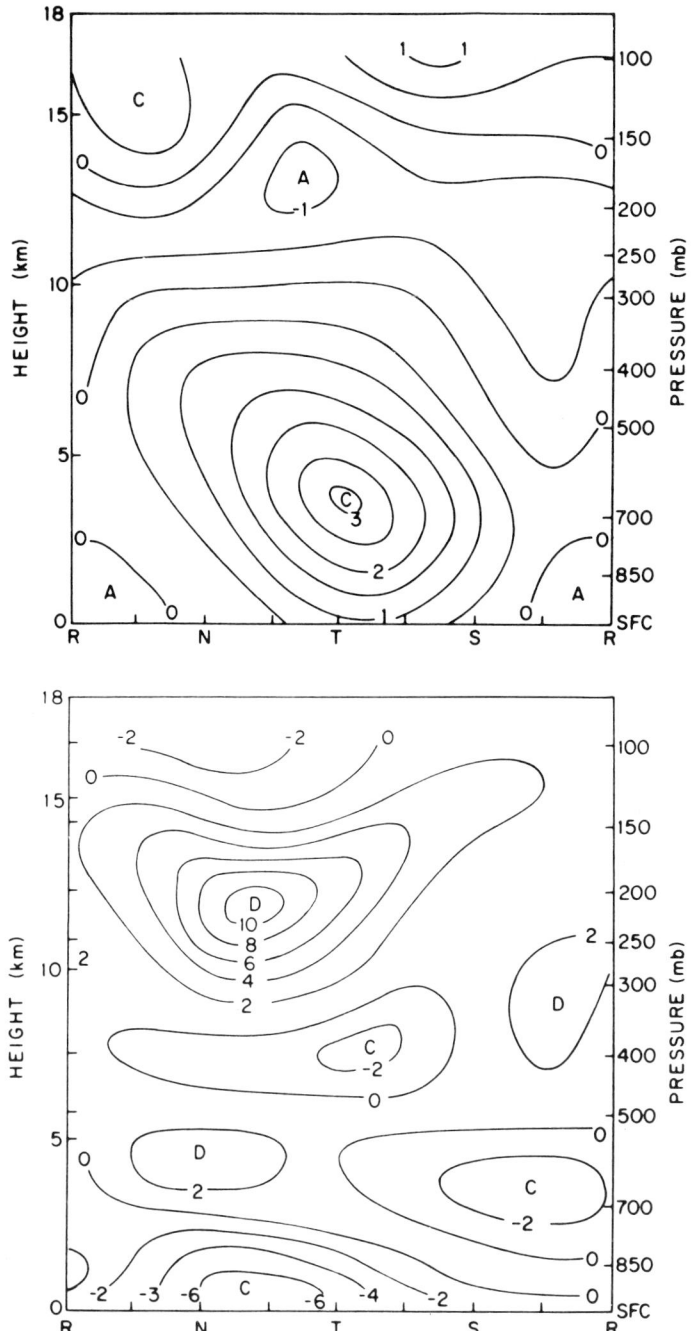

Fig. 8.23. (a) Relative vorticity ($10^{-5} s^{-1}$) and (b) Horizontal divergence ($10^{-6} s^{-1}$) at the central wave latitude (65).

350 CLIMATE AND WEATHER IN THE TROPICS

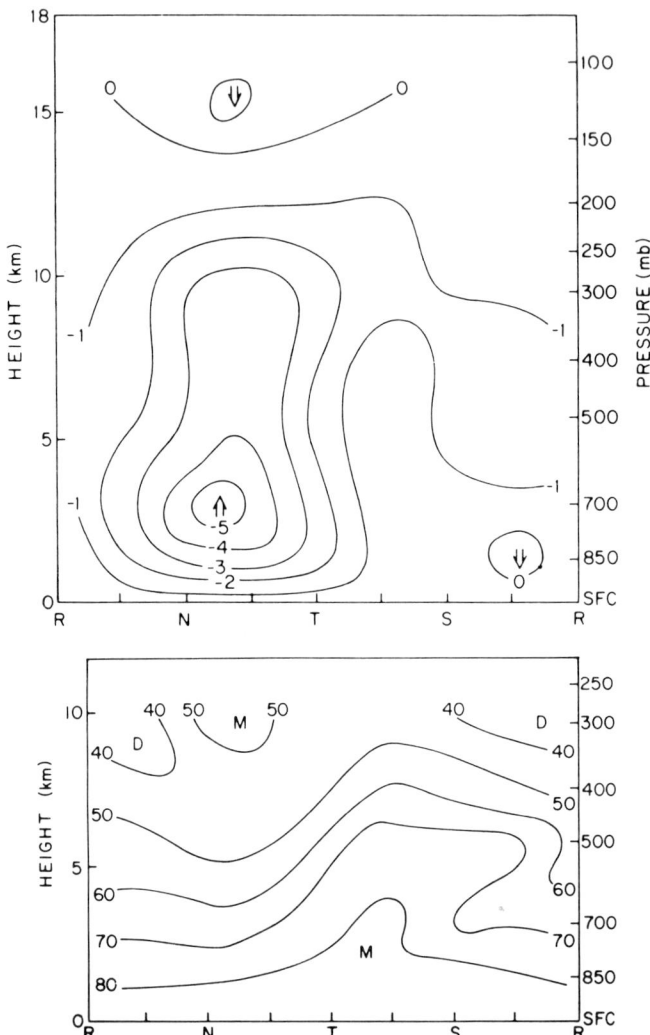

Fig. 8.24. (a) Vertical motion (mb/hour) and (b) Relative humidity (per cent) for the central wave latitude (65).

Energy of motion

As seen earlier, the development of an "outflow anticyclone" or ridge over the lower parts of the waves with high cyclonic vorticity is an essential part of their mechanism of maintenance. The high-tropospheric anticyclonic flow furnishes a means for the exit of the mass ascended in undilute towers; sometimes the anvils extend over long distances from the convective cloud, as much as 100 km and more. An increase in upper anticyclonic vorticity sometimes precedes intense rain events.

In a theoretical study of wave energetics, Estoque and Lin (16) find that the first element in the chain of energy transformation is generation of available potential energy in the upper troposphere by diabatic heating, i.e. by establishing the upper part of the temperature profile of Fig. 8.10. To this internal mechanism should be added a favourable external field of potential energy, mainly the baroclinic trough-ridge arrangement in Yanai's case (Figs 8.19 and 8.20). Often there is anticyclonic flow in the high troposphere west as well as east of a high-level trough, usually at a lower latitude (Fig. 8.25). The convective area near latitude 20° is identical with Yanai's. The lower-latitude cloud is thought to arise as dynamic instability (zero absolute vorticity) develops in the northerly current west of the upper trough as it moves into the col between upper westerlies and easterlies where the 200 mb pressure-height gradient vanishes and essentially inertial motion takes place (75).

Following Estoque and Lin (16), the potential energy generated in the upper troposphere, is transformed into wave energy (through the pressure gradients that have been generated). This includes a low-level fall of pressure often situated in the forward portions of the wave (Fig. 8.11), enabling the wave to overcome frictional dissipation of energy. Also, enhanced energy flow from surface adds to the wave energy over

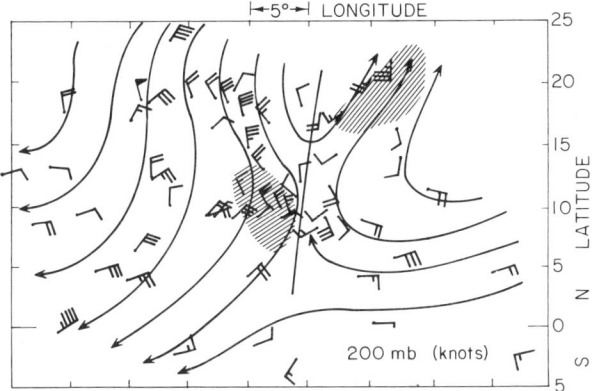

Fig. 8.25. 200 mb winds composited with respect to high-tropospheric trough for five cases in August 1969. Preferred satellite cloud positions are shaded (75).

the ocean (21). In downdraft areas, where low temperatures and specific humidities reach the ocean surface, this transfer can become temporarily very large, especially sensible heat flux. Further studies of the kinetic energy cycle are found in (36, 37, 51) among others.

Wave propagation in entirely steady state occurs, in general, only for short periods. Usually one finds increases and decreases of amplitude and cloud cover. Interaction between low and high altitudes, and between tropical waves and extratropical troughs as in Fig. 8.25, closely governs these changes and must be included in realistic models. For instance, the upper cold trough shown by Yanai (Figs 8.19 and 8.20) did subside releasing potential energy, and this led to a marked intensification of the lower tropical wave. Eventually a hurricane formed.

Reed (66) has computed the energy budget in each of the eight sectors of his composite wave. The types of charts illustrated in Figs 8.21 to 8.24 readily lend themselves to this purpose. In the heat budget over sea, the large terms were precipitation heating *(LP)* and divergence of heat transport (div h_s). Where LP was largest, near the trough, it greatly exceeded radiation balance requirements, so that div h_s also attained its maximum. Conversely, convergence of water vapour had its strongest convergence in that part of the waves. Ocean evaporation was largest in the relatively clear and dry portion near the ridge and least in the rain area and ahead of it. Combining the GATE area with West Africa, the centres of precipitation heating, divergence of sensible and convergence of latent heat shifted westward by one interval as given in Fig. 8.21. Thus, the detailed internal structure of waves can be treated in a budgetary manner as well as the whole wave envelope.

Waves in the stratosphere

Apparently stimulated by the discovery of the 26-month, biennial, stratospheric wind reversal (Chapter 1), Yanai decided to make a detailed study of stratospheric winds available in considerable density during several observation programmes in the western Pacific mainly in the 1950s. Westward-moving long waves were discovered (88) described in detail by Maruyama (41). His abstract concerning the period of (northern) spring and summer of 1958 summarizes the findings as follows:

> Vertical time sections of upper winds show that the wind direction at about 18-24 km height oscillates between southwest and northwest with a period of four to five days. Synoptic analysis shows the existence of the large-scale

wave disturbances moving westward against the westerly flow at a speed of about 2000 km/day. The wave length may be as long as 10,000 kilometers. The phase lines tilt westward with increasing height. Power spectral analysis of meridional wind component indicates that the oscillation is prevailing in the lower stratosphere confined close to the equator. By removing the basic flow through a simple procedure, the disturbances are found to take the form of eddies centered on the equator. Pronounced eddies observed in April 1958 are investigated in particular to determine three-dimensional structure of the disturbances. The westward propagation of the disturbances is apparently owing to the β-effect . . .

When very long waves are mentioned, one immediately thinks of the β-effect in the Rossby long wave formula (Eq. (8.4)). Applying this formula without modification to the data given above, the basic current $U = +32$ m/s, i.e. the direction is correct—from the west, but speed is about double that given by Maruyama on the equator. Any correction limiting the latitudinal width of the wave would very quickly overcorrect through reduction of the dynamic term, also computed by Maruyama. Further analysis is contained in (91, 85, 30); the latter contains a good literature index. It would lead us too far afield to take up the various advanced theoretical concepts involved, but we may mention that one frequency equation (42, 77) envisages a mixed Rossby-gravity wave, which is a rather different approach.

High tropospheric cyclones

Upper and lower troposphere often act seemingly uncoupled for periods of days and more as described earlier. It would be impossible to guess the 200 mb flow from surface or even 700 mb charts, but the reports of weather and especially satellite photographs often provide a clue: where convective cloudiness is missing completely in the rainy season in an area such as the Caribbean, the probability is high that the 200 mb flow is cyclonic. As attested from several long photographic missions in the Pacific Ocean, undertaken by Claude Ronne of the Woods Hole Oceanographic Institution, this also holds true there in many instances (40). On the south side of an upper cyclone, west of the Marshall Islands, very desiccated and elongated anvils, broken off from their low-level stems, covered hundreds of miles; the westerly shear between 500 and 200 mb, at the edge of an elongated trough, was about 15 m/s. These clouds convey the impression of a burnt-out forest. Later, on a leg from Wake Island to Hawaii, clouds die out completely over a distance of 300 miles, apparently under a high-level cyclone.

It is noteworthy that these cloudless areas occur with cold temperatures in the layer 500-200 mb, i.e. in the areas of largest temperature difference between low and high levels. It follows, as already demonstrated from Yanai's case (Figs 8.19 and 8.20), that presence or absence of an outflow mechanism in the high troposphere is the dominant factor for maintaining convection, and not thermal instability. In the core of high-level troughs or cyclones, pressure-heights are at a minimum; any outflow would have to work against the pressure gradient, and that just seems to be impossible for normal convective activity. Upon close inspection, cold core Lows rarely "transform" directly into warm core ones. As in Figs 8.19 and 8.20, a warm centre will start to build perhaps 300-500 km away from the cold core, usually on its eastern side; when the cold dome collapses (90), potential energy is released and a convectively-driven cyclone may spring up at the location of the wave to the east. Here a means exist for the air in cumulonimbus anvils to escape from the scene of convection and to subside somewhere else later.

Propagation of vortex trains

The origin of the upper Lows can often be traced to an elongated trough in the polar westerlies moving eastward with northeast-southwest tilt. The lower portion breaks or "cuts off" (53) and then the centre remains stationary or moves westward, now an isolated centre. Such centres may also form in the equatorial zone and then move toward somewhat higher latitudes (54). The fact remains that they are not isolated vortices at all times. For prolonged periods a series of upper Highs and Lows will march westward across Atlantic or Pacific Ocean, until strong interference from higher latitudes is encountered. The propagation rate, at latitude 15°, is westward at about 8° longitude per day (10 m/s) and the wavelength $(L) \sim 4500$ km. This corresponds to the displacement (C) computed from the Rossby frequency formula for such systems travelling under conservation of vorticity

$$C = \frac{\beta L^2}{4\pi^2}. \tag{8.3}$$

The basic current $U = 0$ and, hence, does not appear. From Eq. (8.3) only westward displacement can be obtained. The case has received little attention, but in reality it occurs often and is important for the evolution of weather systems and the general circulation of the tropics.

When a wave or vortex moves westward against a basic westerly current of, say, 10 m/s, the full formula must be used, i.e.

$$U - C = \frac{\beta L^2}{4\pi^2}. \tag{8.4}$$

The wave length becomes very large, over 6000 km. Nevertheless, 200 mb troughs are observed backing against an upper westwind maximum around latitude 15°N for days in the Atlantic area (Fig. 8.1) with the rate of —8° longitude per day. Relative streamlines and isotherms become parallel; little or no cloudiness may be observed on satellite photos until the outskirts of the troughs are reached.

Maintenance of upper Lows

In the perfect barotropic mode, there would be no problem concerning the maintenance of the upper cold vortices, once aptly described by Rossby as "floating high-level icebergs". Since, however, they appear to dampen even low-tropospheric cumulus activity, at least some vertical motions must be superimposed on the basic barotropic motion. Also, the maximum vorticity at 200 mb, expressed at times as tangential velocity of as much as 25 m/s at a radius of 3-5° latitude from the centre, presumes at least an initial sink of air in the high troposphere. If the vertical motion is down, however, the "iceberg" cannot remain in steady state unless there is strong differential radiation compared to the outskirts which would permit maintenance of the temperature difference. In a fairly cloudless atmosphere, mechanisms as given by Figs 8.3 and 8.4, however, cannot be invoked.

In the absence of any other obvious mechanism it seems necessary to postulate very slight ascent through the high troposphere. There may be no clouds, or there may be very thin, high cirrus in the core observed on a few occasions in such cores at the altitude of jet aircraft. Ascent would lead to a stable thermal layer at its top; such a layer, sometimes called a "false tropopause", indeed occurs (39, 8). Other mechanisms, especially addition of kinetic energy to the high energy ring around the core, would also tend to raise the "iceberg". In this event the vortices are not in fully steady state; some reinforcement of the cold core from the general area of the mid-Atlantic or mid-Pacific troughs occurs through southward or southwestward elongation.

Several individual case analyses for the Caribbean-Atlantic area have been made (38, 8). Particularly instructive is a quote from Carlson (8):

Typically the system extends about 1,200 mi. from north to south and 2,500 mi. east to west. Many of these systems are not detectable in the low-level winds but are often revealed in the low-level temperature field as a cold trough in the easterlies which increases in intensity with height. The lowest level of closed circulation is commonly between 700 and 500 mb. but occasionally the vortex extends well below 700 mb. The maximum temperature variation between the centre of the cold Low and its periphery is found near 300 mb., and the maximum height gradient and wind speed occur near 200 mb., the level where the temperature gradient becomes reversed from that below. Winds are light in the lower-middle troposphere, but the wind speed increases with height and at high levels the strongest winds are found some distance west and northwest of the center. The center of the closed circulation tends to slope with height toward the coldest air which lies northeast of the vortex.

The most notable rainfall and convection occur a few hundred miles from center, *usually in the southeast quadrant* (my italics). Squalls and thunderstorms are often confined to a relatively narrow belt embedded within a much larger cloud cover over the eastern semicircle. Most of this cloud consists of altocumulus and cirrus beneath which the lower cloud forms tend to be suppressed or absent. There is considerable variation in cloud cover and distribution from one disturbance to another. Some systems have been known to be cloud free for a time. The vertical motion pattern is not symmetric but resembles a dipole with centers of rising and sinking motion east and west of the vortex, respectively. Strongest upward motions are naturally associated with the areas of intense convection but both ascending and descending motions are inherent in the baroclinic nature of the system.

The disturbance may either weaken or intensify. In the former instance the decrease in strength is thought to be due to an insufficient source of kinetic energy, while intensification may be promoted by a direct conversion of potential to kinetic energy through release of latent heat along the periphery. On the other hand an increase in the intensity of convection, possibly occurring near the center, may lead to the formation of a warm core easterly wave or vortex and a disintegration of the cold Low.

Erickson (15) has given a clear description of a situation in 1966 when a wave in the easterlies, passing through the Caribbean at about 7° longitude per day, approached a high-tropospheric cold core Low moving northwestward across the southern Bahamas at about 3° longitude per day. On the last days, the northern portion of the wave underneath the cold Low disintegrated and only the southern part could be followed further. On 9 August, 1966 the wave trough was still about 5° longitude east of the upper centre. A cross-section from the southeastern United States to the Lesser Antilles via San Juan, Puerto Rico, affords a good view of the flow and thermal structure north of the Antilles (Fig. 8.26). Only the northern extreme of the wave, which was well marked during passage of the eastern Caribbean, was still visible. The bulk of the troposphere was filled by the cold Low with

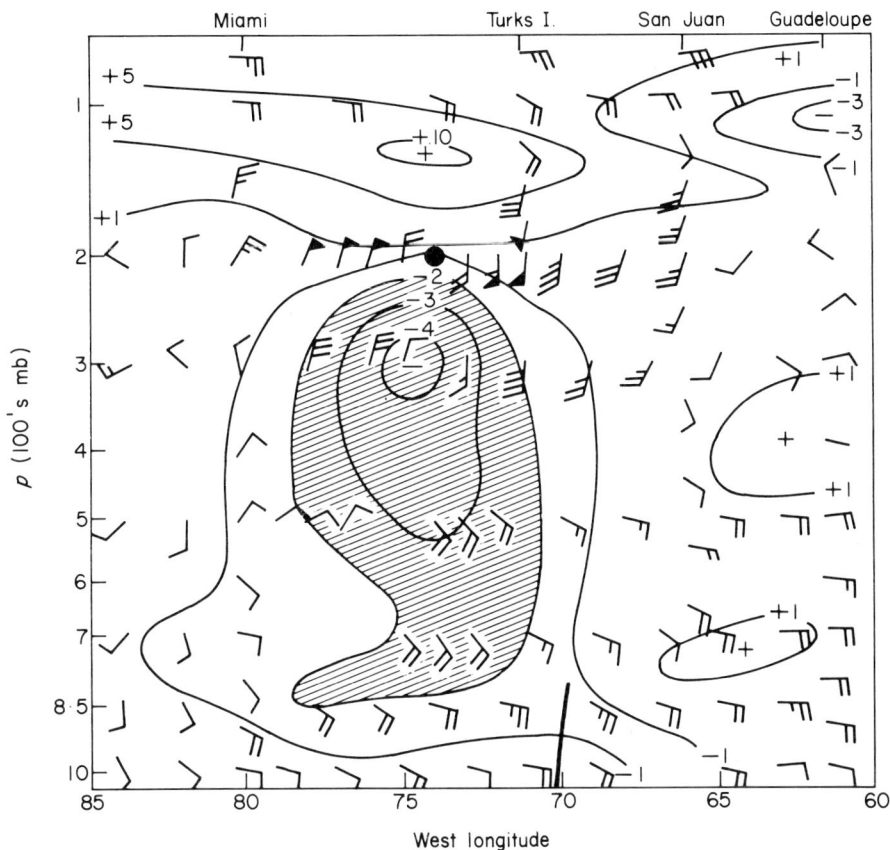

Fig. 8.26. Vertical cross-section of winds (knots) and temperature departures (°C) from the mean tropical atmosphere along a northwest-southeast axis from Florida to the Lesser Antilles intersecting upper Low Centre (heavy dot) on 9 August, 1966. Shaded: Cold core (after 15).

25 m/s winds at the top of the cold layer at 200 mb and topped by a marked stable layer or "false tropopause", a usual accompaniment of such situations which resemble the cold Low structure of higher latitudes (23).

Relative vorticity was at a maximum at 200 mb (Fig. 8.27). Inside the ring of winds with 25-30 m/s, located at about 3° latitude from the centre, it averaged $16 \times 10^{-5} s^{-1}$, compared to $6 \times 10^{-5} s^{-1}$ for the Coriolis parameter, a powerful system indeed. Computed vertical motions showed the principal region of descent under the upper Low. Principal ascent was to the east where the satellite cloud was situated and the flow became anticyclonic, favouring outflow. The computed profile of

358 CLIMATE AND WEATHER IN THE TROPICS

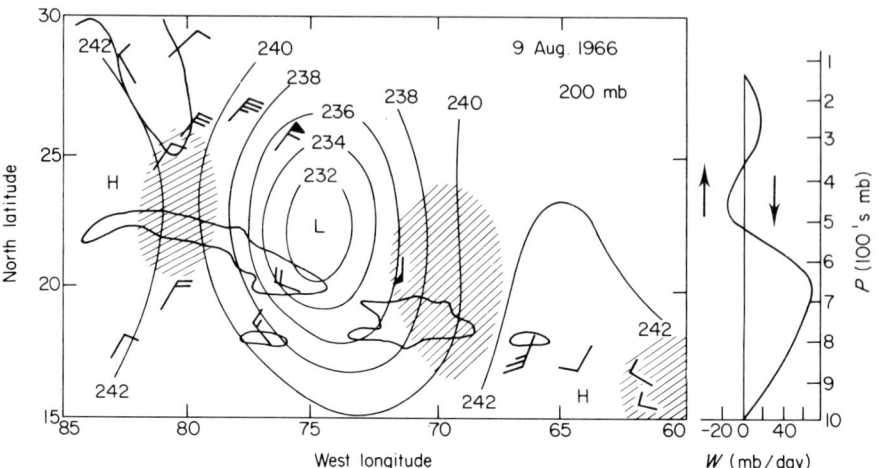

Fig. 8.27. 200 mb wind chart for 9 August, 1966. Heights in 10's of metres above 10 000 m. Shaded: Satellite cloud areas (after 15).

vertical motion did not achieve mass balance, but the case clearly fits with Carlson's and all other evidence showing that a high-tropospheric mechanism must exist to facilitate outflow from undilute towers.

Subtropical cyclones

A variety of upper Low structures occur throughout the tropics. Much effort has been lavished on situations when an upper Low for some reason penetrated to lower levels together with an outbreak of precipitation. The Hawaiian Islands obtain a large fraction of their precipitation from such complex developments; early research has centred there. In contrast to other trade wind areas, the precipitation maximum falls into winter, November to April (Fig. 3.12). Simple cold front passages are likely to be quite dry, but at times the upper trough accompanying these is reinforced through superposition with another trough outside the tropics; sometimes, a cyclone forms near the island, occludes and remains in the area as a deep cyclonic vortex (54, 81). These systems, which have been baptized "subtropical cyclones" are rather different from the uncoupled upper Low with greatest intensity near 200 mb. They are hybrids of many forms intermediate between extratropical cyclones and hurricanes. Hence, description has been rather variable; it has become clear, however, that they occur not just in the cool season, but are encountered widely in the warm or rainy seasons. Old hurricanes that have moved over land also are known to transform into subtropical cyclones at times, producing very heavy rains, for instance, in western Texas.

Simpson (81), lacking enough observation for analysis of individual cases, composited over 20 subtropical cyclones within reach of the Hawaiian Islands and other scattered upper-air stations of the eastern Pacific. He evolved a model of the "kona" storm, the main rain-bearing situation of the islands, so named after the southwestern or kona coast of the Island of Hawaii. Simpson's model of surface wind speed and precipitation in the kona storm is not unlike that seen in some intense waves in the easterlies. The ring of maximum wind is at about 5° latitude from the centre, strongest in the northern semicircle; the rainfall pattern closely follows the wind speed pattern. Inside the core, pressure gradients are weak, and rainfall, if any, is light and falls from altostratus. Rain and wind are heaviest on the east side of the semi-circle.

In an extensive survey, Ramage (61) points out that Simpson's composite contained several species of subtropical cyclones. One subtropical cyclone, for instance, passed the island of Kauai, only miles from the point with the highest rainfall on earth (Chapter 6). Sea-level pressure had fallen to 1004 mb. Heavy precipitation occurred on the forward side of the system with southeast to southwest winds. At the time of front passage with northwest winds, surface temperature dropped from 19-21°C to 13°C and dewpoint from 18°C to 11°C. A marked dry inversion appeared near 800 mb where winds remained southwesterly. Evidently a polar air sector from an initial occluding cyclone was still alive, but, in contrast to high latitudes, where the warm sector is squeezed out of existence with occlusion, at latitude 20° the cold sector has the smallest area; it cannot be forced upward.

Ramage shows a vertical cross-section with rain area 100-300 miles from the centre for a case of early April 1960, when *Tiros* I first gave useful cloud information. Farther inward, clouds were scattered in agreement with Simpson. Beyond 300 miles, upper clouds were largely absent and vigorous subsidence inversion restricted low cloud tops to around 850 mb. The section shows an isentropic dome over the rain area up to 400 mb and warm air in the high troposphere, not the "upper iceberg" discussed earlier. Indeed, this illustration is akin to the earlier rain system model of this chapter. Ramage also speaks of the downdrafts as being responsible for the cold layer below 500 mb and points out that, with the disappearance of the cyclone toward the ground, frictional energy and vorticity dissipation will have very little effect on the cyclone: "Because of its energy exporting character and the insignificant effect of surface friction, a subtropical cyclone apparently does not decay, being usually absorbed by a large-amplitude trough in the polar westerlies." It also may live for one week or longer.

Monsoon depressions

During the 1960s several cyclones were analysed for the area extending from the Arabian Sea to Pakistan and India, of which a detailed account is found in Ramage's textbook (A-15, p. 48 ff). From his illustrations it turns out that these systems have the same character as subtropical cyclones, with cyclonic vorticity maxima near 500 mb, small vorticity at the ground and anticyclonic vorticity in the high troposphere. Similar distributions, with lower temperatures in the equatorial westerlies and higher temperatures in the high-level easterlies, compared to the outskirts, were also found in "disturbances" possessing the characteristics of the monsoon depressions in the Bay of Bengal.

A principal factor which governs all synoptic weather systems of the monsoon period is the high-tropospheric, primary, easterly circulation (Fig. 1.14). In this area and season the primary wind field is so strong and permanent that all synoptic systems disappear between 300 and 200 mb. The monsoon depressions which travel west-northwestward from the Bay of Bengal through north-central India, fall into this class, and they closely resemble subtropical cyclones. An excellent description is contained in (13). Structure in many ways resembles that of Figs 8.21-8.24, except that the cyclonic vortex is deeper and the equatorial westerlies extend to 500 mb and higher at the high latitude (25°), a characteristic also observed, for instance, in the Philippines. The mechanism of formation is related to waves in the high-level easterly jet stream and attendant fields of vorticity advection (32).

For many years, monsoon depressions were thought to originate from impulses coming from the Pacific Ocean via the South China Sea, but the number of such passages, that can be traced easily, is limited. They may be generated partly near the head of the Bay of Bengal itself; a considerable fraction is intensified by energy propagation from the Pacific Ocean with a group velocity based on Rossby waves (34). From a study of weather events over Southeast Asia, one may form the opinion that an impulse is often provided through interaction between extratropical troughs, which travel eastward north of the Himalayas and elongate southward during passage of China east of the mountains. They then come into contact with the equatorial trough zone near the northern border of the South China Sea. Often a wind circulation and an area of convective activity develop in the vicinity of the meeting point.

Onset and interruptions of the Indian summer monsoon

In April and May the rains set in generally over Burma, while they are retarded over India, so that temperatures rise to extremely high values in May. Thunderstorms and squall lines may be generated in the north in association with a high-level perturbation in the pre-monsoon season (33) (cf. also Chapter 4). These appear to differ from squalls in the south at Bangalore (13°N, 77°E) where incursions of the equatorial trough begin in early May (84). Squall lines there attain their maximum in April to July; average frequency is three per month from the 20-year record excepting five in May. During April and May the squalls move predominantly from the sector north to east, thereafter between southwest and northwest, a clear indication of the equatorial trough displacement and the linkage between the squall lines and the trough.

Late in May or early in June, roughly within a three week period, the monsoon then "bursts", often accompanied by violent squalls and occasional intense cyclones on its leading edge in the Arabian Sea and Bay of Bengal. These are anxious weeks for the population since a late monsoon may mean little rain and crop failure. Surface temperature falls with monsoon onset (Fig. 2.17).

Northward progression of the equatorial trough can be traced easily over southern India and, earlier, over Malaya as a sharp shearline between easterlies to the north and westerlies to the south. Beyond 13-15°N, low-level westerlies generally pre-exist, so that the "monsoon trough" can no longer be readily followed. Often its position is deduced from rainfall amounts. After the first burst, which carries the rain roughly to the top of the peninsula, the monsoon advances more gradually toward the Himalayas and northwestern India and Pakistan. In the Indus valley the rainy season may last only a couple of months or less; here the average variability of rainfall from one year to the next is very high (Fig. 3.15). The retreat of the monsoon, beginning in September and often lasting through November, is gradual compared to the advance.

Advance of the monsoon

As early as 1686, Halley stated in the *Memoirs of the Philosophical Society of Great Britain* that the cause of the monsoon is differential heating, a theory that in principle has remained unchallenged. It is not obvious, however, why the monsoon is retarded over India as compared with Burma and with other continents of both hemispheres, and why it subsequently advances in the spectacular manner that has given rise to the term "burst".

In the past, clues have been sought principally in the sea-level pressure distribution over the North and South Indian Oceans. According to Fig. 1.7, a comparable sweep of surface air across the equator does not occur elsewhere. The southerly current is shallow, however, and restricted mainly to the subcloud layer, at least north of the equator. Above 600 m the winds are from the west and even north of west, indicating that the air that reaches India is derived from Africa, containing heavy dust contamination. (Information obtained from photographs taken on reconnaissance flights during the International Indian Ocean expedition by Andrew Bunker, Woods Hole Oceanog. Inst.)

One permanent feature that might help to explain the difference between the onset of the monsoon over India and elsewhere is the channelling effect produced by the height and shape of the Himalaya mountains. In December, a considerable portion of the belt of westerlies circles the southern rim of the Himalayas (Fig. 8.28). In spite of the altitude of the flow shown in this figure, the streamlines follow the contours of the mountains quite well. As prescribed by these contours, a trough lies near longitudes 85-90°E. The winter jet is narrow but intense (Fig. 1.15), and it transports much mass.

In summer (31), an entirely different picture prevails (Fig. 1.14). The westerlies have retreated to the north of the mountains. In the middle troposphere, a trough extends southward across the Arabian Sea, with southwesterly winds turning clockwise over India. This represents a westward displacement of 20° longitude from the wintertime position. The subtropical ridge line lies near 35°N. To the north the westerlies move along the northern boundary of the Himalayas with clockwise curvature.

From this description, Burma is always situated east of the position of the mean upper-air trough, whereas India-Pakistan lies to its west in winter and to its east in summer. Superposition of the high-tropospheric flow pattern and its attendant pressure field on the low-level circulations must have the effect of accelerating the monsoon

SYNOPTIC-SCALE WEATHER SYSTEMS

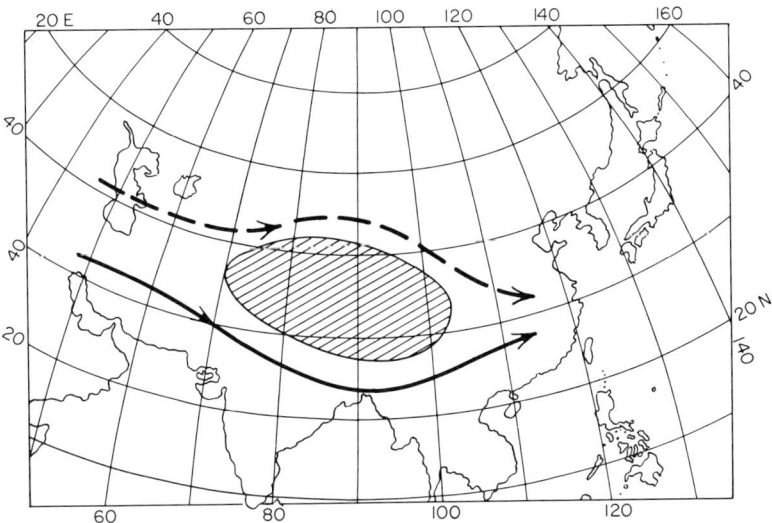

Fig. 8.28. Change in flow pattern in upper atmosphere resulting from displacement of subtropical westerly jet stream to the north of the Himalayas (92).

over Burma and retarding it over India as long as the trough remains in the Bay of Bengal. After it shifts westward, southerly wind components prevail at high levels over the entire Burma-India region and must reinforce the monsoon everywhere. The shift of the trough (92), related to highlevel flow pattern as far as Europe (18, 47), can explain the pattern of advance of the monsoon. This hypothesis permits us to relegate the heat low over northwestern India-Pakistan, often cited as a primary factor in the advance of the monsoon, to a more secondary role. The heat low arises from intense insolation under clear skies over a wide area. However, the clear skies cannot be considered as an *a priori* factor in any monsoon theory (56); it is just another way of saying that the monsoon is retarded.

As the polar westerlies begin to skirt the northern margin of the Asian central mountain massiv, the famous high-tropospheric jet stream in the easterlies, a part of the primary summer circulation, makes its appearance. The roots of the current lie partly over the mainland of Asia itself and partly over the western Pacific Ocean. The origin of the latter branch, very persistent and reliable, is not readily explained; development peaks over India and to its west. Over Africa the current weakens and banks to the right with sinking motion, a mechanism for creation and maintenance of the Sahara desert. The latitude of the easterly axis lies between 10° and 15°N. The boundary

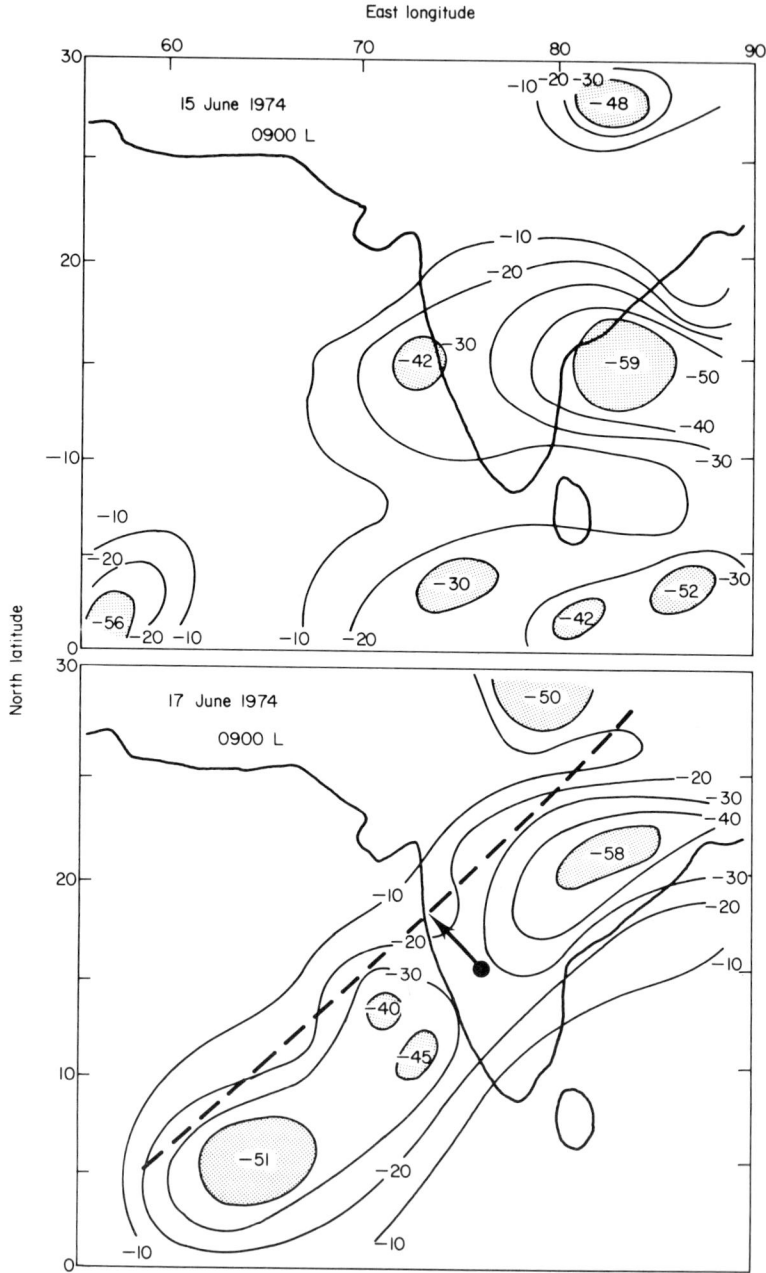

Fig. 8.29. Sequence of four maps showing satellite radiation temperatures (°C) during onset of the 1974 Indian summer monsoon.

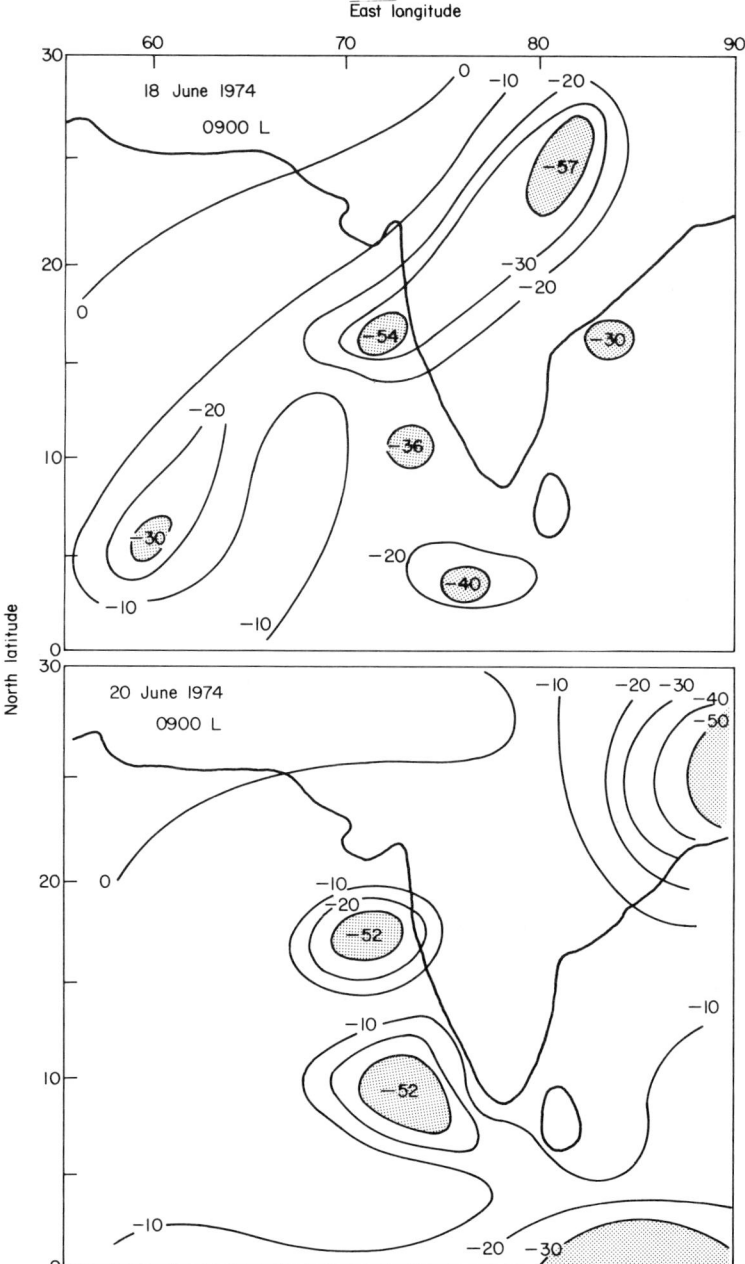

between high-tropospheric easterlies and westerlies lies over or near the southern end of the central mountains. From there horizontal shear of the east wind to the core of the current over southern India follows the distribution demanded by zero absolute angular momentum about an axis fixed in the mountain range (31).

The heated continent with greatly elevated heat source lies to the north, the broad Indian Ocean to the south of the centre of the easterly subtropical jet stream. The seasonal temperature distribution was described in Figs 2.25 and 2.26. The initial rise in upper temperature, at least, must be related to the shift of the westerly subtropical jet stream of winds, since the warming happens well before the onset of rains over northern India. Subsidence on the south side of the subtropical westerlies then overlies the Himalayas. With clear skies, heating of the elevated heat source and upward transport through turbulence combine with the large-scale subsidence to produce general warming of the upper troposphere. In addition, dust transport to the region above 300-200 mb may create a dust layer that intercepts the low-level radiation and acts as a high-tropospheric heat source (Chapter 2).

During the main rainy season, much emphasis has been put on release of latent heat of condensation from huge rainfalls for maintaining the upper temperature field. This hypothesis is not appealing for two reasons. Firstly, the winds at 200 mb blow from east-northeast and, thus, cannot create the very warm air over the mountains, and secondly, precipitation, even in the region of very heavy rains in northeastern India, is intermittent and mostly falls on a few days. Any circulation dependent for its energy on such convection would also be intermittent and variable whereas, in reality, the high-tropospheric winds are the steadiest of the globe.

It is tempting to relate the variable disappearance of the upper westerlies and the onset of the easterly jet stream to the variability in monsoon onset. Rapid weakening of the mean westerly winds during April has indeed been found to be associated with early monsoon, while their (late) persistence indicates delayed monsoon (83). In another study (87) this observation is largely supported—the first burst of the monsoon occurs at about the same time as the largest decreases in zonal wind component do in northern India. However, the early season intensity of the low-level equatorial westerlies is advanced as a second factor. Stronger than average 700 mb west winds at Ghan (1°S, 73°E) representing the lower troposphere favour rapid monsoon development when the high-level westerlies in the north weaken. The interpretation of this finding has not been clarified.

There may be additional complicating factors. Satellite charts, especially those of effective radiation, provide evidence of large, rapid and unsteady fluctuations of circulation, which are different from what may be imagined. Of course, it has been known for a century that the monsoon advance occurs in several pulses and that the early ones subside again for some days before a new thrust gathers. However, such thrusts seen by satellite do not appear to line up with an easterly jet stream or in any way conform to the idea of northward displacement of an east-west line resembling an equatorial trough zone. The period 15-20 June, 1974, may serve as an example (Fig. 8.29). This is rather late in the season for monsoon onset, but within range of the variability to be expected.

At first (15 June), an east-west cloud line of classical appearance straddled India near 15°N, with two intense centres of convection offshore from both coasts. A secondary convergence zone, well marked climatically in Fig. 1.7 and 1.12, was situated farther south. Two days later there had been a merger of these diffuse cloud systems into one long, and very well organized cloud line with northeast-southwest orientation. Such a synoptic feature is not part of previous descriptions. It looks like a large trough extending from higher latitudes into the tropics to the west, with southwest winds and deep convection on the forward side with northward angular momentum transfer. The picture is quite incompatible with adaptation to a concentrated upper east-west current. The line continued its northwestward advance during the next 24 hours. Then it broke down on 20 June; the monsoon burst subsided leaving two strong convective maxima, no doubt one of them a tropical storm, in the Arabian Sea. After a century of research, it appears that knowledge about the monsoon advance and the decisive factors is still rather incomplete.

Fluctuations in the monsoon season

The monsoon was considered an entity largely sealed off from outside influences in the early days of meteorology. The mountains provided a shield in the north supporting a steady monsoon circulation. Actually, little of the air converging in northern India, Pakistan and Bangladesh, as well as in Southeast Asia, escapes into the westerlies. In a limited sense, local mass compensation takes place since the low-level air arriving from the Arabian Sea returns toward it at high levels as an east-northeast wind. However, this does not prevent extra-tropical troughs and the westerly subtropical jet stream from exerting a marked influence on the monsoon. In response to "extratropical forcing", several major monsoon advances to the Himalayas occur in one season; the extensions westward to the Indus valley are controlled by the position and interaction of extratropical troughs passing through Siberia. One can follow these sequences very well with the 1974 satellite radiation data, a year in which two major advances of the monsoon took place toward northwest and when, already in early September, deep convection largely disappeared from the Ganges plain, to revive only sporadically thereafter.

Such behaviour indicates strong dependence on the general circulation to which the so-called "prime mover" is very vulnerable, but it is not the whole story of the well-known and widely studied "monsoon breaks". As in other parts of the tropics, the equatorial (here monsoon) trough goes through a "cycle" of northward advance, breakdown and reformation farther south in the course of several weeks. From a detailed analysis of August 1949 (58) the monsoon trough seen in the 850 mb windfield at first was elongated across the whole peninsula with classical east-southeast—west-northwest orientation. It then shifted gradually northward and eventually became lost in the Himalaya mountains. In the final stage the monsoon westerlies disappeared north of latitude 15° and a new equatorial trough formed right under the centre of the easterly jet stream. The monsoon break was coupled with the northward advance of the equatorial trough into the mountains. The factors underlying such evolution require clarification. They must be investigated in the framework of the "cycle" as a phenomenon encountered in several areas around the globe; the possible role of the Southern Hemisphere winter circulation must be taken into account.

The double convergence zone, one far north in India and the other one spread out in the South Indian Ocean near latitude 5°S (Fig. 8.30), needs to be examined closely in this general context. The evidence for a double convergence zone, so marked in Fig. 1.12, is fully corroborated by the satellite observations. Throughout the season, the secondary convergence zone increases and decreases in intensity just like the primary one. Cold centres organize, travel westward to northward and then disappear. Both northern and southern convergence zones often merge at the equator near 60°E as in the illustration. It has been suggested (pers. comm. Dr. T. M. Krishnamurti) that in seasons when the northern zone is strong, the southern one is weak and vice versa. Whether this holds or not, can be confirmed by increasing amounts of suitable satellite material.

It is surprising, considering the large number of studies of monsoon depressions, that there is no mention of waves in the equatorial westerlies which have a depth of up to 6 km. Rossby waves, at least, might be expected, and even weather-active waves, but the search has been rather fruitless. In 1975, however, satellite cloud analysis and superpressure balloon flights over the Indian Ocean at low levels gave an indication of eastward travelling pulses of the monsoon toward east north of the equator (7).

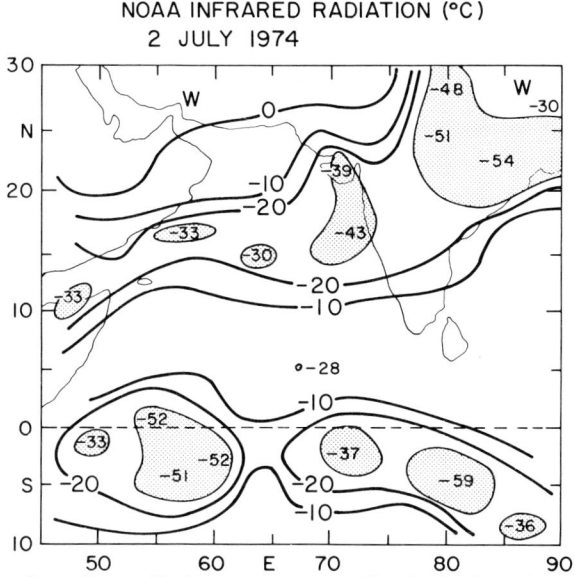

Fig. 8.30. Map of satellite radiation temperatures for the Indian Ocean, 2 July, 1974 (76).

Retreat of the monsoon

The retreat period is, in a sense, a mirror image of the onset. Shift of a single westerly jet stream centre to the south of the Himalayas (68) is sufficient to initiate the disappearance of deep convection from the northern plains. In contrast to the quite spectacular lack of tropical storms in mid-season, these revive quickly in the Bay of Bengal as superposition of troughs in the polar westerlies on the equatorial trough zone, then near 15°N, again becomes possible after the summer interruption (80). This sequence parallels that of the Atlantic (Chapter 10). As long as there is no cold air reservoir and a "sink" for the warm air escaping from a potential hurricane, it cannot develop. As soon as contact with the general poleward meridional temperature gradient is re-established or intensified, the hurricane season resumes.

The Manila episode, 17-21 July, 1972

Meteorological research projects and expeditions only too often find that the conditions they seek to investigate disappear or go to a marked minimum during the duration of the experiment. In particular, measurements have never been taken in a really heavy rainstorm and many generalizations have been made on the basis of very "thin" rainy seasons. Yet it is this class of intense storms whose occurrence or non-occurrence largely decides whether there will be drought, "normal" water supply or flood. The summer of 1972 was marked by a great drought stretching from India across Africa to the Caribbean, but compensation seemed to take place beginning about 90-100°W extending across the Pacific Ocean. The compensation was lively on the western side of that ocean, notably in the Philippines. In the setting of a generally rainy July there occurred a catastrophe in Luzon between the 17th and 21st of the month, when almost one metre (1000 mm) of rain came down, which equals the whole annual precipitation of New York City and nearly twice that of London. It is very interesting to have the chance to analyse such a case in order to assess the mechanisms. It is often said that great rains make great storms, but when really extreme amounts are reported, nothing much occurs in the pressure and windfield in a surprisingly large percentage of the cases. Many hurricanes even are apt to become producers of disastrous rainfalls in their old age over land, more so, or at least as much so, as at the peak of their life (June 1972 in the eastern United States). Why is this so?

The opportunity to study at least some aspects of an extreme event are rare. In the Manila case the weather station kept going, the recording rain gauges disgorged their buckets at record speed and did not stop and, most of all, the weather observers at Clark Air Force Base, north of Manila, kept up their 12-hourly rawinsonde schedule through the whole event with perfection, even though they, too, had one metre of rain at the base. We must congratulate them for their valiant efforts and also the personnel present at Manila† who collected many of the data and preserved them for analysis. All observations were operational, however, as there was no research project and many dearly desired observations, notably radar, are not available.

†Dr. M. M. Obradovich, United Nations Development program consultant, made the Manila observations available. The United States Air Force provided the Clark Air Force Base rawinsonde observations and also rainfall values. As eye witness, Gordon (22) has given an excellent detailed description of the event.

Precipitation

The rain episode lasted from about noon on 17 July, local time, to noon on the 21st. Two Manila recording rain gauge stations—Science Gardens and Port Area—had very similar records. The four day total was 81 cm there, compared to 98 cm at Clark AFB and 115 cm at Baguio in the northern mountains (16°N, 121°E), where one day brought 48 cm. The bucket-type recording gauge records at the Manila stations were divided into 10 minute intervals. Of the total, 40% were without rain. Another 40% had rain up to 1·5 mm/10 min,

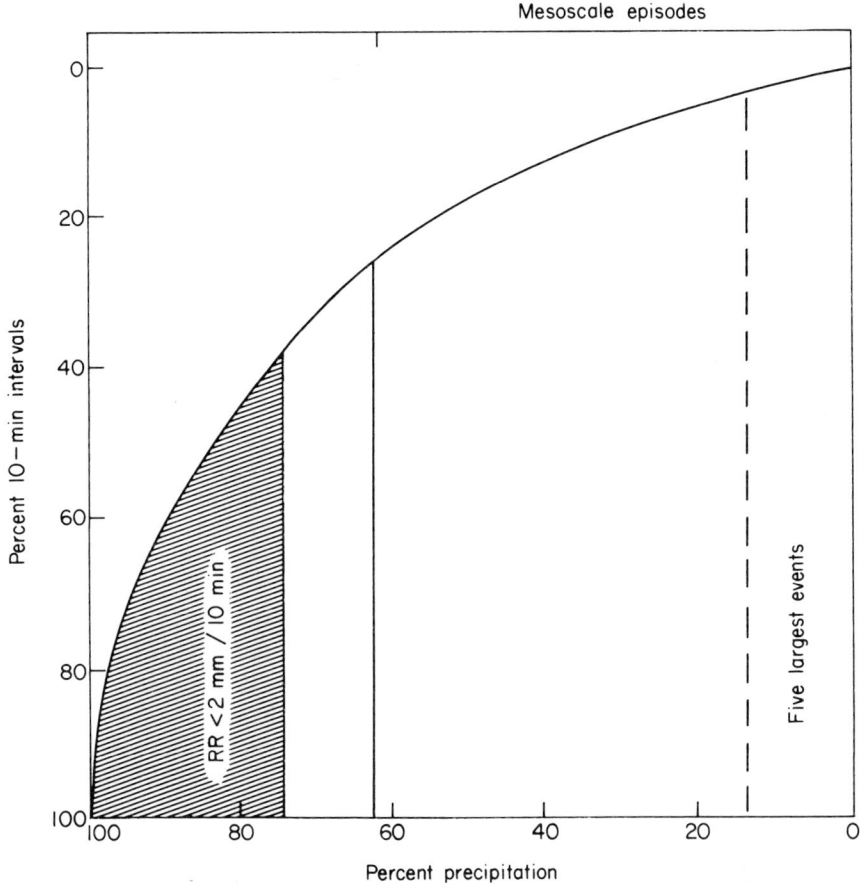

Fig. 8.31. Cumulative per cent precipitation versus per cent time for extreme Manila rain episode, 17-21 July, 1972. Rainfall with less than 2 mm/10 min shaded. Rainless periods not included.

average one mm. This light rain falling between the more intense bursts produced 25% of total precipitation. All higher amounts, yielding 75% of the rain, fell in 20% of the total time.

If we exclude the dry intervals and express time in percentage of 10 minute intervals with rain (Fig. 8.31) we see that the light rain fell in 60% of these intervals. The higher amounts were examined for periods when rain of at least 2 mm/10 min brought 10 mm consecutively. These intervals produced 60% of the rain in 25-30% of the time, and were labelled "mesoscale episodes". Their number averaged four or five per day; precipitation was 35 mm in the mean and duration one hour. These numbers compared so closely with those obtained in other areas that they served to define the "model mesoscale system" described in Chapters 3 and 4. Rain in excess of 10 mm/10 min only occurred seven and ten times, respectively, at the two stations. The highest amount was 21 mm. The five largest events produced 13% of the precipitation in 3% of the time.

For rainfall of 20 cm/day and inflow specific humidity near 20 g/kg, the vertical motion at cloud base will be 10 cm/s, if 10% of the condensed water re-evaporates or escapes from the area with the upper winds. The mean ascent rate to 200 mb would be 20 cm/s, i.e. approximately 10 km/12 hours, and rain would be falling at a steady rate of 1·7 mm/10 min. Such an ascent is obviously too slow. Computing for the mesoscale episodes, upward motion at cloud base is 50 cm/s and mean updraft speed 1 m/s, still very moderate. No doubt one can expect a velocity distribution with some values of several metres per second, but the general outcome is that buoyancy accelerations are not very large, just as nearly all rain fell at fairly moderate rates.

Mean vertical structure

All eight soundings at Clark AFB during the rain period had very nearly the same structure (thermodynamic and wind) so that they can be averaged to yield single profiles. By combining the 0800 and 2000 LT observations, diurnal factors are largely eliminated. The outstanding information (Fig. 2.32) is that temperature anomaly structure (from the Manila July mean) is exactly the same as found in all other such rain areas—cold core in the low levels and warm air relative to the surroundings in the high troposphere! Thus, even the huge rain amounts produced no mechanism for local generation of kinetic energy! In fact, surface wind was below average. Surface pressure was exactly average; the greatest negative height anomalies occurred between 600 and 500 mb. Above 250 mb they became positive and here the greatest outflow should take place.

In contrast to other situations, specific humidity was strongly positive through the whole troposphere, except for the surface level, no doubt due to Clark AFB's location in the lee of the west coast mountain range. For ascent computations, the crosses in Fig. 8.32, representing Manila surface observations, should be preferred. Because of the high moisture θ_e is raised at all levels above 970 mb compared to the mean atmosphere for July at Clark; there is virtually no evidence of downdrafts. Here, the whole troposphere is really lifted as indicated by the arrows in Fig. 8.32 (right side). This is the only difference of substance in thermodynamic structure between this extreme event and the average synoptic system.

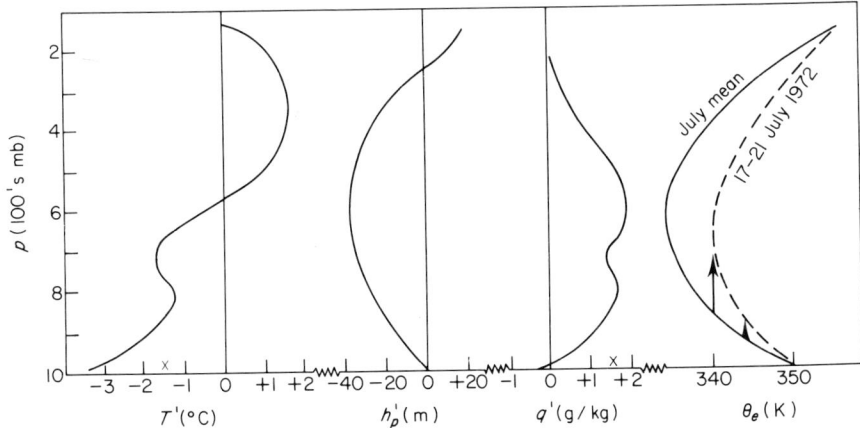

Fig. 8.32. Vertical cross-sections of temperature (°C), specific humidity (g/kg), pressure-height (m) and equivalent potential temperature (°C) deviations from the mean Clark Air Force Base July atmosphere during the heavy rain episode. Crosses mark Manila surface values representative for ascent.

Wind structure

Just as with the thermodynamic observations, those of wind differed drastically from the July mean. Upper winds were westerly through almost the whole troposphere. It must be noted that the July average wind profile is composed of very heterogeneous soundings; Manila is situated sometimes in the southeasterly trades, sometimes in the equatorial westerlies, as the trough zone oscillates across the city. The westerlies in Fig. 8.33 turn clockwise with height on a wide sweep through the troposphere, indicative of geostrophic warm advection everywhere, but no warming materializes in this case during the progression of soundings. Westerlies are strongest at 600-500 mb, just where the temperature anomaly crosses from lower cold to upper warm. Here should be the maximum vorticity, but, unfortunately, we lack a station network for proof.

The meridional component is directed toward north up to 500 mb and toward south higher up. Rather close balance of meridional mass exchange is achieved. There is some indication of a southward shift of an upper trough oriented east-west during the period. At 1200 GMT, 20 July, the pattern of Fig. 8.33 reversed to lower northwest and upper southwest wind. Also, at Baguio the rains started a day earlier and stopped a day earlier than at Manila. There is some suggestion of inward motion from both sides toward a central axis at low levels and outflow at high levels, rather like pictures of a "prime mover".

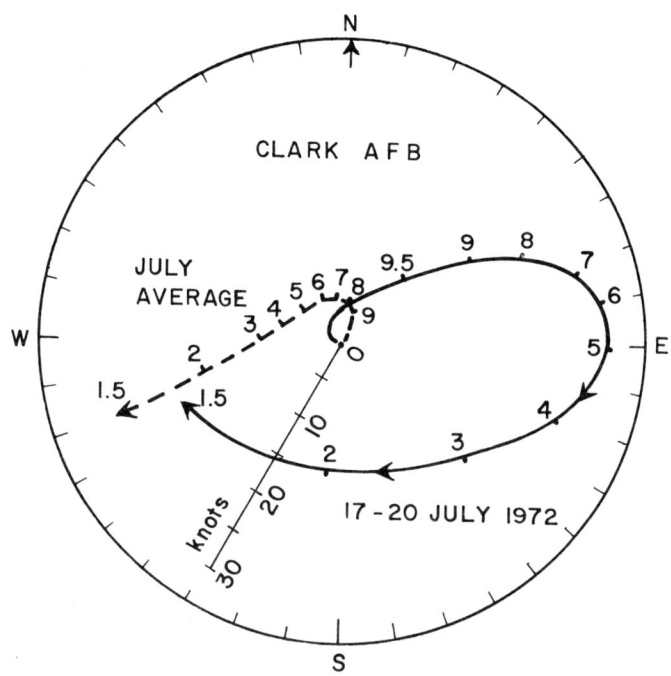

Fig. 8.33. Composite hodograph of the Clark Air Force Base rawin observations during rain episode. Dashed: July average profile.

Synoptic situation

The most puzzling aspect of the Manila episode is the lack of any obvious synoptic core. From Gordon's (22) maps, surface air motion was from southwest across the isobars with lower pressure to the north, a rather common picture. About 2500 km to the northeast was the centre of a typhoon and on the satellite photo the long outflow tail from this distant storm was initially connected with the cloud mass over and around the Philippines. How a wind structure such as that of Fig. 8.33 evolved and how the whole vertical mass circulation system continued for four days, remains unclear. It has been argued that motion of unusual, strong upper westerlies over the coastal mountain ranges was responsible for the situation. Indeed, the clouds as seen by satellite tended to line up north-south over Luzon, with extensions to the South China Sea and the Pacific. No doubt the mountains were a contributing factor, but they can hardly be considered as the key to the situation, especially as the westerlies were at their strongest when the rains ended.

At 200 mb, Gordon's map shows the expected outflow anticyclone so that, if correct, the ascent in the rain episodes was associated with cumulonimbus towers that could reach 200 mb, i.e. undilute towers. Inside the cloud system, the energy transformation was from latent heat to potential energy. The rain, generated along the same moist-adiabatic ascent, occurred without a complementary mechanism that would have allowed shifting of ascent to higher θ_e values. Only through such a mechanism could the mass distribution be altered, possibly generating surface pressure falls and a cyclone. Here lies the main difference between normal and even extreme rain episodes and the hurricane. It is not necessary for extreme rain to occur for cyclogenesis; what matters is that the moist adiabat of ascent shifts to higher values.

On the role of ice processes

The composite Manila sounding (Fig. 8.34) is almost saturated up to 400 mb. The saturation deficit may be instrumental, since the balloons certainly passed through clouds. Alternatively, the presence of salt in the air may have led to condensation at 90-95% relative humidity. Like most soundings in areas of heavy convection (74), this one has a dip of temperature below the freezing level. One may ask whether this dip can be produced through melting of ice and snow particles from above. The rain episode precipitation, excluding the 1 mm amounts, was 14 cm/day; we shall assume that half of the condensate freezes (see below), then the heat required for melting would be $dh = 7 \times F$, where F is latent heat of fusion, or 270 W/m².

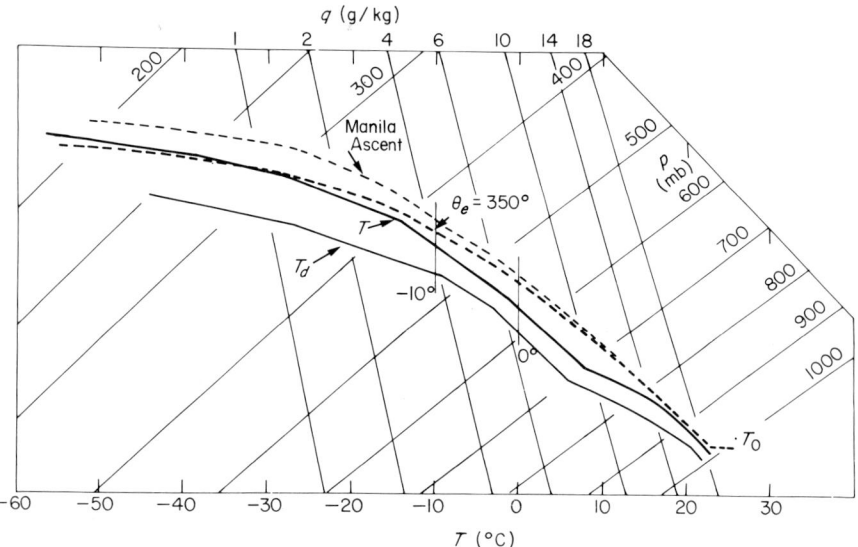

Fig. 8.34. Tephigram of the composite Manila sounding during heavy rain episode. Temperature (T) and dewpoint (T_d) solid. Pseudoadiabatic ascent at θ_e of 350 K, heavy broken. Hypothetical ascent including all energy sources, light broken. T_0 denotes climatic ocean temperature just west of Philippines.

The cooling around 600-700 mb may be assessed as 2°C from the sounding. If this dip is created through the melting process over trajectory distance of, say, 100 km, we can compute whether the required equals the actual thermal advection. The latter,

$$dh_s = c_p u \frac{\partial T}{\partial x} \frac{\delta p}{g}.$$

Here, u is taken as the westwind component, about 10 m/s from the sounding. The pressure thickness may be 100 mb. Given these numbers, $dh_s = 200$ W/m², not quite enough but certainly of the order of magnitude. For exact balance, either the advection must be raised or only 5 g/cm² melted per day. We now turn to the upward velocities.

As just seen, the bulk of the precipitation must be derived from deep convection through undilute cumulonimbus ascents. However, one cannot expect even these clouds to have adiabatic water content at the freezing level, i.e. carry all their water that far. With specific humidity of 20 g/kg at the surface and 5 g/kg at 500 mb, the difference is 15 g/m³ which, from Eq. (4.17), would erode any buoyancy completely. In Fig. 8.34, all energy including the heat of condensed water, is conserved along the curve marked "Manila ascent". The ascent curve begins to deviate from pseudo-adiabatic ascent at $\theta_e = 350$ K at low levels, but the temperature excess of even this ascent over the mean sounding is only 3°C or less, except near 700 mb. This means that buoyancy becomes zero at a water content of no more than 8 or 9 g/m³. For maintaining any upward velocity on the order of 1 m/s, the liquid water content must be reduced to no more than half of the adiabatic value, more likely to a third or 5 g/m³, since subtractions from buoyancy for friction, form drag, etc. must be taken into account. Counting 5 g/kg carried in vapour form beyond 500 mb which freeze or sublimate higher up, minus export of 2 g/kg cirrus outflow and 1 g/kg water vapour from the limited area, the frozen product to be melted on the way down is no more than 7 g/m³. For a melting rate of 5 g/m³/day, computed above, some more water must fall out above 500 mb in unfrozen form. This may very well be the case considering that 500 mb temperature is no less than $-5°C$ with few active freezing nuclei.

An "ice-adiabatic" ascent is used in the Manila sounding from 400 mb upward. At 200 mb the ascent temperature approaches $\theta_e = 360$ K (pseudo-adiabatic) after starting with $\theta_e = 352$ K near the surface. The increment in temperature is 2°C, a large factor in maintaining buoyancy of undilute towers beyond 200 mb to radar echo tops often reported near 150 mb. Ice adiabats probably should be employed in all calculations of deep convection. Upper freezing and lower melting, of course, serve to increase thermal stability; thus, the Manila sounding shows very clearly that it is an atmosphere of near-neutral stability that the heaviest rains can develop. The ice process may be an important factor in vertical energy transport, estimated as high as 25% of required transport in (71). In that event, recycling as illustrated in Fig. 4.27 need not be carried upward beyond the top of the inflow layer (400 mb) for high-tropospheric heat balance.

Conclusion

The limited analysis of the extreme Manila rain episode has provided considerable information. In spite of the flood catastrophe, it rained not much more than half of the time; heavy rainfall rates were confined to 20% or less of the time. Rainfall rates never became very high and vertical motions remained moderate. The rain came in about five major spurts per day, lasting roughly one hour, labelled rain episodes or mesoscale systems embedded in the general storm envelope. Cloud physics processes and the general vertical wind structure may have aided precipitation from the modelling standpoint (52). We can account for the melting of the frozen product coming from the upper parts of cumulonimbus through thermal advection.

The system was cold core up the middle toposphere in marked similarity with rainstorms encountered elsewhere. Pressure-heights were lowest in the middle troposphere, but they rose above the surroundings in the high troposphere leading to an outflow anticyclone. The whole rain area was so solidly saturated that an unsaturated atmosphere capable for generation of downdrafts was not available. Unlike the hurricane, however, the latent heat released could not be used internally for the production of kinetic energy since the system was cold-cored at low and middle levels. A definite synoptic centre or other shape of envelope did not develop. On a larger scale, the whole system was energy-producing since upper outflow occurred at relatively high pressures, but no immediate or important consequences of its existence can be discerned. Some doubts should be raised about theories that only too readily convert latent heat released into kinetic energy of motion.

Squall lines

The Philippine rain was associated with very little wind at the surface, in fact wind speed at Manila was below average. In contrast, some moving systems over land and ocean may be associated with violent weather displays at their forward edge or in some portions of their synoptic envelope. These have been labelled "squall lines" from the name coined for their middle latitude counterpart since in many respects, though not all, the characteristics are similar.

According to Zipser (94):

> Mesoscale systems made up of convective clouds organized in quasi-linear fashion are simpler to analyze than most other mesoscale systems . . . To

avoid confusion [in terminology] the term "squall line" [is defined] to refer to cumulonimbus clouds, organized in linear fashion, associated with a "pseudo-cold front (squall front) at the surface, propagating with considerable speed with respect to the ambient low-level air, in the general direction of the squall wind in the cold air behind the squall front.

The propagation rate of squall lines may be as high as 15-18 m/s. As in middle latitudes they may originate in a synoptic envelope, move forward through this envelope and then die in its outskirts. Thus, squall lines are found to form ahead of the waves in the easterlies described for Africa and the eastern Atlantic, and move to the vicinity of the next ridge downstream where they dissipate. Zipser (93) also noted frequent short duration of synoptic systems when an expanding squall line forms the leading edge and leaves a widening arc of downdrafts and decaying convection behind it. The distinction between "mesoscale" and "synoptic-scale" can become unclear at times when squall lines become very large and powerful.

All synoptic envelopes tend to concentrate their "weather" in about 10% of their area which contains the mesoscale systems. Not all mesoscale systems are squall lines, however, even though most of them have downdrafts, but all squall lines start as mesoscale phenomena and usually end as such with a distinctly shorter life-time than their "mother systems", which may keep spawning new squall lines. Thus, they must be regarded as an extreme kind of mesoscale cloud mass. One distinction may be that they are embedded in dry upper air, permitting very strong downdrafts to develop as in middle latitudes. Suggestive is the earlier observation that most rain falls, and most squall lines start, ahead of wave troughs over Africa although relative humidity in the upper air there is lowest (Fig. 8.24). Further, squall lines always have the low-level indraft along their forward edge because of their rapid motion (Fig. 8.13), amplified when they move from east against low-level equatorial westerlies. This arrangement must be a major reason why squall lines are especially frequent and intense in the Sahel zone of West Africa. Finally, they eject the air rising in cumulonimbi toward their rear, i.e. they move faster than the ambient wind at all heights.

The relative wind structure is understood readily, for instance, when lower easterlies are overlain by upper westerlies and the squall lines move westward with north-south orientation (Fig. 8.13, centre). In West Africa, however, low-level westerlies go over into upper easterlies and even the so-called easterly jet at 200 mb, so that $C - U$ decreases upward for westward moving squall lines there. Perhaps this does not matter as long as $C > U$ even at 200 mb and the anvils are ejected toward the rear. However, cases are on record whereby the squall lines organized themselves so that they moved toward south and even toward east against the general wind field. In middle latitudes, where squall lines preferably occur in connection with westerly jet streams, the anvils in part must move forward and squall lines may move well to the right of high-level winds (14, 48).

A good example of the spreading of a squall line is furnished by Zipser's (93) analysis of an aircraft research mission in the equatorial central Pacific Ocean on 1 April, 1967 (Fig. 8.35). His cross-section at Palmyra (Fig. 8.36) clearly brings out the onset of the squall and subsequent cold downdraft near the surface. The vertical shear from lower easterlies to upper westerlies was so strong (30 m/s) that the inclination of the towers became extreme and eroded already near 500 mb.

Figure 8.36 suggests that high energy is "scooped" at the leading edge of the squall and then ejected backward at high levels. The cold downdraft to the rear has presented some problems when the squall line moves faster than the air at all levels. Then the mid-tropospheric

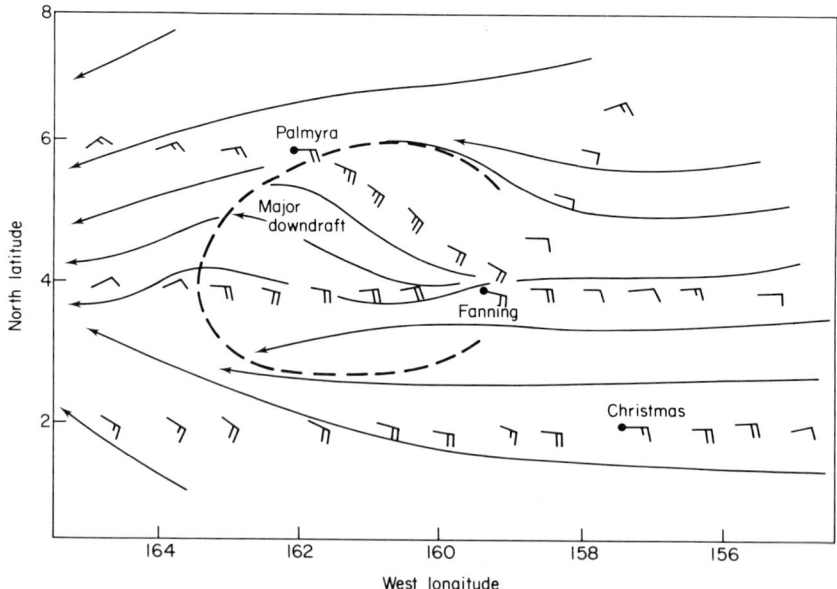

Fig. 8.35. Composite aircraft and rawinsonde station winds in the subcloud layer (150-300 m), streamlines and outline of major downdraft area following squall line (heavy dashed) in the Line Island area during the morning of 1 April, 1967 (local time) (from E. J. Zipser (93)). Figures 8.35 and 8.36 courtesy Dr. E. J. Zipser, Boulder, Colorado.

air, which is accelerated downward behind the squall, for continuity must have been funnelled from the front through the squall, not so easy to comprehend when there is a long squall "front" with a solid row of cumulonimbi. This problem has been treated by Betts (2) and Miller and Betts (44) whose numerical modelling based on Moncrieff and Miller (46) forces them to permit low energy air to cross from front to rear and then descend. The problem of the source of air for the descent is not restricted to squall lines. It occurs, for instance, in the Caribbean case pictured in Figs 4.16 and 4.17, which did not have a sharp leading edge yet overtook low and mid-tropospheric air in relative motion. There, at least, there was no solid wall of cumulonimbus that had to be penetrated. In some cases help comes from the rear side itself, when winds behind the squall pick up enough speed in the middle troposphere, so that they attain forward motion with respect to the squall, as indicated in Fig. 8.36. Upward motion will develop above 500-700 mb, strengthening the anvil or producing altostratus where there is little or no anvil, followed by light rains in the rear part of the system.

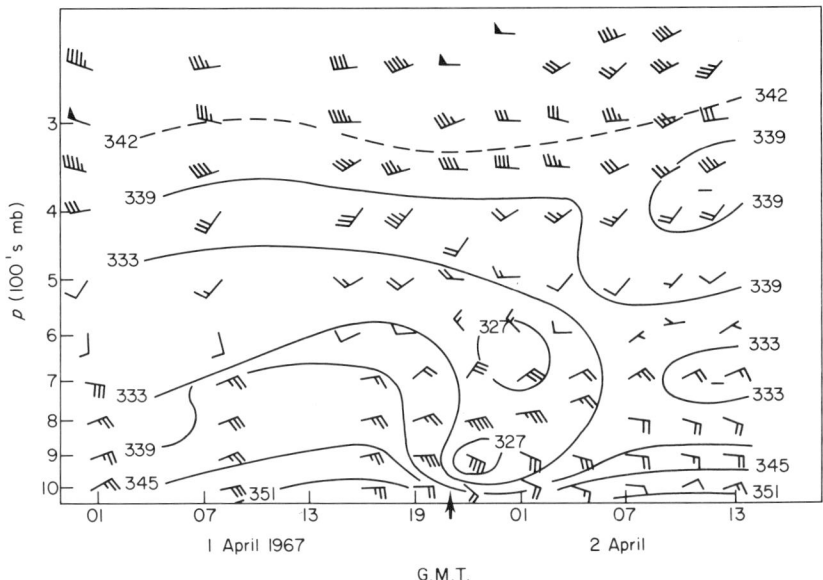

Fig. 8.36. Vertical cross-section of winds (knots) and equivalent potential temperatures (°K) at Palmyra (6°N, 163°W) during passage of squall line shown in Fig. 8.35 (93).

More complications, however, exist. There are the instances of very warm and dry air at some distance behind a mesoscale rain area (again not necessarily a squall line) pictured in Fig. 4.22. Of these, many have been found in GATE (94), and in Venezuelan soundings (2). Betts (2) suggests that the second downdraft is evidence of mesoscale sinking probably driven by the storm dynamics, whereas failure of continuing water supply in a downdraft was given as a possible mechanism in Fig. 4.20. Of course, the drying out must occur over a substantial area, perhaps as large as the mesoscale system, or else upward buoyancy could again develop.

Cool season weather systems

Because of the small seasonal temperature changes of the tropical atmosphere, rainy-season type weather systems can spring up at any time of year when the wind structure becomes favourable. Nevertheless, polar westerlies settle in persistently in most areas, so that monsoon-type rains are precluded. An exception is the western Pacific and the South China Sea where the limit of the polar westerlies is held to 10°N, a downstream effect of the Himalayan mountains; there even tropical storms form and propagate westward in every month of winter.

As the subtropical jet stream passes around the southern rim of the Himalayas in winter (Fig. 1.15) with speeds up to 75 m/s, troughs in this current, also known in other parts of the tropics, have a free path through Northern India toward China. Their track is from northwest near the Caspian Sea area, turning counterclockwise near the head of the Arabian Sea. Areas of rain and wind shift are carried along with the upper troughs. Weakly resembling extratropical cyclones they are known as "western disturbances", as they travel eastward along the southern rim of the Himalayas in northern India. A "block" in the upper air circulation upstream appears to be helpful for vigorous developments (82).

Ramage (60) has picked up these troughs at the exit from India and followed them across southern China to the Pacific coast. On Hong Kong cross-sections trough passages can be followed very regularly (Fig. 8.37). Upper cooling precedes and warming follows the trough axis, seen from the vertical gradient of 24-hour height changes. Wind speed remains below 50 m/s; the jet stream centre lies farther north in China. Precipitation starts in shallow easterlies of the layer surface—850 mb top at the approach of the trough with slow surface pressure falls and high dewpoint (Fig. 8.38). A moist temperature

SYNOPTIC-SCALE WEATHER SYSTEMS

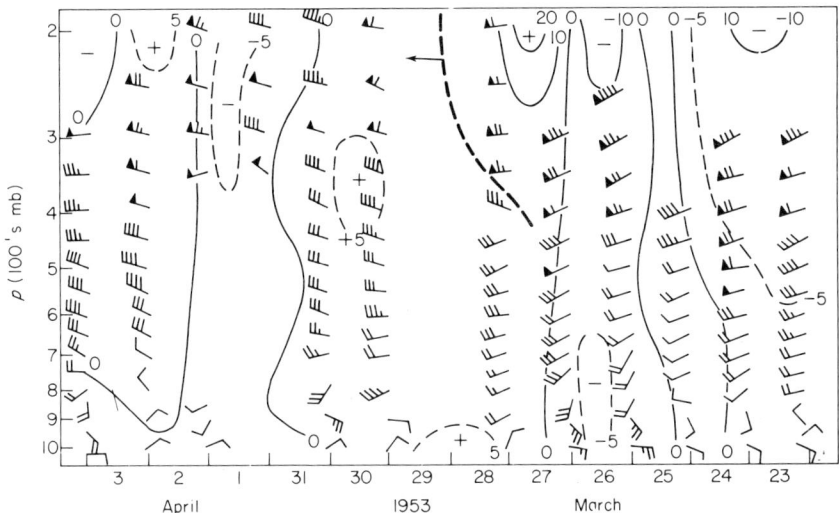

Fig. 8.37. Upper winds (knots) during passage of upper trough (heavy broken line) across Hong Kong, 23 March-4 April, 1953. Also 24 hr height changes (10s m) of isobaric surfaces. From Ramage (adapted after 60), time scale inverted.

inversion with windshift from ENE to SW may remain behind after the axis of the upper trough has passed. Then sporadic rains continue for days under the northwest upper flow with high surface pressure and low dewpoints, as the frontal zone or its equivalent remain not too far south of the coast, perhaps aided by "crachin" development south of the front in late winter and spring (59). The eastward-moving polar trough may take on large amplitude and produce intense weather developments. One such trough passed over the South China Sea on

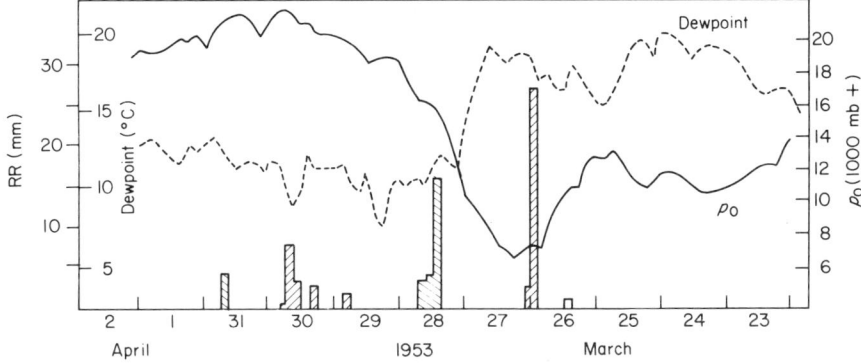

Fig. 8.38. Surface pressure (solid) dewpoint (dashed) and rain events at Hong Kong during passage of polar trough in Fig. 8.37 (adapted after 60).

13 February, 1968, marked at the surface only by lightening of low stratus and weakening of the northeast monsoon. Upon reaching the Chinese east coast, a cyclone developed rapidly north of the Philippines and on 15 February struck Tokyo as a snowstorm of exceptional violence, which paralysed the city. In late winter and spring, the upper troughs from India more often tend to become stationary near 100-110°E leading to copious rains on the polar front extending along the coast line of Southern China.

Quasi-stationary "polar troughs" also occur in the eastern Pacific. On 22 January, 1976, surface easterly isobars curved counterclockwise across the entire trade wind belt almost to the subtropical high pressure centre (Fig. 8.39). An intense and sharply-defined cloud mass seen by the satellite extended along the axis with counterclockwise wind shift. In the layer with most satellite and aircraft winds centred on 250 mb (Fig. 8.40), a very marked northwesterly jet stream approached a long and well-marked polar trough and there lost intensity. A secondary wind maximum developed farther north and east, a common occurrence along deep troughs in the westerlies. The western current gave up its kinetic

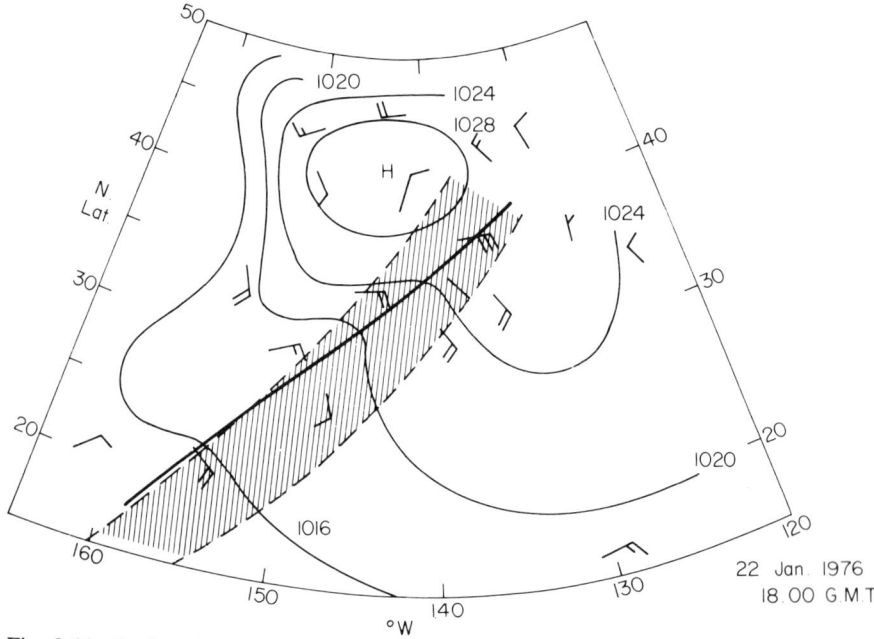

Fig. 8.39. Surface isobars (mb) and winds (knots) for the eastern Pacific Ocean, 22 January, 1976, 1800 GMT. Heavy line: Maximum vorticity in easterlies; shaded: satellite cloud outline.

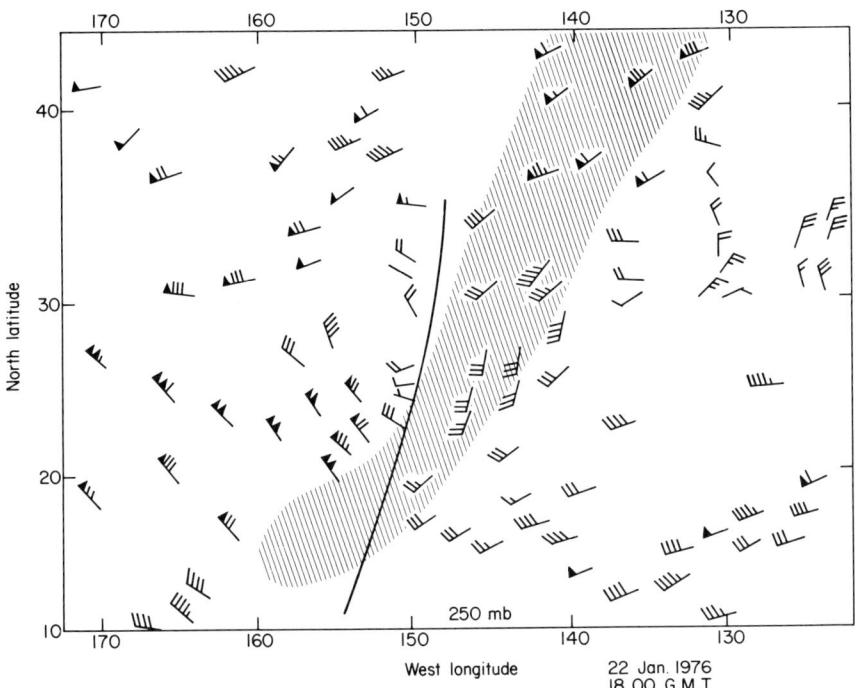

Fig. 8.40. Composite rawin, aircraft and satellite derived wind map centred near 250 mb in the upper troposphere 22 January, 1976, 1800 GMT. Axis of polar trough heavy; satellite cloud outline shaded. Map courtesy Honolulu Forecast Center, National Weather Service.

energy and gained potential energy, deepening the trough and increasing the motion from southwest ahead of it. At higher latitudes to the east, the ascent of warm air produced a new jet stream, curving clockwise.

As far south as latitude 20°N, the western edge of the satellite cloud mass coincided with the upper trough. South of the jet axis, however, the cloud shifted to the west of the trough, somewhat unusual. A subtropical cyclone can split off from such an arrangement. It is difficult to realize much precipitation from polar trough situations, especially when the latter and its low-level reflection move eastward over trades with a marked inversion. Some troughs are strong enough to break the inversion, when equivalent potential temperature decreases upward (Figs 5.19 and 5.20). A cloud mass then appears *in situ* and travels eastward with the troughs. Precipitation is concentrated close to the axis ahead of the low-level cyclonic vorticity maximum; after windshift to northeast, all rain ends rapidly.

Interaction between westerlies and tropics

On the *synoptic* time scale, especially in winter, the tropics are subject to gross domination by intrusions of large-scale polar or *extended* troughs that may reach from latitude 60° to the equator with their enormous energy releases and mass exchanges, coupled with the attending jet streams that form the patterns of long waves in the westerlies. This has been shown beyond doubt already by Cressman (95) with the sparse upper-air observations of 1944-1945. While in the low levels longitudinal symmetry prevails widely, with a north component of the wind in almost every place around the hemisphere (Table 1.1), this is not true in the upper troposphere. There, north and south components of the wind alternate around the globe giving rise to extended southerly or northerly flow over great ranges of longitude. In view of the meridional temperature gradient, these meridional currents operate on this temperature gradient expressed by easterlies decreasing and westerlies increasing upward with height in the troposphere. No single meridian is indicative of the meridional distribution of the pole-to-equator wind component. The tropical atmosphere imparts its energy to the high latitudes within very few and narrow zones of disturbed weather, an old fact made very visible in modern days to the novice by huge narrow meridional tubes of high water vapour transport from lowest to highest latitudes as seen by METEOSAT, the European weather satellite.

In the early days of jet stream exploration in the late 1940s and early 1950s, it was recognized that jet streams tend to go through gradual latitudinal shifts in the course of their life-cycle. Most of these shifts are equatorward and the axis of the current moves to increasing potential temperatures. Thus, the tropics may experience unsettled weather initiated by jet stream "forcing" from above. High rainfall in the equatorial zone has been found correlated with stronger than average jet stream strength in the subtropics (49, good literature index). Under such circumstances, trough extensions from higher latitudes into the tropics and equatorward displacement of jet stream maxima may be more vigorous than with a mean position of the jet stream at quite high latitudes. A proposed feedback from increased equatorial rains through their poleward outflow would reinforce the subtropical jet stream in turn.

On synoptic time scale, cloud masses from South America can make their way across the equator moving at about 5-7° degrees latitude per day, following the Andes and lead to explosive weather events as far north as the Caribbean (Fig. 8.5). Similar northward intrusions are favoured along the east coasts of Australia and Africa.

During GATE they were expected on the western side of Africa through surges of air derived from cold fronts farther south; the network was ready to record and follow these, but nothing happened.

The northward advance of cloud centres from South America should be an event of the southern winter, however, they have been found on satellite sequences in all months. In winter, the initial cold outbreaks can be severe and cause large agricultural damage. One severe outbreak in July 1975 was traced during eight days from Argentina to the Caribbean (55); it killed Brazil's coffee crop of that year.

Surge of the trades

Reverse intrusions of such magnitude from the Northern Hemisphere are not on record, even though the coldest continental air masses are situated there. The monsoon surges of East Asia generally weaken and form a secondary equatorial trough zone north of the equator. Outbreaks of polar air from the United States are violent in the Gulf of Mexico and the Isthmus of Tehuantepec; they then sweep farther southward in the eastern Pacific. Similarly, outbreaks from

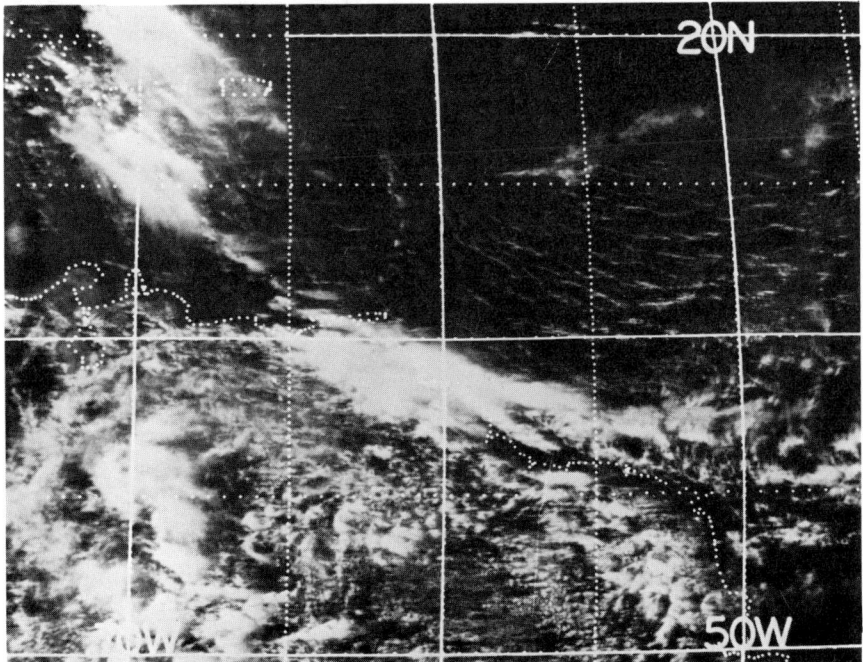

Fig. 8.41. ATS III photo of surge in trades over Venezuela and eastern Caribbean, 24 July, 1972, 1730 GMT.

the North Atlantic will affect the South American coast from Colombia and Venezuela clear to the north coast of Brazil, though. This phenomenon, known as "surge of the trades", should again be confined to winter. Nevertheless, surges occur during summer, emphasizing the role of dynamic anticyclone formation for their origin. Several such surges were encountered in Venezuela during the 1972 experiment, and one of these was exceptionally strong and violent. Prior to 22 July, 1972, an elongated surface Low occupied the subtropics in the Atlantic instead of the normal High. Then, on 23-24 July, sudden and very marked anticyclogenesis took place, coupled with great strengthening of the easterly pressure gradient. Rapid

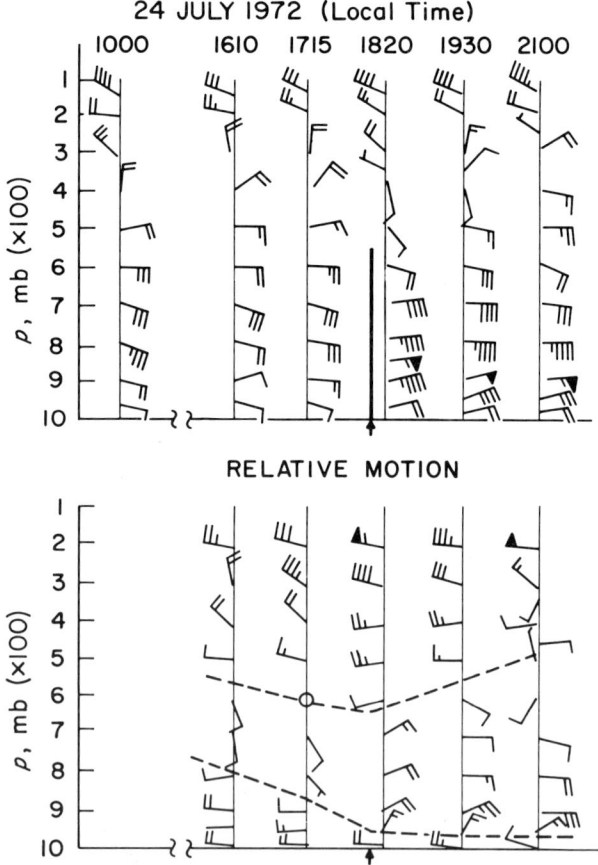

Fig. 8.42. Time section of upper winds (knots) and of winds relative to the moving front of the trade wind surge, 24 July, 1972, taken at Carrizal, Venezuela (9·5°N, 67°W). Approximate time of axis passage marked heavy; dashed lines in lower diagram indicate layer where U-$C > 0$.

acceleration of the easterlies followed over a wide sweep of the Atlantic; during the afternoon of 24 July the leading edge, marked by a very big and elongated sateliite cloud mass followed by thinner streaks of cumulus rows along the wind (Fig. 8.41), arrived over Venezuela. When it reached the meteorological centre at Maracay (10°N, 67°W), afternoon traffic was disrupted as many old palm trees were thrown down, falling across the roads leading from the city centre. Wind speed suddenly rose to 20-30 m/s at 950-900 mb (Fig. 8.42) and gusts of that strength penetrated to the surface.

The period of 24-25 July was the rainiest one experienced in all Venezuela during the 1972 season. Twenty-four hour amounts of 20-50 mm were widespread. Even before the arrival of the edge of the surge many small shower clouds appeared in the sky indicative of convergence ahead of the main system by one hour. The series of almost hourly rawinsonde ascents depicts the great wind violence. As the general surge of the trades advanced toward west, individual squall lines, depicted by radar, moved from east-northeast toward west-southwest at over 13 m/s. The air motion relative to the surge was toward the squall near the ground and it produced most of the heavy rain, but the shear of the wind was extreme in the lowest kilometre; already at 900 mb winds were 27 m/s, twice the propagation rate (Fig. 8.42). The height of the clouds lowered from over 10 to 8 km and even 6 km as the system passed over the radar area. The temperature lapse rate was moist-adiabatic at $\theta_e = 340$ K in almost saturated air.

Five major and several minor surges came into northern Venezuela in the 1972 summer, while there was not a single such event in the 1969 season. Clearly, so large a difference must be related to the difference in mean vertical current structure (Fig. 7.11). In the presence of deep easterlies and nearly barotropic stratification, one of the requirements for squally winds or surges—the large vertical shear—is missing. Strong control of weather by long-term anomalies of zonal current structure from the mean is hereby reaffirmed. At any location, it may take quite a few years before all varieties of synoptic systems have been experienced.

References

(1) Albrecht, B. and Cox, S. K. (1975). The large-scale response of the tropical atmosphere to cloud-modulated infrared heating. *J. Atmos. Sci.* **32**: 16-24.
(2) Betts, A. K. Ref. 8, Chapter 4.
(3) Brooks, C. E. P. Ref. 15, Chapter 4.

(4) Brunt, D. (1921). The dynamics of revolving fluid on a rotating earth. *Proc. Royal Soc.* A, **99**: 397-402.
(5) Burpee, R. (1972). The origin and structure of easterly waves in the lower troposphere of North Africa. *J. Atmos. Sci.* **29**: 77-90.
(6) Burpee, R. (1974). Characteristics of North African easterly waves during the summers of 1968 and 1969. *J. Atmos. Sci.* **31**: 1556-1570.
(7) Cadet, D. and Olory-Togbé, P. (1977). The propagation of tropical disturbances over the Indian Ocean during the summer monsoon. *Mon. Wea. Rev.* **105**: 700-708.
(8) Carlson, T. N. (1968). Structure of a steady-state cold Low. *Mon. Wea. Rev.* **96**: 763-777.
(9) Carlson, T. N. (1969). Synoptic histories of three African disturbances that developed into Atlantic hurricanes. *Mon. Wea. Rev.* **97**: 256-276.
(10) Carlson, T. N. (1969). Some remarks on African disturbances and their progress over the tropical Atlantic. *Mon. Wea. Rev.* **97**: 716-726.
(11) Chang, C. P. (1970). Westward propagating cloud patterns in the tropical Pacific as seen from time-composite satellite photographs. *J. Atmos. Sci.* **27**: 133-138.
(12) Charney, J. G. and Stern, M. E. (1962). On the stability of internal baroclinic jets in a rotating atmosphere. *J. Atmos. Sci.* **19**: 159-172.
(13) Daggupaty, S. M. and Sikka, D. R. (1977). On the vorticity budget and vertical velocity distribution associated with the life-cycle of a monsoon depression. *J. Atmos. Sci.* **34**: 773-792.
(14) De, A. C. (1963). Movement of pre-monsoon squall lines over Gangetic West Bengal as observed by radar at Dum Dum Airport. *Indian J. Meteor. Geoph.* **14**: 37-45.
(15) Erickson, C. O. (1971). Diagnostic study of a tropical disturbance. *Mon. Wea. Rev.* **99**: 67-78.
(16) Estoque, M. A. and Lin, M. S. (1977). Energetics of easterly waves. *Mon. Wea. Rev.* **105**: 582-589.
(17) Flohn, H. (1957). Large-scale aspects of the "Summer Monsoon" in South and East Asia. *J. Meteor. Soc. Japan,* 15th Anniversary Vol: 180-186.
(18) Flohn, H. (1960). Recent investigations on the mechanism of the "Summer Monsoon" of Southern and Eastern Asia. Symposium on Monsoons of the World, India Meteorol. Dept., Delhi, Sec. II: 75.
(19) Freeman, J. C. (1948). An analogy between the equatorial easterlies and supersonic gas flow. *J. Meteor.* **5**: 138-146.
(20) Freeman, J. C. (1968). Concerning a fundamental mechanism in tropical meteorology. *Ann. Meteor.* **1**: 182-185.
(21) Garstang, J. (1967). Sensible and latent heat exchange in low latitude synoptic scale systems. *Tellus* **19**: 492-508.
(22) Gordon, A. H. (1973). The Great Philippine Floods of 1972. *Weather* **28**: 404-415. Comment by H. Riehl (1974). Hot-tower precipitation. *Weather* **29**: 196.
(23) Graves, M. (1951). The relation between the tropopause and convective activity in the subtropics. *Bull. Amer. Meteor. Soc.* **32**: 54-60.
(24) Gray, W. M. (1973). Cumulus convection and larger scale circulations. I. Broadscale and mesoscale considerations. *Mon. Wea. Rev.* **101**: 839-855.
(25) Hantel, M. and Koehne, R. (1975). Die Zweischichtung der tropischen Atmosphäre anhand der "Meteor"-Daten 1965. *"Meteor" Forschungserg.* B, **10**: 21 pp.

(26) Haurwitz, B. and Haurwitz, E. (1939). Pressure and temperature variations in the free atmosphere over Boston. Cambridge: *Harv. Meteor. Stud.* **3:** 74 pp.
(27) Hayden, C. M. (1970). An objective analysis of cloud cluster dimensions and spacing in the tropical North Pacific. *Mon. Wea. Rev.* **98:** 534-540.
(28) Henry, W. K. Ref. 13, Chapter 3.
(29) Holton, J. R. (1971). A diagnostic model for equatorial wave disturbances: the role of vertical shear of the mean zonal wind. *J. Atmos. Sci.* **28:** 55-64.
(30) Holton, J. R. (1972). Waves in the equatorial stratosphere generated by tropospheric heat sources. *J. Atmos. Sci.* **29:** 368-375.
(31) Koteswaram, P. Ref. 27, Chapter 1.
(32) Koteswaram, P. and George, C. A. (1958). On the formation of monsoon depressions in the Bay of Bengal. *Indian J. Meteor. Geoph.* **9:** 9-22.
(33) Koteswaram, P. and Srinivasan, V. (1958). Thunderstorms over Gangetic West Bengal in the pre-monsoon season and the synoptic factors favourable for their formation. *Indian J. Meteor. Geoph.* **9:** 301-312.
(34) Krishnamurti, T. N., Molinari, J., Pan, H. L. and Wong, V. (1977). Downstream amplification of monsoon disturbances. *Mon. Wea. Rev.* **105:** 1281-1297.
(35) Kuhn, P. M. and Suomi, V. E. (1958). An economical net radiometer. *Tellus* **10:** 160-163.
(36) Kung, E. C. (1975). Balance of kinetic energy in the tropical circulation over the Western Pacific. *Quart. J. Roy. Meteor. Soc.* **101:** 293-312.
(37) Kung, E. C. and Burgdorf, H. A. (1978). Maintenance of kinetic energy in large-scale tropical disturbances over the eastern Atlantic. *Quart. J. Roy. Meteor. Soc.* **104:** 393-413.
(38) Lateef, M. A. (1967). Vertical motion, divergence and vorticity in the troposphere over the Caribbean, August 3-5, 1963. *Mon. Wea. Rev.* **95:** 778-790.
(39) Lopez, M. (1948). A technique for detailed radiosonde analysis in the tropics. *Bull. Amer. Meteor. Soc.* **29:** 227-236.
(40) Malkus, J. S. and Riehl, H. Ref. 52, Chapter 4.
(41) Maruyama, T. (1967). Large-scale disturbances in the equatorial lower stratosphere. *J. Meteor. Soc. Japan.* **45:** 391-407.
(42) Matsuno, T. (1966). Quasi-geostrophic motions in the equatorial area. *J. Meteor. Soc. Japan* **44:** 25-43.
(43) Merritt, E. S. (1964). Easterly waves and perturbations, a reappraisal. *J. Appl. Meteor.* **4:** 367-382.
(44) Miller, M. J. and Betts, A. K. Ref. 56, Chapter 4.
(45) Mitsuno, H., Sakai, S. and Hamada, N. (1963). The distribution of clouds around the intertropical convergence zone in the western Pacific. *Geophy. Mag. (Tokyo)* **31:** 685-703.
(46) Moncrieff, M. W. and Miller, M. J. Ref. 57, Chapter 4.
(47) Mothe, P. D. de la and Wright, P. B. (1969). The onset of the Indian southwest monsoon and extratropical 500 mb trough and ridge patterns over Europe and Asia. *Meteor. Mag.* **98:** 145-155.
(48) Newton, C. W. and Newton, H. R. Ref. 60, Chapter 4.
(49) Nicholls, N. (1977). Tropical-extratropical interactions in the Australian region. *Mon. Wea. Rev.* **105:** 826-832.
(50) Nitta, T. and Yanai, M. (1969). A note on the barotropic instability of the tropical easterly current. *J. Meteor. Soc. Japan* **47:** 127-130.
(51) Norquist, D. C., Recker, E. E. and Reed, R. J. (1977). The energetics of African wave disturbances as observed during phase III of GATE. *Mon. Wea. Rev.* **105:** 334-342.

(52) Ogura, Y. and Takahashi, T. (1971). Numerical simulation of the life-cycle of a thunderstorm cell. *Mon. Wea. Rev.* **99:** 895-911.
(53) Palmén, E. (1949). Origin and structure of high-level cyclones south of the maximum westerlies. *Tellus* **1:** 22-31.
(54) Palmer, C. E. (1951). On high-level cyclones originating in the tropics. *Trans. Amer. Geoph. Union* **32:** 683-696.
(55) Parmenter, F. C. Ref. 11, Chapter 7.
(56) Petterssen, S. (1953). On the dynamics of the Indian monsoon. *Indian Acad. Sci. Proc. Sec. A.* **37:** 229-233.
(57) Piersig, W. (1936). Archiv Deutsche Seewarte 54, No. 6. English translation in part: *Bull. Amer. Meteor. Soc.* **25:** 2-17 (1944).
(58) Rahmatullah, M. (1952). Synoptic aspects of the monsoon circulation and rainfall over Indo-Pakistan. *J. Meteor.* **9:** 176-179.
(59) Ramage, C. S. (1954) Non-frontal crachin and the cool season clouds of the China Seas. *Bull. Amer. Meteor. Soc.* **35:** 404-411.
(60) Ramage, C. S. (1955) The cool-season tropical disturbances of Southeast Asia. *J. Meteor.* **12:** 252-262.
(61) Ramage, C. S. (1962). The subtropical cyclone. *J. Geoph. Res.* **67:** 1401-1411.
(62) Rayleigh, Lord (1916). On the dynamics of revolving fluid. *Proc. Royal. Soc. A,* **93:** 148-154.
(63) Reed, R. J. and Recker, E. E. (1971). Structure and properties of synoptic-scale wave disturbances in the equatorial western Pacific. *J. Atmos. Sci.* **28:** 1117-1133.
(64) Reed, R. J. and Johnson, R. H. (1974). The vorticity budget of synoptic-scale wave disturbances in the tropical western Pacific. *J. Atmos. Sci.* **31:** 1784-1790.
(65) Reed, R. J., Norquist, D. C. and Recker, E. E. (1977). The structure and properties of African wave disturbances as observed during phase III of GATE. *Mon. Wea. Rev.* **105:** 317-333.
(66) Reed, R. J. (1979). *In* Structure and properties of synoptic scale wave disturbances in the intertropical convergence zone of the eastern Atlantic. (Thompson, R. M., Payne, S. W., Recker, E. E., Reed, R. J.) *J. Atmos. Sci.* **36:** 53-72.
(67) Regula, H. (1936). Druckschwankungen und Tornados an der Westküste von Afrika. *Ann. Hydrogr. Marit. Meteor.* **64:** 107-111.
(68) Reiter, E. R. and Heuberger, H. (1960). A synoptic example of the retreat of the Indian summer monsoon. *Geografiska Ann.* **42:** 17-35.
(69) Riehl, H. (1945). Waves in the easterlies and the polar front in the tropics. Misc. Rep. 17., Dept. Meteor., Univ. Chicago. 79 pp.
(70) Riehl, H. (1948). On the Formation of West Atlantic Hurricanes. Misc. Rep. 24, Part I., Dept. Meteor., Univ. Chicago.
(71) Riehl, H. and Malkus, J. S. Ref. 27, Chapter 2.
(72) Riehl, H. (1962). Radiation measurements over the Caribbean Sea during the autumn of 1960. *J. Geoph. Res.* **67:** 3935-3942.
(73) Riehl, H. *et al.* Ref. 33, Chapter 3.
(74) Riehl, H. Ref. 21, Chapter 5.
(75) Riehl, H. Ref. 28, Chapter 2.
(76) Riehl, H. and Miller, A. L. Ref. 72, Chapter 4.
(77) Rosenthal, S. L. (1965). Some preliminary theoretical considerations of tropospheric wave motions in equatorial latitudes. *Mon. Wea. Rev.* **93:** 605-612.
(78) Rossby, C.-G. Ref. 29, Chapter 2.

(79) Ruprecht, E. and Gray, W. M. Ref. 77, Chapter 4.
(80) Sen, S. N. (1959). Influence of upper level troughs and ridges on the formation of post-monsoon cyclones in the Bay of Bengal. *Indian J. Meteor. Geoph.* **10:** 7-24.
(81) Simpson, R. H. (1952). Evolution of the kona storm, a subtropical cyclone. *J. Meteor.* **9:** 24-35.
(82) Singh, M. S. (1963). Upper-air circulation associated with a western disturbance. *Indian J. Meteor. Geoph.* **14:** 156-172.
(83) Sircar, N. C. Rai and Patil, C. D. (1962). A study of high-level wind tendency during pre-monsoon months in relation to time of onset of southwest monsoon in India. *Indian J. Meteor. Geoph.* **13:** 468-471.
(84) Soundara Rajan, K. (1961). Squalls at Bangalore. *Indian J. Meteor. Geoph.* **12:** 583-589.
(84a) Wallace, J. M. (1968). Time-longitude sections of tropical cloudiness (Dec. 1966-Nov. 1967).
(85) Wallace, J. M. and Kousky, V. E. (1968). Observational evidence of Kelvin waves in the tropical stratosphere. *J. Atmos. Sci.* **25:** 900-907.
(86) Wallace, J. M. and Chang, C. P. (1969). Spectrum analysis of large-scale wave disturbances in the tropical lower troposphere. *J. Atmos. Sci.* **26:** 1010-1025.
(87) Wright, P. B. (1967). Changes in 200 mb circulation patterns related to the development of the Indian southwest monsoon. *Meteor. Mag.* **96:** 302-315.
(88) Yanai, M. and Murayama, T. (1966). Stratospheric wave disturbances propagating over the equatorial Pacific. *J. Meteor. Soc. Japan,* **44:** 291-294.
(89) Yanai, M. and Nitta, T. (1967). Computation of vertical motion and vorticity budget in a Caribbean easterly wave. *J. Meteor. Soc. Japan,* **45:** 444-466.
(90) Yanai, M. (1968). Evolution of a tropical disturbance in the Caribbean Sea region. *J. Meteor. Soc. Japan,* **46:** 86-109.
(91) Yanai, M., Maruyama, T., Nitta, T. and Hayashi, Y. (1968). Power spectra of large-scale disturbances over the tropical Pacific. *J. Meteor. Japan* **46:** 306-323.
(92) Yin, M. T. (1949). A synoptic-aerological study of the onset of the summer monsoon over India and Burma. *J. Meteor.* **6:** 393-400.
(93) Zipser, E. (1969). The role of organized unsaturated convective downdrafts in the structure and rapid decay of an equatorial disturbance. *J. Appl. Meteor.* **8:** 799-814.
(94) Zipser, E. (1977). Mesoscale and convective-scale downdrafts as distinct components of squall-line structure. *Mon. Wea. Rev.* **105:** 1568-1589.
(95) Cressman, G. P. (1948). On the formation of West Atlantic hurricanes. Mis. Rep. No. 24, Part II, pp. 68-103. University of Chicago.

9

Tropical cyclones†
Structure and Mechanics

A hurricane or typhoon has been an incident in the lives of many. It is not easily forgotten. A big meeting was held in 1977 in Biloxi, Mississippi, east of New Orleans, to commemorate hurricane "Camille"‡ of August 1969 which swept inland there with 900 mb central pressure and a storm wave eight metres high. In the centre were the civil defence people who did wonders during the preparation and clean-up periods, but particularly in the highly critical hours when the storm actually struck. Nothing remains more vivid in memory than the hammering that went on for a whole night in Puerto Rico when a hurricane was announced and 250 000 people boarded up their homes and possessions in San Juan.

Tropical hurricanes have caused the sinking of many ships at sea and countless disasters on shore, but such disasters are not confined to the tropics. Honshu and New England, for example, have been exposed to some of the most violent storms on record.

Since timely warning can make all the difference between enormous loss of life and none, tropical storms were a focal point of interest and

---

†Tropical cyclones with sustained one-minute winds of 74 mph (64 knots) are called *hurricanes* in the Atlantic and eastern Pacific Oceans. The name means "big wind" in the Taino language; its use has expanded to give it a generic meaning. Tropical cyclones of "hurricane" intensity are known as *typhoons* in the western North Pacific Ocean, *willy-willy* in Australia and *baguio* in the Philippine Islands; elsewhere they are called tropical cyclones.

‡The habit of labelling hurricanes with girls' or other names persists; it originates from Stewart's book "Storm" (67).

TROPICAL CYCLONES—STRUCTURE AND MECHANICS

study for many Europeans setting out for distant tropical lands in the age of discovery. Outstanding among them were groups of missionaries who built up a tradition of experience in all hurricane areas. Based on local observations of wind, high cloud drift, colour of sky, sea swell and other features, they were able to perfect warning services whose accomplishments can hardly be guessed at and should never be forgotten. With the advent of weather charts, rapid communications, reconnaissance aircraft and satellites, the role of the missionaries as weather forecasters has declined. The old message "Hoist hurricane warnings." has been displaced by the forecaster who steps before the television camera every few hours to give millions of people the latest news about a storm. Figure 9.1 displays a high point of typhoon activity south of the Japanese islands with two centres as potential threat for the islands. No storm can escape the watchful eye of the satellite above.

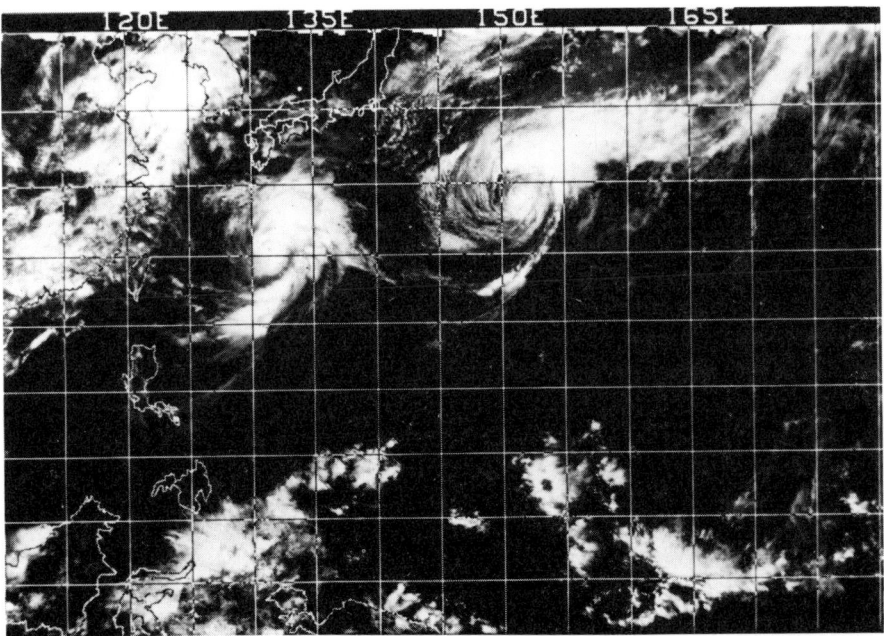

Fig. 9.1. Typhoon pair south of Japan, 2 July, 1976, seen by satellite.

With an expanding horizon of all observations forms, the scientist sees hurricanes as a part of broad atmospheric circulations, influenced in their behaviour by the winds over distant cold lands and in turn reacting on them. Large hurricanes intruding into the westerlies in autumn contribute frequently to the "build-up" of the winter circulation (16).

Yet hurricanes have not had a stable or steady input into the general circulation, at least as far as the history of Atlantic Ocean hurricanes during 1885-1975 is concerned (Fig. 9.2). The mean frequency has been 40 per 5 years. Except for the beginning, activity lay consistently below this mean (30 per 5 years) until 1930 when, in a dramatic reversal from 1930 to 1931, frequency jumped to above average (50 per 5 years) for 40 years until it started to decline again in the 1970s. It is possible, of course, that the Atlantic frequency time curve was countered and cancelled in other oceans, although there is evidence of a world-wide increase in cyclonic activity between 1920 and 1945.

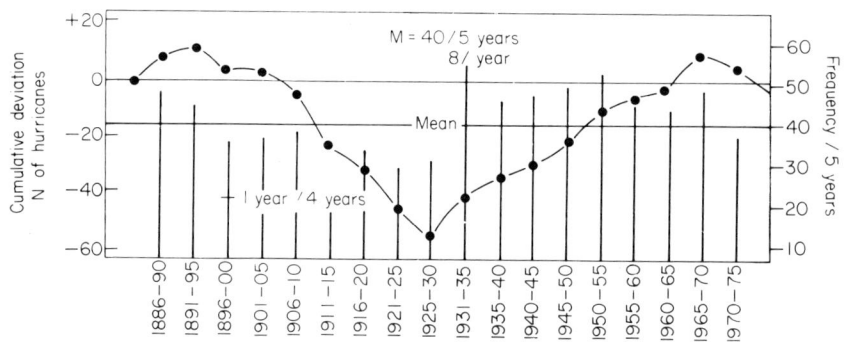

Fig. 9.2. History of Atlantic hurricane frequency in 5-year sums and accumulative deviations from the 90-year mean of 8 hurricanes annually.

Summary of life-cycle

The life-span of tropical cyclones with full hurricane intensity averages out at about six days from the time they form until the time they enter land or recurve into the middle latitudes. Some storms last only a few hours; a few as long as two weeks or even a month. The evolution of the average storm from birth to death has been divided into four stages (14):

(1) *Formative Stage.* Tropical storms form only in or near pre-existing weather systems as described in Chapter 8. Deepening can be a slow process requiring days for the organization of a large area with diffuse winds. It can also be explosive, producing a well-formed eye within 12 hours. Winds usually remain below hurricane force in the

formative stage. Strongest winds are apt to be concentrated in one quadrant, poleward and east of the centre in deepening waves in the easterlies, more variable in the equatorial trough. Surface pressure drops to about 1000 mb.

(2) *Immature Stage.* Not all incipient cyclones become hurricanes. Many have been known to die within 24 hours even though winds had attained hurricane force. Others travel long distances as shallow depressions. If intensification takes place, the lowest pressure rapidly drops below 1000 mb. Winds of hurricane force form a tight band around the centre. The cloud and rain pattern changes from disorganized squalls to narrow organized bands spiralling inward. Only a small area is as yet involved, though there may be a large outer envelope.

(3) *Mature Stage.* The surface pressure at the centre is no longer falling and the maximum wind speeds no longer increasing. Instead, the circulation expands during the mature stage, which may last a full week. Whereas winds of hurricane force may blow within a 30-50 km radius during immaturity, this radius can increase to over 300 km in mature storms. The symmetry is lost as the area of gales and bad weather extends much farther to the right of the direction of motion than to the left (looking downstream).

It is surprising how much the size of mature storms can vary. Even with central pressure less than 950 mb, the radius of some storms is only 100-200 km. If the surface pressure averages 1000 mb over the storm area, the mass will be 3×10^{11}-10^{12} tons. In contrast, storms with similar surface pressure can attain a radius of 1000 km or half the size of the United States and a mass of 3×10^{13} tons, fully two orders of magnitude greater. Only the strongest Icelandic and Aleutian Lows have such a size. In the westerlies, the normal cyclone comprises about 5×10^{12}-1×10^{13} tons.

Mature cyclones also go through irregular periods of increase and decrease, lasting one to three days, usually during interaction with troughs of higher latitudes that extend into the tropics. Short-term wind speed variability in approximately one hour averages over 10% as established by aircraft traverses designed to investigate this point (65).

(4) *Terminal Stages.* The core of hurricanes regularly decreases when they make landfall. Sometimes the whole storm dies within one or two days, sometimes it continues as a type of "subtropical cyclone" giving much rain for days. Over sea, hurricanes have been observed to die, for instance when moving over cold ocean currents as in the northeastern Pacific.

Both over land and sea many cyclones "recurve" into the westerlies and take tracks toward northeast and east in the Northern Hemisphere. The event may be quite unspectacular or it may lead to an enormous blow-up of the general circulation.

Surface structure

Observations taken in tropical cyclones over the years cover all stages of the life-cycle and all storm sizes and intensities. It is no wonder, therefore, that the characteristics deduced by different investigators are not all identical. In some areas, for instance, Japan, most storms are in the terminal phase. In other areas, the immature and mature stages are observed most frequently. These differences must be recognized. Here, we shall be mainly concerned with the mature stage; but we must note immediately that storms level off at very different intensities. One distinguishes on a qualitative basis:

(a) the tropical depression, winds barely touching gale force in one quadrant at most;
(b) the tropical storm, winds of gale, but below hurricane force; the latter is defined as 74 mph or 64 knots;
(c) the minimal hurricane, winds above the threshold in one quadrant only;
(d) the moderate hurricane, winds 80-90 knots around the centre, strongest wind may attain 100 knots and more;
(e) the severe hurricane, strongest winds to 200 knots, the highest observed or inferred from shore damages.

The limits of classes (a) and (b) are fairly generally recognized, though "tropical storm" is often used generically for "hurricane" when enough terms for description are lacking. Hurricane intensity classification has been quite variable, particularly as it is partly dependent on the region considered. In the western Pacific a typhoon may have to show strongest wind of 150 knots to be rated as "severe" or "extreme". Of course, an individual deepening storm may run through the whole sequence (a) to (e), but only a small percentage reaches the ultimate goal.

Because of the close relation between pressure and wind, and because hurricane intensity rating in terms of central pressure is widely accepted, pressure will be the starting point of our description.

Surface pressure

Ordinarily, surface pressure varies little more than 0·3% (3 mb) in the tropics. The central pressure of hurricanes, however, may be 5%, or even 10%, below average sea-level pressure. Sea-level isobars are, therefore, an excellent tool for analysis of tropical storms. Barograph traces at shore stations during hurricane passage, published from all storm areas, especially the Philippines (11, 12), furnish a measure of the pressure field in the core if the rate of movement is known. They all show the same primary features. Figure 9.3 gives a fine view of three storm intensities as observed around the island of Madagascar in the 1959 cyclone season (7). The barograph traces were taken from the records of islands well offshore before the storms made landfall on the main island; the centres passed over or close to the meteorological stations. All three barographs started with pressure of about 1010 mb.

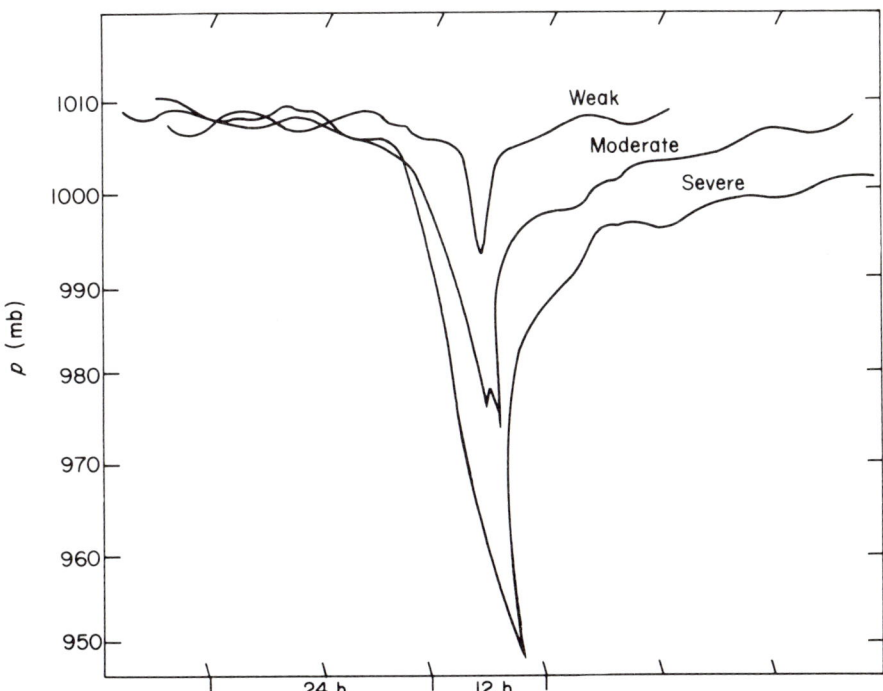

Fig. 9.3. Three barograph traces from Madagascar February-March 1959 illustrating weak, moderate, and severe tropical cyclone (after 7).

Therefore, the pressure depressions shown were 15, 35 and 60 mb. Main pressure drops started near 1005 mb after an initial gradual decline over 12-18 hours. The duration of the core of the weak storm was about four hours; it passed over Madagascar from the west in February, no doubt a rare track at the height of the season. Wind was reported at no more than 15 knots; much stronger squalls should have existed in some part of the perimeter within the concentrated low pressure centre whose diameter is given as 120 km.

The second storm, also in February, on a recurring trajectory toward southeast on the east coast of Madagascar, illustrates the moderate storm. Emphasis is on wind asymmetry; the southern semicircle had the largest expanse of hurricane-force winds. Finally, in the severe case with core duration of about 18 hours, wind speed rose to exceed 100 knots before the anemometer broke down (which happens in many places). In summary, the three traces show the variable degree of storm development very clearly and without having to go to a 900 mb case. Just as clearly the display poses some main problems, such as why do hurricanes level off as observed, and, if Fig. 9.3 is taken as the time history of an individual deepening storm, how does the enormous inner pressure depression develop?

Pressure gradients computed from barograph traces have been measured as 0·5 mb/km to 2 mb/km and more (14), and total pressure drops of 50 mb in 50 to 100 km are barely exceptional. Extreme central pressures are only rarely matched by cyclones over the subpolar oceans. Record pressure near 890 mb has been measured on various known occasions. Particularly reliable are the readings in the Florida Keys on 2 September, 1935, made in a small immature hurricane, and those obtained by Simpson (66) during a descent of reconnaissance aircraft in the eye of a large typhoon.

The wind usually forms a definite ring of maximum velocity at the eye boundary, but the barometer need not stop falling there. Sometimes there is a kink to decreased gradient, and sometimes pressure still falls rapidly inside the eye. Lesser oscillations of variable amplitude and duration are superimposed on the general outline of hurricane barographs. The semidiurnal period is detectable until the steep part of the pressure drop starts (Fig. 9.3). The period of the smaller oscillations, often quite irregular, may vary from seconds to hours; their amplitude may be as much as several millibars. Very short periods may be attributed to the pounding of long hurricane swell on the beaches, wind gusts and the swaying of buildings which house the barograph under the impact of the wind. Oscillations lasting minutes may be related to the varying cloud and rain structure.

In case a hurricane becomes stationary, the pressure profile will, of course, be flat and minimum pressure constant for the duration which, on a few occasions, has lasted 12 to 24 hours.

Low-level winds

By far the greatest damage on shores is caused by wind-produced hurricane tides and waves, and to a lesser degree by the winds themselves. Among the many photographs of hurricane damage, none is more impressive than that of a royal palm with its trunk pierced completely by a long, narrow pine board (68). The pressure that wind exerts on buildings is roughly proportional to the square of the wind speed. With sustained winds over 60 m/s and the associated wind gusts, the number of buildings in any community capable of withstanding the storm decreases quickly. Inhabitants of tropical areas often abandon their homes for the safety of churches, public buildings and other cement structures with steel beams, which have a record of surviving hurricanes quite well up to a point. As the area of extreme wind is usually quite small, it is understandable how storms often wreak complete havoc in one community and leave nearby towns with little damage. Extreme winds of 100 m/s have been estimated in a few instances over sea and deduced from structural damage by engineers over land. Even in an area as large as the Caribbean, such occurrences must be counted in terms of frequency per century.

A composite map of surface or near-surface winds is best made over the sea because of problems regarding surface friction and boundary layer depth over land. Over the ocean, a height of 300 m can be safely regarded as close to the level of strongest winds; most increase of wind speed with height is estimated to be concentrated below 100 m. On one aircraft-measured wind profile in the outskirts of hurricane Eloise (1977) wind speed is given as 20 m/s near 100 m height and at 22 m/s from 200-500 m (52). Above that level it decreased to 14 m/s at 1200 m. Such a decrease, though larger than that observed from aircraft missions flown at several levels, does not appear unusual, but many reliable observations are needed to confirm its existence as typical. Quite generally, the wind at 50 m height already attains 75% of the geostrophic wind (9, 10). Very likely, a similar or even stronger relation holds between $V_{50\ m}$ and V_{max} in hurricanes. In the interior of hurricanes the nominal 50 m level (in the outskirts 30 m), represents the same height as 10-15 m over the undisturbed trades where maximum speed is already nearly attained. The "nominal" level shifts upward in hurricanes if it is to remain above ocean wave height and most of the spray. Over land, the wind profile changes rapidly to one of increased vertical shear as air starts to pass over the near-solid boundary. At a distance of 15 km from shore, surface wind is considerably reduced from that over the ocean, especially in cities, whereas the 100 m wind may not yet be affected.

A good sample of wind observations at 300 m height should describe the low-level wind field closely. The best such sample remains that obtained by Hughes (33) from reconnaissance flights made in the Pacific Ocean in 1944-1945. Once thought an impossible feat, aircraft have penetrated even large and violent typhoons regularly since 1943 with remarkably few accidents. Hughes selected the 84 best flights out of several hundred available. Some of these covered a single stationary storm of moderate intensity; in the stationary case an "axially symmetric" wind field should be most nearly achieved; two-dimensional modelling should be most successful (for coordinate system see Fig. 8.17). Highest wind speed nearly attains 50 m/s (100 knots) mostly due to the large tangential component v_θ (Fig. 9.4). All convergence is concentrated within about 150 km from the centre, very evident in Fig. 9.5, and becomes extreme from a radius of 100 km inward. In the outer area, divergence prevails with magnitude normally encountered in synoptic weather systems. Here, descent must take place and convective activity should be below average. Unusually fine weather preceding a hurricane has proved to be a good warning signal by itself within the trade wind belt. The radius where divergence changes rapidly to intense convergence is often called the "bar" of the storm.

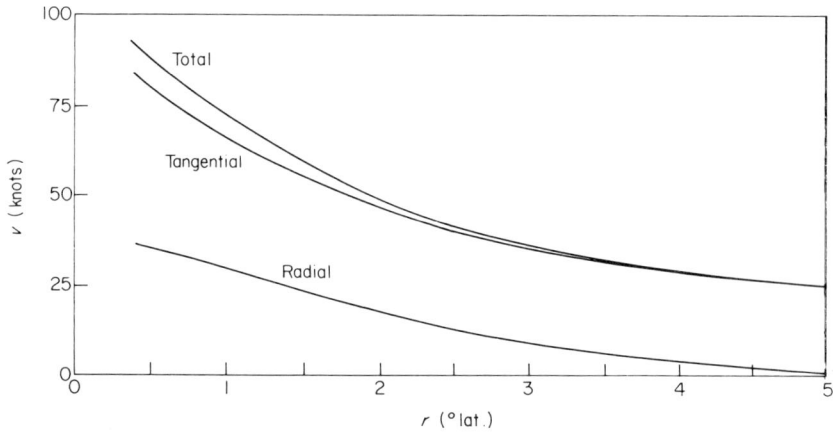

Fig. 9.4. Radial profiles of low-level velocity components for stationary typhoon (33).

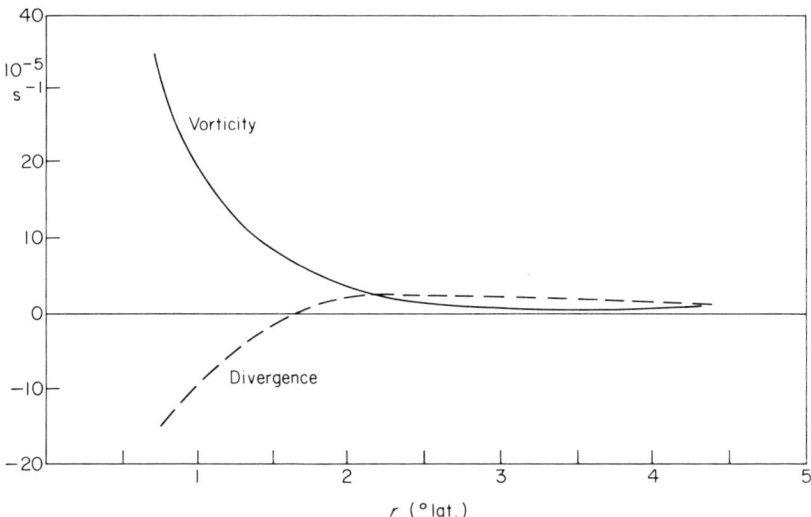

Fig. 9.5. Profiles of divergence (dashed) and relative vorticity (solid) computed from Fig. 9.4 Positive values: divergence or cyclonic vorticity; negative values; convergence or anticyclonic vorticity (33).

Hughes divided all moving storms into three classes; immature, small mature and large mature. Since the wind distribution turned out to be fairly similar except in scale, only the large storms will be discussed. All flights were averaged, using a polar coordinate system centred on the eye; the direction of the storm motion was held fixed rather than geographical directions such as north. This procedure fully preserves the asymmetry in the wind field introduced by the direction of motion. A corresponding Atlantic wind sample gave good agreement (50).

Although averaging always reduced extremes, speeds of 40-45 m/s (100 mph) nevertheless appear in the picture of total wind from the 80 km radius inward (Fig. 9.6). Winds are higher to the right of the direction of motion than to the left. To the right, carrying current and circulation act in the same sense; to the left, they are opposed. This asymmetry is one of the most reliable features of the tropical storm wind field, valid for instance for the Indian Ocean (38) and Japan (34).

Inward spiralling of the streamlines is very marked, especially in the rear quadrants.† There, the radial component of motion is strongest

†Tropical storms are commonly divided into four quadrants orientated with respect to the direction of motion: right front, right rear, left front, left rear.

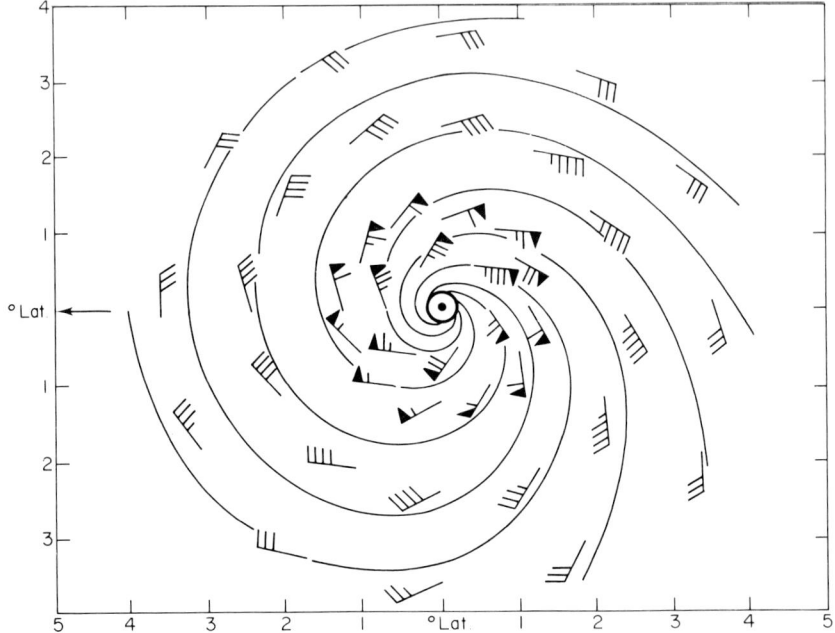

Fig. 9.6. Mean vector wind field and streamlines. Wind speed in knots.

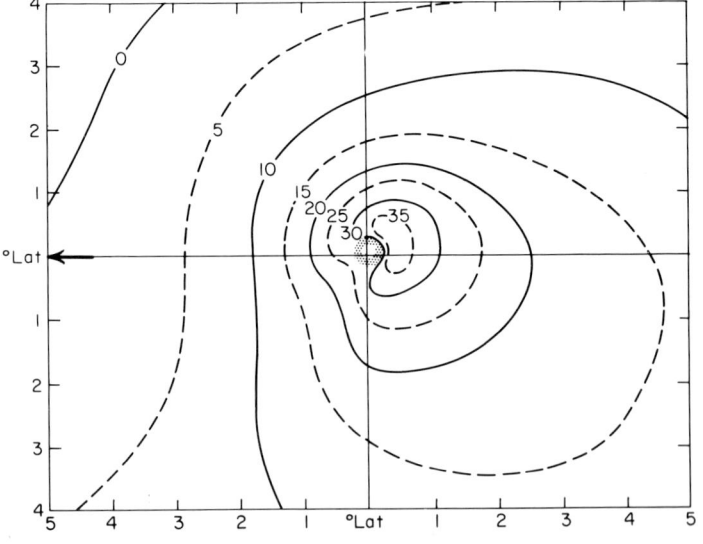

Fig. 9.7. Radial velocity (knots) (33).

(Fig. 9.7). From Figs 9.6 and 9.7 it might seem that the air in the left rear quadrant was approaching the centre most rapidly. Because of the storm movement, however, this is not the case. To obtain a true picture of the inflow, we must subtract the storm motion (average 10 knots) from the total velocity field and then compute the radial velocity relative to the moving centre (Fig. 9.8). We now see that it is the air in the right front quadrant which is really moving most rapidly into the centre, brought out by relative streamlines (Fig. 9.9). Air located in the right front quadrant will move to within 50 km of the centre from a position 200 km away within four hours. In contrast, air approaching from a distance of 400 km in the left rear quadrant requires nearly a day to reach the core. Relative inflow from the forward side is also confirmed for the Indian Ocean (38).

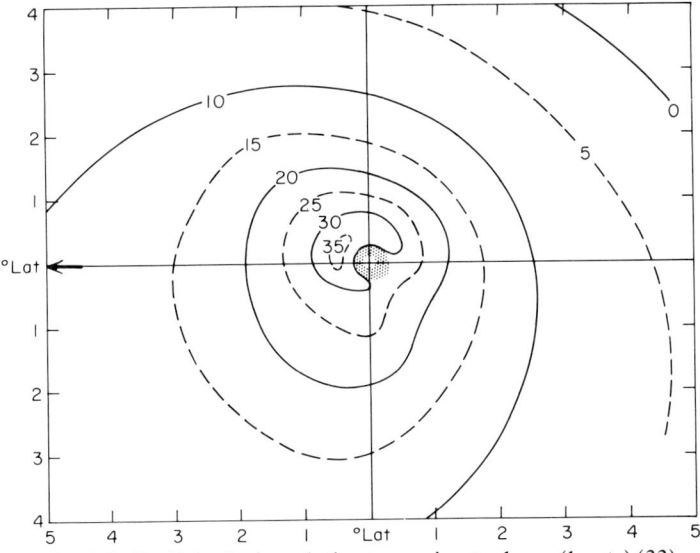

Fig. 9.8. Radial velocity relative to moving typhoon (knots) (33).

In the central area, values of convergence become exceedingly large (Fig. 9.10a), as for the stationary storm. Convergence is strongest in the left rear quadrant, where the satellite picture often shows a huge curving cloud tail. Fully half of the storm is divergent to a distance of about 150 km from the centre in the relative inflow region. Accordingly, prolonged suppression of convective activity (18), followed by the sudden onset of heavy clouds and rain, should be experienced at the approach of a storm, at least while it is moving westward in the trades.

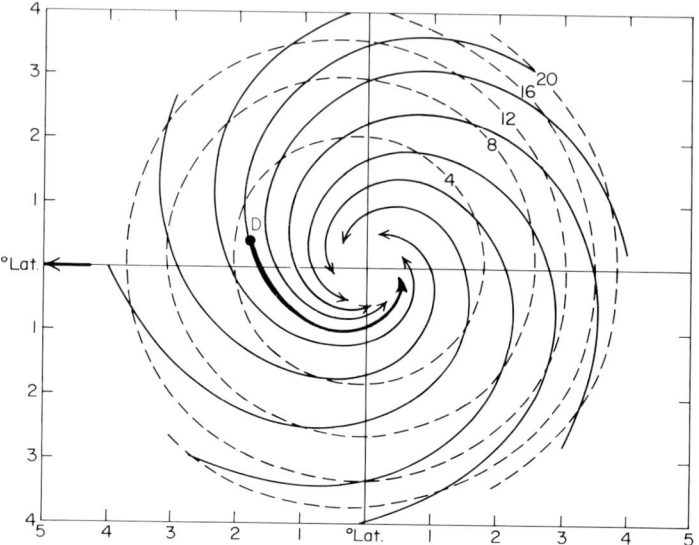

Fig. 9.9. Trajectories relative to moving centre (33). Dashed curves indicate number of hours required to reach a point 30 nautical miles from centre. Heavy curve with point D: trajectory of inflow into hurricane Donna (1960) (see Appendix).

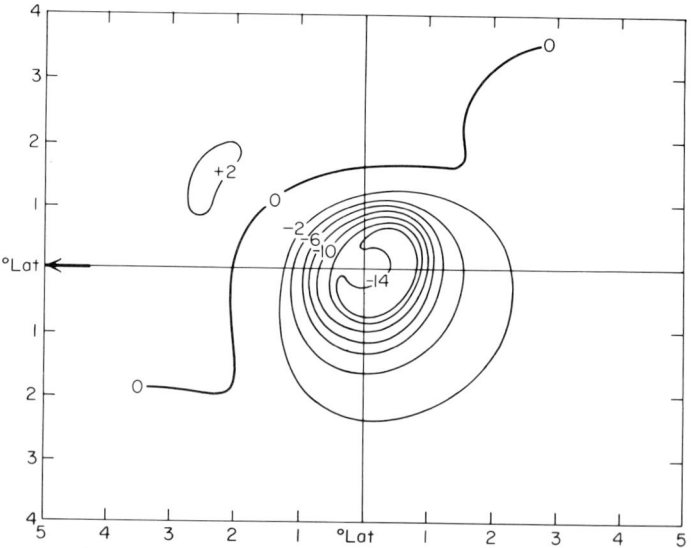

Fig. 9.10a. Field of divergence ($10^{-5}s^{-1}$). Negative values denote convergence (33).

TROPICAL CYCLONES—STRUCTURE AND MECHANICS

Relative vorticity, and therefore most certainly absolute vorticity, are positive everywhere (Fig. 9.10b). Since

$$\zeta_{rel.} = \frac{v_\theta}{r} + \frac{\partial v_\theta}{\partial r},$$

this result is by no means obvious. The rapid inward increase in wind speed represents anticyclonic shear. Opposed is the contribution by the curvature term. Evidently, the latter term predominates, as relative vorticity increases throughout the core to attain values approaching an order of magnitude larger than the Coriolis parameter at latitude 20°. Only inside the ring of maximum wind at, perhaps, a radius of 25 km are curvature and shear both cyclonic and $\zeta_{rel.} = 2\,v_\theta/r$ for solid rotation in the eye. For $r = 25$ km and $v_\theta = 50$ m/s, $\zeta_{rel.} = 2 \times 10^{-3}\,s^{-1}$, a truly extreme value.

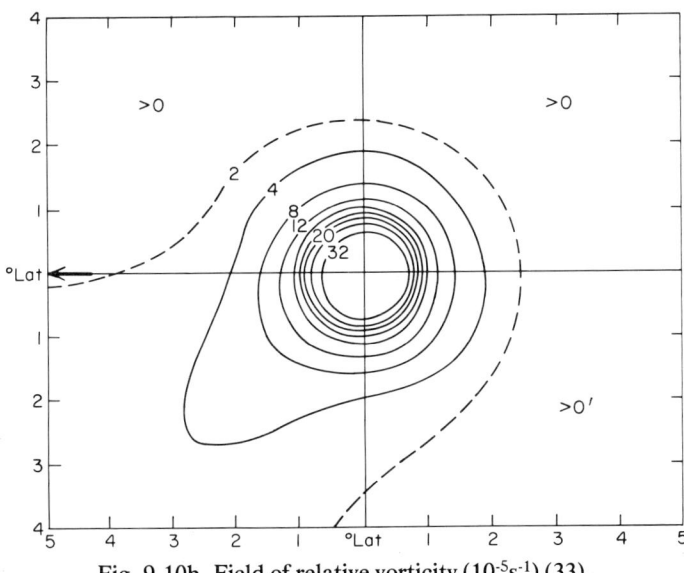

Fig. 9.10b. Field of relative vorticity ($10^{-5}s^{-1}$) (33).

Pressure-wind relations

A standing objective of meteorologists has been the estimation of central wind without having to measure it directly, which is often very difficult or impossible. If central pressure (p_c) or lowest 700 mb height can be measured by aircraft or, potentially, computed by satellite, the cyclostrophic wind relation can be evaluated empirically; though this is not an easy or straightforward task. Various statistical correlations have been undertaken via this route (20, 3). Another study, from 28 years of Pacific typhoon records (2), has led to the formula for maximum wind (V_m)

$$V_m = 3 \cdot 35 (1010 - p_c)^{0 \cdot 644}, \tag{9.1}$$

where V_m is expressed in metres per second. It is pointed out that Eq. (9.1) gives lower speeds than previous regression equations; for instance, with $1010 - p_c = 50$ mb, $V_m = 42$ m/s, the symmetrical part of the windfield. Evidently, frictional retardation is fully included. Maximum hurricane intensity may also be computed via the thermodynamic approach (48).

Temperature and humidity

If air enters the hurricane circulation with the average properties, it should reach condensation at about $p = 970$ mb (Fig. 9.11). Farther inward, a dense fog should cover the ocean, a cloud along the ground, but this is not observed over sea, whereas along land trajectories toward a hurricane off-shore adiabatic expansion does occur; there is no surface energy source. Over the ocean an average ceiling height of 200-300 m remains.

From Eqs. (4.10) and (4.11) a large energy transfer from ocean to air will take place, giving the hurricane a local heat source at the bottom which largely establishes its warm core. From many sources of observation—notably in the Philippines (12) as pointed out by Byers (5)— temperature first falls 2-3°C in the outskirts from adiabatic expansion and rain cooling. Then it *becomes nearly constant* at $T_w - T_{as} = $ 2-3°C. Aircraft-measured temperatures support the isothermal expansion deduced. Even in the eye, at the very low pressure of 900 mb, Fig. 9.26 shows $T_{as} = 26°C$, $q = 20$ g/kg.

All observations support the increase of sensible and latent heat along inflow trajectories. Consider the solid ascent lines of Fig. 9.11. As the air expands adiabatically to 990 mb, temperature falls to 24·4°C, q remains constant and ceiling height decreases to 200 m. Strong air-sea interaction, of course, does not begin suddenly at 990 mb, but assume that this is what happens. Then $T_{as} = $ const $ = 24 \cdot 4°C$ and, if the ceiling remains at 200 m, $T_d = $ const $ = 23°C$. The slope of the T_d line gives the increment in specific humidity, from 18·4 to 20·0 g/kg.

In one rare case when research aircraft penetrated a whole hurricane below and near cloud base (Gladys, 1975) the dotted curves of Fig. 9.11 show temperature and specific humidity changes along the trajectory of Fig. 9.29. Here, most energy increase was accomplished by latent heat increment and decrease of cloud base from

TROPICAL CYCLONES—STRUCTURE AND MECHANICS

700 to 250 m, which shifts the moist-adiabatic ascent from $\theta_e = 348$ K to 358 K, i.e. to a warmer ascent path. The mechanisms of warming the hurricane interior, a difficult subject, will be treated in the Appendix to this chapter.

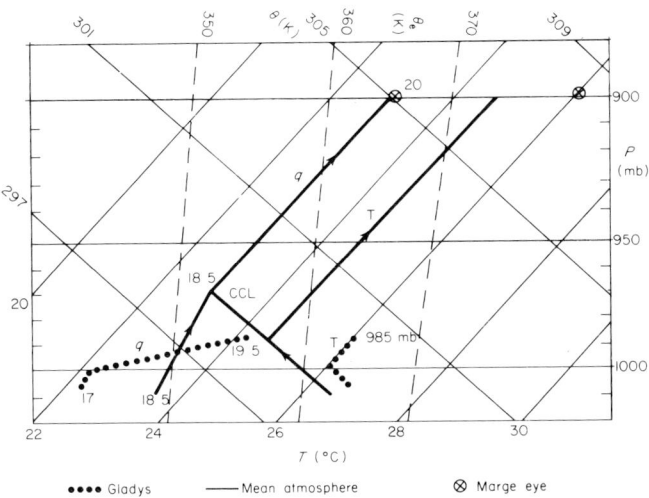

Fig. 9.11. Low-level tephigram showing the variations of temperature and specific humidity for an assumed case (solid line) and as observed in Gladys (1975) (dotted line) along inflow trajectories in the subcloud layer.

Precipitation

Individual rain gauge measurements give only a poor approximation of precipitation in hurricanes. The wind drives rain horizontally and picks up water already fallen to the ground. Even slight topographic features such as buildings, lakes and small hills influence precipitation. Since most records stem from tropical islands and shore stations, usually with mountainous background, reports of rainfall amount and distribution vary widely. In addition, rainfall at any station depends, of course, on its location with respect to hurricane path, intensity and celerity.

Recorded precipitation has been as low as a trace in an October 1941 hurricane, in the Florida Keys, in spite of 60 m/s winds. It has been as extreme as 250 cm, three or four times the average annual precipitation at most middle-latitude cities, during one typhoon passage in the Philippines, dwarfing even the great Philippine flood analyzed in Chapter 8. Amounts up to over 80 cm fell over the eastern slopes of Madagascar during the four days of passage of the severe storm of Fig. 9.3, while west of the central mountain range rainfall was near zero. For the whole month of March the eastern slope had a belt of 100 to 200 cm rain over a distance of 10° latitude (7).

In order to obtain a picture as free as possible from these many local effects one can make a computation. Rainfall, minus evaporation in the air, within a circle of specified radius must equal the difference between inward and outward transport of moisture across the circle plus the moisture source inside the circle, assuming the atmospheric column itself is nearly saturated. Since the outflow takes place in the high troposphere, where the vapour content of the air is very small (Fig. 2.19), it can be readily neglected. The outflow of frozen matter (cirrus), very visible as outward spiral bands on satellite and radar photos, may be assessed as 10% of the condensate.

With these guide lines, precipitation can be computed from the mass inflow (M_r) across a circle with radius r, $M_r = 2\pi r \bar{v}_r \delta p/g$, multiplied by the mean specific humidity $\hat{q}$ over the depth of the layer. Contrary to the calculation at the end of Chapter 4, horizontal eddy transports may here be emitted for a first approximation. This product must be equal to RR, the rainfall per unit area and time over the area, if about 10% of the inflow is subtracted for the ice outflow. After division by πr^2, the rainfall per unit area and time

$$RR = 0.9 \frac{2\bar{v}_r}{r} \hat{q} \frac{\delta p}{g}. \tag{9.2}$$

The caret denotes vertical averaging, the bar, averaging around a circle. The pressure thickness $\delta p/g$, from the discussion in Chapter 4 and especially Fig. 4.26 which applies fully, should not exceed 100 mb; at most, 150 mb. As in the convection model at the end of Chapter 4, convergent low energy air at higher levels must be brought down outside the core and cannot directly enter the upward motion inside, if there is to be a hurricane.

In Fig. 9.12 the computed radial profile of precipitation from the Hughes wind fields is shown together with computed and observed precipitation for hurricane Donna.† The "observed" precipitation was composited around the centre from all reporting Caribbean, Bahamas and Florida stations during its sojourn of several days through this area. It is seen that with this much effort a reasonable coincidence of computed with rain gauge rainfall can be achieved. The Hughes curve falls off more steeply, but such a difference cannot be regarded as serious. Higher values were found for cyclones near India (38). Other computed profiles for individual storms all have rain of

†Fig. 9.14 shows the paths of the hurricanes mainly used in this chapter.

over 50 cm/day near the hurricane eye—which means one centimetre if a station is 30 minutes in this area, a little over two centimetres if it remains an hour, and up to the full amount if the hurricane becomes stationary. Then there is a flood disaster.

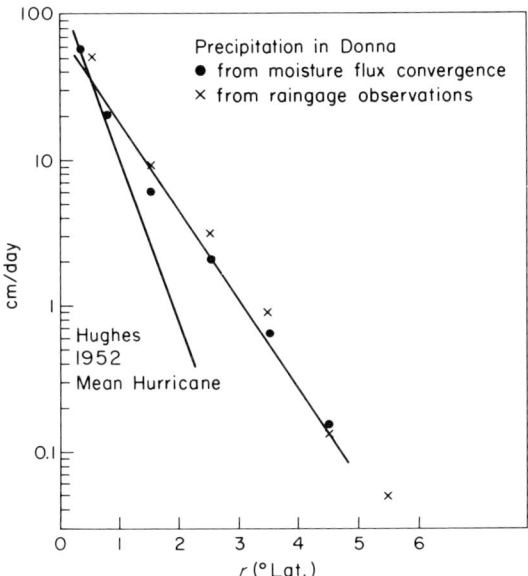

Fig. 9.12. Precipitation profile with radius for mean hurricane, also as observed in hurricane Donna (1960): computed (dots); observed (crosses).

If a storm with the rainfall distribution of either curve moved directly across a station with mean celerity of 6 m/s, storm duration would be two days and total precipitation around 40 cm. Such an average is reasonable and compares favourably with the mean rainfall for hurricanes on the Gulf coast of the United States (8).

Actual rainfall distribution is strongly affected by the asymmetries of circulation. In most hurricanes, clouds and rain extend much farther to the right than to the left of the direction of motion, where indeed it may end abruptly within 60-80 km from the core. Usually, the left front quadrant is safest for aircraft penetration if a storm is moving between west and north. Weather in the left rear quadrant may be much more severe; there, bands of intense convection may extend several hundred kilometres from the storm centre. Satellite photos show these bands as part of the system of spiralling lines of convection.

As tropical storms recurve toward higher latitudes, the precipitation pattern will change. Rainfall becomes concentrated in the right front quadrant of storms moving across the Gulf coast of the United States and elsewhere. This concentration has been ascribed to differences in friction between land and sea, but a change to warm-front type precipitation is often another factor when a cooler airmass is present over land. Drier and cooler air is also often drawn into the circulation from west and northwest. When this happens, the sky clears quickly to the south of a northward-moving centre. The precipitation shield then is situated on the forward side, strongest in the right front quadrant.

Estimates of total latent heat release inside a hurricane envelope indicate LP (integrated) $= 200 - 400 \times 10^{14}$ kJ/day (38, 49), a number to file for estimation of the efficiency of a hurricane in generating kinetic energy. No less than 67% of the latent heat release occurs within the 1° radius, 32% within the 0·5° radius (38). Satellite microwave radiometers are giving considerable detail on rainfall rates in hurricanes, including horizontal distribution and time changes (1). Total energy release agrees closely with the data just quoted for several typhoons measured, while energy release increased by a factor of three during transition from initial to mature stage. The contribution by heavier rains (≥ 0.6 cm/h) grew from 8% to 38% at the same time, and the higher rainfall rates concentrated toward the centre of circulation. This is very valuable information. The water yield of 400×10^{14} kJ/day is 16 km^3/day, equalling the *annual* runoff of the Colorado River—a remarkable comparison!

Surface characteristics of the eye

The eye of hurricanes has held the attention of all who have written on tropical storms from the earliest times. Its name was firmly established long before the days of suggestive satellite and radar photos. Eyes have also been reported from subtropical cyclones and intense extratropical storms. There is a singular point in the middle of every revolving vortex; although most dramatic in the hurricane, a weakening of wind and weather occurs at the centre of every strong circulation. Intense precipitation ceases abruptly at the edge of the eye while the heaviest radar echoes may extend far outward (Fig. 9.13 (a) and (b)). The sky clears at least partly and the sun may appear; as often as not, however, an overcast remains at and above 500 mb, or

TROPICAL CYCLONES—STRUCTURE AND MECHANICS 413

low-level cumuli form one or more separate small spirals within an eye boundary. Observers have described conditions in the eye as "oppressive", "sultry" and "suffocating", but this reaction is psychological and due to the rapid transition from hurricane winds and torrential rains to relative quiet conditions (14). There are also the many pictures of masses of birds settling exhausted on ships' railings.

In a mature hurricane, the eye diameter averages 30 to 50 km. It may attain double that distance in a large typhoon. The eye of two hurricanes passing Tampa, Florida, was large enough for a radiosonde observation to be taken inside (Fig. 9.22). Although the eye is usually pictured as circular, it sometimes becomes elongated;

Helene August 26, 1958
Radar Echoes – 250 mb $T(°C)$

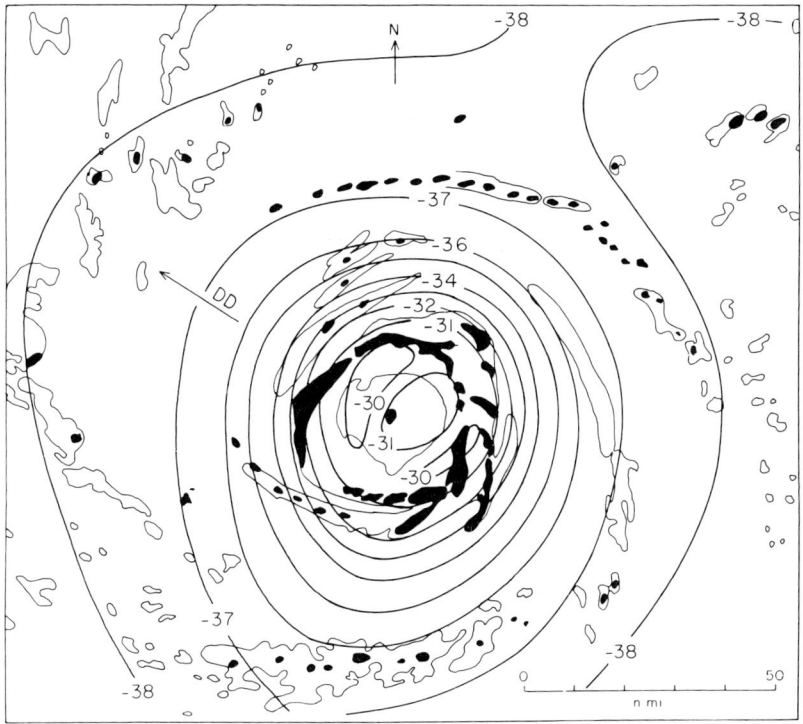

Fig. 9.13. (a) Distribution of hard (solid) and weak (outlined) radar echoes in hurricane Helene (1958) composited from B-47 aircraft radar film and of 250 mb temperature (°C), 26 September, 1958.

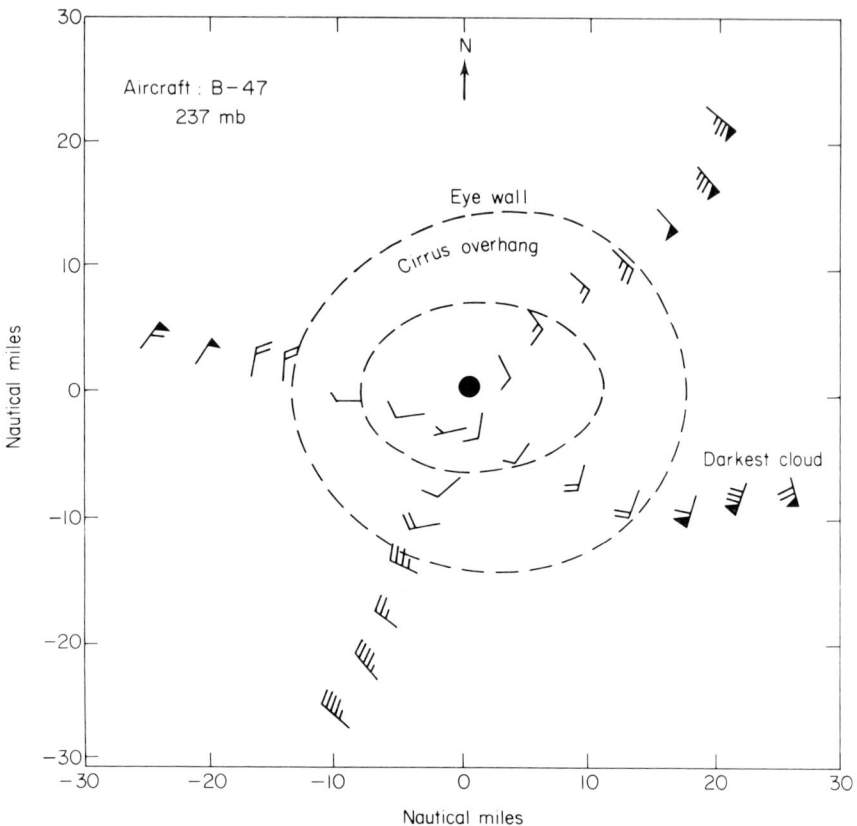

(b) Winds at 250 mb in and near eye, demarcation of eye wall, cirrus overhang and darkest cloud. Plane was in cloud all the time outside eye.

sometimes it is diffuse with a double-structured apperance. Modern observations, especially radar, have proved that an eye does not remain in steady state but is constantly undergoing transformation. In view of the extreme conditions at the edge of the eye this is hardly surprising.

Inside the eye, the surface temperature of the air does not differ from that on the outside though with sunshine it will become somewhat warmer. A few famous cases show rapid temperature rise and humidity drop in the eye at the surface. These observations were once thought to indicate a breakthrough of subsiding upper air to the surface, but they have been largely explained as a result of downdrafts from mountain ranges. Surface wind often gives the appearance of a cartwheel, $v_\theta/r =$ constant, the inner part of a Rankine vortex. Often winds are more variable. If there was a true calm, the air in the eye should move to the rear relative to the moving vortex and rapidly exit there. Continuity would be maintained by descent or by air entering from the front side near the surface; the latter is most probable since humidity is high in addition to temperature. Some experimental beacons have been placed into eyes about two kilometres up. If they stayed inside, and transmitted a signal, the hurricane progress could be followed by monitoring aircraft on the outside without need to penetrate. Interestingly enough, the beacons did not exit in the rear but were contained by the heavy eye wall. *Goes* satellite observations have made this and similar schemes superfluous since they produce a centre position every 30 minutes.

State of the sea

This subject has come up throughout the preceding pages, showing how closely air and ocean are related in hurricanes.

Waves

The extreme wind stress exerted on the ocean surface produces extreme wave group heights in the 20-50 m height range. Detailed discussion of wave generation and propagation, also of tides, may be found in (A-10) and other texts. The waves travel outward in all directions from the storm centre, and they can be spotted, as long sea swell, 1000 km distant and more. Their speed of propagation may be 1500 to 2000 km per day. Since average storm displacement is only one-third of this rate, the ocean waves have provided early warning of an impending hurricane in the past.

Waves generated in the right rear quadrant travel in the direction of the storm. They propagate under the influence of the maximum winds which have relatively little change in direction for a longer time than waves in the other quadrants, and consequently have a comparatively long fetch. Thus, the strongest part of the swell is produced in this part of the storm. When it arrives at a distant coast line, the normal wave frequency, which may be 10-15 per minute, decreases to 2-4 per minute. The direction of the swell indicates the position of the centre at the time the swell was generated, and if this direction remains constant the storm will approach the area directly. If it turns counterclockwise the storm will pass from right to left across the area, as seen by an observer facing the storm. A clockwise turning indicates the opposite movement.

In spite of great ocean depth, the tremendous surface waves in hurricanes initiate vibrations at the bottom of the ocean; these may be recorded by seismographs. Because of their longer period (2-6 s) microseisms due to tropical storms can be distinguished from other vibrations produced, for instance, by local pounding of the surf. Due to the advent of other tracking methods, notably satellites, the use of microseisms appears to have been abandoned.

Among the most severe effects of a hurricane is the damage done along coasts by the ocean waves. Such damage is suffered mainly where land partially surrounds water bodies such as the Bay of Bengal and the Gulf of Mexico. Sustained wind in the right-hand quadrants (in the Northern Hemisphere) piles up water along the shore so that the gravitational tide rises as much as three metres above normal; the high water mark may persist for weeks after a hurricane has passed. Such "storm tides" are of course not peculiar to the tropics; they are a constant danger in all constricted seas, notably the North Sea. High tides permit the long swell to penetrate shore lines and protective sea walls and then break with full force over coastal settlements.

An even greater threat is the so-called "hurricane wave", a sudden rise of the water surging inland with measured heights of as much as seven to eight metres in the Bay of Bengal in November 1970 and in hurricane Camille just east of New Orleans in August 1969; these produce the greatest disasters. Kraus (A-10) has analysed their theory in some detail. Storm waves appear to be best developed in and near the eye, but at times they made landfall well before centre passage. Most descriptions speak of a "wall of water" breaking in from the sea, suggestive of a solitary wave, while others mention series or groups of waves. It has been suggested that the rise of water may be related to the decrease in atmospheric pressure, which would lead to a one metre rise of the ocean surface at $p_0 = 900$ mb.

Over the ocean itself a remarkable change in the wave aspects occurs at hurricane speeds. The waves become organized into narrow straight rows which may be orientated precisely along the wind direction measured by low, overflying aircraft, i.e. wind direction does not change. All foam or spray comes off these lines and between them the ocean appears dark and rather smooth. Some photos have been obtained, notably by the United States Navy. High-resolution airborne radar is expected to give a fuller description of the ocean waves in hurricanes.

Ocean temperature affected by hurricanes

The stirring of the ocean by hurricanes may leave large ocean temperature changes in a hurricane's wake. With intense air-sea exchange high rates of $Q_s = 4000$ j/cm²/day and $Q_e = 8000$ j/cm²/day, total 12 000 j/cm² /day, or ten times and more the normal exchange rate, may be attained. A water column of 30 m depth would cool by 1°C from such exchange. Of course, a hurricane hardly ever remains stationary for a day, so smaller temperature changes should prevail. Nevertheless, Leipper (41) computed total heat loss from the ocean as 20 kj/cm² along the area of hurricane force winds in a Gulf of Mexico hurricane in 1964, which would give a 1°C reduction over a 45 m depth. However, temperature falls as much as 6°C were observed after hurricane passage. Evidently, dynamic processes in the ocean as well as the mean depth of the thermocline must be drawn into the consideration. Other instances of hurricane-related ocean temperature drops are given in (32, 39, 46); one requirement for strong oceanic cooling is slow hurricane motion of less than about 12 knots.

If the thermocline depth is over 50 m, perhaps 100 m, only the 1°C cooling in the upper portion should be observed. Vertical mixing—including internal waves—act to distribute the surface cooling efficiently over the layer. The wind drift should force surface water outward along the path of the cyclone and deposit it at some distance. With a thermocline at 100 m depth, nothing will readily change since the thermocline cannot be expected to be raised easily by 100 m. Leipper's case, as well as others, occurred where the thermocline has depths of 15-30 m. Here, water in and below the thermocline can readily surface and produce large cooling. At some distance the depth of the isothermal layer should increase and the thermocline should sink in compensation.

Leipper found that traces of the cooling could be detected in the Gulf of Mexico for months. In this event, occurrence of further hurricanes would be rendered rather improbable, but, depending on ocean currents and thermocline, the picture may differ (46). Further, in some instances, hurricanes have followed similar paths only one or two weeks apart. Nevertheless, there can be no question, as shown empirically and by modelling ocean temperature differences in hurricanes (54, 62), that introduction of considerably lowered ocean temperatures (2°C) will severely disturb a hurricane as the sensible heat source is cut off. As yet the situation around Australia remains unclarified; north and west of the continent many storms move very slowly for days at full hurricane intensity.

Upper air structure†

Upper-wind structure

One of the earliest controversies regarding tropical storms concerned their vertical extent. Some writers claimed that they disappeared at heights as low as three kilometres, that people standing on mountain tops could look down on the violent cyclonic whirls beneath them. They believed that the cirrus motion remained undisturbed and indicated the direction in which the cyclone was moving—still a good tool in certain circumstances in Florida. Others insisted that hurricanes reach to great heights, 10 km or more, and they thought that cirrus radiated in all directions from the storm and used the change of direction of cirrus movement to track centres.

The answer to this controversy is now well-known. Haurwitz (30) has shown theoretically, and later, through high-level observations,

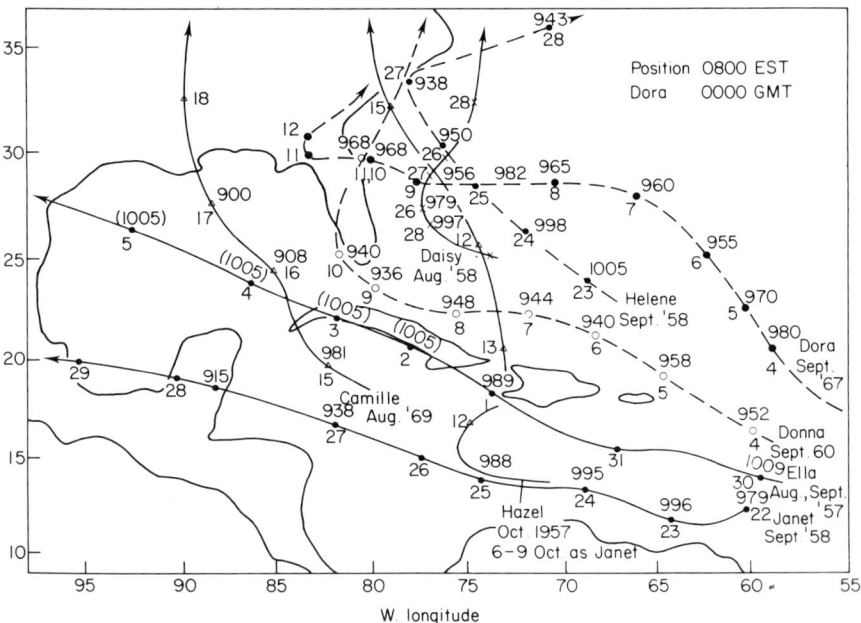

Fig. 9.14. Dates, tracks and central pressure of hurricanes mostly discussed in Chapters 9-11.

†For ease of reference, dates, track and central pressure of the hurricanes mostly discussed in this book are displayed in Fig. 9.14.

empirically, that mature storms extend through the troposphere. Unexpected until the era of rawinsondes was the fact that the high-level flow is anticyclonic, that a clockwise circulation (in the Northern Hemisphere) overlies the low-level counterclockwise wind field at altitudes above 300 mb outward from a radius at which the cyclonic circulation of the core vanishes aloft.

The task of constructing a good, three-dimensional, composite wind field for hurricanes is much more difficult than its determination for the low levels. Upper winds, reasonably, have been sparse in a setting not conducive to releasing and following rawinsonde ascents. Further, the upper circulation is much more variable than the low-level winds; for instance, it depends on how closely a hurricane has approached the polar westerlies. It is best to confine composite diagrams to no more than 6° latitude distance from a centre; even so, the limit is pushed rather far. Jordan (36) made the first valiant attempt at such composition, and in the main her results, carried inward to the 2° radius, have stood up well. With a large increase of upper winds in subsequent years, however, Miller (50) prepared a new composite using selected Atlantic hurricanes. Like Jordan, he shows tangential and radial components for several layers of the troposphere. Tangential components of the two presentations agree well; they are almost constant to 300 mb, then a ring of zero tangential wind encroaches inward to about the 2·5°-3° radius in the high troposphere. Jordan shows v_θ stronger to the right than to the left of the direction of motion; the average should give the steering current. Miller subtracted the steering current and obtained speeds on the left side, which he considers a little too strong (cf. 64). Perhaps subtraction of the whole steering current is an exaggeration. Nevertheless, the conclusion is warranted, that stationary storms are nearly symmetrical and that motion imposes an asymmetry which roughly, but not fully, corresponds to linear addition of a steering current and a vortex (see also Chapter 10).

Figure 9.15 shows averaged vertical v_θ-profiles for hurricane Donna, which happened to pass through most of the dense rawinsonde network of the western Atlantic then in operation. The profiles agree almost exactly with those prepared for a set of composite data (56). The layer closest to the ground cannot be shown on the scale of Fig. 9.15. Already discussed in the last section, it may be added that wind speeds measured on off-shore lighthouses and lightships were virtually identical with the rawin speeds higher up, more proof that nearly all increase of wind with height is very close to the ground in the inner hurricane core. From Fig. 9.15 the wind increases to

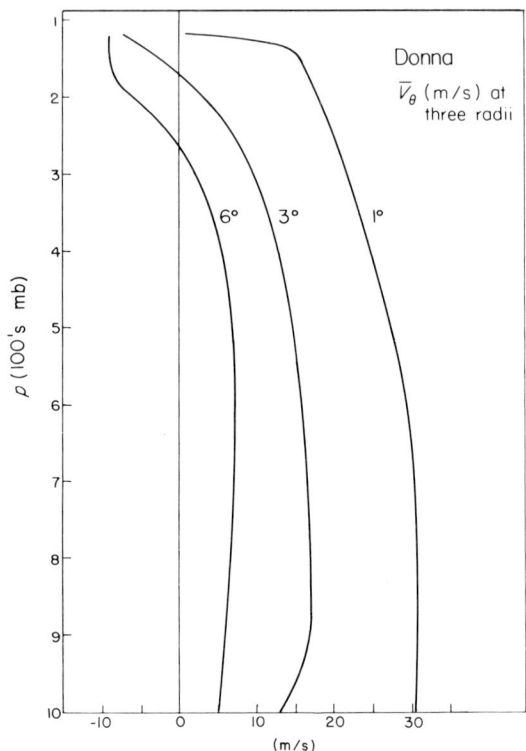

Fig. 9.15. Vertical profiles of tangential velocity (m/s) at three radii in hurricane Donna north of Cuba.

about 900 mb at the 3° radius. However, this part of the profile is uncertain, possibly contaminated by many island mountain effects.

The principal decrease of v_θ does not begin until 400-300 mb. At the 3° radius the circulation becomes anticyclonic at 150 mb, at the 6° radius, 300 to 250 millibars. Figure 9.16, updated from the computations of Jordan (36), is still thought to give a reasonable approximation to $\partial v_\theta / \partial z$ and relative vorticity. The latter becomes negative and approaches the Coriolis parameter in the layer 200-150 mb, suggesting zero or negative absolute vorticity even on constant pressure surfaces.

Together with the tangential velocity, the pressure difference between inside and outside a hurricane is nearly constant as high as v_θ is constant, indicative of uniform temperature on isobaric surfaces (Fig. 9.17). From 500-400 mb upward the pressure difference decreases quickly and becomes negative near 200 mb, when measured

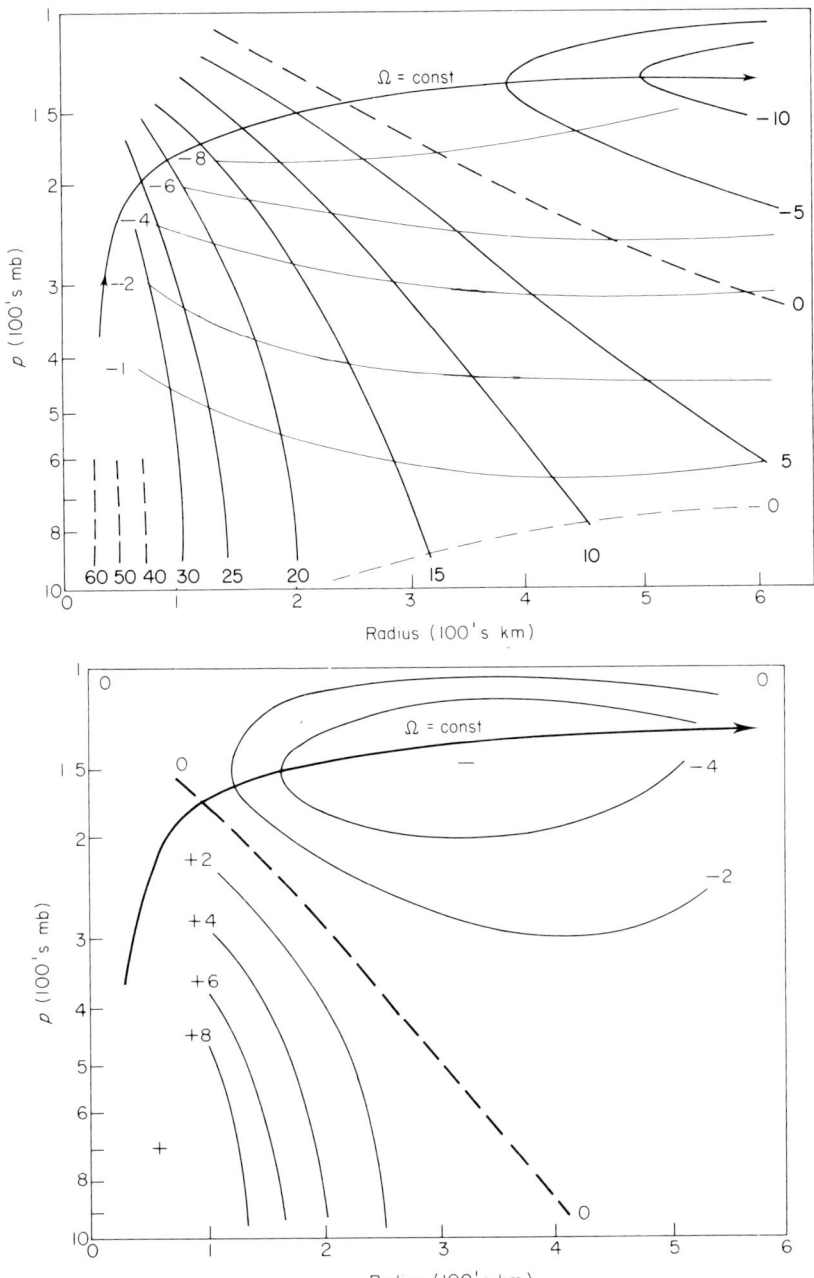

Fig. 9.16. Upper: Tangential velocity component profile (m/s) and lines of equal vertical shear of the tangential wind (m/s per 100 mb) in hurricane Donna. Trajectory for Ω = constant in axially symmetric motion added. Lower: Profile of relative vorticity ($10^{-5}s^{-1}$).

from the radius where tangential velocity is zero in the outflow. The upper pressure difference in a setting of anticyclonic flow is small compared to that in the surface layer. Inward from the radius of strongest wind, pressure-heights fall toward the centre but at a much smaller rate than near the surface, a combined effect of the local heat source within the storm and compression heating due to descent in the eye.

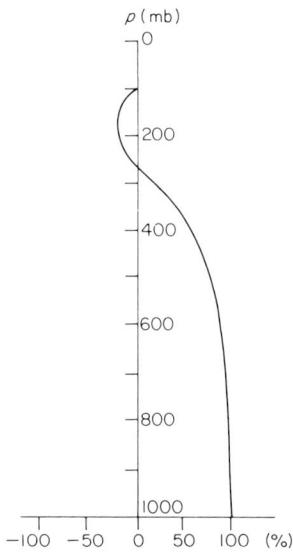

Fig. 9.17. Vertical profile of mean pressure difference between inside and outside of hurricanes, expressed in per cent of the surface difference (58).

The radial component shows maximum inflow in the surface layer and outflow in the high troposphere in all studies. In the middle troposphere, light and variable values of v_r are computed close to the hurricane centre, while in the periphery, net inflow persists as high as 400 mb. The profile for Donna (Fig. 9.18) is quite typical; it agrees with Miller's results and inflow profiles found in various other hurricanes (cf. also Fig. 10.8). Such net inflow, from all previous deductions, must descend, otherwise a severe brake would be imposed on the hurricane engine with entrance of mid-tropospheric, low-energy air into the centre. From the v_θ-profiles it is clear that the mid-tropospheric air cannot contract too far inward. Otherwise v_θ should increase strongly with height—abstraction of absolute angular momentum in the surface layer due to friction but inward motion at

TROPICAL CYCLONES—STRUCTURE AND MECHANICS

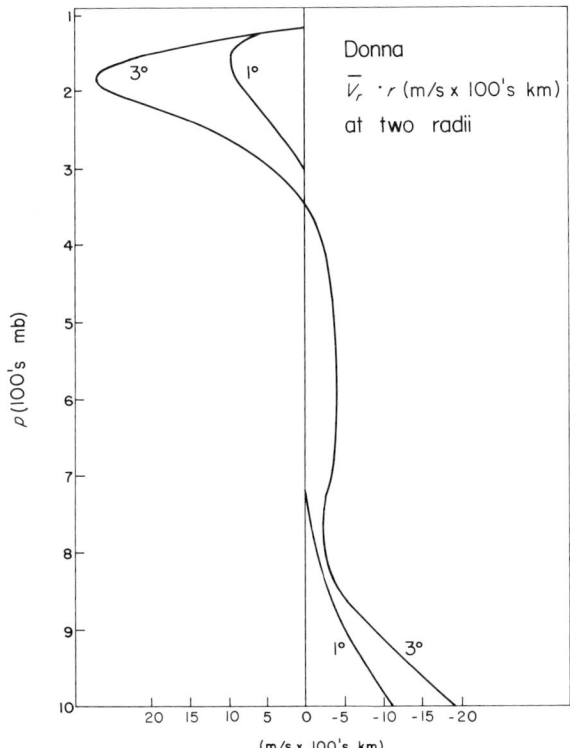

Fig. 9.18. Vertical profiles of $\bar{v}_r r$ (m/s × 100 km) at two radii in hurricane Donna.

nearly constant momentum higher up. Nevertheless, following Ooyama (54), it is the slight contraction of the inflow during descent which accounts for the fact that cyclonic rotation is encountered in the deep, mid-tropospheric layer as far out as the 6° radius. The descent must be pictured as facilitated by the downdraft mechanism along the outer spiral bands where convective turbulence may be intense.

Outflow charts prepared by Miller and computed for Donna are presented in Fig. 9.19. The principal difference is the lack of inner cyclonic motion in Miller's composite, possibly related to the fact that his chart extends as high as 16 km. Neither diagram indicates the very strong northeasterly to easterly outflow with speeds of 25 to 30 m/s often found south of Pacific typhoons moving west at low latitudes. The principle of all these structures remains the same but no complete similarity in detail should be sought. A detailed construction of a mean typhoon approaching the Japanese Islands from south (34) shows much similarity with the foregoing; the cyclonic circulation

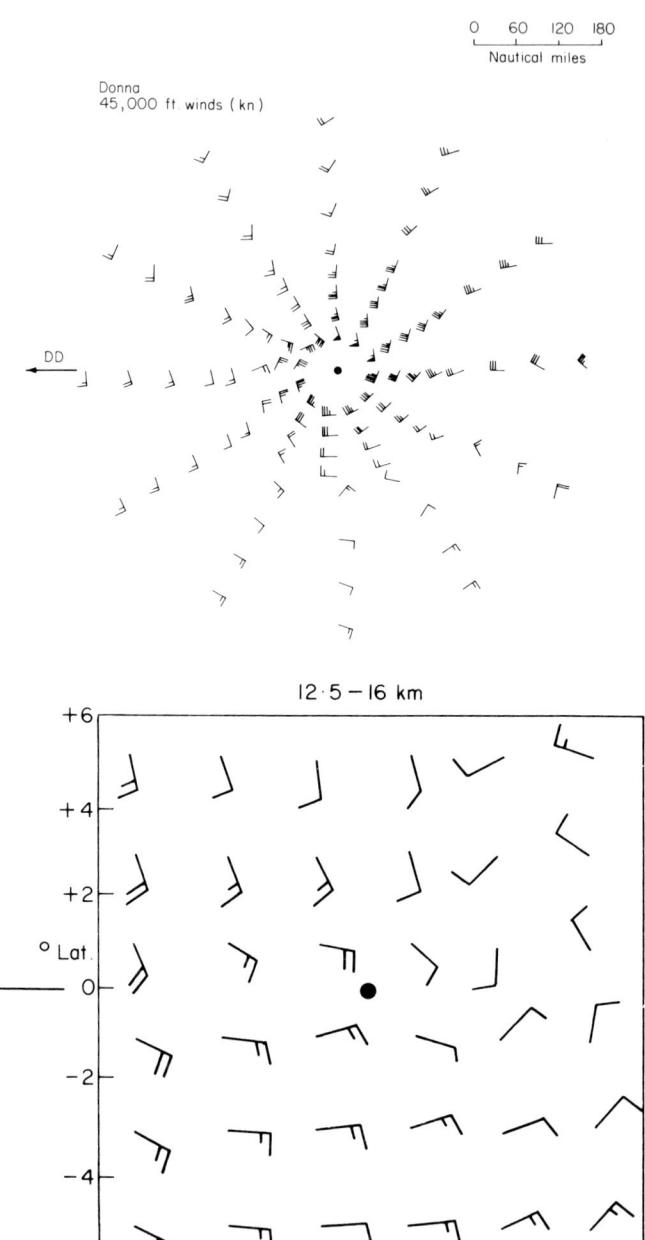

Fig. 9.19. Upper: Windfield at 13 700 m for hurricane Donna between the 1° and 6° radii (knots). Lower: The composite windfield from 12·5 to 16 km in knots (50).

covers a wider area, possibly because of the higher latitude and late stage of the storms composited.

The hurricane circulation weakens and disappears toward 100 mb, and easterlies already prevail at this height. These strengthen upward considerably to 50 mb (Fig. 9.20) and attain 25 m/s at 25 mb (29). Of course, there is no information within the hurricane core, but one would think that any strong perturbation there would be transmitted as far as stations on the Florida coast. On the basis of the charts shown, the assumption of a 100 mb top appears justified.

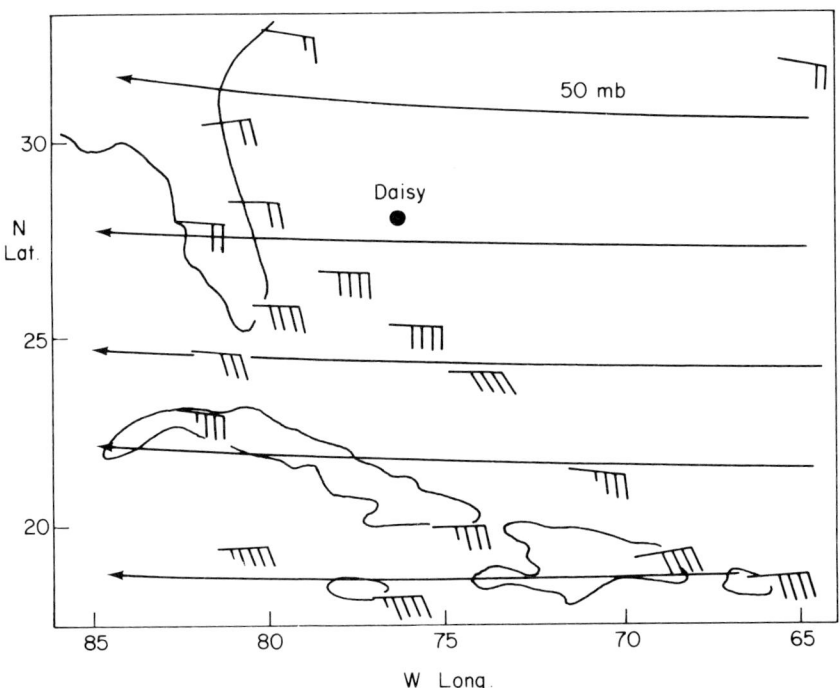

Fig. 9.20. Winds at 50 mb over the West Atlantic-Florida region during hurricane Daisy on 26 August, 1958 (29).

Several other features of importance have been enumerated by Shea and Gray (64): the slope of the ring of maximum wind is small and a function of intensity; vertical wind shear is much smaller in deepening than in filling storms; in intense storms the maximum wind occurs closer to the centre than in weak storms. Also noted, but not generally agreed, are that the maximum wind occurs within the eye wall cloud area—at times it is well outside or there is no eye wall, and that with increasing latitude the maximum winds occur farther away from the storm centre than at low latitudes, implying a widening of the eye. This appears very reasonable because of the increasing influence of Coriolis force with latitude, i.e. gradient replaces cyclostrophic wind. However, according to G. Bell, Royal Observatory, Hong Kong, eye radius is independent of latitude.

Cloud systems

Although the cloud sequence varies considerably from one storm to the next, accounts of many eye witnesses of the precursory signs are remarkably uniform. On the day before storms which move west to northwestward the weather is often unusually fine (subsidence) and the barometer above normal. Arrival of the long sea swell is the first warning, then the barometer slowly begins to drop (Fig. 9.3) and the wind may start blowing from an unaccustomed direction. In the low levels, normal convective activity is suppressed and amount and depth of cumuli are below average. In the high troposphere, a cloud sequence begins which is closely akin to that observed during warm-front approach in middle latitudes. Cirrus makes its appearance, followed by cirrostratus, altostratus and altocumulus. Then several brief periods of cumulus congestus with showers may be experienced as the outer convection spirals approach and pass. The barometer falls more rapidly, and finally a dark wall of clouds approaches—historically called the "bar" of the hurricane. With its arrival the full force of the hurricane is unleashed.

Good agreement exists between the low-level cloud sequence and the field of divergence (hence also, of low-level vertical motion) computed in Fig. 9.10. Now we may add that the description of the upper clouds accords with the observations and deductions about the winds aloft. A narrow current, at first ascending almost vertically, then flowing rapidly outward in the upper troposphere, will reproduce exactly the reported cloud sequence.

Satellite photos

Out of the immense number of photogenic satellite pictures, we have already seen the photo of hurricane Anita (Fig. 7.8). Since the satellite looks on the scene from above, the whitish spiral (indicating cold temperatures) and almost circular cirrus mass around the eye that peaks out, are the upper outflow. The bands of darker colour (warm), evident especially over the southern United States, belong to the low-level inflow bands. Satellite films clearly show where the motion is inward and outward. Again the reader is urged to avail himself of any chance to see such films; they are even carried often as part of regular television programmes. An older picture is that of Fig. 9.21 which looks like the genie out of a bottle. Here the outflow toward northeast is evident where the typhoon has established contact with the polar westerlies, whose own cloud lines are visible west of the hurricane outflow. Inferences may be drawn from such pictures about maximum hurricane wind speed (21). For further illustrations, see Figs 9.1 and 10.1.

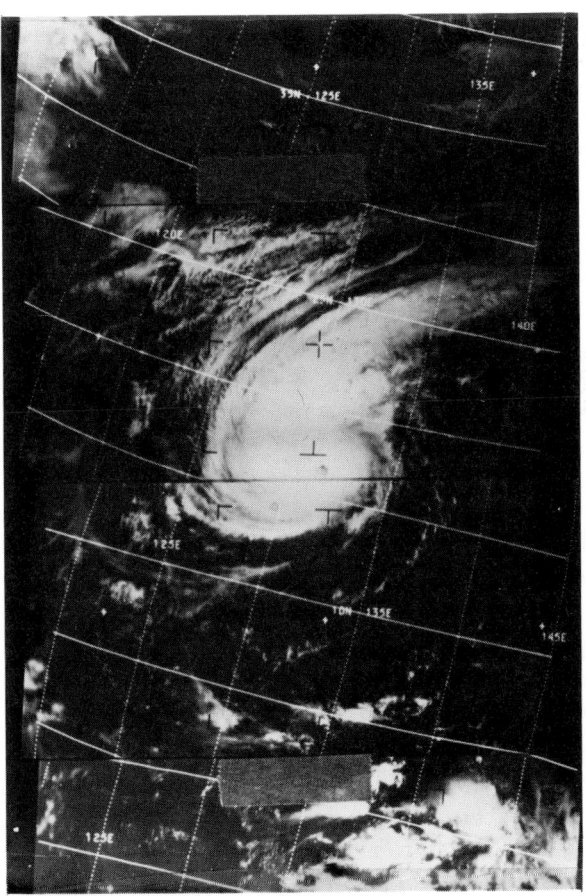

Fig. 9.21. Satellite photo of typhoon centred near 17°N, 132°E, on 15 November, 1967. Eye is discernible.

Radar bands

Observers' accounts have long indicated that the fury of a hurricane does not persist constantly. Extreme violence alternates with periods of comparative lull, even in the core, but little analysis of these oscillations was possible until the advent of radar led to a remarkable discovery (45, 69). The central storm area is not filled with rain-sized (1mm+) particles uniformly through the troposphere. Rather, we find narrow, elongated bands of enhanced intensity, interspersed with areas where the echo is feeble or absent (Fig. 7.10). Some bands have their roots hundreds of kilometres from the centre. From there, they spiral inward, though the organization of the spirals has been noted to emanate from the centre outward (63)! Radar, of course, sees the incoming spirals; the RHI picture can be interpreted to show increasing cloud height along their length.

Caution is needed in the interpretation of radar photos, not only because of the curvature of the earth (see Chapter 7) but also because of beam attenuation in very heavy rain even from 10 cm apparatus. The far side of a hurricane cloud photo beyond the eye nearly always looks truncated, which gives an unrealistic picture of unsteadiness, but real unsteadiness, nevertheless, occurs. Bands "push in" and then die away after having reformed the eye. In the eye ring the ascent in undilute cumulonimbi attains 5-10% of the ring area, not more than in synoptic weather systems but, nevertheless, in organized concentration (44). The balance of space is filled with weak echoes from light rain from anvils. The towers may extend somewhat above 150 mb after the energy addition along the earth's surface. They push the air above it upward, perhaps as far as the tropopause, producing pileus clouds which sit like a cap or layer cake on top of the solid towers reaching up.

Slope and walls of the eye

One of the finest views of the hurricane, described vividly by Simpson (66) and others, is the cloud panorama which surrounds the observer flying at high altitudes inside of the eye wall. The author experienced such a view only once, but the sight was unforgettable. Flying at 300 and 200 mb in hurricane Donna, the airborne radar formed a beautiful and very clear ring around the plane. One could see the low part of the eye filled with two separate spirals curling inward. Higher up, the eye was clear and one could see a great discrepancy between north and south sides. The southern semi-circle was filled with solid cirrus of milky appearance in the sunlight, topping out at 250 mb; the sheet was completely smooth and without waves or protrusions seen from above. The eye wall had a large cirrus overhang sloping into the eye (like Helene, Fig. 9.13); huge masses of ice cascaded inward from the solid sheet, an enormous icefall, creating the descent in the eye. The ice first accelerated the downdraft and then evaporated. On account of very high stability and warm temperatures, however, the lower part of the descent must have been forced. From tritium measurements (53) in other cases it has been estimated that descent contains a stratospheric admixture of about 5% of the mass in the eye at low levels, drawn from the layer 125-100 mb, which is entirely possible.

On the north side of the Donna eye, to the right of the direction of motion, there was a huge glaciated iceberg with a very hard and sharp outline towering far above the aircraft at 200 mb—a stunning sight. It appeared to be slightly receding with height, i.e. sloping outward. This veritable iceberg contained all ascending mass and upon penetration it proved violent. As well as an enormous updraft, there was a thunderstorm going on and what appeared to be a heavy mixture of ice and water struck the plane. The pilot backed into the eye after one minute (about half penetration) since the Navy jet was only marginally stressed for an encounter of such extremity which contrasts sharply with many smooth and non-turbulent penetrations in the middle troposphere. Thunderstorms are not supposed to be part of hurricanes, but in another case, hurricane Helene, thunderstorms were also encountered on several crossings of the eastern eye wall near 850 mb on 25 September, 1958.

In summary, the eye wall appeared vertical everywhere up to 300-200 mb. Of course, a slight widening compared to the low levels was possible. Higher up, an increasing outward slope of the huge mountain of cloud above the aircraft was visually detectable.

Thermal structure

Observations taken since the development of radiosonde instrumentation have confirmed the classical idea that air in the interior of tropical storms is less dense than that in the surroundings. Aircraft experiments, notably those of the United States National Hurricane Research Project, have greatly added to the knowledge of thermal structure as well as all other aspects of tropical storms. The drawback of aircraft missions is that, while they give great detail on constant pressure or level surfaces, even a multiple plane mission can only explore a few levels in the vertical on one day, and the analyst must interpolate between levels.

Vertical eye wall

One of the best explored hurricanes was Daisy of 1958 which gradually moved northward east of Florida (Fig. 9.14) within easy reach of the hurricane research planes based in southern Florida. In the mature stage, surface pressure was near 950 millibars inside the eye and strongest wind was 50 meters per second. For this time Fig. 9.22 shows a composite sounding at the 20 mile radius (curve C), together with the mean tropical atmosphere (A), undilute ascent at 350 K with the surface properties of the inflowing trade wind air (B) and the hurricane eye sounding taken at Tampa, Florida, in October 1944 (D).

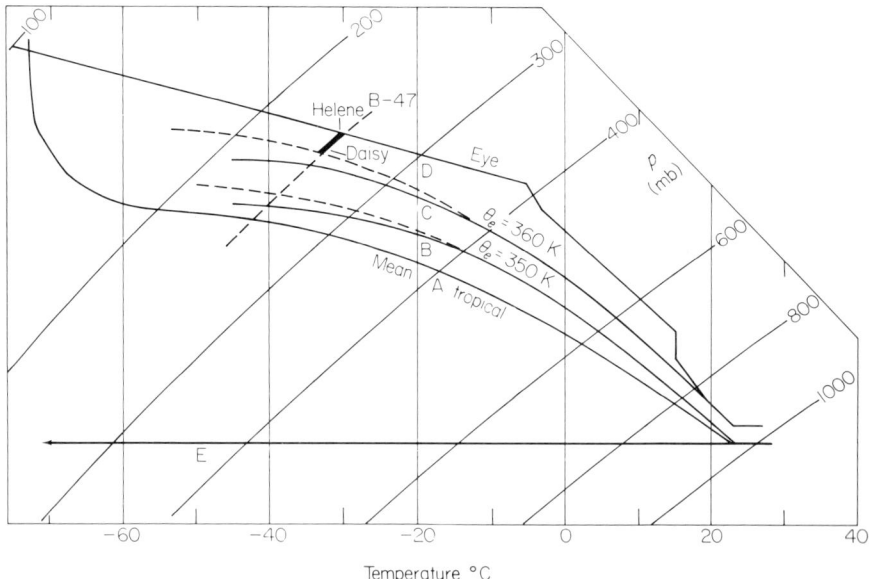

Fig. 9.22. Tephigram showing: (A) mean tropical atmosphere (Fig. 2.19); (B) ascent at $\theta_e = 350$ K; (C) ascent at 360 K after isothermal expansion to surface pressure of 960 mb; (D) eye sounding at Tampa, Florida, in October 1944.
Dashed: modification of ascents (b) and (c) along ice adiabats.
Heavy solid: range of aircraft-observed temperatures in rain area outside eye in two hurricanes.

The mean tropical atmosphere, which largely forms the outer thermal envelope around a hurricane, is the coldest sounding. Ascent B shows the temperature profile that can be called to life by undilute ascent at 350 K. If measured along the vertical, it is hydrostatically compatible with sea-level pressure of 1000-995 mb. Another increase of the same magnitude is achieved with isothermal expansion of the inflow to 960 mb and increase in specific humidity of 3 g/kg. This ascent, with $\theta_e = 360$ K in the low levels, is hydrostatically compatible with the 960 mb pressure if it occurs vertically. Even in this moderate hurricane, at least half of the temperature gradient vital for the functioning of the hurricane machine is created within the system itself, no doubt a major reason why hurricanes are rare. In extratropical cyclones a pre-existing temperature field becomes concentrated and its potential energy is reduced with the growth of the cyclone. In the tropics, the cyclone must produce the temperature gradient and then also maintain and defend it.

Above about 400 mb both inside soundings begin to depart noticeably from the pseudo-adiabatic ascent. Many measurements of high quality and importance were made by B-47 aircraft of the United States Air Force flying near 250 mb; these permit an excellent description of high-tropospheric structure. The nominal θ_e rises by 5 or 6°C, just as in case of Fig. 8.34. It is highly probable that the heating mechanism described there, largely from release of latent heat of fusion, is also fully operative here and produces the departure of the soundings on the warm side. An abnormally cold layer underneath the zero degree isotherm has been regularly encountered in hurricanes and, as in the Manila case of Chapter 8, has been interpreted as being due to melting ice and snow.

Both in hurricanes Helene (Fig. 9.13) and Daisy (Fig. 9.23) a very fine cross was flown by the US Air Force B-47 aircraft. In Helene, the warm temperatures of $-30°$ and $-31°$C, based on very careful evaluation[†], occurred both inside and outside the eye, marked heavy in Fig. 9.22. Thus, ascent temperatures in the rain area and eye descent temperatures were identical at 250 mb; it so happens that eye sounding "D" also fits. In Daisy θ_e increases inward from 350 K ($T = -39°$C) to 365 K ($T = -32°$C) in the inner ascent area, here defined by the ring of strongest winds. Within 8 to 13 km from the best centre position winds suddenly become very light. Here $T = -32°$C, almost as warm as in Helene. To confuse things, however, the film shows the plane to have remained continuously in cloud during both eye passes,

†By Margaret Chaffee, formerly at Woods Hole Oceanographic Institution.

one of which was a "bulls" eye, not an easy feat in a large 400-knot airplane when the eye diameter is so small!

Figure 9.23 indicates obviously that the ascent, therewith the eye wall, did not slope outward. Actually, the eye encountered by aircraft underneath was 32 km wide, hence larger than at 250 mb. It is, therefore, quite correct to say that the ascents of Fig. 9.22, although drawn following the ascent path of air, also represent a vertical sounding. With some assumption about the top of the clouds and assessing the top of the hurricane as 100 mb, from the last section, the sea-level pressure can be calculated from the sounding. Such a computation was carried out for a range of θ_e values (43) (not including the ice

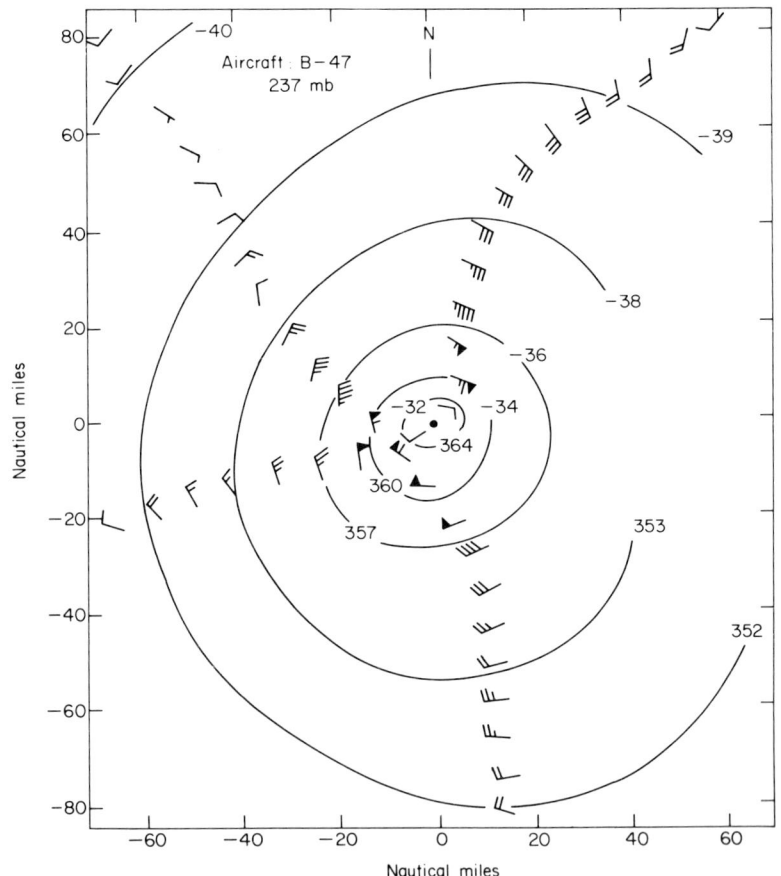

Fig. 9.23. Winds (knots), temperature (°C) and θ_e (°K) observed by B-47 aircraft in hurricane Daisy at maturity, 27 August, 1958. Eye wall is 10 km from centre dot.

phase) and it was found that in the range from 950 to 965 K an approximately linear relation between θ_e and p_c held, given by

$$dp_c = -2 \cdot 5 \, d\theta_e, \tag{9.3}$$

down to $p_c \sim 950$ mb. Accordingly, a 25 mb surface pressure drop should be found between the two hurricane rain area ascents of Fig. 9.22, not very different from what was observed. The above relation was tried out for eight storms where eye-wall temperatures could be identified from B-47 flights, several in different stages of development. The slope constant in Eq. (9.3) was 2·56 over the range 350-368 K, very close to the formula (61). Probably the large values reflect the upper warming due to ice phase. Except at the lowest θ_e values and highest central pressures, the B-47 measured θ_e values were higher than those of the inflow layer by up to 5 or 6°C, again emphasizing the likelihood of massive release of latent heat of fusion.

In summary, it appears that the ascent of air at progressively higher θ_e-values going toward the eye produces net tropospheric warming from which the surface pressure field can be deduced. From Fig. 9.17 this warming on constant pressure surfaces becomes noticeable only above approximately 500-400 mb. If the ascent of low-level air only occurs in 5 to 10% of the inner convective area (44), a question remains how the heat is diffused from the ascent so that it leads to an inner area warming. Heat diffusion from rising cumulonimbi has been deduced for non-hurricane synoptic systems. It could play the same role in the hurricane in the upper layers where, coupled with the accompanying wind shear, the undilute tower becomes vulnerable to erosion. In spite of the unfavourable saturated setting for development of downdrafts, Gray (24) has calculated an extensive draft structure from aircraft missions at lower altitudes and these would also be found at high levels. Therefore, descent between ascending towers has been suggested as a mechanism for raising upper temperatures, as well as outward diffusion of heat from the sinking air in the eye. In the latter event, air would have to exit from the eye where it is shown entering in Fig. 9.24; the mass appears to be much too small to accomplish large temperature increases for an area an order of magnitude larger. Dry downdrafts also are difficult to accept; even though intense radar echoes are concentrated in small areas (Fig. 9.13(a)), the intervening space nevertheless is filled with thick cloud from satellite and aircraft observations, not the setting for drying and warming downdrafts. It

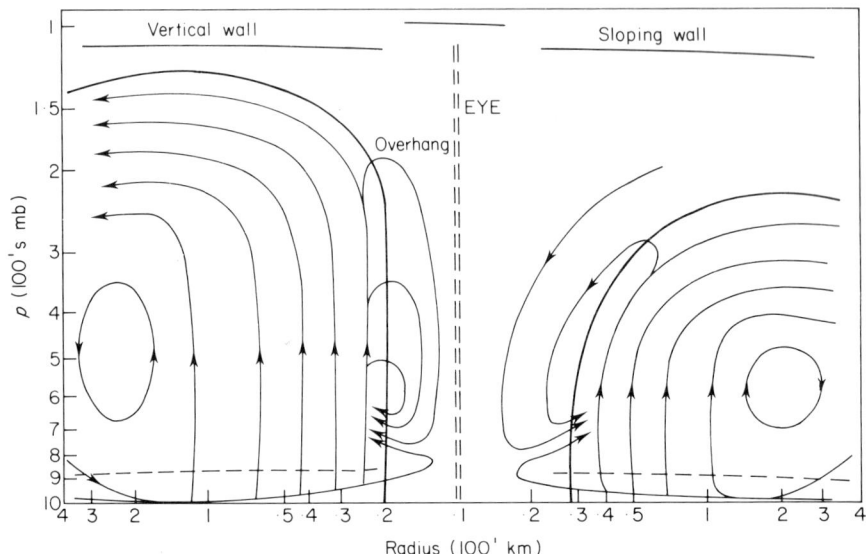

Fig. 9.24. Sketch illustrating hurricane with vertical and sloping walls. Radius logarithmically spaced. Dashed: top of main inflow layer.

appears that there is an unresolved problem and that erosion from towers leaning with increasing slope at high levels may be the most viable mechanism.

The possibility of temperature increase due to internal dissipation of kinetic energy was considered in hurricane Daisy (60). Much warming could not be accomplished by a reduction of wind speed from, say, 50 to 30 m/s which would yield 0·8°C, but persistent energy reduction of that magnitude following a hurricane would have the order of the radiational cold source and should, therefore, be retained in calculations.

Sloping eye wall

Figures 9.22 and 9.23 show the essential thermodynamic transformations in hurricanes with solid eye wall. They demonstrate the importance of the energy flux from the ocean in increasing θ_e. The changing thermodynamic ascent paths affect the mass field, give rise to a surface pressure gradient and therewith to generation of kinetic energy. So far we have not considered the position of Haurwitz (30) who believes surface pressure is determined from warming due to descent above a sloping eye wall. The wall would be shaped like a surface barograph trace. As its slope increases downward, the depth of the layer with adiabatically warmed air increases so that surface pressure—still assuming a 100 mb top—falls correspondingly (Fig. 9.24, cf. also Fig. 15.9 by Palmén and Newton, A-13). The precipitation profile of Fig. 9.12 could no longer fully hold, as most rain would occur farther outward. Such cases do happen with surface pressure profiles and winds very similar to those of the hurricanes with a vertical eye wall (19, 31).

One hurricane with central pressure near or below 960 mb which remained without closed eye wall for at least four days was Dora of 1964 as it slowly approached northern Florida from the southeast (65). Research missions at four levels up to 600 mb were carried out on 5 September, when the storm was moving very slowly so that much of the relative inflow took place in the eastern sector, somewhat different from Fig. 9.8. No particular problem was expected on approach from the southeast until the navigator began to complain that there was no eye wall and he had to work very hard navigating the aircraft to the centre. Normally, the pilot sees the clear eye ring on his radar screen and can manage the plane easily without navigational assistance.

The key illustration of Fig. 9.25 shows the RHI radar display from 30 n.mi. to the left to 30 n.mi. to the right of the aircraft situated in the middle of the diagram. Radar return remained largely below

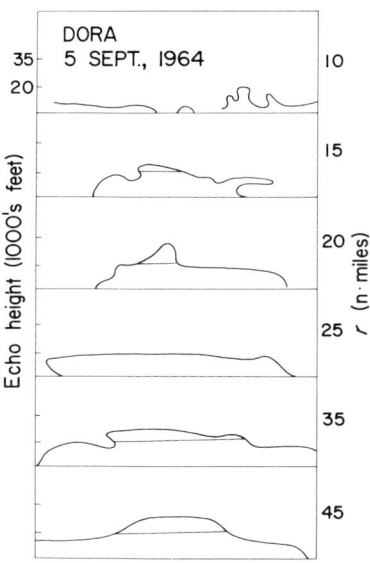

Fig. 9.25. Radar RHI cross-sections along relative inflow trajectory into hurricane Dora, 5 September, 1964. Echoes extending about 6100 m shaded.

6000 m near the centre where only a few isolated cumulonimbi could be detected. Echo tops remained at or below 9000 m all the way out, even though a composite echo picture shows a well-organized spiral pattern (65), evidently at lower levels only. A well-organized ring of maximum wind was present, with speeds above 50 m/s in the southeast quadrant and of over 45 m/s all around the storm. Pressure dropped 20 mb from the 64 km to the 32 km radius and then another 13 mb inside the eye with winds decreasing inward! It appears that most of the pressure fall of the inner region must be related to descent along and inside a sloping eye wall. Data were imperfect and no solution is apparent to questions as to location of the ring of strongest wind with largest pressure gradient continuous on both inside and outside of the 45-50 m/s isotachs. Further, with the large drop of central pressure inside the eye, how can descent with outflow be maintained against the pressure gradient force? It appears that knowledge of hurricanes and their mechanisms remains quite imperfect, even though almost identical pressure and windfields are observed for the two types of Fig. 9.24.

In spite of Dora's dry appearance in the inner area on 5 September, this did not stop her from delivering 25-40 cm of rain once the centre had entered northern Florida. Further, on 7 September, the storm was the subject of an interesting experiment designed to measure natural short-term variability over 30 to 60 minutes, described in some detail by Sheets (65). The meso-turbulence turned out to be large in wind speed, up to 20% and more on the hourly time scale, cloud and eye wall formations as well as the other elements. Such information is of more than casual interest for statistical evaluation of cloud seeding experiments in hurricanes. Additionally, Dora displayed further "non-standard" mean features on that day. Maximum wind ranging from 30-45 m/s had moved outward to the 50 nautical miles radius while such eye wall as could be found in the northwest quadrant was at 10-15 nautical miles radius. In spite of the variability of mean structure and of considerable short-period "turbulent" changes, the whole hurricane envelope proved very stable, as has been found in other cases as well. Except for the detailed aircraft missions, no one would have guessed the short and medium-time changes which did happen.

Temperature and moisture in the eye

Descending motion in the eye must be invoked not only in storms such as Dora to explain central pressure. Hydrostatic computations along

TROPICAL CYCLONES—STRUCTURE AND MECHANICS

even the warmest moist ascents will not yield the very low pressures in the centre of numerous hurricanes. Since the time of the first penetration of hurricanes by aircraft, large and sudden temperature rises have been encountered at the inner edge of the eye wall (curve D, Fig. 9.22). Jordan (37) has summarized temperatures (from dropsondes, aircraft descents and a few radiosondes) as a function of central pressure (Table 9.1). In each of his categories the range is large, though smallest at the surface, indicating the mitigating effect of the ocean. In the mean, surface temperatures are almost constant —isothermal with pressure—but a slight rise of 2°C to the lowest pressures is indicated.

Table 9.1. Mean temperature data (°C) for the eye of tropical cyclones
(The classifications (I-VI) are based on surface pressure, Jordan (37))

	I	II	III	IV	V	VI
Surface Pressure (mb)	998-980	979-965	964-950	949-939	924-906	901-883
No. of Cases	14	15	13	4	17	10
Surface	25·2	25·7	25·0	26·5	26·7	27·1
900 mb	21·7	22·9	23·2	24·5	26·2	
800 mb	17·6	19·3	19·3	20·7	22·9	24·5
700 mb	12·7	14·7	15·8	18·2	18·9	22·1
600 mb	7·1	8·8	11·0	14·0		
500 mb	2·4	0·7	3·5	6·0		

Maximum and minimum temperatures (°C) in the sample of data used in computing the mean data shown above

	I	II	III	IV	V	VI
Surface	27-24	30-24	26-24	27-26	30-24	29-23
900 mb	24-19	29-21	25-22	25-24	29-24	
800 mb	20-16	23-16	22-16	23-19	25-20	29-21
700 mb	18-10	17-12	18-14	21-16	22-14	27-17
600 mb	12-04	11-06	14-09	15-11		
500 mb	0-(—5)	4-(—3)	06-01	08-03		

One really extreme eye sounding, not included in the table, is that of typhoon Marge of 1951 (66). Here, a large temperature inversion was present between 850 and 750 mb (Fig. 9.26), sea-level pressure 900 mb. The inversion base decreases from 1·2 km above the ocean at about 980 mb to 500 m or less at the lowest pressures (35). This is the layer often filled with cloud spirals.

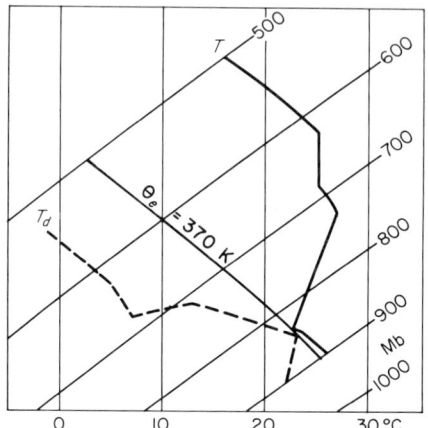

Fig. 9.26. Tephigram of dropsonde observation in typhoon Marge near 20°N, 136°E on 15 August, 1951 (66). $\theta_e = 370$ K added.

Of high interest are the moisture values reported in eyes, which permit computation of θ_e values. Figure 9.26, for instance, indicates specific humidity of 20 g/kg at sea level, 19 g/kg at 800 mb, 12 g/kg at 700 mb and still 8·5 g/kg at 600 mb. These are very high values in the "eye" setting. It turns out that $\theta_e = 370$ K, almost constant, in Fig. 9.26. In the Tampa eye sounding of 1944, humidities were high to 600 mb, then decreased to 1 g/kg at 350 mb. θ_e was constant, in this case at 360 K, at least to 300 mb. Higher up, the temperature curve deviated to potential temperatures of 370-380 K near 100 mb. How can these structures be understood?

If the eye, once formed, retained the same mass, then there should be observed a wind exactly equal to the storm velocity. If the eye is calm, except perhaps for rotary motion of the v_θ/r = const type, air should all the time enter the eye on the forward side and exit in the rear. Neither of these two extremes occurs, but eyes are never in steady state; they undergo short periods of intensification and decay. Horizontal penetration of rain bands into the centre with high moisture and liquid water content will, on descent to lower levels, produce eye soundings like that of Fig. 9.22, as long as θ_e in the ascending rain area and θ_e in the eye are equal (42). If the descending air retains its high angular momentum, it may be able to move outward against the pressure gradient near the surface if there is an imbalance of forces. As just seen, this solution was unlikely for hurricane Dora which had very light winds near the centre but a strong inward pressure gradient.

How does the approximately constant $\theta_e = 370$ K of Fig. 9.26 originate? As Fig. 9.11 indicates, inward mass movement of the surface air through the pressure gradient at constant temperature and dewpoint will lead to $\theta_e = 370$ K at 900 mb. Therefore, it is entirely possible for the inward mixing process to occur. No very large vertical displacements are required except for the possible slight admixture of substratospheric air. Considering water vapour constant during descent—without knowledge about evaporation of water drops—saturated air with 8·5 g/kg and $\theta_e = 370$ K would enter at 475 mb to produce the 600 mb T and T_d of Fig. 9.26. The origin for the air located at 700 mb is at 575 mb. At 800 mb, almost no vertical motion need occur, and the downdraft must be forced. For a typical mid-tropospheric temperature difference of 10°C and virtual temperature difference of 9°C about 30 g/m^3 of water would be needed for minimal buoyancy downward from Eq. (4.17); this value is clearly out of the practical range. A dynamical mechanism, capable of explaining the descent in the eye, has not been advanced.

Another potential source of heating in the eye, well removed from contact with the ocean, is internal dissipation of kinetic energy. Figures 9.13 and 9.23 at 250 mb, and many eye penetrations at lower elevations confirm the sudden drop in wind speed and, therefore, of kinetic energy, as the eye is entered. Outside air entering the eye must lose its energy of motion in part through frictional interaction which will lead to warming unless kinetic energy in the eye is to rise with time. Complete dissipation of the energy of a 50 m/s wind would produce a rise of 2·5°C/g or, if the admixture of high-energy air is, say, one-tenth of the eye air, the eye temperature rise will be 0·25°C. Knowledge is insufficient to pursue the matter beyond this point, but internal dissipation has been found to be a large residual in kinetic energy balances (60, 51, 47) in moderate hurricanes and may play an enhanced role in severe storms. One problem is the disposition of absolute angular momentum, which must be conserved. It can be reduced by flow toward higher pressure during descent or by lateral mixing through the very sharp momentum gradient at the edge of the eye. In any event, the angular momentum of the mean flow cannot be "dissipated".

Dynamic relations

The hurricane was seen early as a Rankine vortex with velocity profile given by $v_\theta/r =$ const inside the ring of maximum wind and $v_\theta r =$ const outside (13). Later, the outer profile was modified to

$v_\theta r^x =$ const, where the range of x is about 0·4 to 0·6. From modelling (43) and observations (60, 61) $x = 1$ is too large. This slope of the v_θ-profile is not always completely constant (31) and eventually $v_\theta \to 0$ in the outer area. A completely precise value of x is not to be expected from simple linear modelling, especially since hurricanes undergo frequent intensity changes; a value of 0·5 appears to be a good average.

From detailed low-level wind and pressure charts based on aircraft reconnaissance, the maximum wind turns out to be strongly super-gradient (28, 40); this is to be expected since the inward radial motion does not stop precisely at the radius where the cyclostrophic balance is fulfilled. Asymmetries of pressure-wind relations around the hurricane are thought to be due to differences in trajectory curvature resulting from hurricane displacement (40). While a stationary hurricane is more symmetrical than a moving one, calculation of all terms in the radial equation of motion 9.5 along aircraft traverses reveals residual imbalances which must be due to internal friction. The latter balances 25-30% of the pressure force (23), a theme that has been further developed (24-26). Further, radial and vertical modelling using the velocity components and the vertical motion has been carried out (6, 15, 17, 61). A simple two-dimensional model is presented in addition to the numerical modelling efforts in Chapter 11 so the subject need not be pursued further here.

Budgets of momentum and energy

Absolute angular momentum

Definitions

The equations for horizontal motion in polar coordinates, neglecting lateral friction, are:

$$\frac{dv_\theta}{dt} + \frac{v_\theta v_r}{r} + fv_r = -\frac{1}{\varrho}\frac{\partial p}{r\partial\theta} + \frac{1}{\varrho}\frac{\partial \tau_{\theta z}}{\partial z}. \tag{9.4}$$

$$\frac{dv_r}{dt} - \frac{v_\theta^2}{r} - fv_\theta = -\frac{1}{\varrho}\frac{\partial p}{\partial r} + \frac{1}{\varrho}\frac{\partial \tau_{rz}}{\partial z}. \tag{9.5}$$

It should be restated that the radial component v_r is defined as positive outward. In the literature the definition has been variable; most terms will change sign if it is defined as positive inward. Equation (9.5) becomes the cyclostrophic equation if all terms except the pressure force and the centripetal acceleration are discarded. Equation (9.4) is transformed to a momentum equation by multiplying with the radius r. Then the left side becomes $r\, dv_\theta/dt + v_\theta\, v_r + r\, fv_r$. Now

$$r\frac{dv_\theta}{dt} + v_\theta v_r = \frac{d}{dt}(v_\theta r).$$

Further, $v_r = dr/dt$, so that the last term

$$rfv_r = rf\, dr/dt = f\, d/dt\,(r^2/2) = d/dt\,(fr^2/2) - (r^2/2)\, df/dt.$$

Then

$$\frac{d}{dt}\left(v_\theta r + \frac{fr^2}{2}\right) - \frac{r^2}{2}\frac{df}{dt} = -\frac{1}{\varrho}\frac{\partial p}{\partial \theta} + \frac{r}{\varrho}\frac{\partial \tau_{\theta z}}{\partial z}. \tag{9.6}$$

Historically, the component of the earth's angular momentum per unit mass about the vertical axis of the hurricane has been defined by

$$\Omega = v_\theta r + \frac{fr^2}{2} = \text{const.} \tag{9.7}$$

This equation is a special case of Eq. (1.1) for a vortex centre at any latitude, hence also identical with Eq. (8.1). The momentum definition will be valid if every term of Eq. (9.6) is individually zero. If the equation is integrated over the storm volume, the pressure term will vanish by definition, since pressure integrated around a closed curve must come back to itself. The frictional term will by no means vanish in the surface layer where momentum is transferred from air to ocean. In the upper troposphere, in spite of vertical turbulent fluxes (25, 26), trajectory calculations show Eq. (9.7) to be a good approximation, at least for a symmetrical storm. Finally, the second term on the left side of Eq. (9.6) must also vanish, i.e. one sees the hurricane occurring at constant latitude, no doubt valid for very small storms but quite unwise for large ones with size of middle latitude cyclones and more. For instance, for a storm centred at latitude 15° with 5° radius, the Coriolis parameter varies by a factor of two around the periphery between latitudes 10° and 20°. In a line integral much will depend on whether the inflow is mainly from the equatorial zone at low, or from the polar side at high, Coriolis parameter. Given the definition used earlier, namely that $df/dt = \beta v$ in Cartesian coordinates (mixing coordinate systems) one often sees southerly inflow in the Northern Hemisphere and also southerly outflow in the high troposphere as in Fig. 9.19 (a) and (b). Such flow would seem to act as a substantial brake on the hurricane—low earth's momentum inflow on the equatorial and high outflow on the polar side. Care is needed, however, since the hurricane motion must also be taken into account. With a poleward component of track, the usual event, this component may well counteract the apparent advection of earth's momentum, reducing this effect to zero or even reversing its sign. This rather important matter has received inadequate attention in the literature in all its facets, so that it cannot be pursued further here. However, concern is sufficient to discuss the momentum budget only as far as the 3° radius.

Momentum budget

For budget purposes, the left side of Eq. (9.4) is transformed to a flux equation by integrating over the hurricane volume. One then obtains the transports $F_{\Omega(radial)}$ at any radius and $F_{\Omega(vertical)}$ between two radii (56):

$$F_{\Omega(radial)} = -\frac{2\pi r^2}{g} \left[\int\int_{p_2}^{p_1} \bar{v}_\theta \bar{v}_r \, dp + \int_{p_2}^{p_1} \overline{v'_\theta v'_r} \, dp + \frac{fr}{2} \int_{p_2}^{p_1} \bar{v}_r \, dp \right] \quad (9.8)$$

$$F_{\Omega(vertical)} = -\frac{2\pi}{g} \left[\int\int_{r_1}^{r_2} \tilde{v}_\theta \tilde{\omega} \, r^2 \, dr + \int_{r_1}^{r_2} \widetilde{v^*_\theta \omega^*} \, r^2 \, dr + \frac{f}{2} \int_{r_1}^{r_2} \tilde{\omega} r^3 \, dr \right] \quad (9.9)$$

Here, $\omega \equiv dp/dt$. Further, the surface transport to the ocean

$$F_{\Omega(surface)} = 2\pi \int_{r_1}^{r_2} \tau_{\theta,0} r^2 \, dr \quad (9.10)$$

where $\tau_{\theta,0}$ is the stress at the nominal height of 'ship's deck level" discussed earlier. This height must rise inward from 10 m height if ocean waves and most spray are to remain below it; a thorough analysis of this problem has not been made.

In Eq. (9.8) the first term denotes transport by the mean ageostrophic circulation $\bar{v}_r$ in Figs 9.5 and 9.18, the second tranport by the deviations from symmetry going around the periphery in Figs 9.6 and 9.19 and the third term transport of the earth's rotation assuming constant Coriolis parameter. Momentum budgets have been featured very strongly in hurricane research because of differences in interpretation concerning the relative importance of the three terms.

TROPICAL CYCLONES—STRUCTURE AND MECHANICS

The third term must vanish at any radius as long as small net mass transports indicative of deepening or filling of the hurricane are omitted. These net fluxes, though utterly important, cannot be approached via the budgetary route, since they are orders of magnitude smaller than the accuracy of flux determination. However, even if the third term makes no contribution to radial transport, its convergence or divergence between radii in lower and upper atmosphere will give rise to vertical transport, a matter that will come even more strongly to the fore in computing the momentum budget for the earth's general circulation in Chapter 12. The second term, the "eddy" flux, has been proposed as a potential mechanism for "forcing" a hurricane through strong momentum inflow in a limited sector of the storm's periphery only.

Figure 9.27 shows the budget for a steady state mean hurricane based on data as discussed earlier in this chapter (56). Other budgets

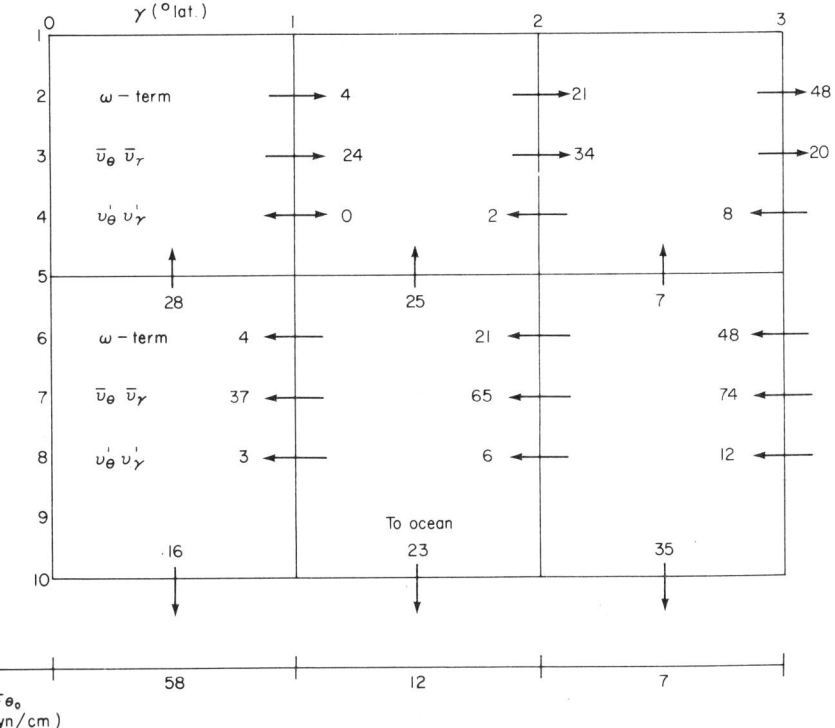

Fig. 9.27. Budget of absolute angular momentum (10^{22} g cm^2 s^{-2}) for average hurricane (56). Bottom: tangential surface stress (dyn cm^{-2}).

are contained in (31), among many. Most net flux is accomplished by the mean momentum term, especially through the 1° radius. The "eddy" term is inward but small there. Its importance increases outward; at the 3° radius it is directed inward both in inflow and outflow so that its contribution is 26% of total transport. In budgets extending outward as far as the 10° radius, no outer boundary has been found where momentum transport ceases, a difficult problem for modellers who wish to obtain a "closed" hurricane.

Calculations in individual hurricanes vary as to the role and sign of the eddy term. From Fig. 9.19(a) a result quite similar to that of Fig. 9.27 was computed at the 1° radius: mean outflow of 20 units and eddy inflow of 2 units. In hurricane Daisy (60) results were variable; at the height of the storm eddy outflow was computed outward to the 150 km radius. In a circumnavigation of hurricane Helene on October 25, 1958, near 500 m height the mean term with $\bar{v}_\theta = 28$ m/s and $\bar{v}_r = -11$ m/s near the 1° radius was dominant with inflow of 12 units as in Fig. 9.27 for a pressure thickness of 100 mb, i.e. rather close to the inflow given for 300 mb in the mean budget. The eddy term is often due to a wave-number "one", i.e. the asymmetry introduced by superposition of vortex and steering current. In this case the first harmonic of v_θ and v_r differed by 90° in phase, i.e. one variable was zero when the other had a maximum or minimum. Therewith the whole correlation is zero. Transport by small-scale eddies computed from the one minute variations (3-mile scale) also was insignificant, justifying the omission of lateral stresses in the momentum budgets. Since Helene was deepening rapidly at the time of the line integral flight, the increase in intensity must be related clearly to the increase in mass circulation with deepening, which appears to be generally observed. Figure 10.2 shows this growth for hurricane Daisy; on 26 August, 1958, the day comparable for 25 September for Helene, $\bar{v}_r$ was only 6 m/s (12 kn), just half of the Helene value.

It appears to be fair to conclude that the contribution by the eddy term is unsystematic and not necessarily related to hurricane growth or intensity at small radii (59). The eddy momentum transport does increase outward in Fig. 9.27, however, and one can picture that at large radii it may predominate (57), especially as no adequate momentum source from the sea can be found within hurricane envelopes, if the boundary is defined as the limit of low-level cyclonic flow. Exactly at those large radii, however, the variable Coriolis parameter must be taken into account for a convincing theory.

It is worth noting that another term is missing from the flux diagram, i.e. the one arising from the motion of the system, which can be taken into account by introducing winds relative to the moving storm as just considered for the variable latitude effect. For instance, $v_{\theta\,\text{rel}} = v_\theta - C \sin \theta$. Computation of a momentum budget in relative coordinates showed that resultant inflow was reduced by 20% when propagation was included and that the eddy term remained negligible (22).

The surface stress (Eq. (9.10)) is determined readily by the difference in momentum transport between two radii in Fig. 9.27, if no momentum is lost to the stratosphere. Further, recalling Eq. (4.9), surface stress can be related to V^2, the total velocity or, in case of the tangential component here considered, to $v_{\theta 0} V_0$, where V_0 is total scalar speed at the reference level. Often, the relation between surface stress and radius is depicted in illustrations; more interesting perhaps is the relation between the total stress τ and V_0 given in Fig. 9.28 for

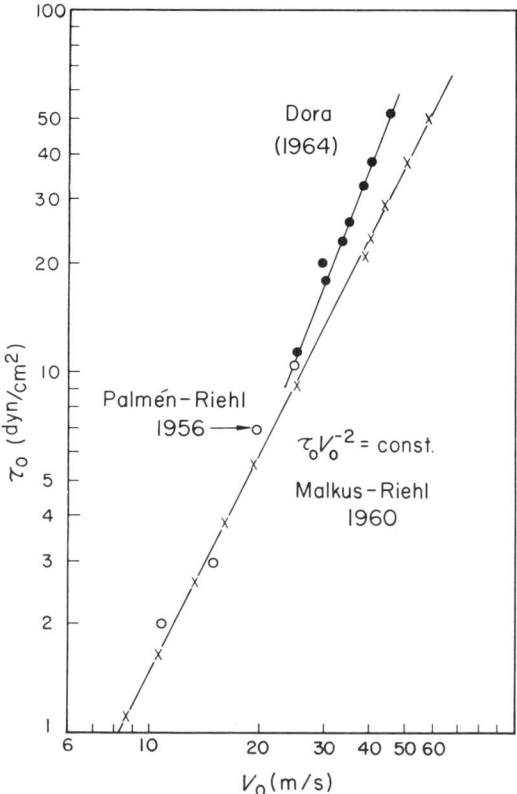

Fig. 9.28. Surface stress against surface wind speed for two hurricanes and one model storm.

the mean hurricane, for a model storm (43) and again for hurricane Dora without closed eye wall showing a very similar stress profile in spite of apparently different energetics. The drag coefficient C_d, not measured directly in hurricanes, can be computed from Eq. (4.9). Usually, C_d increases inward with wind speed, for instance from 1·3 to 2·1 × 10^{-3} from the 3° to the 1° radius in Fig. 9.27. Such information has been used for developing the regression equation (Garratt A-5).

$$C_d \times 10^3 = 0·51\ V^{0·46}, \tag{9.11}$$

where V is expressed in m/s. The coefficient would rise from an average oceanic value near 1·5 × 10^{-3} to 3·5 × 10^{-3} at 50 m/s, evidently produced by breaking waves at very high wind speeds and ocean spray thrown up into the air. The original meaning of C_d is rather lost under these circumstances, but it still provides a nominal reference and check on calculations which no one has been disposed to abandon.

Kinetic energy

In a steady-state storm import of kinetic energy through a cylindrical boundary plus production in the volume inside this boundary must equal the energy transfer to the ocean and/or dissipation within the atmosphere. Both the import through a boundary such as the 600 km radius (55) and time changes inside are found to be negligible compared to production and frictional loss. Close to the centre advection becomes an important term on account of the convergence of mass flow. Production of kinetic energy decreases inward with the decrease of total mass flow, even though production per unit mass continues to rise to the eye wall. More will be said about kinetic energy release in the following section.

The efficiency (ε) of a thermal engine in converting heat to mechanical energy can be defined as the ratio of the mechanical energy produced to the heat released. For the mean hurricane the kinetic energy produced was computed as 150 × 10^{18} ergs per second or 0·36 × 10^{12} kW·h/day between the 0·25 and 6° radii (56). The latent heat released was 13·3 × 10^{12} kW·h per day (cf. 38, 49). Thus, $\varepsilon = 3\%$, a somewhat higher value than is usually estimated for the general circulation and extratropical cyclones but still very small indeed, indicating that the mechanism for energy release resides in the central area with local oceanic heat source and not in the advection of large masses of water vapour into the system from without. Most rain is used for the change of ascent from curve E in Fig. 9.22 to curve B. This fraction, about 80%, produces no mass change, hence no kinetic energy. To subtract this part, $\varepsilon \sim 13\%$ depending on which ascent curve is used, is far more realistic.

In terms of energy production by man, the fraction of heat released which is converted to kinetic energy is very large. The electric power production of the United States is about 1×10^{10} kW·h per day; hence, the mechanical energy generated in hurricanes is 36 times as large (200 times in 1957).

Ground dissipation of kinetic energy will be discussed at the end of this chapter. Internal dissipation in the free atmosphere attains the magnitude of advection or production in the high-speed core, say from about the 80 km radius inward depending on storm size. We have found that such dissipation may be important for the thermal structure of the eye, but potentially it can also raise the temperature of the outflow, so that a hurricane may be warmed 1-2°C above 300 mb, not negligible as the same magnitude as that derived from the freezing process is obtained.

Thermal and latent heat energy

This budget introduces several factors so far not considered: radiation, ventilation of a hurricane by outside air, the source of the water vapour and the efficiency in converting potential to kinetic energy. As noted in the case of the Manila flood, no kinetic energy need be produced by enormous precipitation as long as the condensation takes place along the same moist-adiabatic ascent paths and does not affect the mass field. Hence, there is no valid basis for quoting huge releases of latent heat in hurricanes as a reason for the existence of such storms.

Release of potential energy

An interesting comparison exists in the latent and sensible energy budgets of hurricanes Daisy (60) and Helene (51) which were brought to identical units by Palmén and Newton (A-13) for the budget inside the 1° radius (Table 9.2). During the latent heat release the atmosphere inside the cyclones is raised vertically and acquires a large amount of potential energy. The export of $gz + c_p T$ greatly exceeds the import. One can also say that the excess is closely proportional to the potential temperature difference. The inflow of latent heat, if little water vapour plus cirrus escapes, almost cancels this net export. If complete energy balance was obtained, the air would rise on the same thermodynamic path throughout the hurricane, given by curve (B) of Fig. 9.22. Latent heat energy is transformed to potential energy. Kinetic energy production will take place only if the starting sonde is that of the mean tropical atmosphere. The production of kinetic energy through release of potential energy (W) is given by

$$W = -\frac{R}{g} A \int \widetilde{\omega^* T^*} \, d \ln p \qquad (4.31)$$

or the simplified version of Eq. (4.32). Positive generation would occur if the ascent is along curve (B) and the descent along curve (A) of Fig. 9.22. Such release can lead to tropical storms with sea-level pressure near 1000 mb, but there will be no hurricane.

Table 9.2. Radial energy balance (10^{12} kj/s)[a]

	Daisy (60)	Helene (51)
Net latent heat inflow	+34·1	+31·9
Flux of latent and sensible heat from sea	+ 3·4 (10%)	+ 6·2 (19%)
Import of kinetic energy	+ 0·4	+ 0·2
Total energy source	37·9	38·3
Less net export of $(c_p T + gz)$	−36·9	−37·4
Balance for radiation cooling	1·0	0·9

[a] After Palmén and Newton (A-13), transports through the 1°radius.

Within the mature hurricane the latent plus sensible oceanic energy source shifts the ascent path to successively higher energy, curve (C) in Fig. 9.22. From Table 9.2 the oceanic source was 10% of the import of latent energy in Daisy; in Helene as high as 20%. Still the added energy is a small percentage of the inward transport—really within limits of knowing this transport. The source is highly important, however, for the production of kinetic energy; the correlation in Eq. (4.31) increases with baroclinity from the hurricane interior outward. The area on the thermodynamic charts between the curves (A) and (C) is double that between (A) and (B). Thus, for $p_c = 960$ mb, kinetic energy production is greatly enhanced. As p_c falls to still lower values the positive area in Fig. 9.22 continues to increase until a ceiling is reached at the ascent of $\theta_e = 370$ K.

Radiation

In terms of budget, the energy inflow plus source exceeds the export. The difference must be balanced by the radiation cooling which turns out to have very reasonable values as computed on the bottom line of Table 9.2. There are, in addition, very interesting details inside the 1° radius. Black-body radiation computed from the top of the cirrus shield observed in hurricane Daisy (60) decreased markedly inward from $-1\cdot2$°C/day beyond the 60 km radius to only $-0\cdot9$°C/day in the eye wall where the clouds are considerably higher and colder. One of the ways by which a hurricane can protect the warm inside temperatures is by raising the height of the upper outflow. It is not without significance that the height of the outflow is observed to rise as a storm deepens.

Ventilation

The three-dimensional budget permits an estimate of the constraint exercised upon a hurricane through "ventilation", i.e. inflow on one side and outflow on the other without net convergence or divergence:

> We may postulate in addition a small vertical wind shear, at least up to 500-400 mb, and this is often obtained. Vertical columns will then remain a unit, and the moisture aloft will not be blown away. A strong wind shear resists building and maintenance of trade cumuli and thunderstorms; cyclogenesis is probably affected in the same way. The only alternative would be a deep moist layer over an area so large that the vertical shear did not matter. This situation is rare. It follows that a basic current with strong vertical shear in the lower layers would be an obstacle; as yet this point has not received the attention it deserves. "Tropical Meteorology" (1954) H. Riehl.

Since then, Gray (27) has elaborated and also found small wind shear a condition for hurricane formation and maintenance. Numerous other authors have reached the same conclusion.

Examples are plentiful that wind shear acts as a brake. In Chapter 8 we noted the correlation between the occurrence of tropical cyclones in the North Indian Ocean and the appearance of high-tropospheric westerlies south of the Himalayas. We can now add that the large shear from equatorial westerlies to high-level easterlies contributes to the absence of tropical cyclones at the height of the monsoon season (27). Gray has conducted a worldwide study relating vertical wind shear and cyclone incidence. Aircraft patterns flown in hurricanes have revealed that some ventilation of cores is taking place and that considerable increments in the oceanic heat source would be needed to counteract a strong cooling and drying constraint from outside. In hurricane Daisy (60) detailed computations were made which are somewhat complex; the reader is therefore referred to this publication. In summary, ventilation did exercise a constraint on the hurricane at maturity on 27 August, 1958. This constraint is sometimes compared to that exercised by ground friction applied to the v_θ-component which has a tendency to retard the rotation rate. Here, we take quite a different viewpoint: only through the frictional transfer of momentum to the ground do we obtain the vertical wind shear needed to maintain the interior temperature field in quasi-equilibrium through the cyclostrophic thermal wind relation. Thus, we regard the friction not as a constraint but as a highly essential part of the mechanism which permits the hurricane to exist at all—in marked contrast to a large number of synoptic systems in which no such frictional mechanism is operative and which never become tropical storms. (See Chapter 11 for further discussion.)

Hurricanes as closed systems

The moisture budget may be used to determine whether a hurricane can exist in isolation from the thermal viewpoint. One can calculate the area of a thermally self-contained hurricane assuming descent against radiational cooling on the outside. In the first law of thermodynamics the change in energy may be given by $c_p\,(dT/dt)_{\text{rad}}$. Then

$$c_p\left(\frac{dT}{dt}\right)_{\text{rad}} = c_p\frac{dT}{dt} - \frac{1}{\varrho}\frac{dp}{dt}. \tag{9.12}$$

For steady temperature inside a closed area $dT/dt = w\,\partial T/\partial z$ and $dp/dt = w\,\partial p/\partial z = -\varrho g w$ with use of the hydrostatic equation. Inserting in Eq. (9.12)

$$c_p\left(\frac{dT}{dt}\right)_{\text{rad}} = c_p w\frac{\partial T}{\partial z} + gw.$$

Solving for the vertical motion

$$w = \frac{\left(\frac{dT}{dt}\right)_{rad}}{\frac{\partial T}{\partial z} + \frac{g}{c_p}}, \qquad (9.13)$$

where g/c_p is dry-adiabatic lapse rate. Given $(dT/dt)_{rad} = -1°C/day$ and $\partial T/\partial z = -0.6°C/100$ m, $w = -250$ m/day. For mass continuity at 700 mb

$$\int \varrho w \delta A = \int_0^{2\pi} \int_{1000}^{700 \text{ mb}} v_r \, r \delta\theta \, \frac{\delta p}{g}, \qquad (9.14)$$

where A is the area of descent beyond the 2° radius. On the right side of the equation, $v_r = -5$ m/s at $r = 2°$ (Fig. 9.18) so that the mass inflow $M_r = 19 \times 10^{12}$ g/s. Solving for A with the foregoing data, $A = 6.6 \times 10^6$ km^2 and the radius of the limit of descent $r = 1450$ km $= 13°$ latitude. This value is at the high end of the range of observed hurricane radii.

We next determine whether climatically estimated evaporation rates in the tropics can furnish the water needed for precipitation in the hurricane. From Fig. 9.12, average precipitation is about 10 cm/day inside the 300 km radius for Donna; the outer portion may be neglected. Therewith the required water is 28×10^{15} g/day. Dividing by the area inside the 13° radius, the average evaporation $E = 0.42$ g/cm^2/day, about 25% more than is to be expected from the annual evaporation profile of Fig. 3.1. However, there are large portions of the tropical oceans where evaporation is about 0.4 cm/day; further, using the curve marked "Hughes" in Fig. 9.12, the precipitation is about 25% lower than for Donna, and the condition that $E = P$ is met given the large outer radius. However, neither Donna nor most of the typhoons making up the Hughes curve were nearly that large. Assuming an average radius of 7.5° latitude the area of the hurricane is only one-third the maximum area and will not be able to furnish the required evaporation. Hence, from the moisture budget view, the normal hurricane must be an open system drawing on water vapour evaporated well beyond its limits and in return exporting approximately equal potential energy through high-level outflow. The observed limits on hurricane area size are not understood.

References

(1) Adler, R. F. and Rodgers, E. B. (1977). Satellite-observed latent heat release in a tropical cyclone. *Mon. Wea. Rev.* **105:** 956-963.
(2) Atkinson, G. D. and Holliday, C. R. (1977). Tropical cyclone minimum sea-level pressure/maximum wind relationship for the western North Pacific. *Mon. Wea. Rev.* **105:** 421-427.
(3) Bell, G. J. (1961). The estimation of surface wind speeds in tropical cyclones. *Bull. Amer. Meteor. Soc.* **42:** 382-383.
(4) Browner et al. Ref. 1, Chapter 6.
(5) Byers, H. R. (1944). "General Meteorology". McGraw-Hill Book Co., New York.
(6) Carrier, A. F., Hammond, A. L. and George, O. D. (1971). A model of the mature hurricane. *J. Fluid Mech.* **47** (pt 1): 145-170.
(7) Chaussard, A. and LaPlace, L. (1959). Les perturbations dans le sud-ouest de l'Ocean Indien. *Météorologie* (4) **56:** 323-366.
(8) Cline, I. M. (1926). "Tropical Cyclones." McGraw-Hill Book Co., New York.
(9) Defant, A. and Ertel, H. (1942). Der thermodynamische Wirkungsgrad der Atmosphäre. *Ann. Hydrogr. Maritim. Meteor.* **70:** 161-168.
(10) Defant, A. and Defant, F. (1958). "Physikalische Dynamik der Atmosphäre." p. 288. Akad. Verlag, Frankfurt a/M., 527 pp.
(11) Deppermann, C. E. (1937). Wind and rainfall distribution in selected Philippine typhoon. Bureau of Printing, Manila.
(12) Deppermann, C. E. (1939). "Some characteristics of Philippine typhoons." Bureau of Printing, Manila.
(13) Deppermann, C. E. (1947). Notes on the origin and structure of Philippine typhoons. *Bull. Amer. Meteor. Soc.* **28:** 399-404.
(14) Dunn, G. E. (1951). Tropical cyclones. *In* "Compendium of Meteorology." (T. F. Malone, Ed.) pp. 887-901. American Meteorological Society, Boston.
(15) Durst, C. S. and Sutcliffe, R. C. (1938). *Quart. J. Roy. Meteor. Soc.* **64:** 75.
(16) Erickson, C. O. and Winston, J. S. (1972). Tropical storm, mid-latitude, cloud band connections and the autumnal buildup of the planetary circulation. *J. Appl. Meteor.* **11:** 23-36.
(17) Estoque, A. (1962). Vertical and radial motions in a tropical cyclone. *Tellus* **14:** 394-402.
(18) Fett, R. W. (1964). Aspects of hurricane structure: new model considerations suggested by TIROS and project mercury observations. *Mon. Wea. Rev.* **92:** 43-60.
(19) Fett, R. W. (1968). Some unusual aspects concerning the development and structure of typhoon Billie—July 1967. *Mon. Wea. Rev.* **96:** 637-648.
(20) Fletcher, R. D. (1955). Computation of maximum surface winds in hurricanes. *Bull. Amer. Meteor. Soc.* **36:** 247-250.
(21) Fritz, S., Hubert, L. F. and Timchalk, A. (1966). Some inferences from satellite pictures of tropical disturbances. *Mon. Wea. Rev.* **94:** 231-236.
(22) Gangopadhyaya, M. and Riehl, H. (1959). Exchange of heat, moisture and momentum between hurricane Ella (1958) and its environment. *Quart. J. Roy. Meteor. Soc.* **85:** 278-287.
(23) Gray, W. M. (1962). On the balance of forces and radial acceleration in hurricanes. *Quart. J. Roy. Meteor. Soc.* **88:** 430-458.

(24) Gray, W. M. (1965). Calculation of cumulus vertical draft velocities in hurricanes from aircraft observations. *J. Appl. Meteor.* **4:** 463-474.
(25) Gray, W. M. (1966). On the scales of motion and internal stress characteristics of the hurricane. *J. Atmos. Sci.* **23:** 278-288.
(26) Gray, W. M. (1967). The mutual variation of wind, shear and baroclinity in the cumulus convective atmosphere of the hurricane. *Mon. Wea. Rev.* **95:** 55-73.
(27) Gray, W. M. (1968). Global view of the origin of tropical disturbances and storms. *Mon. Wea. Rev.* **96:** 669-700.
(28) Gray, W. M. and Shea, D. J. (1973). The hurricane's inner core region, II: thermal stability and dynamic characteristics. *J. Atmos. Sci.* **30:** 1544-1560.
(29) Groening, H. U. Ref. 20, Chapter 1.
(30) Haurwitz, B. (1935). The height of tropical cyclones and the eye of the storm. *Mon. Wea. Rev.* **63:** 45-49.
(31) Hawkins, H. F. and Rubsam, D. T. (1968). Hurricane Hilda, 1964. I. Genesis, as revealed by satellite photographs, conventional and aircraft data. *Mon. Wea. Rev.* **96:** 428-452; II. Structure and budgets of the hurricane on October 1, 1964, **96:** 617-636; III. Degradation of the hurricane. **96:** 701-707.
(32) Hazelworth, J. B. (1968). Water temperature variations resulting from hurricanes. *J. Geoph. Res.* **73:** 5105-5123.
(33) Hughes, L. A. (1952). On the low-level wind structure of tropical storms. *J. Meteor.* **9:** 422-428.
(34) Izawa, T. (1964). On the Mean Wind Structure of Typhoon. Japan Meteor. Agency, Meteor. Res. Inst., Tech. Note No. 2. 19 pp.
(35) Jordan, C. L. (1952). On the low-level structure of the typhoon eye. *J. Meteor.* **9:** 285-290.
(36) Jordan, E. S. (1952). An observational study of the upper-wind circulation around tropical cyclones. *J. Meteor.* **9:** 340-346.
(37) Jordan, C. L. (1958). The thermal structure of the core of tropical cyclones. *Geophysica* (Helsinki) **6:** 281-297.
(38) Koteswaram, P. and Gaspar, S. (1956). The surface structure of tropical cyclones in the Indian area. *Indian J. Meteor. Geoph.* **7:** 339-352.
(39) Landis, R. C. and Leipper, D. F. (1968). Effects of hurricane Betsy upon Atlantic Ocean temperature, based upon radio-transmitted data. *J. Appl. Meteor.* **7:** 554-562.
(40) LaSeur, N. E. and Hawkins, H. F. (1963). An analysis of hurricane Cleo (1958) based on data from research reconnaissance aircraft. *Mon. Wea. Rev.* **91:** 694-709.
(41) Leipper, D. F. (1967). Observed ocean conditions and hurricane Hilda, 1964. *J. Atmos. Sci.* **24:** 182-196.
(42) Malkus, J. S. (1958). On the structure and maintenance of the mature hurricane eye. *J. Meteor.* **15:** 337-349.
(43) Malkus, J. S. and Riehl, H. (1960). On the dynamics and energy transformations in steady-state hurricanes. *Tellus* **12:** 1-20.
(44) Malkus, J. S., Ronne, C. and Chaffee, M. (1961). Cloud pattern in hurricane Daisy, 1958. *Tellus* **13:** 8-30.
(45) Maynard, R. H. (1945). Radar and weather. *J. Meteor.* **2:** 214-226.
(46) McFadden, J. D. (1967). Sea-surface temperatures in the wake of hurricane Betsy (1965). *Mon. Wea. Rev.* **96:** 299-302.
(47) Merceret, F. J. (1976). The turbulent microstructure of hurricane Caroline (1975). *Mon. Wea. Rev.* **104:** 1297-1307.

(48) Miller, B. I. (1958). On the maximum intensity of hurricanes. *J. Meteor.* **15:** 184-195.
(49) Miller, B. I. (1958). Rainfall rates in Florida hurricanes. *Mon. Wea. Rev.* **86:** 258-264.
(50) Miller, B. I. (1958). The Three-Dimensional Wind Structure around a Tropical Cyclone. Natl. Ocean. Atmos. Admin. USA, Natl. Hurricane Res. Proj. Rep. 15. 41 pp.
(51) Miller, B. I. (1962). On the Momentum and Energy Balance of Hurricane Helene (1958). Natl. Ocean. Atmos. Admin. USA, Natl. Hurricane Res. Proj. Rep. 53. 19 pp.
(52) Moss, M. S. and Merceret, F. J. (1976). A note on several low-layer features of hurricane Eloise (1976). *Mon. Wea. Rev.* **104:** 967-971.
(53) Östlund, H. G. (1967). Hurricane Tritium II: Air-Sea Exchange of Water in Hurricane Betsy. Report of Inst. Marine Sci., Univ. Miami. 23 pp.
(54) Ooyama, K. (1969). Numerical simulation on the life-cycle of tropical cyclones. *J. Atmos. Sci.* **26:** 3-40.
(55) Palmén, E. and Jordan, C. L. (1955). Note on the release of kinetic energy in tropical cyclones. *Tellus* **7:** 186-188.
(56) Palmén, E. and Riehl, H. (1957). Budget of angular momentum and energy in tropical cyclones. *J. Meteor.* **14:** 150-159.
(57) Pfeffer, R. L. (1958). Concerning the mechanics of hurricanes. *J. Meteor.* **15:** 113-120.
(58) Riehl, H. Ref. 70, Chapter 8.
(59) Riehl, H. (1961). On the mechanism of angular momentum transport in hurricanes. *J. Meteor.* **18:** 113-115.
(60) Riehl, H. and Malkus, J. S. (1961). Some aspect of hurricane Daisy, 1958. *Tellus* **13:** 181-213.
(61) Riehl, H. (1963). Some relations between wind and thermal structure of steady-state hurricanes. *J. Atmos. Sci.* **20:** 276-287.
(62) Sadler, J. C. (1964). Tropical cyclones of the eastern North Pacific as revealed by TIROS observations. *J. Appl. Meteor.* **3:** 347-366.
(63) Senn, H. V. and Hiser, H. W. (1959). On the origin of hurricane spiral rain bands. *J. Meteor.* **16:** 419-426.
(64) Shea, D. J. and Gray, W. M. (1973). The hurricane's inner core region. I. Symmetric and asymmetric structure. *J. Atmos. Sci.* **101:** 1544-1564.
(65) Sheets, R. C. (1968). The structure of hurricane Dora (1964). Nat. Ocean. Atmos. Admin. USA, Tech. Memo 83. 64 pp.
(66) Simpson, R. H. (1952). Exploring the eye of typhoon Marge, 1951. *Bull. Amer. Meteor. Soc.* **33:** 286-298.
(67) Stewart, G. R. (1941). "Storm." MacMillan Co., Canada. 349 pp.
(68) Tannehill, I. R. (1942). "Hurricanes." Princeton University Press.
(69) Wexler, H. (1947). Structure of hurricanes as determined by radar. *Ann. New York Acad. Sci.* **48:** 821-844.
(70) Wu, J. (1974). Evaporation due to spray. *J. Geoph. Res.* **79:** 4107-4109.

Appendix

Discussion of local heat source

The most important information gathered from the observations of thermal structure in hurricanes is that the ascent path of air shifts to increasing values of θ_e in the interior; a baroclinic thermal structure is produced through which energy of motion is generated*. Early in the chapter it was shown that air expands isothermally and specific humidity increases at constant dewpoint in the surface layer going inward given constant cloud base height. Then the first law of thermodynamics can be written

$$\frac{dh}{dt} = c_p \frac{T}{\theta}\frac{d\theta}{dt} = -\frac{1}{\varrho}\frac{dp}{dt} = -RT\frac{1}{p}\frac{dp}{dt}. \qquad (9.15)$$

Upon integration between $p = 1000$ millibars and p_{central}

$$p_c = 1000\left(\frac{T}{\theta_c}\right)^{c_p/R}, \qquad (9.16)$$

where θ_c is central potential temperature (56). Here is a second constraint on surface pressure, in addition to the hydrostatic one which is closely fulfilled (30). Only when both constraints are consistent and yield the same value of surface pressure, can a hurricane exist.

The increase in θ_c has been ascribed to strong sensible heat flow Q_s from the ocean when wind speed is high and $T_w - T_{as} = $ const at about 2-3°C compared to 0.5°C in the undisturbed trades. However, in going through the whole pressure difference 1000 to 900 mb, sensible heat increases by 8 J/g and latent heat by 4 J/g, whereas one is normally accustomed to see a small ration of Q_s/Q_e. Of course, this ratio is not directly comparable to the ratio $c_p\,d\theta/L dq$, following a trajectory, since sensible heat transport up cannot penetrate far beyond cloud base, whereas turbulent flux of latent heat can continue unhindered there. However, mechanisms other than Q_s could be of importance, such as turbulent downward transport of sensible heat

---

*In hurricane exploration near latitude 35°N, however, much larger temperature variations were encountered which indicate the intrusion of different air masses (40).

across cloud base, radiation, evaporation from falling rain and downdrafts. At high velocities, heating due to internal dissipation of kinetic energy could contribute to downward heat transport.

Hurricane Gladys (1975)

It is not possible to evaluate all these potential contributors. However, an aircraft reconnaissance mission in hurricane Gladys gives the opportunity to determine to what extent Q_s satisfies the heat budget and also the first law of thermodynamics. This flight was conducted in the sub-cloud layer and near cloud base; missions in earlier years were conducted higher up in the clouds where release of latent heat does not permit evaluation of the Q_s contribution. Figure 9.29 shows a representative sample of wind reports. We shall consider the shaded area drawn backward from r_2 at the maximum wind of 47-48 m/s to r_1 with 20 m/s which, in view of slow storm motion, may be considered as a trajectory over the hour which the air needed to cover the 110 km distance.

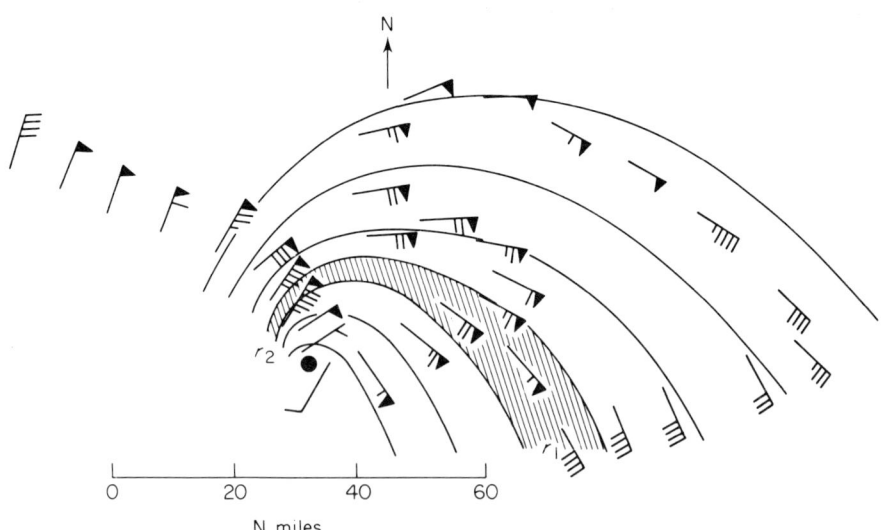

Fig. 9.29. Windfield in hurricane Gladys on 29 September, 1977 (knots) explored by aircraft of Natl. Ocean. Atmos. Admin., USA with aircraft flying below or near cloud base (150-300 m height).
Shaded: trajectory for computations.

Depth of layer with Q_s

We shall consider Q_s as given by Eq. (4.10) without the second term, since this added heat source is transported outward with the water vapour. In Eq. (9.15) we may express

$$\frac{dh}{dt} = -\frac{1}{\varrho}\frac{\partial F_{hz}}{\partial z} \tag{9.17}$$

so that after integration

$$\int \frac{dh}{dt} \varrho\, \delta z = -(F_{h\,\text{top}} - F_{h\,\text{bottom}}) = Q_s$$

if the only energy source is situated at the surface. Then

$$\varrho c_p\, c_D\, (T_w - T_s) \int V\, dt = (p_{r_1} - p_{r_2})\, dz. \tag{9.18}$$

The pressure difference is observed as 1000-987 mb = 18 mb. On the left $\int V dt = l$, the trajectory length, using $T_w - T_{as} = 2°C$ from observations, $dz = 360$ m in Eq. (9.18), very near cloud-base, but extending to 500 m in holes.

Energy balance along the trajectory

In the first law of thermodynamics the source will be assumed as Q_s. The pressure term may be expanded using the horizontal kinetic energy equation and the hydrostatic equation, as derived in textbooks. The local pressure change integrated over the trajectory is zero, since it contains both portions with falling and with rising pressure. After integration with time from r_1 to r_2 and over dz,

$$\frac{\varrho\, c_p\, c_D}{\varrho \delta z}(T_w - T_{as})l = c_p(\hat{T}_{r_2} - \hat{T}_{r_1}) +$$

$$\left(\frac{\hat{V}^2}{2}\right)_{r_2} - \left(\frac{\hat{V}^2}{2}\right)_{r_1} + \frac{c_D}{\delta z}\int \hat{V}^2\, \delta l + D - \overline{K'w'} \text{ term at top.} \tag{9.19}$$

In the term for ground friction $\int \hat{V}^3\, dt = \int \hat{V}^2\, dl$ as used before for $\int \hat{V}\, dl$ at constant C_d. From gust probe measurements, the upper $\overline{K'w'}$ flux is small. Then calculation gives

Sensible heat source	1·8
Kinetic energy gain	0·9
Frictional loss	1·0

It is seen that balance is very nearly obtained. At least from 1000 mb inward (Fig. 9.11), the expansion is isothermal. There is no residual for D. If the latter is rated as $5 \cdot 10^{-3} m^2/s^3$, it will be 10^{-2}J over the trajectory, hence 1% of the energy transferred per gram to the ocean, i.e. immeasurable. Gain of kinetic energy and frictional loss are equal as also found in modelling (61). The fraction of kinetic energy gain is very high; normally it is only a small difference between production and dissipation. Thus, seen in this light, the hurricane appears as a very effective engine. The details of the dissipation below 50 m are not known, except that hurricanes do generate ocean currents and waves. The action of spray assuming much of the air's kinetic energy may play an important role, suggested also by laboratory experiments (70). It may be added that, from the budget calculations, $c_p \, d\theta/L dq = 0 \cdot 21$ whereas $Q_s/Q_e = 0 \cdot 13$. A fraction of the latent heat gain is transported by turbulence through cloud base. But the major part is absorbed into and carried by the mean motion.

We return now to Fig. 9.9, where the relative trajectory is shown for surface air given by lighthouse observations for hurricane Donna (1960). These observations permit carrying out the same calculations just presented. In this instance of a strong hurricane $p_c = 960$ mb; $v_\theta)_{r_2} = 60$ m/s; $v_\theta)_{r_1} = 25$ m/s; $T_w \, T_{as} = 2°C$ as before. One obtains for a four-hour trajectory (joule/gram):

Sensible heat source	4·5
Gain of kinetic energy	1·5
Frictional loss	4·5

In this instance, balance is not achieved; the sensible heat source (16% of the latent heat source) is insufficient to balance the kinetic energy terms. This is not surprising since, from Eq. (9.19) the ratio Q_s/Friction $= c_p \, (T_w - T_{as})/\bar{v}^2$. Since $T_w - T_{as} = $ constant, Q_s must fail to meet the energy balance above a threshold value of V_{max} which is near 50 m/s.† For still higher speeds on emust postulate downward heat transport through cloud base, since the isothermal expansion is maintained. Measurements do not exist.

---

†Parallel reasoning has been employed to determine the maximum probable storm and precipitation yield for reservoir and spillway design in the Venezuelan Andes (Ref. 32, Chapter 3).

10

Tropical Cyclones Formation and Movement

Tropical cyclones principally develop in the following areas and seasons.
(1) Tropical North Atlantic Ocean;
   a. East of the Lesser Antilles and the Caribbean east of 70°W, July to early October,
   b. North of the West Indies, June to October,
   c. Western Caribbean, June and late September to early November,
   d. Gulf of Mexico, June to November,
(2) North Pacific Ocean off the west coast of Central America, June to October;
(3) Western North Pacific Ocean, May to November, some storms in all months;
(4) Bay of Bengal and Arabian Sea, May to June and October to November;
(5) South Pacific Ocean west of 140°W, December to April;
(6) South Indian Ocean;
   a. Northwestern coast of Australia, November to April,
   b. West of 90°E, November to May.

No record of a fully developed hurricane exists for the South Atlantic and the South Pacific east of 140°W. In the north-central Pacific a few near-hurricanes or minimal hurricanes have made their appearance. Seasonally, the mean latitude of storm formation moves poleward in the first half of the season and then retreats equatorward. Early and late season cyclones form mostly in the belt from 5° to 15° in the Northern Hemisphere, at the height of the season between 10°

and 25°. In the Atlantic, cyclogenesis between 25° and 30°N is fairly common. The poleward limit for true tropical hurricane formation there is about 35°N. Extratropical cyclones with winds of hurricane force occur several times per winter over the northern and southern oceans, but they will not be treated here.

Because of open observing networks, many tropical storms passed over the oceans without being detected during the early years of meteorology. This follows from the fact that in several regions "frequency" has risen with increasing density of observing stations, demonstrated most obviously by the records for Australia (39). There, "frequency" rose from five to ten for two years around 1910 to 20 to 25 around 1960 when, in the satellite era, the curve levels off. Statistics are most trustworthy for the Atlantic Ocean, where ship traffic has been heavy and island stations numerous for many years. Thus, there is justification in publishing Fig. 9.2 which is based on an average of eight storms per year. In other ocean areas, satellite photography has upset frequency tables so badly that it is best not to quote numbers

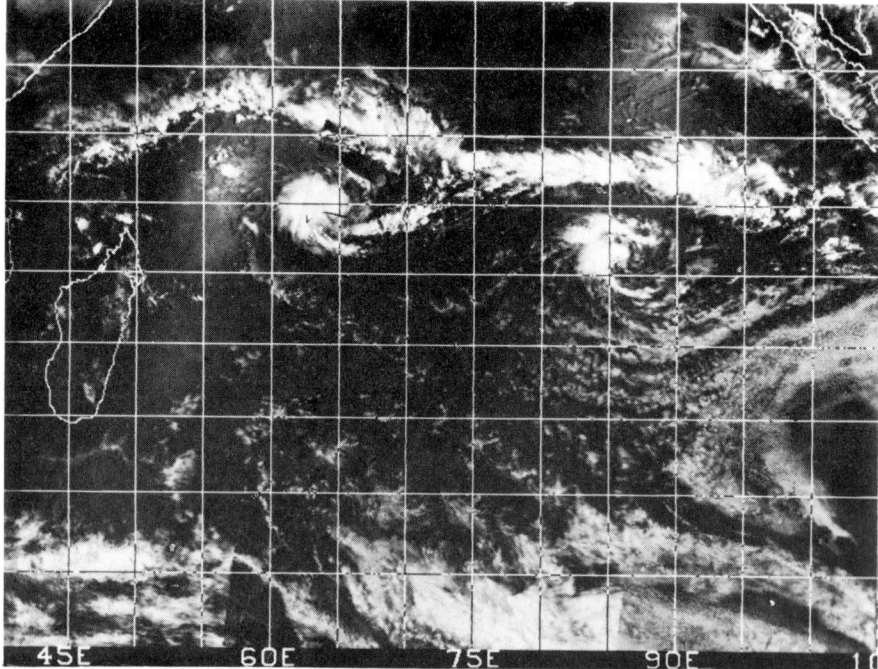

Fig. 10.1. (a) Satellite photo of the South Indian Ocean 18 February, 1977 showing formation of two tropical cyclones 7-10° latitude south of east-west cloud line marking equatorial trough.

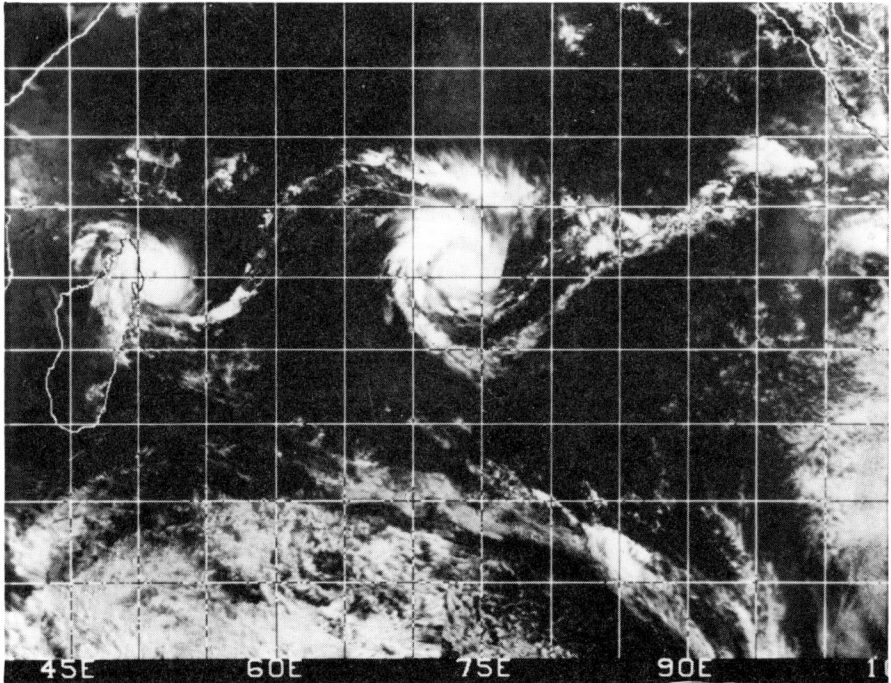

Fig. 10.1. (b) The fully mature storms, eyes visible, four days later. The western one is making landfall on northern Madagascar. No rotation of the centres occurred as they remained over 20° longitude apart.

until new frequencies for at least one decade are published from the satellite record. Especially in the eastern Pacific and in the South Indian Ocean, area frequencies have been grossly underestimated. There may, however, have been a trend toward increasing cyclone frequency in all oceans prior to the advent of satellites (44). One thing remains certain: tropical cyclones are very rare compared to extratropical ones. One used to quote 50 hurricanes per year as being the world average. Even if the true number is closer to 100, this is still a tiny amount compared to formation of at least ten extratropical cyclones per day or several thousand per year. Further, about 20 out of the 50 annual hurricanes were found to occur in the northwestern Pacific (40%). This ratio may decrease; northern Pacific typhoons often are large and many strike land, while those of the South Indian Ocean may come and go without being noticed—except by satellite (Fig. 10.1). A detailed climatology of typhoon frequencies in the western Pacific has been prepared including the frequency of lowest pressure in each storm (3).

Some observations during hurricane formation

Even in the satellite era it remains of value to cite indications of an impending development.

Pressure

Sea-level pressure falls more than 3-3.5 mb/day or reaches values of 5 mb or more below normal for the area or season. Caution must be exercised because of the large-scale rapidly travelling pressure waves mentioned in Chapter 7. Significant pressure fall occurs in small areas only.

Winds

In the trades, easterly winds with speeds 25% and more above normal in a limited area should be suspected, especially if the flow curves cyclonically. Any wind with direction ranging from south through west to north, when it should normally be easterly, is a danger signal.

Weather

Steady rain at several adjoining stations (in contrast to shower type precipitation) or unusually heavy precipitation with cirrostratus and altostratus overcast are frequent precursors of storms.

These are indications for the general public. The meteorologist can compute surface inflow across a circle centred on a suspicious area, given enough ship or low-level pilot balloon observations. The use of satellite-derived, low-level winds is questionable; early inflow may be concentrated in the lowest kilometre. Mass inflow increases as a storm deepens (38), a very dependable history (Fig. 10.2). Veering of wind with height confines the convergence to the lowest kilometre (27). Circle inflow computations using surface winds should be repeated on every map following a suspicious area. Inflow increasing to, say, 1 or 2 m/s around the 4-5° radius should be cause for alarm.

In the initial stages winds tend to be strongest near, perhaps below, cloud base and decrease with height whether or not deepening to hurricane intensity ensues (53). Potential development or re-development of a centre perhaps no more than 100-200 km from the Gulf of Mexico and Florida coasts of the United States is a frequent and critical event. Aircraft monitoring at 600 m height, the cloud base in the trades, is a definite requirement in such situations. Flights in the middle troposphere may show nothing, and then there may follow an unpleasant surprise.

A forecaster should use the tools just enumerated as they can be very helpful. Simply finding the equatorial trough cloud band on the satellite photo will not suffice. Hurricanes develop in elongated troughs well off the centre of the convergence zone on its poleward side (33, 27). Figure 10.1 illustrates this event for two simultaneous storm formations in the South Indian Ocean. Considerable sophistication is needed to find the right spot for development early if good surface maps are lacking. The second photo shows both storms at maturity four days later; they indeed became well developed. The thin trail of cloud between them is a remnant of the old equatorial trough cloud band.

Variability of hurricane intensity

The detailed wind structure even of a mature hurricane undergoes approximately a 10% variability on the time scale of one hour. A whole hurricane, as large and intense as Donna (1960), experiences variations of 10-30 mb in central pressure over a period of two or three days. Extreme pressures, such as 900 mb, appear to hold for no more than 12-24 hours, although there may be a second recurrence (20, 32, 69). The innermost portion of the windfield, especially, which depends heavily on inward penetration of mass is very sensitive and subject to strong alterations. The eye tends to intensify and then weaken as successive strong radar bands push in from the outside and reform a

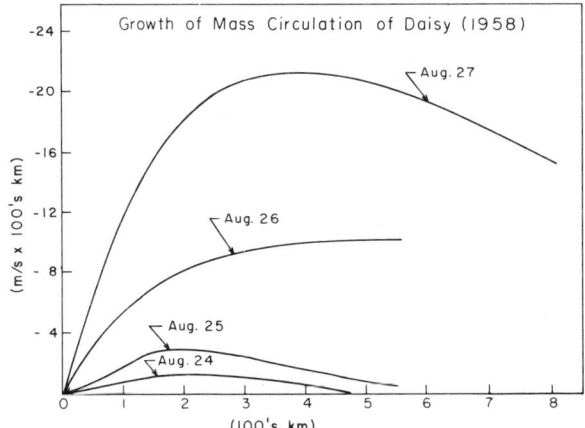

Fig. 10.2. Growth of mass circulation of hurricane Daisy (65).

strong eye, followed by decay. Thus, aircraft reconnaissance often encounters the peaked profile of strongest wind on one day and on the next a larger area, some tens of kilometres along the radius, where v_θ is nearly constant, i.e. the wind profile is in a transition stage and not in "steady-state" mode. Nevertheless, the hurricane envelopes hold together remarkably well and their pumping up and down can be readily followed, except when strong interference from the outside occurs. A hurricane diameter of no less than about 3000 km is required for the circulation to be capable of acting as an isolated vortex for energy balance. Such cases do happen, but they are very rare. The normal hurricane with perhaps half this diameter or less is always exposed to outside influences, most of which come from higher latitudes.

Formation of hurricanes

Discussion of research through computer modelling will be found in the next chapter. It remains worthwhile, however, to give a qualitative survey of some of the principal ideas on hurricane formation, intensification to great strength, and transformation in the late stages. The problem, of course, involves the unsteady state. There is every indication that two-dimensional considerations fail, though they may represent the mature storm well. For these reasons, as well as the fact that the observational base in cyclogenetic situations well out over the oceans is usually unsatisfactory, the subject of formation has tenaciously resisted solution. Nevertheless, some generally valid rules can be cited with confidence before entering into the problem areas.

General requirements for cyclogenesis

(1) In the first place, one can identify high-tropospheric monthly or seasonal flow and temperature fields, which tend to favour or suppress hurricane generation. For the Caribbean these are discussed in Chapter 12: a warm upper anticyclone is present when hurricane development is high; a cold upper trough minimizes their occurrence. The difference in pattern (Figs 12. 17-18) is large and unmistakeable; similar gross distinctions may be found in other hurricane regions. When these patterns are well developed, they serve as a good guide to the probability of hurricane formation, while they persist.

TROPICAL CYCLONES—FORMATION AND MOVEMENT

Five other conditions, necessary but not sufficient, must be fulfilled before the question whether or not a hurricane will form need be considered.

(2) The atmosphere must be capable of permitting deep convection to occur. Some thermodynamic structure resembling the mean atmosphere of Table 2.1 must be present rather than an atmosphere with trade wind inversion or a deep, dry upper layer. However, actual occurrence of deep convection has been found not to be a requirement in the early stages.

(3) Ocean temperature must be high enough (relative to air temperature) to permit interface transfer of enough latent heat and, to a smaller extent, sensible heat so that undilute ascent can take place from the subcloud layer. Usually, 26-27°C is considered a threshold value in the deep tropics (49). Regional charts showing the positions where an incipient centre turned hurricane agree well with this threshold. Palmén (49) shows the regional distribution of 300 mb temperature difference between observed values and those obtained through adiabatic ascent from the surface for the Atlantic. In February, the difference is barely zero in the hurricane belt; in September, the ascent is warmer by 5°C in the hurricane region, a very large difference (cf. end of Chapter 2). In winter, surface layers become slightly cooler while upper temperatures may rise enhancing thermal stability. Drying of low-level air in winter is another large factor.

(4) A pre-existing concentration of cyclonic vorticity in the low to middle troposphere serves as "initial disturbance". Hurricanes always form from initial vorticity fields described in Chapter 8 (15). Since these usually carry cloud systems when conditions 2) and 3) are met, they are readily identified and tracked on satellite photographs. All locations of potential development can be pinpointed with the satellite information on every day around the tropics of the whole globe—a notable technological achievement.

(5) The angular momentum of the air drawn into a central region to become a revolving vortex depends in good measure on the component of the earth's rotation $\omega\sin\phi$ about any prospective hurricane axis. Thus, no hurricane forms on the equator. Generally, cyclogenesis occurs between latitudes 5° and 25°, but incipient hurricanes have been encountered within 2° of the equator, in the western Pacific and northeastern Indian Ocean, with sufficient frequency to warrant the suggestion that "the limiting effect of the Coriolis parameter may have been overrated" (8). The size of these cyclones initially is very small. The forecaster may place a circle with

2° radius around suspicious areas close to the equator and try the $\overline{v'_\theta v'_r}$ correlation discussed under the heading Momentum Budget in Chapter 9. If the Coriolis effect is indeed overrated, the eddy transfer of angular momentum must show up with large values.

(6) Vertical shear of the horizontal wind should be small near a prospective hurricane (but not necessarily small in neighbouring interacting systems) in order for ventilation not to disperse any incipient warm air accumulation. With preference the initial disturbance should have upper anticyclonic outflow; this condition is normally satisfied (Chapter 8).

One factor whose importance has remained in doubt, are sea surface temperature anomalies. While the role of cold ocean currents in weakening and suppressing hurricanes has been firmly established, the same cannot be said about the role of local "hot spots" or even the Gulf Stream for hurricane formation and intensification. Of course, from Eq. (9.18) any ocean temperature anomaly that would increase $T_w - T_{as}$, say from 2° to 3°, would be a favourable factor. However, in the formative stages there are other influences of overriding importance, so that the "signal-to-noise" ratio for sea surface temperatures is weak and not convincing. A map of locations where tropical cyclones reached hurricane intensity in the period 1901-1957 (81) reveals no relation even with the normal sea surface temperature gradient, apart from general concentration in the western half of the ocean. No maximum is found over and around the Gulf Stream; the warmest area, the Gulf of Mexico, has had decades when it was free of full hurricanes and a larger number of tropical storms did not manage to achieve hurricane intensity (63).

Nevertheless, a positive correlation between sea surface temperature and intensification and track of tropical cyclones has been noted (19) as well as variations in cyclone intensity with changing temperature of the underlying surface along the track (51, 52). In the eastern Atlantic, enhanced ocean temperatures in 1966 were coupled with intensification of waves in the easterlies coming out of Africa while, with lower temperature, such deepening did not occur in 1968 (10).

On the seasonal scale, Ramage (56) suggests that ocean temperature, lowered by strong evaporation in the monsoon season in the seas around India, prevents hurricane-force cyclones from forming in the months with strongest monsoon. This text lists three criteria why they should not occur: lack of interaction with the polar westerlies and the poleward meridional temperature gradient (Chapter 8), large vertical shear of the horizontal wind (Chapter 9) and lowered ocean temperatures. Are all the separate influences working together, and what is their relative importance? It is clear that in-depth studies are needed rather than the statistical treatment of single factors.

Long period trends

On the five-year time scale and longer, when variations in the general circulation may be expected to control ocean temperatures closely, the trend in hurricane frequencies of Fig. 9.2 could be related to ocean temperature changes on the order of 0·5°C in 15 years (63) Correlation between sea-level pressure and vorticity was negative in the western Atlantic as found in other studies. Thus, cold ocean temperatures and anticyclonic vorticity of the surface flow combine to act as a brake on hurricane development, whereas cyclonic relative vorticity of the mean flow favours formation, as noted also in the Southern Hemisphere (23).

Long period shifts of the belt of westerlies affects hurricane frequency. With the summer geostrophic westerly maximum at 50°N at 700 mb and weaker than average westerlies farther south, hurricane development is favoured (5). Five-year charts of sea-level pressure from the mean corroborate this finding. In summer in the Atlantic Ocean, geostrophic easterly flow was below average prior to 1930 and then changed to above average, just when the sharp rise in hurricane frequency took place (Fig. 9.2).

The criteria (2) to (6) listed above are found over the tropical and subtropical oceans so regularly and for such a long time each year that hurricanes should be the rule rather than the exception over the oceans if the criteria were sufficient. Yet the fraction of initial vorticity maxima that intensifies lies between one and ten per cent, closer in most years to one per cent though there are exceptions, such as August-October 1950 in the Atlantic. We now proceed to further investigation of potential mechanism in low and high troposphere.

Low and middle troposphere

Warm core

The synoptic systems of the lower troposphere very frequently have a cold core up to about 500-400 mb (Chapter 8). This cold core must be "transformed" to a warm core for hurricane development, or the warm core may spring up in another part of a synoptic envelope. Yanai, in particular, has laid stress on this point (77, A-25). He sees the transformation as occurring through enhanced convection which establishes warm temperature anomalies first in the layer around 400 mb, a slow process resulting from a small difference between condensation heating and cooling due to ascent. Relative motion between a lower cyclonic envelope and an upper ridge may accomplish the same warning more easily (60). Whatever the precise mechanism, the mesoscale updrafts and downdrafts depicted in Figs 4.20 and 4.21, must be eliminated, especially the lower right-hand quadrant of Fig. 4.21. Thereby the transport of low-energy air to the surface (a strong hurricane preventative!) is stopped and the arrangement of *point sinks* for the converging air may go over into ascent over several per cent of the synoptic convergence zone as found in the final hurricane. Thereby a solidifying cloud mass would be created, capable of developing the fully mixed subcloud layer which has been observed (46). Enhanced transfer of latent and sensible heat cannot be invoked in the initial phases, except if all hurricane formation is referred to oceanic "hot spots", since wind speed has not yet risen, nor has $T_w - T_a = 2°C$ become established in any inflow. This is the great disadvantage compared to extratropical cyclone development which starts by concentrating an existing temperature gradient in a narrow zone and, hence, occurs frequently and rapidly.

Even when a warm core tropical "depression" has been established, hurricane formation occurs only rarely. At peak strength the depression may have developed a vertical temperature structure above itself which corresponds to vertical ascent of air at $\theta_e = 350$ K, the energy of the normal outside atmosphere in the mean tropical atmosphere (curve (B) in Fig. 9.22). Assuming undisturbed conditions in the high troposphere, the surface pressure is computed to be about 1000 mb, and this is a threshold value than cannot be exceeded unless a heat source in the interior becomes established. There are many instances, when incipient depressions flare up to such intensity, only to subside again in short order (4, 9, 74). The uniform conclusion has been that "strong concurrent large-scale forcing" is a requirement for deepening to hurricane strength.

Convective instability

Thermodynamic processes, such as release of "convective instability of the second kind" (11, 47), discussed in more detail in Chapter 11, should have their fullest chance to be effective in tropical depressions whereas the pressure-rainfall (satellite cloud) relations of Figs 7.1 and 7.2 do not appear favourable. The problem posed is relevant indeed:

TROPICAL CYCLONES—FORMATION AND MOVEMENT

Why do cyclones form in a conditionally unstable tropical atmosphere whose vertical thermal structure is apparently more favourable to small-scale cumulus convection than to convective circulations of tropical cyclone scale? It is proposed that the cyclone develops in a kind of secondary instability in which existing cumulus convection is augmented in regions of low-level horizontal convergence and quenched in regions of low-level divergence. The cumulus and cyclone-scale motions are thus to be regarded as cooperating rather than as competing—the clouds supplying the latent heat energy to the cyclone, and the cyclone supplying the fuel, in the form of moisture, to the clouds.

Later it is said that

> ... the amplification of the disturbance is due to the surface frictionally induced convergence of moisture and liberation of latent heat in the centre of the cyclone (11).

In principle, the same question is asked as in Chapter 8: why do the normally observed synoptic weather systems break down into something like 100 narrow disorganized chimneys of convergence rather than form one well-organized vortex? The modelling leans heavily on the generally known rise of mass circulation with growth of the hurricane depressions (Fig. 10.2) which certainly concentrates the moisture supply. Empirically, the major event of starting the hurricane has already occurred when the mass circulation becomes unified in the direction of a central core. It would appear difficult to ascribe this early departure from the normal synoptic scattering to ground friction. Further, the problem remains that a superabundance of initial vortices either does not develop or fades away after one day or less. Nor does the hurricane of necessity form at the location of initial convergence and vorticity concentration, as made very evident by the 5-10° latitude difference between equatorial convergence zone and locus of cyclone developments in Fig. 10.1.

Turning of vortex tubes

A potential dynamical factor through the turning of horizontal vortex tubes to vorticity about the vertical axis is given in Cartesian coordinates by

$$\frac{\partial \zeta}{\mathrm{d}t} = \ldots + \frac{\partial u}{\partial z}\frac{\partial w}{\partial y} - \frac{\partial v}{\partial z}\frac{\partial w}{\partial x}. \tag{10.1}$$

The second term especially has been investigated (59). Given a slow-moving wave with ascent concentrated east of a wave axis, $\partial w/\partial x$ is positive and, hence, $\partial v/\partial z$ must be negative for a positive contribution to ζ. Such a slope of the axis toward the convergence zone with height is often found, since the rain area is cold up to 500-400 mb from Chapter 8. In case of a fast-moving wave with concentration of "weather" west of the axis, $\partial w/\partial x$ is negative across the axis, so that $\partial v /\partial z$ must be positive; again the axis should slope toward the cold rain area with height. Both cases, therefore, are structured so that a vorticity increase in Eq. (10.1) is favoured. Given a vertical shear of v of 5 m/s/5 km (not large) and a horizontal gradient of vertical motion of no more than 4 cm/s/200 km, the contribution to ζ will be $2 \times 10^{-5} s^{-1}$, of half a Coriolis parameter near latitude 17°/day. With a sharp edge of the precipitation area the magnitude of the term can easily rise to $1 \times 10^{-4} s^{-1}$ per day. The trouble is that the effect tends to increase the cyclonic vorticity of cold core precipitation areas. In a warm cyclone it will, conversely, inhibit development, and this no doubt is one severe constraint on the intensification of warm waves.

Rate of cross-isobar flow

The rate of increase in kinetic energy through the work done by the pressure force is exceptionally high inside the mature hurricane, whereas in the surrounding trades pressure and frictional forces are nearly balanced. If, in a given system, the cross-isobar angle of flow should suddenly increase, the production term will increase relative to the frictional retardation, at least temporarily. The existing pressure gradient is used effectively for actually producing increases in kinetic energy on short trajectories from high to low pressure as sketched in Fig. 10.3. In contrast, the dashed trajectory indicates a case where air converging toward an initial centre may move inward at nearly constant kinetic energy on a long path, with pressure and frictional forces in approximate balance. In a one-dimensional inflow model the maximum wind increased from 55 to 80 m/s, when the inflow angle changed from 20° to 25° (40). Such a change will occur, for instance, when a narrow and elongated wedge of high pressure advances into the tropics just east of an initial depression (59). Then the inflow angle may temporarily rise to as much as 90° leading to a rapid initial energy increase. However, the large inflow angle is not maintained, so that any high-wind area may again fade away quickly.

In summary, we have noted several thermodynamic and dynamic mechanisms that may affect the lower troposphere in the initial stages of hurricane formation. However, we have not arrived at a hurricane criterion, partly, perhaps, because some of the mechanisms appear to

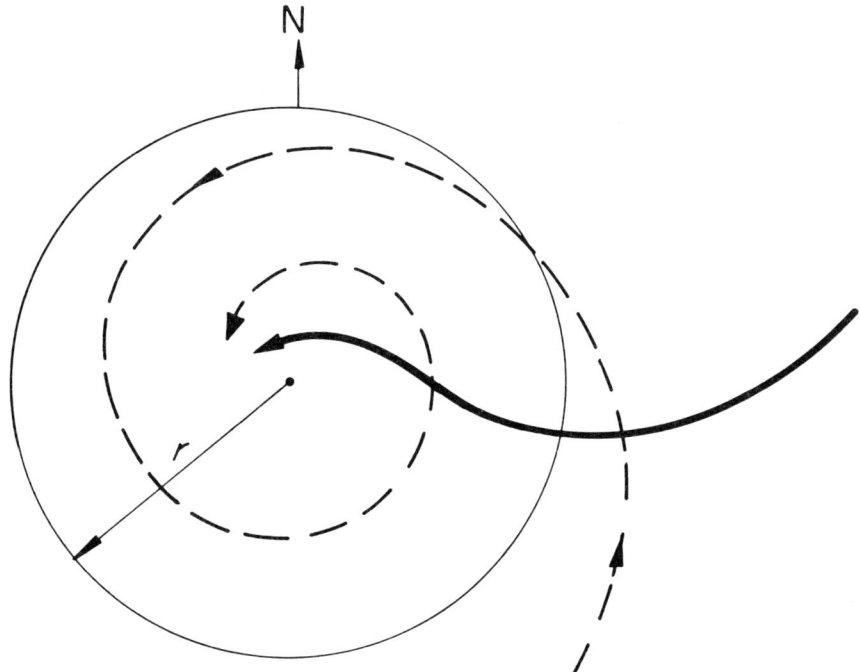

Fig. 10.3. Sketch of surface inflow into initial disturbance at small and large crossing angles at outer periphery.

be conflicting. In addition to internal criteria, external "forcing" through changing the inflow angle and through eddy angular momentum transport has been proposed. However, many incipient cyclones have been simply engulfed and obliterated by momentum intrusions such as surges of the trade. Surface air mass contrasts, a major factor in extratropical cyclone development, similarly act to destroy incipient centres and even major typhoons when relatively cold air penetrates to the centre. The effect of such intrusions is the main reason why a frontal hypothesis of hurricane formation analogous to that outside the tropics will not work. Any ascent of even slightly cool air with, say, $T = 24°C$ and $q = 16$ g/kg, $\theta_e = 342$ K will drop the tropospheric $\hat{T}_v$ below that of the mean tropical atmosphere. Energy would have to be expended to raise such air and the warm interior temperature field could never form.

The high troposphere

Because of the tendency for low and high troposphere to be decoupled in tropics (30) one should be prepared to find the formative tropical cyclone in a variety of high-level environments. However, Fett (17) notes that:

> the upper-level structure of the formative cyclone of moderate intensity is grossly similar to that of the storm at maturity . . . (1) an upper-level trough or shear line which preceded the low-level trough and which moved with the storm system [toward west]; and (2) an upper-level ridge or anticyclone which was generally superimposed over the convective area east of the lower-level trough or vortex. As has been found with the mature storm, strong subsidence occurred under the upper trough and coincided with a cloudiness minimum area in advance of the major convective cloudiness and west of the trough axis. Upward vertical motion with heavy convective activity was suggested under the upper-level ridge or anticyclone to the rear [east] of the lower-level trough.

It so happens that one of Fett's major cases presented in vertical time section is that of the maps in Figs 8.19 and 8.20.

This description is accurate and most valid in the Caribbean area south of latitude 20°N. For hurricanes forming farther north, and in the Southern Hemisphere also at lower latitudes, the upper trough is the equatorward tail of a trough in the polar westerlies. Cyclogenesis then is near the subtropical ridge line in the high troposphere. Further, intense convection associated with the surface system, while often observed, is not required in the early stages (18, 76). However, the model provides a sound base for analysis of hurricane-generating mechanisms involved.

Criteria for barotropic instability of the anticyclonic ridge or vortex have been developed (71, 36), summarized and extended by Alaka (1). Consider an anticyclonic vortex initially in gradient wind balance, i.e. Eq. (9.5) holds with $dv_r/dt = 0$. The vortex is also assumed to be frictionless. If now a disturbance is introduced by increasing the pressure gradient, dv_r/dt will become positive and may increase without limit depending on the following inequality:

$$\left(f + \frac{2v_\theta}{r}\right)\left(f + \frac{\partial v_\theta}{\partial r} + \frac{v_\theta}{r}\right) < 0. \tag{10.2}$$

The second bracket in this expression is the absolute vorticity of motion around the vortex. Following Sawyer (71) instability will develop if the first term is positive but the absolute vorticity negative.

However, no cases with negative absolute vorticity as initial condition have been observed (62). Even in the mature hurricane absolute vorticity at best approaches zero value in the outflow but does not exceed it (Fig. 9.16(b)). Another possibility is that of the expression in Eq. (10.2) being exactly zero. Then the outlfow will continue with a rate depending on the initial pressure perturbation as long as the increased pressure gradient remains alive. The theory of the stability of barotropic (76) and baroclinic (48) circular vortices has been greatly extended. Altogether, however, consideration of such vortices is somewhat unrealistic. Outflow patterns in general look like those of Fig. 9.19; the mass exit is concentrated in one quadrant for the storm, and this holds for the initial stages.

A more promising approach is consideration of baroclinic release of kinetic energy, not baroclinic instability of a current, rather the circumstances attending particular synoptic systems of limited areal extent. There are two sources of energy: (1) warming through condensation and (2) importation of a cold layer representing high potential energy into the tropics. We shall first take up condensation heating.

In former days, upper anticyclones were regarded mainly as "dynamic highs"; their warm core is derived from subsiding motion. This model, plus the observation that tropical rain areas are largely cold-cored, made any case for direct conversion of latent heat energy released to kinetic energy seem rather hopeless. Since then, upper warm cores with outflow from undilute towers above lower cold cores were observed, which alters the situation. Consider Fig. 10.4, a 250 mb chart from the southeastern United States and surroundings with enough data to permit reasonably reliable analysis (64). An elongated, narrow area of intense rain extended for several days from the southeast across the Gulf of Mexico. Precipitable water of the air columns was 6 g/cm^2, a value seldom exceeded in large-scale convection. The upper troposphere was warm relative to the outskirts and already at the standard level of 250 mb, in general use at the time, maximum pressure-heights were found above the rain area. Accelerated outflow from the rain area toward southeast and south is clearly evident. The energy cycle goes from latent heat to potential energy to kinetic energy. The small jet stream centre east of Florida does not attain the full strength computed from transit through 60 m along the indicated trajectory. Speed should be 35 m/s rather than 25 m/s. However, the chart is below the level of strongest wind; more than 30 m/s was attained at 200 mb, and further, the wind of 25 m/s is greater than the geostrophic wind computed as 15 m/s and also in excess of the gradient wind. In any event, the occurrence of the energy

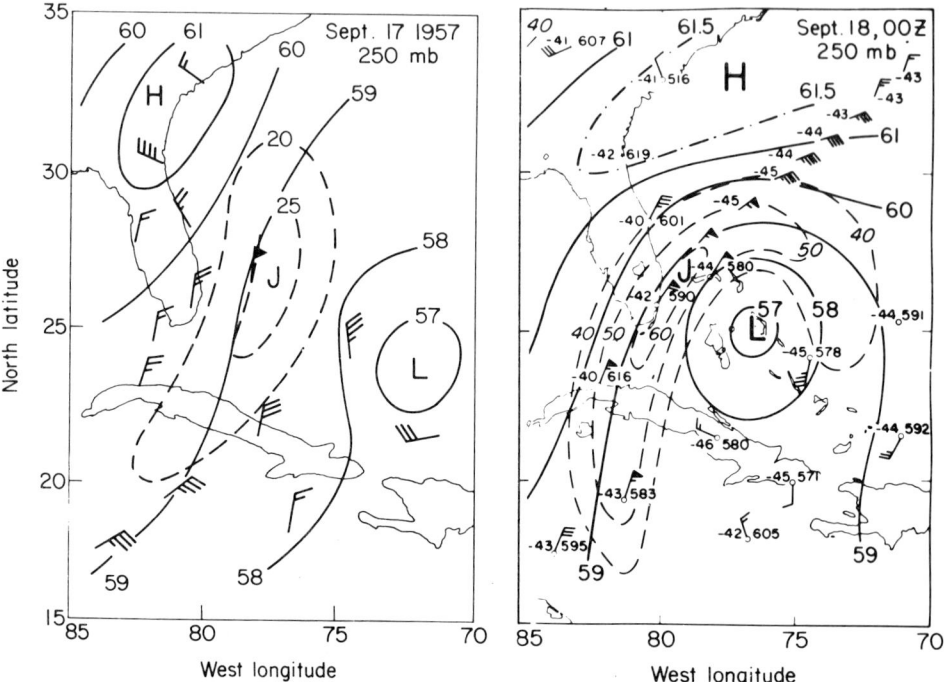

Fig. 10.4. 250 mb chart, 17 September, 1957, 00 GMT. Contours in hundreds of feet, base 30 000 feet (30 m intervals between contours). Isotachs (dashed) in m/s (64).

Fig. 10.5. 250 mb chart, 18 September, 1957, 00 GMT. Includes isotachs and B-47 aircraft flight data east of Florida (64).

transformation cycle was demonstrated and has been found since with increasing frequency as observation networks and special observing programmes have led to greatly increased data density.

Considerable intensification took place on the following day (Fig. 10.5) as an upper Low from the West Atlantic moved westward toward Florida and, through the relative motion, impinged on the stationary rain band there. We see a form of superposition of two weather systems and potential intensification of the energy cycle through the juxtaposition of two energy sources. Winds at 200 mb at Miami attained 36 m/s. Before pursuing this subject further, let us view one striking example of a tropical cyclone in the Gulf of Mexico.

Hurricane Ella (1958)

In going over the literature on hurricane formation one of the pitfalls is that often there is a description of cases where one factor thought favourable verifies; especially when a storm intensifies near an oceanic high temperature anomaly. The bulk of all situations where this does not happen receives no such publicity. In this chapter the methodology has been to emphasize the negative aspect; this course will be continued with one gross example which contains strong warning against oversimplification of any kind.

Prominent in this book and in the literature have been the hurricanes of 1958 when an advance in the knowledge about such storms was achieved through the peak operation of the US National Hurricane Research Project. Hurricane Ella, the spectacular nonperformer of that year, however, has fallen into oblivion, though it was thoroughly explored by research aircraft.

This hurricane formed at the end of August in the eastern Caribbean (Fig. 9.14); it then committed the mistake of making landfall on eastern Cuba and travelling the length of that island with the eye on the south shore. Usually, a centre will back off to the nearest oceanic refuge, but not this time. When intercepted by hurricane research aircraft on 2 September, an extensive rain area was encountered in the right semi-circle, especially the right rear quadrant. Temperature in the rain area did increase from 8° to 13°C at 730 mb going toward the centre, still typical of hurricane structure. At that level the windfield showed $v_\theta/r =$ constant out to almost $r =$ 200 km, where speed was 25 m/s. Such a profile, indicative that the inner hurricane core has died, has been observed in various tropical storms. South of the centre of what was then called "tropical storm" the sky was clear.

A composite map of high-tropospheric rawinsonde and flight observations shows anticyclonic outflow and no ring of cyclonic v_θ around the centre (Fig. 10.6) though present at low levels. Of course, there was no eye wall; the origin of the diverging winds appeared to be over the rain area southeast of the centre. There the high-level aircraft was in cloud and turbulence at 237 mb, and temperatures were highest corresponding to $\theta_e = 350$ K; in the west, temperature was 2-3°C lower. All of these features may be regarded as typical of hurricane diminished to tropical storm, but they also indicate that the storm was alive and the upper circulation anticyclonic enough to expect regeneration as it passed off the east end of Cuba into the Gulf of Mexico with the highest sea surface temperatures of the whole Atlantic hurricane area and as high as are found anywhere. The event was heavily monitored, but nothing happened! Ella went out to sea and travelled the whole length of the Gulf of Mexico to the southern Texas coast without deepening while the research aircraft were exploring and all radar sets along the coast were incessantly trained toward the centre and its progress.

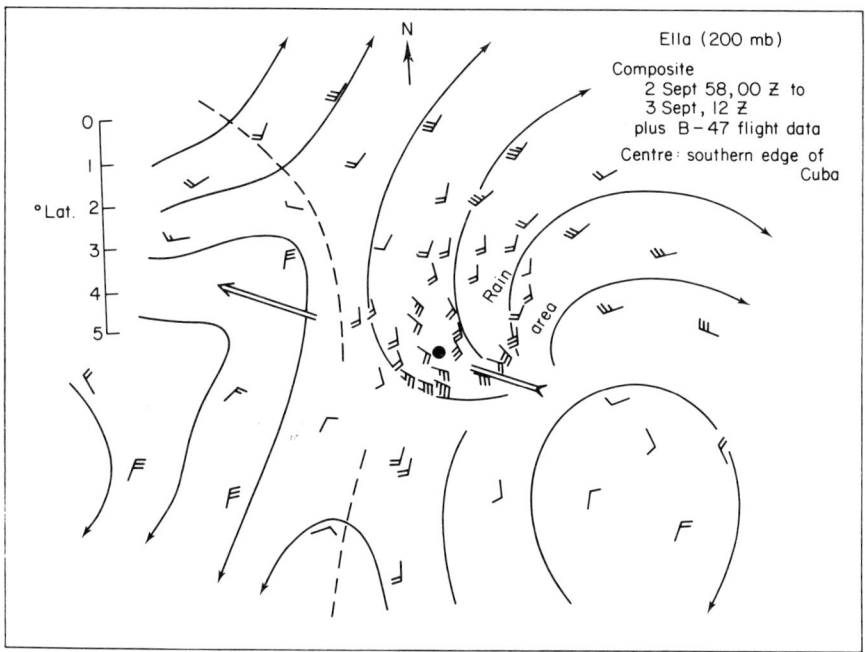

Fig. 10.6. 200 mb chart for hurricane Ella when located on central Cuba on 2 September, 1958. B-47 flight data and rawinsonde winds over 36 hours composited. Winds in knots. Relative latitude scale shown.

Figure 10.7 shows the ex-hurricane in the middle of the Gulf. Surface pressure was below 1006 mb and reported wind speeds 40 knots. The centre was well formed, with the typical asymmetry due to crowding of wind and isobars to the right of the direction of motion. The cloud and rain shield occupied the northern semi-circle. Upon breaking out of cloud near the centre no further clouds whatsoever

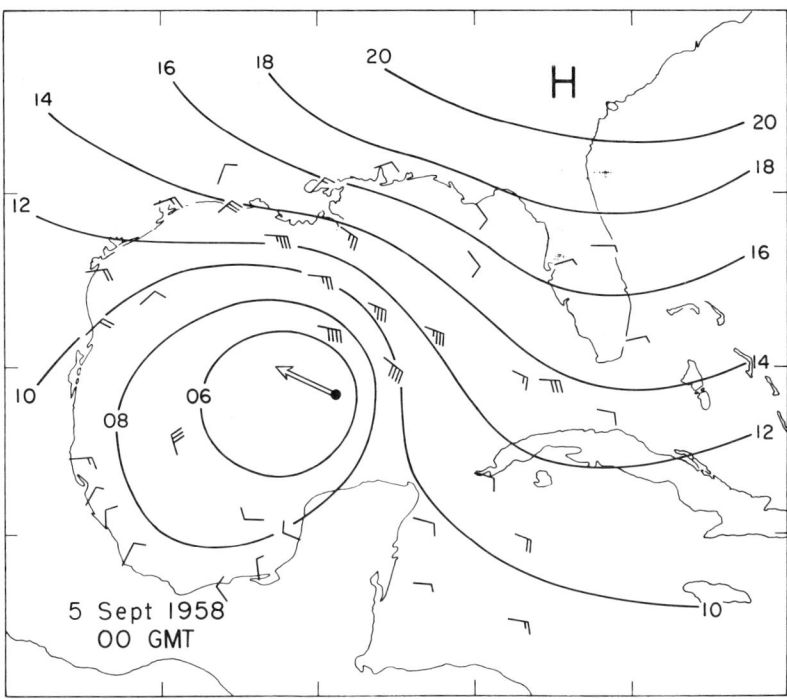

Fig. 10.7. Surface isobars and winds as Ella passes central Gulf of Mexico on way to Texas coast without reintensification.

could be seen from the aircraft toward south. Conditions were excellent for frictional convergence, yet the net mass flow inward integrated around the Gulf of Mexico (Fig. 10.8) was much less than two days earlier. Even if the slight surface outflow is dismissed as unrealistic, the fact remains that net inflow was small and concentrated in the middle troposphere at low θ_e, in strong contrast with the low-level increase in mass inflow with hurricane growth (Fig. 10.2). At 200 mb, anticyclonic flow was vigorous around the entire Gulf (Fig. 10.9). In the northeast quadrant, but only there, tangential wind speed equalled $-rf/2$. Evidently, surface frictional convergence, high sea surface temperatures, small vertical wind shear and large, vigorous upper anticyclone were jointly unable to put renewed life into the tropical storm that once had been a hurricane and was violating all "known" principles when it refused to redevelop. There were no obvious negative factors, such as low-level cold air inflow.

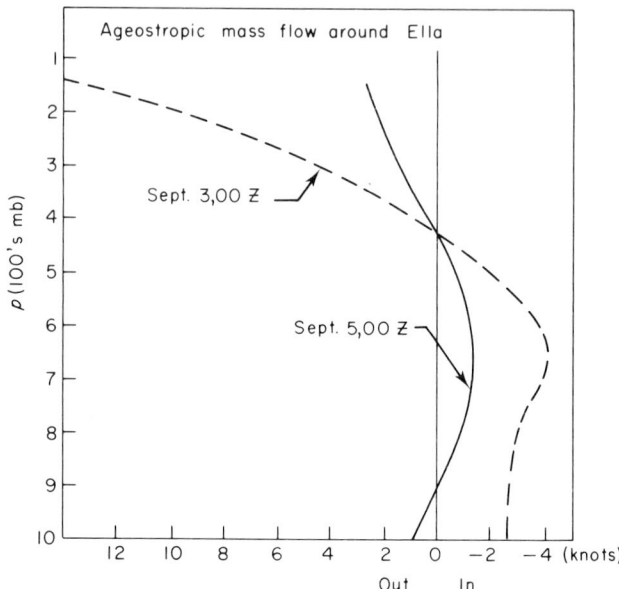

Fig. 10.8. Ageostrophic mass flow in the layer 1000-700 mb on 3 September, 1958, across a circle with 5° latitude radius; and on 5 September as integrated around the Gulf of Mexico.

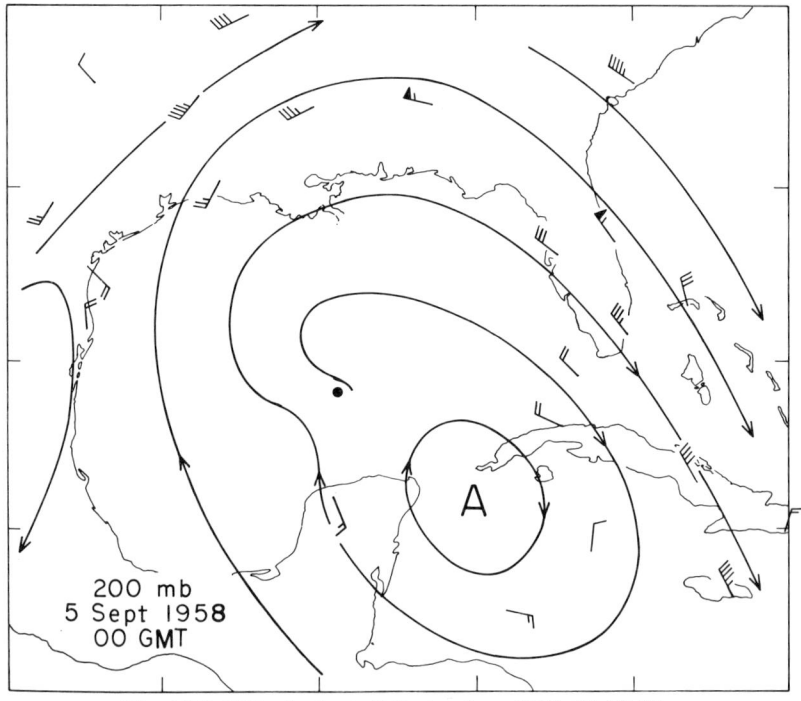

Fig. 10.9. 200 mb chart, 5 September, 1958, 00 GMT.

Release of potential energy
Evidence

The major reason for showing Ella in some detail is the demonstration that generation of kinetic energy through condensation heating, which took place widely across the Gulf of Mexico, could do no more than maintain the tropical storm in steady state. The missing element of the situation must be release of potential energy brought to the vicinity of the centre from the outside, for instance as shown in Figs 10.4 and 10.5. Various synoptic patterns can be encountered that lead to juxtaposition of an incipient centre and a limited reservoir of cold air in the middle and upper troposphere capable of sinking with attendant intensification of the circulation between the two sources of energy: advance of the lower latitude portion of an extratropical trough toward the periphery of the tropical depression or wave (13, 16, 35, 37, 57, 59, 62); a pre-existing deep trough in the tropics, very similar (Figs 8.19 and 8.20) (78); a cold reservoir seen as trough preceding a tropical wave or small storm (17); initial upper level cyclones in the mean Pacific upper-air trough (Fig. 1.14) (69); and the suggestion of energy transmission westward from a deep trough in Mid-Pacific (54). In fact, just about everyone who has dealt synoptically with the formation problem, has reached the same conclusion in principle, in contrast to theoretical modellers.

A really gross case occurred in August 1969. A small, recently-formed hurricane was rounding the western tip of Cuba (Fig. 10.10(a)) while a very large and deep upper Low, extreme for the season, occupied the South-Central states of the United States. This Low had been present for several days in quasi-steady state. As Camille started north in the Gulf of Mexico, the cold Low collapsed, the cold temperatures disappeared as the cold mass sank, leaving behind only a narrow trough of high cyclonic vorticity in which the principal sinking motion was concentrated, hence converging at 200 mb (Fig. 10.10(b)). Camille deepened rapidly to 900 mb (!), a pressure held until landfall, while an intense outflow anticyclone formed on its eastern flank. Many similar developments, less extreme, can be cited. For instance, in the first days of October 1949, a deep upper-air trough over the southern United States and northern Gulf of Mexico simply vanished, while an incipient centre in the southern Gulf acquired hurricane intensity with little lag (61). Another instance of intensification from minimal to extreme intensity was hurricane Janet of September 1955. After travelling westward for several days in steady state as a minimal hurricane behind a westward-advancing,

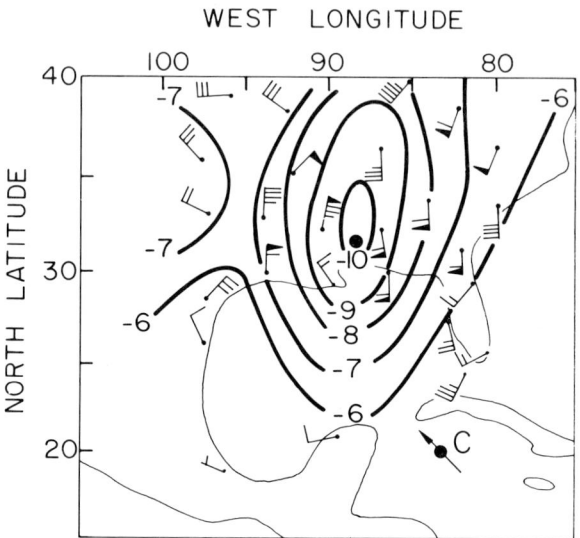

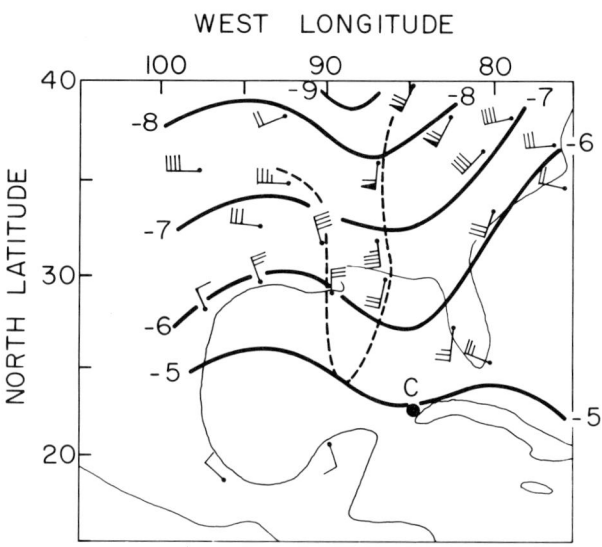

Fig. 10.10. 500 mb isotherms (°C) and 200 mb winds over the southeastern United States during northwest travel of tropical storm Camille in western Caribbean: (upper) 15 August, 1969, 00 GMT; (lower) one day later.

high-level trough just as described (17), this trough collapsed as it came into contact with another large upper anticyclone over the Gulf and western Caribbean. Extreme deepening of Janet ensued. It passed over Swan Island destroying the key weather station. A reconnaissance aircraft was lost; later the hurricane made landfall on the Mexican coast with measured central pressure of 914 mb and winds estimated at 100 m/s from the destruction. The occasional extreme re-intensification of hurricanes moving out of the tropics (discussed below) just shows the baroclinic energy transformation in a somewhat different setting.

Further comments

The energy release mechanism, which is set in "motion" by the sinking cold domes, is again that of Eqs. (4.31) and (4.32). One may think of the upper portion of curve (A), the mean tropical atmosphere, in Fig. 9.22 as being replaced by air about 3 to 5°C colder. Then Eq. (4.31) is evaluated from the difference between curve (B) and the altered curve (A). The potential energy gradient is increased; therefore, the work done by a vertical mass circulation of the same strength as before will approximately double. Some numbers will be quoted in connection with hurricane Hazel for which Palmén (50) has made detailed computations.

It is suggested that the foregoing has pointed to the missing link and mechanism in the evolution of hurricanes and in secondary growth of small hurricanes to extreme intensity. However, the end of the story has not yet been reached. Juxtaposition of an incipient centre with a source of potential energy should be accepted as necessary, but there are well-documented instances when movement of an upper trough over and through a tropical circulation has caused it to weaken or disappear. This is the situation when hurricanes are found to die over the open sea. The ventilation of the upper portion of warm convective areas by cold air advected by a trough will remove the warm centre and destroy generation of kinetic energy from condensation heating. The critical difference between these cases and those examined before is that the cold dome sinks in the latter and does not sink in the other situations. We herewith arrive at the further requirement; the structure of cold domes must be such as to cause them to sink upon approach to an incipient tropical centre, and the cold air itself must enter the tropical circulation no more at high than at low elevations (62).

Herewith we arrive at a very difficult point. One can hardly say that the tropical system itself has the capability of initiating the collapse of cold domes. In the Camille situation, the hurricane evidently was the incidental beneficiary of a major event in the lower middle latitudes. No obvious reason has been found why the cold dome, stable for three days, suddenly collapsed. Given a high-tropospheric cyclonic vortex in gradient wind equilibrium, the most obvious cause leading to sinking is movement to lower latitudes. Then the Coriolis force diminishes and eventually becomes unable to sustain the cold dome against the inward acceleration by the pressure gradient force. Prolonged southeastward motion of deep cold domes across the United States often can be directly linked to rapid formation of intense east coast cyclones in winter when the latitude of a dome may have decreased by ten degrees.

One would expect this mechanism to be especially reliable in the tropics because of the large change of Coriolis parameter with latitude. Indeed, cases of collapse probably due to this mechanism can be cited; but it does not always happen. The dome approaching Florida in Fig. 10.5, for instance, moved into the southwestern Caribbean retaining its intensity. Another mechanism would be for the dome to lose angular momentum when a trough in the westerlies impinges on the polar side. Qualitative decision in these cases is very difficult. It would seem, however, that a baroclinic, numerical prediction model should be capable of giving a correct prediction since no special features of condensation or sub-grid events are involved. Concerning the equatorial extensions of polar troughs, the jet stream on the western side frequently does not cross the trough but subsides with anticyclonic turning, while a new velocity maximum develops on the eastern side (Fig. 8.40). Such a structure has the required energy release mechanism built in and should provide a most favourable setting for intensification when it comes in contact with a tropical disturbance. In contrast, open troughs through which the high-tropospheric air passes without slowing down or subsiding, should be the kind that destroy rather than intensify a tropical centre. Thus, in the trough situations qualitative decisions are more readily made than in the cold Low cases. It appears that extension of numerical prediction models in the direction of exploring external, rather than internal mechanism of formation should have an excellent chance of solving the residual difficulties attending the at one time untractable formation problem.

Maximum intensity of hurricanes

In general, hurricanes need three days to grow from incipient to mature state. As long as surface conditions remain favourable, one might expect them to level off at about the same intensity since the low-level characteristics of the inflow differ only a little over the western parts of the tropical oceans. However, nothing could be farther from the truth. The peak intensity of hurricanes ranges from a bare minimum of 64 knots to 150-200 knots in the same region, with all intermediate intensities represented. Size, as mentioned earlier, increases with increasing latitude until the subtropical ridge line is reached. The difference is attributable to the same factor just discussed, that is, the release of potential energy through interconnection between a hurricane and its surroundings at large. A hurricane confined to the deep tropics and without outlet for the outflow to the mid-latitude westerlies generally retains minimal intensity. The solenoid field available for release of potential energy may be given by a temperature difference of only 1-2°C. As soon as a channel is opened into the westerlies and the main latitudinal temperature gradient, one finds the intensity of hurricanes increasing rapidly. The channel may be opened by the usual poleward displacement of hurricanes, then maximum intensity is attained just before recurvature at the subtropical ridge (66) as long as ocean temperature does not decrease. On other occasions the middle latitude long waves, through development of huge amplitude, will reach deep into the tropics, especially when the autumnal build-up of baroclinity takes place in high latitudes. A connection between maximum intensity and ocean temperature has also been claimed (42).

Late hurricane stages

There is a wide range of terminal phases which a hurricane may undergo. A few die over water under a non-sinking cold Low or trough in the high troposphere, as has just been discussed. Most often, storms keep on a west to northwest track until they make landfall, or they recurve into the polar westerlies. While over the ocean, central wind speed drops by about 50% over 16° latitude after recurvature in the Pacific (66), apart from cases with strong redevelopment. Hurricanes making landfall always decrease in central wind speed and pressure rises. This filling used to be ascribed to strong frictional retardation over land, an explanation that no doubt has merit when a hurricane crosses mountain ranges. A severe hurricane in 1928, for instance, passed with its centre along the whole length of the central mountain range of Puerto Rico (Fig. 6.7); in nine hours of traverse, pressure rose over 30 mb (73). The maximum wind of typhoons crossing the Philippine mountains decreases by one-third (7). Further, the boundary layer structure will change over land with reduction of wind near the solid surface while, say, the 100 m wind may at first remain unaffected.

However, the general weakening of hurricanes over land, especially flat land like Florida, is due to the sudden cutting off of the oceanic energy source under the hurricane, which makes it impossible to maintain the thermal field (34). Miller (43) has shown the removal of the heat source to be the correct answer in a very detailed study of the filling of hurricane Donna over southern Florida in September 1960.

Hurricanes approaching land in autumn also encounter the sudden invasion of polar-type air into the forward quadrants. Such intrusion has destroyed even huge typhoons advancing against the China coast. Often a storm will turn direction unexpectedly and quickly, gliding parallel to the coast or even drawing away. This happens along the coast of the southeastern United States (see Helene of September 1958; Fig. 9.14) (2). After passing Cape Hatteras going northeastward, the danger of cold air entering again recedes since the land curves away toward northwest.

After one or two days over land a surprisingly large number of hurricanes acquire a structure similar to that of subtropical Lows. The cyclonic bottom portion is greatly weakened, but the capability to draw low-level air inward and produce heavy to very heavy precipitation persists, aided of course when even a mild, extratropical type of

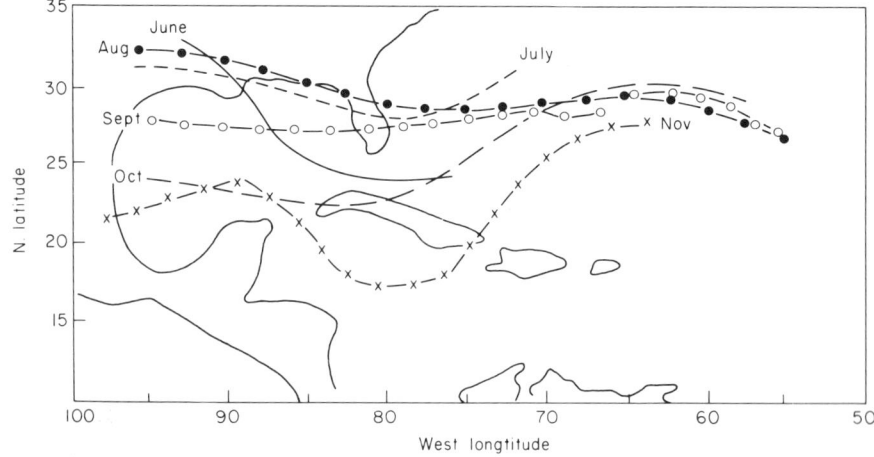

Fig. 10.11. Average latitude of recurvature of Atlantic hurricanes in the months of the season (12).

TROPICAL CYCLONES—FORMATION AND MOVEMENT   485

cyclogenesis is going on. From Texas to the northeastern United States some 50 cm of rain and more may fall in 24 to 36 hours over a substantial area and bring on some of the great flooding disasters that cause damages far beyond those of the hurricane stages of the systems. The mechanisms by which the rainfall capability remains so high are not obvious, but it is a good principle to be prepared for severe weather even when an old hurricane arrives after two or three days over land. Camille of August 1969 weakened substantially as she moved up the Mississippi, but then she turned east and produced a long, narrow belt of disastrous rain with well over half a metre of water across the whole Appalachian mountain range almost to the Atlantic coast (28). Precipitation of as much as 15 cm has been delivered by an old hurricane as far north as Chicago.

Renewed deepening in middle latitudes

The synoptic situation present over a continent at the time of landfall greatly affects the future fate of a hurricane. Sometimes, there is a dynamic anticyclone to the north, and under these circumstances, a hurricane that has already drifted northeastward to sea, may be reversed in its direction and re-enter land with marked weakening. At times the hurricane arrives from south and southeast when a deep trough approaches from the west, and a cyclogenetic situation is in the offing in any event, to which the hurricane then lends further and, at times, spectacular impetus. Such situations are frequent around the Japanese Islands (41, 45, 72). Best known is hurricane Hazel (Fig. 9.14), which drifted as a small storm through the eastern Caribbean and then rapidly deepened north of the Antilles when it changed course to northward. A spectacular outflow anticyclone had already formed two days before landfall and it became the ridge of the long wave pattern in the westerlies over the Atlantic. Deepening continued all the way as the hurricane approached the South Carolina coast from southeast. At that time a deep trough in the westerlies with closed centre and 500 mb temperatures near $-30°C$, very cold for mid-October, was moving eastward through the Middle West. Cyclogenesis was expected in any event; the relative approach of trough and hurricane produced a development of extraordinary intensity. The surface centre—one may argue whether it was the same one or a redevelopment—accelerated to forward speed of 30 m/s on a course slightly west of north and produced a band with 15-20 cm rain about 300 km wide in its 12-hour traverse from the Carolinas to southern

Canada. With such heavy rain amounts in a very short time, a precipitation catastrophe ensued in several areas, at its worst near Toronto. Winds at 300 mb were accelerated from 30-40 m/s on the day before to 80 m/s and possibly more after the juncture of the two systems.

Palmén (50) who developed Eq. (4.31) for this occasion, computed the release of kinetic energy as 19-20 $\times$ 10^{10}kW; the export of kinetic energy from the area which contained the intense energy transformation had about the same value whereas friction was only one tenth; thus, most of the energy produced was exported. Potential energy was restored by the release of latent heat which was an order of magnitude larger with 156 $\times$ 10^{10}kW; the energy conversion attained the unusually high value of over 10%. The generation is equal to the average importation of kinetic energy from the tropics into the middle latitudes in winter through the subtropical jet stream. Two or three cyclones of the magnitude of Hazel, operating constantly, would suffice to maintain the kinetic energy of the whole hemisphere from 30° poleward against frictional dissipation. Here occurred one of nature's great upheavals, seldom realized in such magnitude.

Motion of hurricanes

Since the days when monthly surface isobars could first be drawn and tropical cyclone tracks charted with some accuracy, meteorologists have known that these tracks roughly parallel the mean isobars around the subtropical anticyclone. This led to the concept of steering (24, 29). Considering the shape of some recurving tracks, some writers have gone so far as to predict hurricane displacements by assuming parabolic or hyperbolic path shapes. Unfortunately, such simple solutions rarely give a satisfactory answer. In Fig. 10.12, selected tracks from September hurricanes of more than three days' duration have been superimposed. Although an undertone suggestive of parabola or hyperbola is undeniable, all kinds of track shapes appear. These are shown in model form in Fig. 10.13. Tracks (a-c) are the most frequent ones encountered; d-f are examples of uncommon displacements. Paths that lead out of the tropics are called "recurving" paths, a somewhat inaccurate but well-established name.

Early and late season cyclones often move poleward from their inception, since they form close to the subtropical ridge. This happens especially in the Southern Hemisphere where the westerlies over the

TROPICAL CYCLONES—FORMATION AND MOVEMENT 487

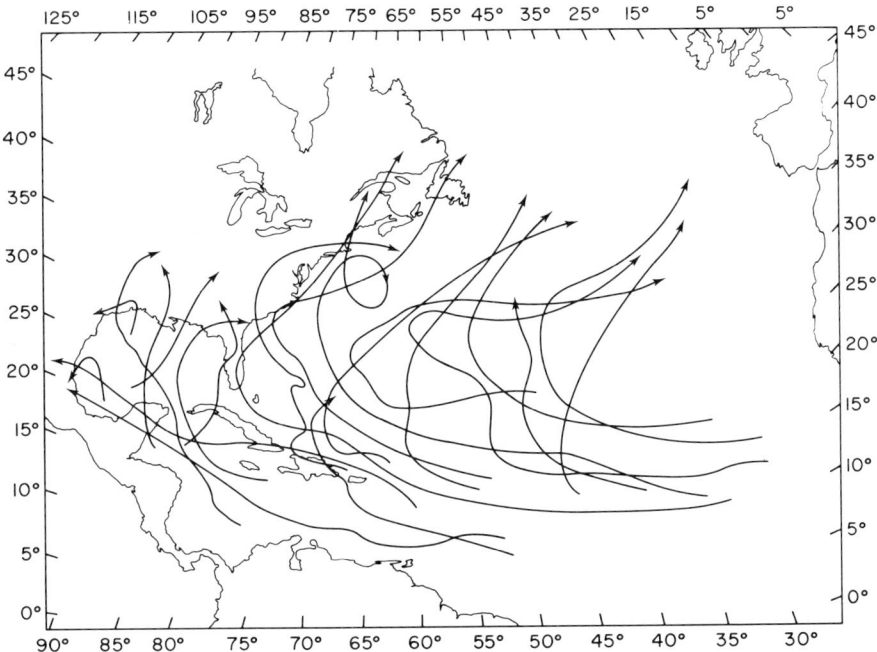

Fig. 10.12. Selection of Atlantic hurricane tracks of more than three days' duration, 1950-1963.

oceans are on average closer to the equator than in the Northern Hemisphere (22). However, this does not apply to the South Indian Ocean or to typhoons propagating westward in the far western Pacific in winter south of a stable, subtropical, high-tropospheric ridge near 10°N.

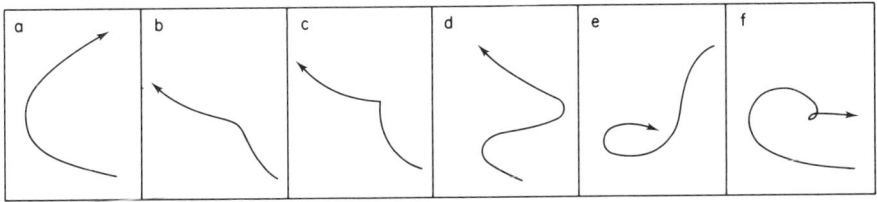

Fig. 10.13. Models of hurricane tracks.

Steering

Track (a) is the simple path around the subtropical anticyclone. Tracks (b-d) show a hurricane bypassed by a trough on the poleward side for the Northern Hemisphere. In (b) the storm is pulled up slightly, but the trough is moving fast toward east or has small amplitude or both, so the hurricane soon comes under the influence of the upper winds to the rear of the trough and curves back toward west. In model (c) the influence of the trough is much stronger. A cusp point is reached on the track where indeed it is uncertain what the storm will do. In such moments the limit of determinacy of atmospheric happenings appears to be reached. Small random impulses may cause the hurricane to turn back toward west, but in other cases recurvature out of the tropics continues after halting for perhaps as much as a day. Finally, in (d) the hurricane does recurve at first, but then there is a drastic rearrangement in middle latitudes; the trough causing recurvature loses its amplitude and is replaced by an upper anticyclone, so the hurricane again turns toward west.

Internal forces

All of the foregoing tracks as well as those of Fig. 9.16 have one thing in common—a tendency to move poleward. This tendency, shared by cyclones at all latitudes, indicates propulsion by an "internal" force in addition to the external one described. As suggested by Rossby (67, 68) poleward acceleration of cyclones and equatorward acceleration of anticyclones can arise from the variation of the Coriolis parameter across the width of a storm, but it is difficult to measure this acceleration. The internal force can accomplish an appreciable fraction of total displacement (14). This should especially hold for socalled supertyphoons with a diameter large enough not to be subject to steering influences; these, indeed, generally take a steady path toward westnorthwest. A second internal force arises from superposition of steering current and vortex which is non-linear due to the v_θ^2/r term in Eq. (9.4) and will cause sinusoidal oscillations about a mean storm path (79).

Interaction of vortices

In hydrodynamics, the interaction between vortex pairs has been the subject of analysis for many years. Vortices either attract or repel each other; they rotate about a centre of gravity located on the straight line

TROPICAL CYCLONES—FORMATION AND MOVEMENT

or great circle connecting them. The position of this centre depends on the relative mass of the vortices. If they are equal, they rotate about a point midway between them, if not, the ratio of the two masses determines the location of the point. In addition to mass, the relative intensity of circulation within each vortex influences the position of the centre of rotation.

Interaction between vortices has been studied especially by Fujiwhara (21) and Haurwitz (31). Consider two cyclones as in Figs 9.1 and 10.1. The eastern centre will exert an equatorward-directed force on the western one through its circulation. In turn, the western centre exerts a poleward-directed force on the eastern one, thus, the axis connecting the centres will rotate counterclockwise with time. The interaction usually starts only when the centres are no more than 15° longitude apart. In Fig. 10.1, with vortex separation of 20° longitude, rotation typically did not occur. Figure 10.14 illustrates a pair in a position very similar to that of Fig. 9.1 south of Japan. The vortices

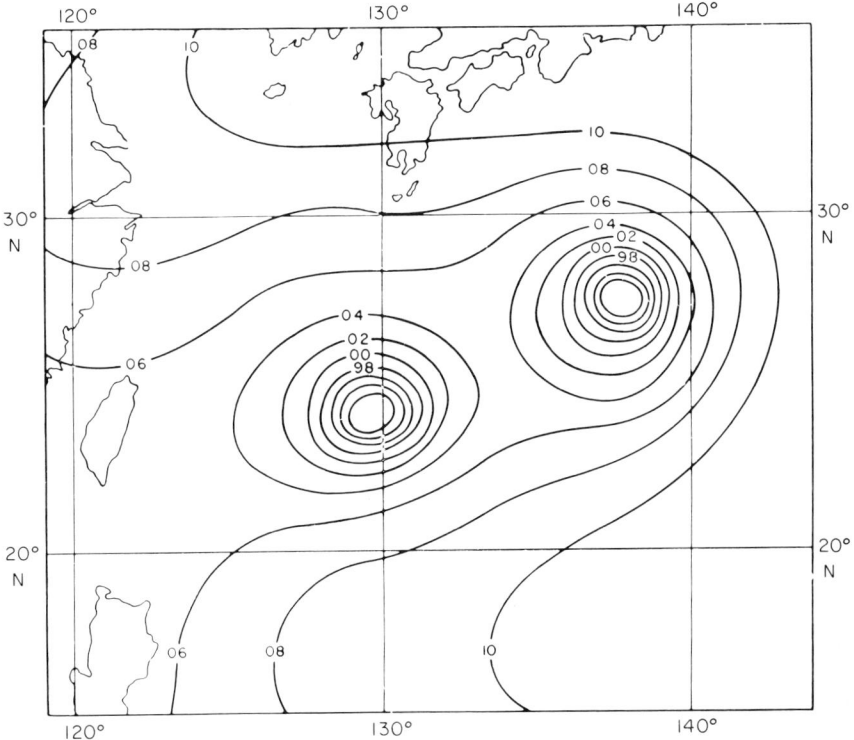

Fig. 10.14. Surface chart for the western Pacific, 25 August, 1945 (31).

were of equal size and intensity. From their track and that of the midpoint between them (Fig. 10.15) larger outside influences affected the entire hurricane pair. If the displacement of the midpoint is subtracted, we obtain the relative motion of the vortices and then the interaction is quite clear. The storms rotate about each other and are mutually attracted. As they come closer together, their influence on each other grows and the relative rotation rate increases from 7° to 30° in 12 hours, a considerable factor in prediction.

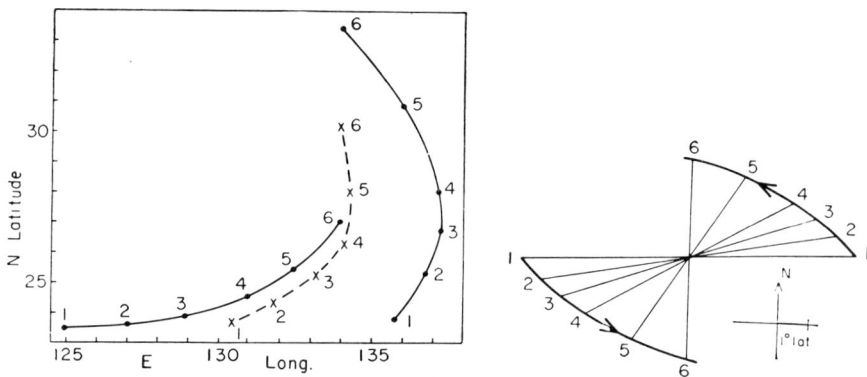

Fig. 10.15. Left: Track of typhoons of Fig. 10.14 (solid) and of centre of rotation (dashed) for six 12-hour intervals beginning 24 August, 1945.
Right: Relative motion of typhoons.

Certainly, it is not easy to find many cases so clear-cut. Outside systems with much more mass usually dominate, and their influence must be included, so one arrives at a multiple interaction problem that requires computer facilities for solution.

Although the interaction principle has been applied mainly to hurricane pairs, this may not be its only use. Consider a pair consisting of a cyclone and an anticyclone. If the mass of the anticyclone is very great compared with that of the cyclone, we observe a quasi-stationary high and a storm that travels clockwise around its periphery; this is another way of looking at the steering current. The situation is not unlike that of the system earth-moon. Both bodies rotate about a common centre of gravity, but the mass of the earth is so much greater than that of the moon, that the centre of rotation lies in the interior of the earth.

Supertyphoons with diameter up to 3000 km in the Pacific have the same size and mass as the subtropical anticyclones. Then the typhoons influence the anticyclone; if the latter is situated north of the cyclone, it is propelled westward. Since the High, in turn, drives the typhoon westward, we may see the pair advancing westward over a large portion of the Pacific Ocean as a couplet. In the end, either the High is broken down by the westerlies or the typhoon touches land.

Forecasting

Only a brief historical survey will be given here, since computer prediction is certain to take over.

Hurricane forecasting has proved to be very difficult, the location and timing of landfall as well as the expected intensity and the associated threat from the ocean on the shore. With modern communication facilities, ships at sea can evade nearly all hurricane centres except those accelerating rapidly out of the tropics. The last major reported disaster at sea was that of a German school sailing vessel intercepted in Mid-Atlantic in 1957 by a hurricane which took an unusual course somewhat south of east after recurvature.

For prediction on shore, there is the severe problem that the area of heavy damage is usually small compared to the best efforts of limiting the "area of uncertainty" in centre prediction. With regard to "timing of arrival", there is no harm if a hurricane arrives a little later than anticipated. The danger lies in its coming early when its forward motion suddenly accelerates.

In the last days of August 1954, a severe hurricane struck New England. After moving northward slowly from 26 to 30 August, it accelerated and moved from Cape Hatteras to Long Island in just 12 hours. Ten days later another hurricane followed almost the same path; it also greatly accelerated but it moved just slightly more toward east and, thus, bypassed New England at sea to strike the State of Maine and southeastern Canada. Short-term prediction for the general public was good and the slightly more eastern path of the second centre was recognized. For securing industrial facilities and ports, two or three days' warning of a hurricane may be required. No such accuracy as yet can be remotely achieved.

In early forecasting history, meteorologists sought to establish the "steering current", which serves well as long as a hurricane is in the deep tropics unaffected by troughs in the westerlies with long equatorward extensions. Problems arise when a choice has to be made between several of the model paths in Fig. 10.13, notably when hurricanes are approaching Florida and the other southeastern states from the east-southeast. The question has to be asked, will the hurricane recurve? A startling example of this happening came in 1958 (Fig. 9.14) when hurricane Helene almost entered and then suddenly turned to skirt the coastline. Dora of 1964 was nearly two days off Jacksonville, Florida, looking the situation over. Then she gradually resumed a westward path and became the first hurricane on record to strike northern Florida from the east. Donna (1960) was a "regular" performer which had the distinction of striking in succession all three land areas of the United States most subject to hurricanes. Daisy (1958) remained off-shore.

With the establishment of rawinsonde networks in the tropics, it became evident that hurricanes would recurve if the westerlies of an impinging trough from the west reached to the latitude of the storm at 500 mb (58). Surprises due to a storm first curving up, then suddenly turning west again, would hereby be avoided. A condition, of course, was that the trough maintained its structure and intensity. With an increasing file of case histories, computation of the 24-hour path from 500 mb charts with a determination of the net forces acting on the boundary of hurricanes was attempted. The internal force was taken into account by allowing statistically for enhanced meridional movement.

This type of approach was further developed at the National Hurricane Center, Miami, in the form of statistical forecasting models. Dynamical methods (70) were included too, as well as a mix of climatological historical analogues; previous track and persistence of deviations from the seasonal mean always remained key tools. Although slow improvement in short term prediction skill resulted, all forecasting methods (including the computerized ones) suffer from rapid loss of skill with increasing forecast length. Predictions at times are too ready to recurve a hurricane. This happened in Camille of 1969 which the forecasts steered into the northeast Gulf, whereas the hurricane developed a great anticyclone on the eastern side as it deepened rapidly and, thus, maintained its advance in the general direction of New Orleans (Fig. 9.14). Rapidly deepening centres often make a small left turn, in the Southern Hemisphere, a right turn.

Various additional forecasting methods have been proposed. On the long-term basis an outlook of probable paths is given by the long waves in the westerlies. Deviations of mean position and amplitude show where preferred tracks out of the tropics are most likely to occur (5), and thermal patterns appear again. In the seas around India, steering along the thermal wind at 500 mb is preferred to the actual wind (25). Tracks tend to follow warm water anomalies (75, 19); successive tracks tend not to follow the same track on account of a wake of cold water left behind by the first storm (6). However, various exceptions to the latter hypothesis have been noted in the Pacific Ocean (55) and similar tracks occur in the Atlantic as well. It is difficult to incorporate all these suggestions quantitatively in forecast routines. The concept of high tropospheric steering, regularly employed in the United States Hurricane Warning Service† with success in certain selected types of general flow patterns, has been revived with Pacific data (26). In view of the average high-tropospheric wind fields (Fig. 9.19) it has proved impossible to develop a general technique based on high-level flow alone; hence, computations based on the lower troposphere up to 500 mb were widely developed and used. One can conclude that all efforts in hurricane prediction have not converged toward a single, generally applicable scheme and that, in spite of an oversupply of forecasting aids of one form or another, the forecaster in the end still has to "look out of the window" and make a personal decision.

There is little expectation that further pursuit of forecasting improvements with any method short of full iterative models on high-speed computers will do much to close the gap between existing skill and forecast accuracy and advance warnings demanded. Even the numerical prediction faces the problem of the cusp point or loop on a track, when essentially there is indeterminacy. On the other hand, the rapid acceleration of the two 1954 hurricanes should be more readily predictable since mid-latitude circulations are largely involved. The opinion has been advanced that it may be easier to modify than to predict hurricanes, and the potential truth of this assertion cannot be denied.

†By Grady Norton, first chief hurricane forecaster.

References

(1) Alaka, M. A. (1958). Dynamics of upper-air outflow in incipient hurricanes. *Geophysica* (Helsinki) **6**: 133-146.
(2) Arakawa, H. (1959). Coast Effect on Typhoon Movement. Pap. Meteor. Geoph. 9, pp. 123-126. Meteor. Inst. Tokyo.
(3) Arakawa, H. (1963). Typhoon Climatology as Revealed by Data of the Japanese Weather Service. Proc. Seminar Tropical Cyclones, Tokyo. pp. 31-36. Japan. Meteor. Agency.
(4) Astling, E. G., Daggupaty, S. M. and Leslie, K. R. (1972). A diagnostic study of a non-deepening disturbance in the Caribbean Sea. *J. Appl. Meteor.* **11**: 1305-1317.
(5) Ballenzweig, E. (1959). Relation of long-period circulation anomalies to tropical storm formation and motion. *J. Meteor.* **16**: 121-139.

(6) Brand, S. (1971). The effects on a tropical cyclone of cooler surface waters due to upwelling and mixing produced by a prior tropical cyclone. *J. Appl. Meteor.* **10:** 865-874.
(7) Brand, S. and Blelloch, J. W. (1973). Changes in the characteristics of typhoons crossing the Philippines. *J. Appl. Meteor.* **12:** 104-109.
(8) Brunt, A. T. (1969). Low latitude cyclones. *Aust. Meteor. Mag.* **17:** 67-90.
(9) Carlson, T. N. Ref. 9, Chapter 8.
(10) Carlson, T. N. (1971). An apparent relationship between the sea-surface temperature of the tropical Atlantic and the development of African disturbances into tropical storms. *Mon. Wea. Rev.* **99:** 309-310.
(11) Charney, J. G. and Eliassen, A. (1964). On the growth of the hurricane depression. *J. Atmos. Sci.* **21:** 68-75.
(12) Colón, J. (1953). A study of hurricane tracks for forecasting purposes. *Mon. Wea. Rev.* **81:** 53-66.
(13) Cressman, G. P. (1948). On the Formation of West Atlantic Hurricanes. Misc. Rep. 24, Part II, pp. 68-103. Dept. Meteor., Univ. Chicago.
(14) Cressman, G. P. (1951). The development and motion of typhoon Doris (1950). *Bull. Amer. Meteor. Soc.* **32:** 326-333.
(15) Dunn, G. E. (1940). Cyclogenesis in the tropical Atlantic. *Bull. Amer. Meteor. Soc.* **21:** 215-229.
(16) Erickson, C. O. (1967). Some aspects of the development of hurricane Dorothy (1966). *Mon. Wea. Rev.* **95:** 121-130.
(17) Fett, R. W. (1966). Upper-level structure of the formative tropical cyclone. *Mon. Wea. Rev.* **94:** 9-18.
(18) Fett, R. W. Ref. 18, Chapter 9.
(19) Fisher, E. L. (1958). Hurricanes and the sea-surface temperature field. *J. Meteor.* **15:** 328-333.
(20) Fortner, L. E. (1958). Typhoon Sarah, 1956. *Bull. Amer. Meteor. Soc.* **39:** 633-639.
(21) Fujiwhara, S. (1921). *Quart. J. Roy. Meteor. Soc.* **47:** 287; *ibid.* (1923). **49:** 75.
(22) Gabites, J. F. (1963). The Movement of Tropical Cyclones. Proc. Seminar on Tropical Cyclones, Tokyo. pp. 159-164. Japan. Meteor. Agency.
(23) Gabites, J. F. (1963). The Origin of Tropical Cyclones. Proc. Seminar on Trop. Cyclones, Tokyo. pp. 53-58. Japan. Meteor. Agency.
(24) Garriott, E. B. (1895). *Mon. Wea. Rev.* **23:** 167.
(25) George, C. A. (1953). Thermal thickness patterns and tropical storms. *Indian J. Meteor. Geoph.* **4:** 279-290.
(26) George, J. E. and Gray, W. M. (1977). Tropical cyclone recurvature and non-recurvature as related to surrounding wind-height fields. *J. Appl. Meteor.* **16:** 34-42.
(27) Gray, W. M. Reference 27, Chapter 9.
(28) Haggard, W. H., Bilton, T. H. and Crutcher, H. L. (1973). Maximum rainfall from tropical cyclone systems which cross the Appalachians. *J. Appl. Meteor.* **12:** 50-61.
(29) Hann, J. (1875). *Z. Öst. Ges. Meteor.* **10:** 81.
(30) Hantel, M. and Koehne, R. Ref. 25, Chapter 8.
(31) Haurwitz, B. (1951). The motion of binary tropical cyclones 1951. *Arch. Meteor. Geoph. Biokl.* (A) **4:** 73-86.
(32) Holliday, C. R. (1977). Double intensification of typhoon Gloria, 1974. *Mon. Wea. Rev.* **105:** 523-528.

(33) Hubert, L. F. (1949). High tropospheric westerlies of the equatorial west Pacific Ocean. *J. Meteor.* **6**: 216-224.
(34) Hubert, L. F. (1955). Frictional filling of hurricanes. *Bull. Amer. Meteor. Soc.* **36**: 440-445.
(35) Klein, W. H., (1957). The weather and circulation of June 1957. *Mon. Wea. Rev.* **85**: 208-220.
(36) Kleinschmidt, E. (1951). Grundlagen einer Theorie der Tropischen Zyklonen. *Arch. Meteor. Geoph. Biokl.* (A) **4**: 53-72.
(37) Koteswaram, P. and George, C. A. (1957). The formation and structure of tropical cyclones in the Indian Sea area. *J. Meteor. Soc. Japan* **75**: 309-322.
(38) Krueger, D. W. (1959). A relation between the mass circulation through hurricanes and their intensity. *Bull. Amer. Meteor. Soc.* **40**: 182-189.
(39) Lourensz, R. S. (1977). Tropical Cyclones in the Australian Region, July 1909 to June 1975. Australian Govt. Publish. Svc., Canberra. 111 pp.
(40) Malkus, J. S. and Riehl, H. Ref. 43, Chapter 9.
(41) Matano, H. and Sekioka, M. (1971). On the synoptic structure of typhoon Cora, 1969, as the compound system of tropical and extratropical cyclones. *J. Meteor. Soc. Japan* **49**: 282-296.
(42) Miller, B. I. Ref. 48, Chapter 9.
(43) Miller, B. I. (1964). A study of the filling of hurricane Donna (1960) over land. *Mon. Wea. Rev.* **92**: 389-406.
(44) Milton, D. (1974). Some observations of global trends in tropical cyclone frequencies. *Weather* **29**: 267-270.
(45) Mohri, K. (1956). On the characteristic structure in the mid-tropospheric westerlies accompanied with recurved typhoons. *Geoph. Mag.* (Tokyo) **27**: 237-247.
(46) Moss, M. S. and Merceret, F. J. Ref. 52, Chapter 9.
(47) Ooyama, K. Reference 63, Chapter 4. Also: Ooyama, K. (1971). Convection and convective adjustment, a theory on parameterization of cumulus convection. *J. Meteor. Soc. Japan* **49**: 744-756.
(48) Ooyama, K. (1966). On the stability of the baroclinic circular vortex: a sufficient criterion for instability. *J. Atmos. Sci.* **23**: 43-53.
(49) Palmén, E. (1948). On the formation and structure of tropical hurricanes. *Geophysica* (Helsinki) **3**: 26-38.
(50) Palmén, E. (1958). Vertical circulation and release of kinetic energy during the development of hurricane Hazel into an extratropical storm. *Tellus* **10**: 1-23.
(51) Perlroth, I. (1962). Relationship of central pressure of hurricane Esther (1961) and the sea surface temperature field. *Tellus* **14**: 403-408.
(52) Perlroth, I. (1967). Hurricane behavior as related to oceanographic environmental conditions. *Tellus* **19**: 258-268.
(53) Pike, A. C. (1962). Some Observations on Low-Level Wind Variation in the Vertical in Tropical Cyclones. Atmos. Sci. Tech. Paper 29, Dept. Atmos. Sci., Colorado State Univ., Fort Collins. 8 pp.
(54) Ramage, C. S. (1959). Hurricane development. *J. Meteor.* **167**: 227-237.
(55) Ramage, C. S. (1972). Interaction between tropical cyclones and the China Seas. *Weather* **27**: 484-493.
(56) Ramage, C. S. (1974). Monsoonal influences on the annual variation of tropical cyclone development over the Indian and Pacific Oceans. *Mon. Wea. Rev.* **102**: 745-753.

(57) Ramage, C. S. (1974). The typhoons of October 1970 in the South China Sea: intensification, decay and ocean interaction. *J. Appl. Meteor.* **13:** 739-751.
(58) Riehl, H. and Shafer, R. J. (1944). The recurvature of tropical storms. *J. Meteor.* **1:** 42-54.
(59) Riehl, H. Ref. 70, Chapter 8.
(60) Riehl, H. (1948). On the formation of typhoons. *J. Meteor.* **5:** 247-264.
(61) Riehl, H. and Burgner, N. M. (1950). Further studies of the movement and formation of hurricanes and their forecasting. *Bull. Amer. Meteor. Soc.* **31:** 244-253.
(62) Riehl, H. (1950). A model of hurricane formation. *J. Appl. Phys.* **21:** 917-925.
(63) Riehl, H. (1956). Sea surface temperature anomalies and hurricanes. *Bull. Amer. Meteor. Soc.* **37:** 413-417.
(64) Riehl, H. (1959). On production of kinetic energy from condensation heating. *In* "The Atmosphere and the Sea in Motion." (B. Bolin, Ed.) pp. 381-399. Rockefeller Institute Press.
(65) Riehl, H. and Malkus, J. S. Ref. 60, Chapter 9.
(66) Riehl, H. (1972). Intensity of recurved typhoons. *J. Appl. Meteor.* **11:** 613-615.
(67) Rossby, C.-G. (1948). On displacements and intensity changes of atmospheric vortices. *J. Marine Res.* **7:** 175-187.
(68) Rossby, C.-G. (1949). On a mechanism for the release of potential energy in the atmosphere. *J. Meteor.* **6:** 163-180.
(69) Sadler, J. C. (1976). Typhoon Intensity Changes in the Pacific Stormfury Area. Dept. Meteor. Univ. Hawaii. 13 pp.
(70) Sanders, F. and Burpee, R. W. (1968). Experiments in barotropic hurricane track forecasting. *J. Appl. Meteor.* **7:** 313-323.
(71) Sawyer, J. S. (1947). Notes on the theory of tropical cyclones. *Quart. J. Roy. Meteor. Soc.* **73:** 101-126.
(72) Sekioda, M. (1970). On the behavior of cloud patterns as seen on satellite photographs in the transformation of a typhoon into an extratropical cyclone. *J. Meteor. Soc. Japan* **48:** 224-234.
(73) Stone, R. G. (1940). A note on the distortion of the hurricane San Felipe (II) by the mountains of Puerto Rico. *Trans. Amer. Geoph. Union* **21:** 254-255.
(74) Smith, C. L., Zipser, E. J., Daggupaty, S. M. and Sapp, L. (1975). An experiment in tropical mesoscale analysis. *Mon Wea. Rev.* Part I. **103:** 878-892. Part II. **103:** 893-903.
(75) Tisdale, C. F. and Clapp, P. F. (1963). Origin and paths of hurricanes and tropical storms related to certain physical parameters at the air-sea interface. *J. Appl. Meteor.* **2:** 358-367.
(76) Yanai, M. (1961). Dynamical aspects of typhoon formation. *J. Meteor. Soc. Japan* **39:** 282-309.
(77) Yanai, M. (1961). A detailed analysis of typhoon formation. *J. Meteor. Soc. Japan* **31:** 187-214.
(78) Yanai, M. Ref. 90, Chapter 8.
(79) Yeh, T. C. (1950). The motion of tropical storms under the influence of a superimposed southerly current. *J. Meteor.* **7:** 108-113.
(80) Tannehill, I. R. (1956). "Hurricanes." Princeton Univ. Press, 308 pp.
(81) Dunn, G. E. and Miller, B. I. (1960). "Atlantic Hurricanes." Louisiana State Univ. Press, Baton Rouge, La. 326 pp.

11

Numerical Hurricane Prediction

by
Ferdinand Baer

Introduction

One of the most significant events to affect meteorology was the development of the high-speed digital computer. It was the good fortune of meteorology to have the inventor of the computer, J. von Neumann, apply his interest to the problem of weather forecasting and efforts to utilize this new tool started in the late 1940s (7) Since that time the literature has mushroomed with applications based on computer usage. Although the pilot efforts at prediction were focused on the middle latitudes—the area where the greatest numbers of people are affected by weather—it was not long before all aspects of global weather came under computer study.

The practical benefits inherent in utilizing computers for predicting weather are undeniable. With its ability to handle large volumes of data and make formidable calculations quickly, the computer can give the meteorologist guidance on preparing forecasts for public information. The presumption is, of course, that such guidance material is both useful and adequate. To reach this stage, substantial research was, and continues to be, necessary. Thus, we come to the second function of the computer in meteorology: to provide an understanding of the behaviour of the atmosphere as a fluid in motion over the earth. To achieve reliable forecast capability, substantial work is essential and indeed sizeable amounts of scientific expertise have been applied.

Studies of the tropics utilizing computers were inevitable and have become commonplace. Initial hesitation was, perhaps, based on lack of data, since initial values play a prominent role in numerical forecasting, as well as the recognition that some influential physical factors in tropical motions are difficult to treat. These latter factors include the effect of organized convection, the relative importance of divergent motion in generally weak flow, and the impact of air-sea exchange. Nevertheless, given the overwhelming size of the tropics, their interaction with mid-latitude flow and their general long-term effects on the overall fluid behaviour soon brought them into the realm of computer study.

Inevitably interest focused on the tropical hurricane, the most prominent event in the tropics, which not only has strong social implications due to its intensity and potential for damage, but which creates the most scientific curiosity because of its complex physical structure and evolution. Attention in numerical forecasting has focused on the development of the hurricane from a weak vortex through its mature stage, testing theories which are particularly applicable to the tropical vortex. It should be pointed out that hurricane development is only part of the story being studied by computer application. The evolution of the depression from an easterly wave is not yet understood. Hurricane movement and its environmental interaction have been studied, but are not as well understood as the development problem; nor are numerical forecasts as yet reliable. Clearly, since models of the hurricane are inadequate to give repeatedly successful forecasts for given conditions, the possibility of modifying them through effects in computer models simulating activities in the real atmosphere is premature. Some computations of this type have, nevertheless, been attempted.

Although the hurricane has a dominant impact, it is not the only aspect of numerical prediction in the tropics. Large-scale models can forecast the propagation of easterly waves, and the general circulation, which knows no boundaries, may be rooted prominently within the tropical zone. Numerical models of the general circulation have been successful since the growth of computer capacity. Finally, meteorologists are becoming more aware of the impact which oceans, with their high heat capacities, have on climate. Since the largest area of ocean surface is in the tropics, comprehending the interaction between ocean and atmosphere becomes essential to both physical understanding and a modelling and prediction potential.

By what method does one approach the problem of numerical prediction with emphasis on tropical flow? There are at least two accepted procedures with a third mode reflecting their combination:

(1) *Dynamical*, a deductive procedure prognosticating from a

given state of the atmosphere by specifying non-linear partial differential equations and integrating those equations with time. Besides understanding the physics which explains the system, this procedure requires substantial approximation and computational skill together with powerful computers;

(2) *Inductive* procedure, generates prediction formulae from past observational information. Such formulae may or may not depend upon current data, but require an unavailable abundance of past records for forecast accuracy;

(3) *Dynamical-statistical* forecast methods, may use information from a dynamical model together with the inductive procedure.

Whereas the dynamical approach may provide information on the behaviour of the fluid system as well as yielding predictive information, the statistical approach is primarily oriented toward forecast skill. Lacking sufficient data to provide statistical predictions without large variance, it is evident why most research effort has gone into studying tropical forecasting via the dynamic route.

The success achieved in developing a tropical prediction method, frequently denoted as a "model", can clearly be established by comparing the model's output (or forecast) with the observations which have been outlined in the previous chapters. Thus, one may ask about evolution of thermal and flow fields as well as moisture distribution in the forecast domain and compare them to carefully analysed statistics. If concern is for a particular forecast period, model output must be compared to observations. Detailed observations are frequently not available in the tropics or, if they are, they do not have the required density to check predictive models. Thus, recourse to statistical or composite data must be available. This model-checking procedure is applicable to the methods introduced above.

This discussion will focus on the dynamical approach, consistent with the evolution of numerical prediction of tropical flow. The physical assumptions underlying this approach are embodied in the Navier-Stokes equations for fluid flow, an equation for fluid mass conservation, a thermodynamical statement usually expressed by defining entropy changes, a relation between the variables of state and conservation equations for mass elements embedded in the fluid which undergo independent variations, such as change of phase. These equations should form a consistent set, together with necessary boundary and initial conditions, to allow for a prediction of the future state of the fluid. Although much research effort expended on this problem has been focused on the hurricane, the prediction of tropical flow or the general circulation, it involves the same procedure if a dynamical approach is applied. Our capacity to deal with the approach as outlined above, as well as that which has actually been achieved, will be presented. Hopefully the historical evolution of our accumulated knowledge will become evident from what follows.

Dynamical method

The general equations necessary to carry out a forecast have just been outlined, and must match in number the dependent variables. For the most comprehensive problem, the dependent variables include two horizontal motion variables denoted by the vector $\mathbf{V}$, vertical motion, pressure, density, temperature and at least one moisture variable, although some sophisticated models include separately cloud moisture and rain water. The equations are applicable to general circulation modelling, hurricane modelling, vortex development, wave propagation and many other fluid flow applications. Unfortunately, specification of the equations does not provide an immediate solution and the equations must be approximated to suit the event for which they will be used. It is not feasible nor practical to contain all scales, both in time and space, within a single calculation. One cannot hope to predict the evolution of individual cumulus cells on the scale of one to five kilometres, with a lifetime of 30-60 minutes, together with the hurricane which encompasses several thousand kilometres and may endure for a fortnight. Thus actual prediction systems for each scale range may appear quite different from one another.

The general equations are as follows. If we let the three-dimensional velocity vector be given as $\overline{V} = \mathbf{V} + w\vec{k}$ with $\vec{k}$ unit vector normal to the horizontal surface, the motion equations become

$$\frac{d\overline{V}}{dt} + 2\overline{\Omega} \times \overline{V} = -\frac{1}{\varrho}\nabla p - \nabla \Phi + \overline{F}. \tag{11.1}$$

Here $\overline{\Omega}$ represents the earth's rotation vector and Φ is the geopotential. Clearly $\overline{F}$ incorporates the forces (internal) acting on the fluid and may be interpreted as the Reynolds' stresses for the given scale of interest.

Mass conservation is represented by the equation of continuity

$$\frac{d\varrho}{dt} = -\varrho \nabla \cdot \overline{V} \tag{11.2}$$

which, if combined with Eq. (11.1) yields the momentum form of the motion equations, a popular representation in the literature.

$$\frac{\partial \varrho \overline{V}}{\partial t} + \nabla \cdot \varrho \overline{V}\overline{V} + 2\overline{\Omega} \times \varrho \overline{V} = -\nabla p - \varrho \nabla \Phi + \varrho \overline{F}. \qquad (11.3)$$

For thermodynamics we state the first law in terms of the potential temperature θ which is related to entropy through the relation $S \equiv c_p \ln \theta$,

$$\frac{d\theta}{dt} = \frac{\theta}{c_p T} \frac{dQ}{dt} + F_\theta, \qquad (11.4)$$

where dQ/dt reflects nonadiabatic heating and F_θ incorporates the thermal stresses acting in the fluid.

Finally, we include a statement on water vapour conservation, noting that additional relations for water substance may be added if desired.

$$\frac{dq}{dt} = E - P + S_t + F_q, \qquad (11.5)$$

where again F_q incorporates the vapour stresses, and S_t represents the water storage in clouds or rain. It should be noted that we presume the moist atmosphere to behave as a perfect gas, such that

$$p = \varrho RT. \qquad (11.6)$$

Application of scale analysis allows both simplification and alteration of Eqs. (11.1) to (11.6). Perhaps the most frequently used result of scale analysis is the hydrostatic approximation. To a high degree of accuracy for scales on the order of hurricanes and larger (6, 43, A-6) the vertical pressure gradient is in balance with the geopotential gradient, i.e.

$$\frac{\partial p}{\partial z} = -\varrho \frac{\partial \Phi}{\partial z} = -\varrho g \qquad (11.7)$$

which makes the vertical velocity a diagnostic variable since Eq. (11.7) must replace the vertical equation in (11.1). It should be noted that if one wishes to include in a prediction model the effects of cumulus evolution on their own scale (approximately 1-5 km) Eq. (11.7) is not particularly appropriate and the equation for vertical acceleration must be maintained. This is precisely what has been done in a detailed account of hurricane development (59), wherein the explicit growth and decay of individual convective cells was allowed.

Utilization of the hydrostatic approximation allows for a coordinate transformation in the vertical such that pressure and geopotential reverse their roles as dependent and independent variables, respectively, with the transformation given by Eq. (11.7) (12). Of particular value from this transformation is the representation of continuity

$$\frac{\partial \omega}{\partial p} = - \nabla \cdot \mathbf{V}, \qquad (11.8)$$

with ω denoting the substantial change of pressure with time. For simplified boundary conditions, this equation removes the prognostic requirement on the mass field. A further transformation in frequent use is the σ-coordinate, where σ is defined as

$$\sigma \equiv \frac{p}{p_s} \quad \text{or} \quad \sigma \equiv \frac{p - p_t}{p_s - p_t}. \qquad (11.9)$$

Here p_t and p_s refer to the top pressure and surface pressure of the model. The convenience of this representation lies in the fact that normalized surfaces can be defined despite topographic intrusions. However, an equation must be specified for the prediction of surface pressure as it changes with time (42).

With the hydrostatic assumption and pressure coordinates, Eq. (11.1) may be represented by Eq. (11.7) and the two-dimensional motion equation,

$$\frac{\partial \mathbf{V}}{\partial t} + \mathbf{V} \cdot \nabla \mathbf{V} + \omega \frac{\partial \mathbf{V}}{\partial p} + f \vec{k} \times \mathbf{V} = - \nabla \Phi + \overline{F}_h + \overline{F}_p, \qquad (11.10)$$

where $\overline{F}_h$ and $\overline{F}_p$ are the two-dimensional stress vectors representing the horizontal and vertical frictional forces, respectively. The velocity vector may now be transformed into a rotational and divergent part. Thus,

$$\mathbf{V} = \vec{k} \times \nabla \psi + \nabla \chi, \qquad (11.11)$$

where ψ represents a stream function (rotation) and χ represents a velocity potential (divergence). Prediction equations in these two variables may be developed by taking the vertical component of the curl and the divergence of Eq. (11.10) respectively. We introduce this procedure because it easily incorporates the quasi-geostrophic approximation, an assumption which states that, since divergence is sufficiently small, its time variation may be neglected. The resultant divergence equation yields a balance (boundary value problem) between the fields of vorticity (stream function), divergence and geopotential. Justification for this approximation has been given in terms of scale analysis (6) and in terms of energy conservation (27). Equation (11.10) is denoted as the "primitive equation". The quasi-geostrophic equations are represented simply by a prediction equation for vorticity and a balance condition from the divergence equation. This system, the quasi-geostrophic system, has substantial historical precedence in numerical forecasting and many of the earlier hurricane development models used it. Since the assumption presupposes small divergence, it was essential that these models did not include forcing, which would stimulate divergent motion. For completeness we now write the quasi-geostrophic equations, neglecting all reference to divergence in the balance equation.

$$\frac{d\xi_a}{dt} + \omega \frac{\partial \xi_a}{\partial p} + \xi_a \nabla \cdot \backslash \mathbf{V} + \vec{k} \cdot \nabla \omega x \frac{\partial \backslash \mathbf{V}}{\partial p} = \vec{k} \cdot \nabla x (\bar{F}_h + \bar{F}_p)$$

$$- \nabla \cdot \xi_a \nabla \psi + \tfrac{1}{2} \nabla^2 (\nabla \psi)^2 = - \nabla^2 \Phi$$

$$\xi_a \equiv \nabla^2 \psi + f \quad ; \quad \frac{d}{dt} \equiv \frac{\partial}{\partial t} + \backslash \mathbf{V} \cdot \nabla . \tag{11.12}$$

Note that in Eq. (11.12) the laplacian operator is planar. Hurricane models based on these approximate equations may be found in (8, 22, 39, 41, 56, 57). Since gravitational instabilities are inhibited by the balance assumption, the effects of gravitational excitation which are inherent in moist convection must be incorporated into the model by parametric means. Convective parameterization has already been discussed in Chapter 4 but we shall consider it further.

Let us now consider the choice of horizontal coordinates in the surfaces of either pressure, σ, or geopotential, Φ. This choice clearly depends on the problem to be solved. Should we wish to calculate the general circulation over the entire globe, spherical polar coordinates (latitude/longitude) are clearly in order (51). If we wish to compute the flow evolution in the tropics over a large longitudinal domain, we could map the domain onto a Mercator projection and use Cartesian coordinates in the plane. This expansion has been used in many studies (35, 32, 20). Finally, when the hurricane development problem is considered, since the hurricane has such similarity to a vortex, cylindrical coordinates have been chosen. Under this expansion, the hurricane eye is placed (and usually remains) at the centre of the coordinate system with the radial coordinate increasing outward and the angular coordinate rotating about the axis at fixed radius. We will not enumerate all the models which use this system, since it is so common, but we shall note that most modellers presume the hurricane (or the corresponding vortex) to be axially symmetric. This somewhat unrealistic assumption not only allows the modeller to focus on dominant features which excite the hurricane, but allows him to analyse with considerable simplification. This is so because symmetry assumes independence from the axial coordinate, reducing the independent space variables to one in the horizontal surface. If we define that variable as r, the radial distance from the axis of rotation, a balanced set of symmetric equations for the hurricane may be derived from Eqs. (11.1) to (11.6) and appears as follows, noting that v_r represents the radial velocity and v_θ the tangential velocity

$$\frac{\partial v_\theta}{\partial t} + v_r\left(f + \frac{\partial v_\theta}{\partial r} + \frac{v_\theta}{r}\right) + \omega\frac{\partial v_\theta}{\partial p} = F_{th} + F_{tp}$$

$$v_\theta\left(f + \frac{v_\theta}{r}\right) = \frac{\partial \Phi}{\partial r}$$

$$\frac{\partial \theta}{\partial t} + v_r\frac{\partial \theta}{\partial r} + \omega\frac{\partial \theta}{\partial p} = \frac{\theta}{c_p T}\frac{dQ}{dt} + F_\theta$$

$$\frac{\partial q}{\partial t} + v_r\frac{\partial q}{\partial r} + \omega\frac{\partial q}{\partial p} = E - P + S_t + F_q$$

$$\frac{1}{r}\frac{\partial}{\partial r}(rv_r) + \frac{\partial \omega}{\partial p} = 0. \qquad (11.13)$$

Because the hurricane is normally found at low latitude, the Coriolis parameter (f) is frequently considered constant. This is a poor assumption since it is just at low latitudes that the percentage change of the Coriolis parameter with latitude becomes very large. An attempt to assess the effect of variation in f by calculating in the β-plane has been made (29).

Having now established the equations which form a consistent set for determining the future state of a model atmosphere, we must determine how the model is forced by stresses and heating, as well as prescribing boundary conditions, which may also involve forcing. We shall begin by outlining ways in which forces may be represented, highlighting the variety of assumptions made in different models.

Internal stress

Many of the models developed to predict tropical flow and/or hurricanes incorporate some form of internal stress. These terms evolve from the Reynolds' stresses due to Reynolds' averaging of the basic equations. We have represented these terms by the general symbol $\bar{F}$ in Eqs. (11.1), (11.4) and (11.5). In all cases K-theory (cf., Eq. (4.8)) has been applied in one form or another. This theory derives from analogy with molecular motion so that the subgrid scale correlations are related to the large scale by a functional coefficient, generally denoted as an eddy diffusion coefficient. We shall show several of the possible ways in which these terms may be referred to the prognostic variables. It should be noted that vertical fluxes, either at the surface or in the boundary layer, are essential to all models so that momentum, heat and moisture may be brought into the atmospheric system from the underlying surface. However, in the earlier, balanced models (9, 39) internal diffusion was not significant. Because of their balanced nature and the limitations this imposes on the growth and propagation of high frequency gravity waves, the scale-dependent damping of these diffusion terms was not required for stable predictions. Interestingly, for the earliest hurricane development models involving the primitive equations (18), the dissipative character of the stresses (although included) was insufficient to prevent gravitational instability on the smallest resolvable scales.

The stresses may be broken up into two parts; one in the horizontal surface denoted into a subscript h and one out of the surface (in the direction of increasing geopotential) denoted by the subscript p. Clearly the three dimensional vector $\bar{F}$ will have a vector component in the surface and a scalar component normal to the surface. Since most models incorporate the hydrostatic assumption, we shall not discuss the latter stress. The two-dimensional stress vector $\overline{F}$ as well as those stresses for heat and moisture are then given as,

$$\overline{F} = \overline{F}_h + \overline{F}_p$$

$$F_\theta = F_{\theta h} + F_{\theta p}$$

$$F_q = F_{qh} + F_{qp}, \tag{11.14}$$

where

$$\overline{F}_h = F_{hr}\vec{r} + F_{h\lambda}\vec{\lambda} = F_{rx}\vec{i} + F_{hy}\vec{j}$$

$$\overline{F}_p = F_{pr}\vec{r} + F_{p\lambda}\vec{\lambda} = F_{px}\vec{i} + F_{py}\vec{j}. \tag{11.15}$$

In the latter set we have expanded the horizontal vectors in both cylindrical coordinates (utilized in symmetric vortex models) as well as Cartesian coordinates utilized in the Mercator projection. Following K-theory,

$$\overline{F}_h = \nabla \cdot (K_h \nabla \backslash \mathbf{V})$$

$$\overline{F}_p = g \frac{\partial \overline{\tau}}{\partial p}, \tag{11.16}$$

where ∇ is a surface operator and $\overline{\tau}$ is the vertical shearing stress vector. We shall defer discussion of surface stress to the section on boundary conditions. By analogy the expressions for heat and moisture diffusion are,

$$F_{\theta h} = \nabla \cdot K_\theta \nabla \theta \quad ; \quad F_{qh} = \nabla \cdot K_q \nabla q$$

$$F_{\theta p} = \frac{\theta}{c_p T} \frac{\partial m_\theta}{\partial p} \quad ; \quad F_{qp} = \frac{\partial m_q}{\partial p}$$

$$m_\theta \equiv K_{\theta p} \frac{\partial \theta}{\partial p} \qquad m_q \equiv K_{qp} \frac{\partial q}{\partial p}. \tag{11.17}$$

The eddy diffusion coefficients listed in Eqs. (11.16) and (11.17) take on a variety of values in the many models which have been proposed.

There is, however, one more coefficient which is related to $\bar{\tau}$ and defined

$$\bar{\tau} \equiv K_M \frac{\partial \mathbf{V}}{\partial p}, \text{ (see Eq. (4.8))} \tag{11.18}$$

The simplest approximation would be to make all coefficients constant and equal. Indeed, in some cases the constant has been chosen equal to zero. At the other extreme, the coefficients have been chosen as non-linear functions of the prognostic variables. Most studies of these coefficients have been focused on the momentum diffusion coefficient, K_h. That the coefficient depends on the flow deformation *(D)* as well as the smallest scale increment in the model (Δs) has been conjectured (55). Thus,

$$K_h = a + b (\Delta s)^2 |D|, \tag{11.19}$$

where in *Cartesian coordinates* the deformation would be written as

$$D^2 = \left(\frac{\partial u}{\partial x} - \frac{\partial v}{\partial y}\right)^2 + \left(\frac{\partial v}{\partial x} + \frac{\partial u}{\partial y}\right)^2. \tag{11.20}$$

This formulation has been used in (2, 50). An alternate and simpler linear form was adopted (3) as

$$K_h = C_1 |\mathbf{V}| + C_2. \tag{11.21}$$

The horizontal coefficients for heat and moisture, K_θ and K_q are often set equal to the momentum coefficient K_h for lack of a better choice. However, the relationship between these coefficients suggest that their proportionality depends on the flux Richardson number (25).

The variety of representations for the vertical flux coefficient utilized in models and defined by Eq. (11.18) is as broad as that for K_h. The Rossby-Montgomery approach (55) suggests that this coefficient be dependent on mixing length l and the vertical wind shear. The mixing length may change with height, and in many cases is rapidly reduced to zero outside the boundary layer. In functional form, we have

$$K_M = l^2 (p) \left|\frac{\partial \mathbf{V}}{\partial p}\right|. \tag{11.22}$$

Mixing length has been used as a function of some vertical structure function $\mu(z)$ as well as the density *(1)*. More commonly *l* is established as a linearly decreasing function of the increasing vertical coordinate. Because this coefficient may be stability dependent, a stability factor has been added (50) to a formula comparable to Eq. (11.22), giving representations both to stable and unstable configurations. This stability parameter is characterized by the ratio of the vertical shear of virtual equivalent potential temperature to the magnitude of vertical wind shear. The corresponding vertical coefficients for heat and moisture may or may not be directly related to this momentum coefficient.

Heating

Reference to Eq. (11.4) will indicate that non-adiabatic heating of the atmospheric system has been represented by a term labeled dQ/dt. This term should include radiative heating as well as other sources. Those other sources are dominated by convection, both wet and dry. Indeed it is found in the tropics that convective heating is so pronounced that very few models include radiative heating and cooling. We shall consequently not discuss this heating mechanism although the interested reader will find a comprehensive treatment of how such heating may be incorporated into a model in (53).

The evidence of convection, as indicated in Chapter 4, is apparent in cumulus activity, and that activity is on a small horizontal scale (1-5 km). If our models were able to stretch over an unlimited scale range, the cumulus clouds could be modelled and forecast as part of the physical system, and their heating effects would impinge directly on the large scale. Since this is at present not generally feasible, convective heating in the thermodynamic equation is somewhat akin to the Reynolds' stresses in the momentum equations, representing the bulk characteristics of small-scale thermal processes. Whereas the parameterization of momentum diffusion is characterized by K-theory, convective parameterization is somewhat more involved insofar as the stability characteristics are more overwhelming. We have outlined parameterization schemes in Chapter 4. We shall here indicate how they may be incorporated into an atmospheric model.

In the earliest calculations involving developing hurricane models (18, 58) the devastating effect of convective instability on model prediction was overlooked and the consequences of latent heat release was included in the models in the form,

$$\frac{dQ}{dt} = -L\omega \frac{\partial q_s}{\partial p} \quad \text{if } \omega < 0. \tag{11.23}$$

This procedure provided thermal energy to the system but did not guarantee gravitational stability. Since the unstably-rising particles are confined to the convective scale, the instability was manifest on the smallest resolvable model scale and rapidly eroded the calculation. Linear stability analysis supported these results.

Although convection takes place on the small scale of the cumulus, it is apparent that when many such cells go through their cycle in a contiguous domain, the atmosphere organizes this convection in some sense and makes it appear as a force on a larger scale; at least this is an undeniable observation from developing hurricanes. This observation and the need for a parameterization scheme on the large scale led to the development of the CISK (Conditional Instability of the Second Kind) hypothesis (40, 9). By this mechanism, frictional convergence in the boundary layer delivers the moisture flux which is carried aloft and provides the necessary release of latent heat to warm the atmosphere above the boundary layer. The vertical distribution of this heating is adjusted so that gravitational instability is avoided, but energy input effects are maintained and on a scale considerably larger than the cumulus. In its simplest form, the heat released is proportional to the vertical velocity at the top of the boundary layer, the convergence of mass into the column, the fraction of the column which may be in cloud, and some measure of the saturated moisture variable, usually its value at cloud base. As originally prescribed (9),

$$\frac{dQ}{dt} = \frac{gL}{q_{sc}} \frac{dq_s}{dp} I. \tag{11.24}$$

where I represents the convergence of moisture into an atmospheric column and clearly incorporates the boundary layer vertical velocity (ω_{bl}), q_s is the saturated specific humidity and q_{sc} its value in the top of the friction layer. The vertical velocity was explicitly incorporated (39) as well as a structure parameter η describing how heat was vertically distributed, by the formula

$$\frac{dQ}{dt} = \eta \frac{\partial \bar{\theta}}{\partial p} \omega_{bl}. \tag{11.25}$$

A particular defect of these schemes was their inability to terminate the heating process so long as moisture flux continued. Since hurricanes do reach a steady-state level before decaying, some mechanism was required in the models to simulate this process. It was hoped, of course, that the simulation process would also characterize nature. The problem was successfully modelled (22) by showing that entrainment was going on at all levels of the cloud; and that mixing with the environment was gradually bringing the environment into thermal equilibrium with the cloud structure. Once equilibrium took place no exchange of heat was possible and the efficiency of convection was depleted. This process was introduced into the modelling relations by making heating proportional to the difference in cloud temperature and environmental temperature (22). Thus, in addition to the factors of moisture flux and fractional rate of cloud mass in the rising column, the heating was given as

$$\frac{dQ}{dt} \propto (T_{\text{environment}} - T_{\text{cloud}}), \qquad (11.26)$$

(see also Chapter 4, Eq. (4.28)). This form of parameterization became very popular and has been used with many variants since its development. The procedure has been modified (23) to include shallow convection as well as the deep convection. A still more complex theory has been developed (5) wherein a large population of clouds all detraining at different levels has been described. Further, a scheme, whereby an arbitrary cloud model may be used to supply energy in a parametric sense for the heating function dQ/dt has been proposed (2).

A parallel view to convective parameterization, evolving not from a need to organize convection on the hurricane scale, but still intended to avoid gravitational instability has been presented (30). The scheme is applicable both to wet and dry convection but there is clearly no moisture release in dry convection. The process is difficult to spell out in equation form since it involves testing for lapse conditions. During a prediction with the model equations chosen, some form of Eqs. (11.1) to (11.6), the lapse rate is sensed. If it exceeds the critical (dry-adiabatic for dry and moist-adiabatic for saturated) the instability is converted to heat energy and latent energy, if available. The adjustment is made at all necessary levels but the thermal adjustment of the entire column is constrained such that potential energy of the column remains unchanged. In this way gravitational instability is avoided and heat is carried upward.

Boundary conditions

A variety of boundary conditions may be found in tropical modelling literature, each model having its own peculiarities; however, the variations are all based on some fundamental assumptions. The interested reader can consult various models as represented in the literature to savour the individuality of the modellers.

The boundary conditions to be specified depend on the choice of domain and the relevant differential equations to be solved. We may say that the dependent variables and possibly their derivatives must be specified on the lateral boundaries at all levels, as well as the top and bottom of the model over the entire surface. Let us consider first the boundary conditions at the top of the atmosphere. Most models will be represented to some fixed pressure $p_{top} \leqslant 100$ mb, where p_{top} may go to zero. For the condition at this level, almost all models impose no mass flux through the surface such that, in pressure coordinates, $\omega_{top} = 0$. This condition has the advantage that gravitational waves are inhibited from travelling along the surface. On the other hand, some studies suggest that the condition allows for trapping of vertically propagating energy by setting up standing modes in the fluid. If these modes are forced by physical processes, resonance may ensue, setting up instabilities which may not have a real physical basis. Since this phenomenon has not been diagnosed in the experimental models under review (although it may well exist), modellers have not included an absorption layer at the top of the fluid. Together with the flux condition on ω, most models state that the temperature at the top remains fixed, as well as the moisture variable which is frequently set to zero. For the horizontal flow variable, the velocity and its gradient may either be made to vanish or some specified flow along the surface may be provided. These conditions must hold for all time.

Lateral boundary conditions depend crucially on the nature of the domain. For a general circulation type model (GCM) which is hemispheric or global, one needs only periodicity in longitude and regularity at the poles. A statement of symmetry or antisymmetry would be adequate at the equator for a hemispheric model. On the other hand for a regional domain, fluxes, flow and physical conditions must be specified on the boundaries. Two possibilities exist and are exploited. For the hurricane model (symmetric or antisymmetric), the boundary conditions are set for all time. For regional models which may or may not contain hurricanes, boundary conditions may be calculated as a function of time from an additional model, which includes the model for which boundary conditions are sought as a subset. This latter procedure is characterized by model nesting.

Let us consider the hurricane vortex problem as an example for setting fixed boundary conditions. In this model, represented by Eq. (11.13), the outer boundary is defined by some radius, r_c. One may now define an open or closed system. For a closed system, no mass exchange occurs across the interface at r_c. Thus, this boundary acts as a rigid wall and the normal velocity $v_r(r_c) = 0$. If, furthermore, we do not want any divergence at this surface, the gradient of momentum in the radial direction must also vanish, i.e. $\partial/\partial r \, (rv_r) = 0$ at $r = r_c$. Finally, the tangential velocity (v_θ) must be specified, as well as the temperature and moisture. It has been found that if r_c is not substantially far from the centre of the vortex, say $r_c < 1500$ km, these fixed conditions may play a significant role in the vortex evolution.

Open conditions allow for mass flux into the region, but the radial velocity must be defined. One could, as has been done, presume some gradual inflow into lower layers and outflow aloft. It has been found, however, that outflow can and should be calculated from trajectories inside the model domain. Again for open conditions, temperature and moisture must be established although they may also be advected out in regions of outflow and specified from internally predicted values.

For the vortex model, symmetric or not, conditions must also be specified at the axis, $r = 0$. For continuity and simplicity, most models use the conditions,

$$v_r = v_\theta = \frac{\partial T}{\partial r} = \frac{\partial q}{\partial r} = 0$$

at $r = 0$. Except for the bottom surface, the conditions so far specified close the system.

We come now to the conditions at the earth's surface. These conditions are complicated by the fact that there is exchange of momentum with the solid earth (or ocean) as well as transfer of heat and moisture from both the earth and ocean. Since the ocean represents a changing fluid as complex as the atmosphere, one should in principle have an interactive interface between two fluids whose properties are continually changing. Fortunately, oceanic changes occur on time scales much larger than those of interest to tropical modellers, so that the properties of the ocean may be held fixed (as are those of the solid earth) during the prediction period of any model considered.

Specification of moisture flux into the atmosphere from the ocean, if oceanic temperatures exceed 27°C, has been shown in previous chapters to be crucial to vortex evolution, so that great care must be

NUMERICAL HURRICANE PREDICTION 513

taken in the representation of surface boundary conditions. The atmospheric boundary layer is a very complex region, and detailed prediction of its properties applicable to numerical prediction models has been discussed (11). However, this level of sophistication has not penetrated tropical models. Indeed the closest approach to boundary layer theory as applied to the hurricane development problem is to be found in (25), which considers the boundary layer from a Monin-Obukhov framework, defining friction velocity, friction potential temperature and a frictional moisture variable all based on specification of surface conditions.

The mass flux at the surface is clearly zero except for moisture, and thus most models use the condition $w = dz/dt = 0$ at $z = 0$. This is frequently translated to $\omega = 0$ at surface pressure. This latter assumption has the advantage of inhibiting surface gravity waves, but it is a very bad assumption in a vortex where the pressure gradient can become as large as 1 mb/km—then $\omega_0 \to 1$ mb/min. In conjunction with this condition, the bulk properties of the boundary layer are calculated in models by specifying the other fluxes from the surface. Some of the early balanced vortex models (9,39) ignored the details of the changing boundary layer and simply specified the vertical velocity at the top of the boundary layer as the required condition for modelling in the free atmosphere. This vertical velocity was established by frictional convergence in the boundary layer and calculated on the assumption of "Ekman pumping" (8); it may be defined as

$$\omega_0 = \alpha \left(\frac{K_M}{f}\right)^{\frac{1}{2}} \zeta_{gs}$$

where α is a factor based on surface cross isobaric flow, K_M is the diffusion coefficient discussed earlier, and ζ_{gs} is the geostrophic vorticity at the bottom of the model. This boundary condition clearly allows for mass and moisture flux into the model and is used in the convective parameterization considered earlier.

Most models which include the boundary layer specify fluxes at the model's surface. These fluxes, representing momentum, heat and moisture respectively (see Eqs. (11.17) and (11.18)) are made to depend on the drag coefficient C_D, the magnitude of the surface wind $|\mathbf{V}_s|$ and characteristic values of the surface and air respectively. Thus, we find, in general,

$$\overline{\tau}_s = C_D |\mathbf{V}_s| \mathbf{V}_s$$

$$m_{\theta s} = -C_D |\mathbf{V}_s|(T_w - T_{as})$$

$$m_{qs} = -C_D |\mathbf{V}_s|(q_w - q_{as}). \qquad (11.27)$$
$$(4.9\text{-}4.11)$$

In these relations, V_s, T_{as} and q_{as} represent surface (or near surface) model values which are predicted, and variables denoted by subscript w correspond to sea-surface values. Characteristic values used in models are $T_w = 27°C$ and $q_w = 23\text{-}24$ g/kg^{-1}. In some models the surface moisture and heat fluxes are specified for all time, so that q_{as} and/or T_{as} need not be forecast to establish the values of m_{qs} and $m_{\theta s}$. How these quantities are propagated up from the surface, given their values from Eq. (11.27), has been discussed under the heading of internal stresses.

Although most models use a constant surface drag with the characteristic value $C_D = 3 \times 10^{-3}$, some variation has been attempted (see also Chapter 4). The coefficient has been made proportional to the surface wind (Eq. 9.11), (A-5). In another study (50) more boundary layer detail was incorporated, utilizing the logarithmic layer such that

$$\left(\frac{C_D(z)}{C_{D10}}\right)^{\frac{1}{2}} = \frac{k_0}{(C_{D10})^{\frac{1}{2}} \ln\left(\frac{z}{z_0}\right) + k_0},$$

where the drag coefficient is strongly height dependent. Furthermore, C_{D10} represents the coefficient at ten metres, which depends linearly on the wind at that level through a formula similar to that of (35). These subtleties probably do not play a major role in the prediction capability.

Numerics

Having a consistent set of equations with suitable boundary conditions does not automatically provide a solution. Unfortunately, the system with which we are dealing is very complicated, involving non-linear, partial differential equations for which known analytic solutions do not exist. It is, therefore, essential to reduce the system of equations to computational form, thus allowing us with the aid of a computer to produce a forecast.

Several procedures have been devised to aid in this transformation. Perhaps the most common method is denoted as the finite difference procedures. This technique, heavily used in tropical modelling, requires transformation of the continuous geometric domain over which the given equations describe physical events to a network of points, called grid points. These points, which may or may not be equally spaced over the domain, replace the continuum as the location of values for the dependent variables; for the forecast problem these include the flow variables, pressure, density, temperature and moisture variables. Thus, we may conceive of each pressure surface as containing an $N \times M$ network of points, with P such surfaces. This yields a domain of $N \times M \times P$ points (clearly finite) by which the domain is defined. If for descriptive purposes we represent the domain in Cartesian coordinates (not essential or even generally done) any point in the domain will have its location given as $(x,y,p) \rightarrow (i\Delta x, j\Delta y, k\Delta p)$ where $1 \leq i \leq N$, $1 \leq j \leq M$, $1 \leq k \leq P$. If the differential operators in the prediction equations are now defined on this set of points, and the time domain is converted to a finite set of points in time, the entire system is converted to a "finite difference" system. In general, one gets a very large matrix system (the dependent variables at the grid points being the matrix elements) to extrapolate by some suitable method into the future.

Other means for solving non-linear systems are included under expansion procedures. The spectral method involves an expansion of the dependent variables in a truncated set of known space dependent functions with unknown, but predictable, time dependent coefficients. These functions, when introduced into the prediction set Eqs. (11.1) to (11.6), will yield a set of ordinary non-linear equations in time alone, which may be forecast by finite difference means. This procedure has had some success for general circulation forecasting wherein the expansion functions are known as surface spherical harmonics; the approach has been reviewed (28). Another development which may prove useful in solving difficult systems is the finite element method wherein once again a grid of points is chosen to represent the domain, but local functions (local to each point) are defined from which derivatives may be taken. Application of the procedure to meteorological prediction may be found in (10).

We shall focus attention on the finite difference approach for tropical modelling. Although the transformation to finite difference equations formally allows for a solution, we must assure that the solution contains some measure of reality. Thus, we might expect that our choice of finite-differencing will provide a consistent, stable and convergent solution. If the equations are not consistent, the process is meaningless. If the calculations are not stable, the solution will eventually grow without bound and become meaningless. If the solutions to the finite difference equations do not converge to the solutions for the continuous problem, as the mesh of points increases by decreasing their distance from one another, the solutions will not be applicable to the problem in question. Thus, great care must be exercised in the finite difference procedure. These issues have been studied (44), but we cannot treat them in detail.

Nevertheless, it should be pointed out that errors must arise in converting from continuous to finite difference systems. This is evident from the fact that the solution will not be known at all points, but simply on a countable subset of points. Since in the interval between any two points at which a solution is known there exists an infinity of undefined points, the error problem is substantial. Furthermore, all studies of these problems indicate that the most serious errors occur on the smallest scale. By this we mean that any quantity which varies from point to point will have its greatest error over its shortest variation. If we define the interval between points along one space axis as Δx, this error in any quantity, say $T(x)$, will be in $T[(i + 1)\Delta x] - T(i\Delta x)$. Moreover, because of the non-linear transfer of amplitude from one scale to another, and because the limiting scale cannot be smaller than $2\Delta x$ in the finite difference system, "aliasing" causes its principal pile-up of error on the smallest ($2\Delta x$) scale. We must bear these error sources in mind.

A large number of finite-difference schemes have been devised to represent differential operators, each with its own properties. For application to the meteorological forecast problem, these devices have been reviewed in (19, 33). Two schemes have been selected as most popular for hurricane models, although they have also been used in the more general tropical prediction models. For convenience of reference, let us define a few of these operators in their most simple terms.

1. Centred difference in space (not cylindrical coordinates):

$$\frac{\partial u(x)}{\partial x} = \frac{u[(i + 1)\Delta x] - u[(i - 1)\Delta x]}{2\Delta x} . \qquad (11.29)$$

2. Upstream advective difference:

$$v\frac{\partial u}{\partial x} = \frac{|v(i\Delta x)|}{\Delta x}\left\{u(i\Delta x) - u[(i-1)\Delta x]\right\} \text{ if } v>0$$

$$= \frac{|v(i\Delta x)|}{\Delta x}\left\{u(i\Delta x) - u[(i+1)\Delta x]\right\} \text{ if } v<0. \qquad (11.30)$$

3. Forward difference in time:

$$u[(n+1)\Delta t] = u(n\Delta t) + \Delta t\, F(n\Delta t). \qquad (11.31)$$

4. Centred difference in time (leap-frog):

$$u[(n+1)\Delta t] = u[(n-1)\Delta t] + 2\Delta t\, F(n\Delta t). \qquad (11.32)$$

In Eqs. (11.29) and (11.30), Δx represents the distance between any two adjacent points in the domain, and may represent any coordinate direction. Although the function (u) may clearly depend on all three spatial coordinates and time, this reference has been suppressed for convenience. Similarly, in Eqs. (11.31) and (11.32), the quantity, u, is also a function of the space variables. Moreover, F is a non-linear function of the dependent variables, evaluated at the time $t = n\Delta t$, where Δt is the smallest increment of time at which a new value may be extrapolated. Finally, u may represent any of the dependent variables of the system Eqs. (11.1) to (11.6) to be forecast.

Historically, Eq. (11.29) in space has been combined with Eq. (11.32) in time to yield forecast equations which have been tested. The stability and truncation properties of the schemes are tested by applying the finite difference operators to the differential system, then linearizing and solving the linear equations. Since the linear differential equations also have known solutions, the solutions to the finite-difference equations may be compared to them. By this procedure, errors of the approximation have been defined. Based upon this procedure we note that small scales have the largest errors. Analysis of the leap-frog scheme (Eq. (11.32)) shows that the difference equation is of higher order than the differential equation, leading to computational solutions, which oscillate with period of $2\Delta t$. The leap-frog scheme is amplitude conserving (for the physical mode) but has serious phase error problems. Because of its non-damping character, small-scale errors may become overwhelming. This can be offset in the system because the diffusion terms are themselves dissipative, with their largest damping on the smallest scales.

Hurricane modellers have found the combination of Eqs. (11.30) and (11.31) to be more suitable to their prediction models. This system, upstream advective and forward in time, proves on linear analysis to be stable and damping the shortest scales. Thus, effects such as "aliasing", which amplify the short scales, are contained by this damping. Non-advection type terms in the equations are treated by centred differencing. Alternative schemes of greater complexity and sophistication have also been utilized. These include the energy conserving schemes (4) and the Kuri-box method (24). There is no clear evidence that one method is outstandingly superior to another, provided stability requirements are imposed in the calculations.

Of particular interest is the variety and originality exercised by modellers in reducing their domain to grid form. For the vertical domain, we find σ surfaces, pressure surfaces, geopotential surfaces and homogeneous layers. In some cases the boundary layer is simply specified, in others numerous surfaces on which predictions are made penetrate it. Some models have equally spaced surfaces in the vertical, others choose a more dense representation in the lower atmosphere, with increased spacing aloft. Thus, we find that Ooyama (41) requires only a few homogeneous layers, whereas Yamasaki (60) has 25 layers extending to 20 km with a minimum spacing of 600 m below 9 km. Between these extremes one finds models ranging from two levels to 13 layers.

The horizontal structuring is equally as varied. Consider first the symmetric hurricane model described by Eq. (11.13) which incorporates only one horizontal dimension. Modellers have solved this system over a variety of regions, from domains with r_c as small as 400 km to those with $r_c = 3400$ km. Experience indicates that the domain should exceed 1000 km, and this is indeed the range in which most modellers work. Their choice is often constrained by computer capacity, an important issue but not appropriate for discussion here. The minimum distance between points, here denoted as Δr, ranges from 400 m (60) to 30 km and more. Since many modellers feel that the small-scale action in the hurricane takes place in the vicinity of the vortex centre, many have also generated non-constant grids. For these grids, the distance between points expands as one goes outward from the axis of rotation. Considerable variety may be found in the process. For some models Δr is continually expanding as a function of r (22), whereas for others the change comes discretely (2). The most compelling argument for reduced grid size is based on the scale at which physical processes in the model act. Since convection acts on scales smaller than most models can resolve, convective parameterization has been utilized to transfer this effect to larger scales.

For three-dimensional models, gridding must be set in each two-dimensional surface. If only the hurricane is involved, a simulated circular region (25) or a rectangular region (3) may be chosen. Since these models represent the equatorial region, they are generally projected onto a Mercator map with Cartesian coordinates. Three-dimensional models, if sufficiently large in domain, may not only predict the hurricane development, but may also predict its movement, as well as its interaction with the environment. Indeed such a model could predict the evolution of the tropical environment with or without an incipient hurricane. Since the hurricane is the predominant tropical event, three-dimensional modellers have recognized the scaling requirements for its evolution as did two-dimensional modellers when they gradually reduced the grid increment to compensate for small-scale effects near the hurricane centre. When applied to three-dimensional models this procedure has led to a process called "grid nesting". The nesting process is defined as follows. An overall domain is chosen and represented by a grid with given Δx and Δy (say in Cartesian coordinates), yielding a matrix of $N \times M$ points. In some subdomain of this region, preferably overlying the hurricane, another grid is specified on which the increments are $\Delta x' = s\Delta x$, $\Delta y' = s\Delta y$ where s is fractional so that perhaps two or three points lie between the points of the coarse grid if s is a half or one-third. This procedure may be continued such that a third subgrid might lie inside the second with yet smaller grid spacing. Note the similarity of this approach to the gradually expanding grid in two-dimensional models.

The finite-difference equations which represent the model of interest are defined on each of the grids as chosen above and may be integrated on them. However, the fine scale subgrids must have their boundary conditions defined as a function of time from the grid of the next lower resolution. This must be done by interpolation, since a coarse grid will not have all the information needed by a finer grid. If each resolution system is integrated (predicted) separately, the only interaction of the grid systems will be from coarser to finer through boundary conditions. Such calculations have been performed. However, direct interaction between the grid systems is possible if at different times during the integration the local values on the finer grid are averaged and transferred in some fashion to a representative point in the local domain of the coarse grid. Experiments with this procedure may be found in (13, 32, 26). A final elaboration on the process has been pursued by Jones (16) wherein he defines conditions under which the fine grid moves in time within the confines of the coarse grid. By this mechanism the motion of the hurricane may be determined and the fine grid may be focused over the vortex domain.

Models of the large-scale tropical flow on the hemispheric scale utilize numerical processes of the type outlined above, which are similar to those applied in general circulation models with somewhat larger minimum resolution, i.e. large basic Δx, Δy. These models can see only the gross properties of hurricanes and their impact on the circulation.

It should be noted, finally, that the choice of Δt, the minimum increment over which the time extrapolation is made, depends on the choice of the space increments used, as well as the nature of the prediction equations. Since linear theory will predict the maximum value of Δt which will allow a computationally stable calculation, most modellers solve for the linear properties of their equations and utilize a value of Δt somewhat smaller than stability indicates. In most cases this choice works well for the non-linear equations, and for the many models discussed involving hurricane prediction, the value successfully used is close to $\Delta t \approx 60$ seconds. Some few models will allow a value slightly larger, whereas models with very fine space resolution may require $\Delta t \leq 10$ seconds. Experimentation (29) with the semi-implicit scheme (46) indicates allowance for a substantially larger time step, but the appropriateness of this procedure to hurricane prediction is not sufficiently documented.

Initial conditions

The various models described above are all designed for predicting the future state of the tropical atmosphere, be it the hurricane, equatorial wave or general circulation. However, since they represent an initial value problem, the predicted state depends upon the initial conditions. As the various models are designed for different purposes, initial conditions vary.

The symmetric vortex models are primarily designed to test model forcing and its ability to develop an incipient hurricane. Thus, they uniformly develop their calculations from theoretical, rather than observed, initial conditions, although these initial conditions do represent a developing hurricane. Although there is some variety in the choice of these conditions, the similarities among the models' initial conditions are apparent. Most utilize the mean hurricane season sounding data for vertical temperature and moisture distribution (17). They also begin with a vortex which is both radially and vertically structured. This vortex decreases upward and has its maximum near 900 mb in the range from 50 to 250 km from the axis of rotation, with the most frequently used radius approximately 200 km. At this radius the maximum tangential velocity ranges from 5 to 18 m/s. This velocity decreases exponentially both inward and outward from the radius of maximum. The pressure field is generated from the balanced (gradient) wind relation, and the meridional circulation is determined from the thermal wind relationship. These conditions are sufficient to initiate model calculations.

An interesting experiment (60) began with no motion, but with a temperature and moisture distribution. Since this model did not parameterize convection but was on a small enough scale to be nonhydrostatic, buoyancy perturbations were initially introduced at regular intervals to initiate convection and develop the vortex.

Three-dimensional models were designed for varying purposes and thus utilize a wide variety of initial conditions. Those models, which were to simulate hurricane development, but without the symmetry constraint, utilize initial conditions very similar to those for symmetric models as discussed above. Models (32, 35) were tested with observed data intending to forecast a real atmospheric situation. One was tested on data from two hurricanes and on a non-developing storm (35), while the other (32) utilized observations preceding the development of hurricane Isbell (1964).

It is evident that initialization difficulties with observed data arise in the tropics as they do in middle latitudes. Those models for which observations were used as initial conditions received substantially massaged conditions. Initialization problems are compounded in the tropics by lack of data as well as weaker balance conditions, although applicable techniques have been discussed and tested (20, 36). Despite these efforts, prediction results utilizing observed and adjusted initial conditions must be interpreted relative to the limitations and shortcomings of the initial data.

Steady-state models

The discussion in dynamical modelling tacitly assumed that predictions are derived by the solution of an initial value problem. Although most of the research in tropical prediction has followed this approach, much valuable information as to the nature of the steady hurricane can be deduced by solving the steady-state system. The steady-state assumption of ignoring time variations in the predictive equations presupposes weak time variations and/or yields the asymptotic solution to the time varying equations. One particularly interesting example of this technique (61), based on an earlier one-dimensional model (62), and which uses cylindrically symmetrical equations (two-dimensional) will now be described. Only the case with vertical eye wall will be considered (Fig. 9.24). Surface pressure is defined through the thermodynamic processes following the trajectory and hydrostatically through the vertical ascent given a top of the hurricane at 100 mb. In the steady, mature hurricane, the gradient thermal wind equation for the tangential component may be expected to be valid to a high approximation. This involves the vertical shear of v_θ; but this shear, in trajectory calculations, depends on the frictional transfer of momentum from air to ocean. We wish to determine the constraints, so that a comparable solution for wind and thermal fields can exist and permit a steady-state storm.

Wind structure

The applicable equations for horizontal motion in polar coordinates with axial symmetry are given by the first two of Eqs. (11.13). Noting that $v_r = dr/dt$ and $df/dt = \beta v$ where v represents the northward component of velocity, neglecting F_{th} and setting

$$F_{tp} = \frac{r}{\varrho} \frac{\partial \tau_{\theta z}}{\partial z}$$

(see 11.16), we may replace the first of (11.13) by a momentum conservation condition,

$$\frac{d}{dt} \Omega = \frac{\beta r^2 v}{z} + \frac{r}{\varrho} \frac{\partial \tau_{\theta z}}{\partial z} \tag{11.33}$$

$$\Omega \equiv v_\theta r + \frac{fr^2}{2}. \tag{11.34}$$

Note that Ω represents the component of angular momentum per unit mass about the hurricane's vertical axis, and will be conserved if no stress is acting and the variation of f is weak or negligible. Friction must be considered for the inflow when the air is in contact with the ocean; in the outflow internal friction may exist, but conservation of momentum will be assumed for the two-dimensional model. The first

NUMERICAL HURRICANE PREDICTION

term on the right of Eq. (11.33), which represents the sweep of northerly or southerly winds through the hurricane, has usually been neglected. Since β, although nearly constant, is very large in the tropics, such neglect forces one to restrain momentum analysis to the inner storm area, surely not exceeding the 3° radius. In the belt 10-20° latitude, for instance, the Coriolis parameter changes by a factor of two. Thus, extension of calculations to the 5° radius for a centre located at 15° can lead to major errors in momentum budgets and conclusions drawn therefrom (Chapter 9). Bearing this limitation in mind, we shall adapt the definition of Ω (Eq. (11.34)) within 300 km of the centre.

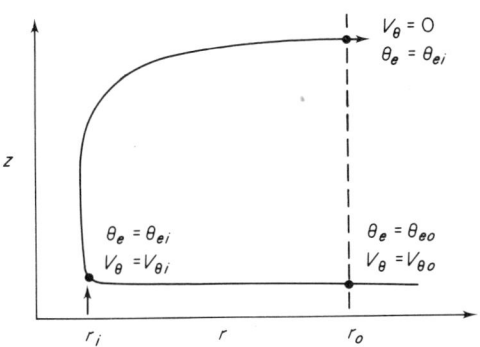

Fig. 11.1. Illustrating coordinate system and distribution of variables, for steady-state hurricane (61).

Concern will be limited to the interval between the centre and the radius r_o where, in the outflow, $v_\theta = 0$ (Fig. 11.1). The central hurricane intensity is to be specified from constraints at r_o. Integrating inward along the outlfow, it follows directly from the assumption of constant Ω that

$$\frac{fr_o^2}{2} = v_{\theta i} r_i + \frac{fr_i^2}{2}. \tag{11.35}$$

Usually, $fr_i \ll v_{\theta i}$, so that

$$r_o = \left(\frac{2}{f} v_{\theta i} r_i\right)^{1/2}. \tag{11.36}$$

For the inflow we use (11.33) with the above considerations,

$$\frac{d\Omega}{dt} = \frac{r}{\varrho} \frac{\partial \tau_{\theta z}}{\partial z} . \tag{11.37}$$

Integrating over a depth δz, at the top of which $\tau_{\theta z}$ is assumed to vanish,

$$\int \frac{d\Omega}{dt} \varrho \delta z = -r\tau_{\theta s} , \tag{11.38}$$

where $\tau_{\theta s}$ is the surface tangential stress component, considered evaluated at the nominal height of 30 m above the ocean surface where the ocean waves plus spray stop. Since the long history of hurricane observations the world over shows that from the ring of maximum wind outward a relation of the form $Vr^x =$ const applies, the outer part of a Rankine vortex, we adopt such a relation. Additional background is furnished by momentum and energy balances in hurricane Daisy and in model storms (62) indicating that frictional transfer of energy to the surface is independent of radius in concentric rings of increasing radius so that $\overline{v_\theta^3 r^2} =$ const and, for the symmetrical case, $\overline{V} r^{2/3} =$ constant. Alternatively one may assume that potential vorticity is conserved in the inflow requiring a distribution of surface stress

$$r\tau_{\theta s} = \text{constant} \tag{11.39}$$

This relation has been verified empirically in the inner hurricane area but not outside the 1-2° radius where $\tau_{\theta s} r^{1.5} =$ const is more applicable. With the stress definition of Eq. (4.9), $v_\theta^2 r =$ const or

$$v_\theta r^{1/2} = \text{constant} . \tag{11.40}$$

This form is preferable to that containing V since it can be used directly in the component equations. The exponent 0·5 represents a rough mean value of coefficients ranging from 0·3 to 0·7, the latter being valid only in asymmetrical storms with high maximum wind. From Eq. (11.40)

$$\zeta_{rel} = \frac{v_\theta}{r} + \frac{\partial v_\theta}{\partial r} = \frac{1}{2} \frac{v_\theta}{r} .$$

NUMERICAL HURRICANE PREDICTION

At the radius of maximum wind, r_1, the inner boundary conditions demand that $v_{\theta i}$ must be the same for inflow and outflow. Evaluating the constant in Eq. (11.40) at $v_\theta = v_{\theta i}$, $r = r_i$, and combining with (11.36), the radius of innermost penetration is given by

$$r_i = \frac{r_c^3 f^2}{(4v_{\theta 0}^2)}, \qquad (11.41)$$

where $v_{\theta 0}$ is the tangential component of the inflow at r_0. The associated maximum tangential speed is

$$v_{\theta i} = \frac{2 v_{\theta 0}^2}{f r_0}. \qquad (11.42)$$

Figure 11.2 shows central speed and radius determined from these outside parameters for a given latitude. Similar diagrams can be prepared for other latitudes. In general one finds, in accordance with many observations, that r_i increases and $v_{\theta i}$ decreases as a storm enters the middle latitudes. From Fig. 11.2 the higher the tangential speed is at a given r_0, the higher also is the central speed and the smaller is r_i. The larger r_0 is for given $v_{\theta 0}$, the smaller is the central speed.

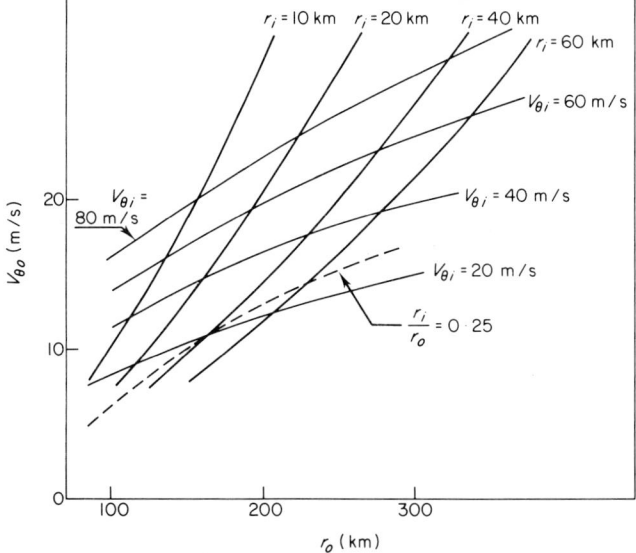

Fig. 11.2. Maximum speed ($v_{\theta i}$) and width of eye (r_i) as function of r_0 and $v_{\theta i}$ at latitude 28°.

In principle, Eqs. (11.41) and (11.42) solve the problem of determining the core intensity from external constraints. Figure 11.3 provides an example. It has been pointed out that r_0 may be difficult to determine even when upper aeroplane observations are available, since it depends on a sloping trajectory. However, inverting the problem, r_0 may be established from measured central parameters. This value of r_0 can then be compared with that obtained at, say, the 250 mb flight level to see whether a close correlation exists. Of even more interest is the determination of r_0 from satellite-derived winds and comparison of observed and computed central values. The satellite would need to see not only the upper outflow to determine r_0, but also enough of the lower inflow to estimate $v_{\theta 0}$. In many cases, breaks in the upper clouds afford such a view of the low-level spirals.

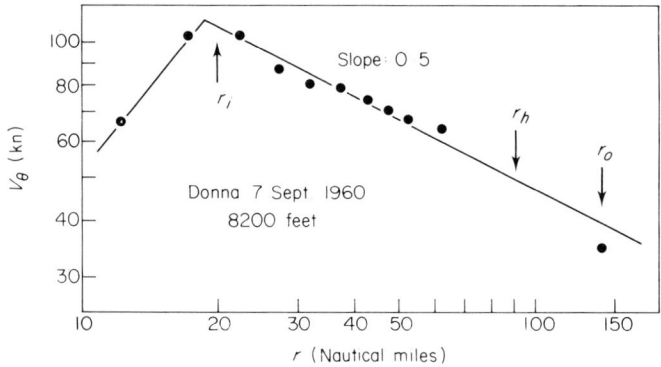

Fig. 11.3. Radial profile of tangential velocity, hurricane Donna, 7 September, 1960. For track and intensity of Donna see Fig. 9.14.

Kinetic energy computations

In the outflow, assumed steady and frictionless, the Bernoulli theorem of Eq. (7.2) is valid for generation of kinetic energy, if the storm is also stationary. Then, in isobaric surfaces, $\mathbf{V} \cdot \nabla h_p = dh_p/dt$, whereas in moving systems $\mathbf{V} \cdot \nabla h_p = dh_p/dt - \partial h_p/\partial t$; the local change will be very large in the hurricane interior. In the stationary case the total pressure (or h_p) change determines the kinetic energy production (positive or negative). Thus, for the outflow,

$$\frac{v_{\theta i}^2}{2} = g(h_0 - h_i)_u, \tag{11.43}$$

where $(h_0 - h_i)_{upper}$ is the height difference referred to a constant pressure surface and the hydrostatic change associated with ascent is excluded.

For the inflow radial balance of forces from the second equation of (11.13) may, to rather good approximation, be expressed by the cyclostrophic relation

$$\frac{v_\theta^2}{r} = g\frac{\partial h}{\partial r}. \tag{11.44}$$

Also $v_\theta^2 = v_{\theta i}^2 r_i/r$ from Eq. (11.40). Thus,

$$v_{\theta i}^2 \frac{r_i}{r^2} = g\frac{\partial h}{\partial r}.$$

Since h is a function of r alone we may integrate between r_i and r_0. Then

$$v_{\theta i}^2 \frac{r_0 - r_i}{r_0} = g(h_0 - h_i)$$

and, for $r_i \ll r_0$

$$v_{\theta i}^2 = g(h_0 - h_i)_{lower}. \tag{11.45}$$

This is a highly important result, to be compared with Eq. (11.43). It takes twice as large a height gradient in the surface inflow than in the outflow to obtain the same kinetic energy. If $v_{\theta 0}^2$ cannot be neglected, the difference would be greater, but in any moderate or strong hurricane this neglect is justified. It follows that *about half of the generation of kinetic energy by pressure forces is actually used to increase the kinetic energy of the inflow; the other half maintains the kinetic energy against ground friction.* The rate of kinetic energy generation is very high, perhaps the highest to be found anywhere. It cannot be expected to be valid far beyond r_0; in the trades it is correct to state that pressure and frictional forces almost cancel and that net generation of kinetic energy, positive or negative, is a very small residual (Chapter 5).

Equations (11.43) and (11.45) permit us to relate central wind speed to the mean virtual temperature field generated inside the hurricane. The thickness between the isobaric surfaces of inflow and outflow D

$\equiv h_u - h_l$ and the change in thickness $D_i - D_0 = (h_0 - h_i)_l -$
$(h_0 - h_i)_u$. Since $(h_0 - h_i)_l = 2(h_0 - h_i)_u$ from (11.43) and (11.45),
$D_i - D_0 = (h_0 - h_i)_u$. Introducing the hydrostatic equation differentiated between constant pressure surfaces,

$$\frac{\hat{T}_{vi} - \hat{T}_{vo}}{\hat{T}_{v(m)}} = \frac{\hat{D}_i - \hat{D}_o}{\hat{D}} = \frac{(h_o - h_i)_u}{\hat{D}}. \tag{11.46}$$

Inserting Eq. (11.43),

$$v_{\theta i}^2 = 2g \frac{\hat{D}}{\hat{T}_v} \delta \hat{T}_v \tag{11.47}$$

or approximately,

$$v_{\theta i} = 27 \, \delta \, \hat{T}_v^{0.5}. \tag{11.48}$$

when $\hat{D} = 100$ km, $\hat{T}_v = 270$ K and v_θ is expressed in metres per second. Obviously an upper limit on hurricane intensity is imposed by the square-root relationship. The temperature difference must be 4°C for a 50 m/s moderate hurricane and 16°C for an extreme 100 m/s storm. The latter value, very difficult to realize, approaches the temperature difference between polar and tropical air entering normal middle latitude cyclones. Thus, in hurricanes, as contrasted to extratropical storms, the efficiency in conversion of other energy forms to kinetic energy is quite superior.

Thermal wind

The temperature difference across the hurricane can also be determined from the thermal wind equation assuming its validity. Differentiating Eq. (11.44) with respect to pressure,

$$\frac{2v_\theta}{r} \frac{\partial v_\theta}{\partial p} = \frac{\partial}{\partial r} \frac{\partial (gh)}{\partial p}. \tag{11.49}$$

Using the hydrostatic equation the radial temperature gradient is introduced:

$$\frac{\partial v_\theta^2}{\partial \ln p} = -R \frac{\partial \hat{T}_v}{\partial \ln r}. \tag{11.50}$$

Table 11.1. Radial distributions in model hurricane

r km	v_θ in m/s	v_r in m/s	α deg	$\tau_{\theta,s}$ dym/cm^2	$v_r r$ in 10^5m^2/s	div$_2$ V 10^{-5}s^{-1}	W 6% area m/s	v_θ out m/s	ΔV_θ vertical m/s	$\Delta \hat{T}_v$ cumul. °C
33	60	8.0	7	90	2		7	60	0	5
50	47	10.0	12	55	5	40	4	40	10	2.0
70	39	11.5	16	38	8	24	3	26	13	0.94
90	34	12.7	19	28	11	17	2	18	15	0.55
110	30	12.3	22	23	14	13		14	16	0.33
150	25	11.7	25	16	18	9.5	1.5	9	16	0.15
200	21	10.0	28	10	20	6.5		4	17	small
250	18	8.0	30	8	20	0		0	18	

This equation was evaluated using the v_θ profiles for inflow and outflow given in Table 11.1 and the column marked Δv_θ between 900 and 200 mb. We note that $\Delta \hat{T}_v$ across the hurricane should be 4·1°C, almost 1°C less than computed from Eq. (11.48). Approximately half a degree can be added if the Coriolis parameter is included in the outer portions of the hurricane beyond $r = 100$ km, so that essentially the temperature differences from kinetic energy and thermal wind considerations are equal.

The vertical shear arises from the fact that in the inflow $\Omega_o = v_{\theta o} r_o + fr_o^2/2$, while in the outflow the first term is zero by definition at r_o. Thus, the momentum given up to the ocean over the inflow trajectory is $v_{\theta o} r_o$. If this frictional momentum transfer did not exist, the vertical shear would be zero in the two-dimensional model. The thermal wind would be zero and a balanced hurricane could not exist. *Thus, it is essential for the correct amount of momentum to be transferred to the ocean to establish and maintain the temperature field needed for the kinetic energy generation.* This is one of the basic requirements for the occurrence of hurricanes; one can readily see now why other synoptic rain systems, which die out toward the ocean in association with their cold core, cannot efficiently generate kinetic energy and fall apart into an assortment of mesoscale rain areas, even though a general upper outflow mechanism exists.

Radial motion, outflow

The detailed radial distribution of temperature gradient as determined by Eqs. (11.48) and (11.50) differs somewhat. In the core the vertical shear of v_θ is small and demands little support through the temperature gradient, while in the outskirts the reverse is true. Thus, there is an imbalance. Radial acceleration must be added, especially in the outflow which is directed toward greater pressure heights, at least until r_0, and only beyond this radius toward lower heights; thus,

$$\frac{dv_r}{dt} = \frac{v_\theta^2}{r} - g \frac{\partial h_p}{\partial r}. \tag{11.51}$$

A mechanism must exist for air to be accelerated outward from the core against the pressure-height field. The mechanism from Eq. (11.51) is due to v_θ^2/r which is most conveniently regarded as a "centrifugal force" here. This force must exceed the inward-directed pressure force in order for an outward acceleration to occur. In our steady model

NUMERICAL HURRICANE PREDICTION

$\mathrm{d}v_r/\mathrm{d}t = v_r \partial v_r/\partial r = \partial/\partial r \, (v_r^2/2)$. Substituting in Eq. (11.51) and integrating,

$$\left.\frac{v_r^2}{2}\right]_{r_2} - \left.\frac{v_r^2}{2}\right]_{r_1} = v_\theta^2 \, \delta \ln r - g(h_{r_2} - h_{r_1}) .$$

The centrifugal term is computed with the data in Table 11.1. The pressure force is computed from Eq. (11.43). Near the centre, the centrifugal force greatly exceeds the pressure force, so that v_r is computed as high as 17 m/s at 50 km, outflow angle 20°. From there v_r decreases outward. At $r = 70$ km, it has dropped to 12 m/s, outflow angle 30°, the same v_r value as in the inflow. Farther out, v_r continues to decrease, but the outflow angle increases, reaching 90° at $r = r_0$ by definition.

Radial motion, inflow

If we allow for acceleration through non-linear advection in Eq. (11.33), the left side of the equation, when divided by r, becomes $v_r \, (f + v_\theta/r + \partial v_\theta/\partial r) = v_r \, (f + \zeta_{\mathrm{rel}}) = v_r \, (f + \tfrac{1}{2} v_\theta/r)$. Integrating Eq. (11.33) over the depth of the inflow layer, utilizing this acceleration,

$$\int \frac{v_r}{2r}(2fr + v_\theta)\varrho \Delta z = -C_d v_\theta^2/\cos \alpha .$$

Assuming mean values over the depth of the layer

$$\hat{v}_r = -\frac{C_d}{\Delta z} 2r \frac{\hat{v}_\theta^2}{(2fr + \hat{v}_\theta)\cos \alpha} . \tag{11.52}$$

Evaluation at $r_0 = 250$ km and latitude 20° ($f = 5 \times 10^{-5}\mathrm{s}^{-1}$) yields the following values:

$v_{\theta 0}$ (m/s)	$-v_{r0}$ (m/s)	$-\alpha$ (deg)
5	1·0	11
10	3·6	20
15	7·4	26
20	11·2	29
25	15·6	31

Thus, the radial wind component and, presumably, also the mass inflow increase strongly with the tangential component and the absolute angular momentum about the hurricane axis. The inflow angle increases at low v_θ and then becomes steady near 30°. This is a rather high value; more commonly one observes inflow angles in the range from 15° to 25°, averaging 20° within 200 km from the centre. For any given outer boundary condition $-v_r$ at first increases, then decreases inward (Table 11.1). The variation is smaller than obtained in some other models (62) and there is no inner boundary condition requiring $v_{ri} = 0$. Rather the inflow goes through the eyewall with decreasing speed; earlier we have seen such inflow as a requirement to explain the eye low-level observations. Inflow angle and mass transport per unit layer thickness, however, decrease very strongly, the latter is only 7% of the outer inflow at the eye boundary (Table 11.1). Computed horizontal divergence rises to $-4 \times 10^{-4} s^{-1}$, rather more than in the mean storm of Fig. 9.11. For a layer thickness of one kilometre the same column in the Table gives the vertical motion at 1 km height in centimetres per second. If the inflow as far out as 100 km goes up only in hard radar bands and "hot towers" (Chapter 9)—still an area ten times larger than in ordinary synoptic systems—the vertical motion (next column) becomes large rapidly as the core is approached; however, vertical mass transport between concentric rings of air remains constant to the 100 km radius. Only a small fraction of the air approaching the core is able to penetrate into its centre, but this small fraction is correlated with total mass inflow and central pressure. It rises with increasing hurricane intensity (Fig. 10.2), and therewith also the production of kinetic energy through the generation term $\varrho \, d \, KE/dt = -g \, \overline{v_r \, \partial h_p / \partial r}$. Here, the long bar has been used since, from Chapter 10, it is the push of inflow in one sector rather than a general increase around the whole periphery which is the effective link in the deepening of hurricanes.

The analysis just concluded indicates, together with the direct information of Table 11.1, how the simple steady-state approximation to the dynamical equations may be utilized in a non-prognostic way to yield both information on the physical characteristics of the system, and those reasonable values to which a prediction should conform. Further, this presentation, as well as Chapter 10, contains strong suggestions on the mechanisms through which generation and weakening of hurricanes is initiated and carried out.

Statistical method

We have already seen the requirements for producing a dynamical forecast in the tropics and with emphasis on hurricane prediction. An alternate approach which may yield useful guidance information with considerably less synoptic input, but at a cost of greater dispersion in forecast accuracy, may come by statistical means. This mechanism utilizes historical information in an analogue or recursive sense to give a probability statement for the motion and possibly intensification of a hurricane. Numerous programmes to achieve this goal have been devised with charmingly creative acronyms (at least by contrast to dynamical models) (15).

The analogue methods do not utilize synoptic data directly. They rely on abstracting properties of the hurricane situation such as its location, moisture, time of year and intensity, and with this information they search the historical record of hurricane tracks for similar past occurrences. In processing this information, the computer generates probability ellipses based on bivariate, normal distributions to identify expected tracks which represent the forecast. Weighting may be added to the analogues based on their frequency of occurrence (14); these models require access to large data banks of tropical storm tracks if they are to succeed.

Forecasts based on regression equations may or may not include analogue information and may or may not incorporate synoptic data. Some may indeed utilize output from dynamical prediction models in their formulation. These models are represented by a regression equation which, incorporating predictors, establishes the future state of the hurricane. These predictors are sought, by statistical means, from an assortment of information available both about the past history of the event in question (statistical sample) and its current state. By utilization of significance tests, these predictors are identified, classified and sorted as to utility. Serious difficulties arise in this process, however. Models which utilize analogue data in regression equations are denoted "simulated analogue methods". These methods also use location, motion, time of year and intensity of the storm as predictors, and search analogues for climatology and persistence to establish their regression equation.

Whereas the simulated analogue formulation does not utilize synoptic data, the statistical synoptic method does so. Examples of this technique are the US National Hurricane Center NHC 67 and NHC 72 models (34, 37). Here, observed height data and gradients of the storm may be incorporated into the sampling information in determining the relevant predictors for the regression equation, as well as climatology given from analogues. In addition, weather patterns and steering currents may also be correlated with the motion and included in the predictive equation.

Finally, we consider the class of models which incorporates in its statistics the output of dynamical models and which are termed statistical-dynamical models. These models include some information from a numerical prediction which clearly implies a thorough use of synoptic data. MOHATT (45) uses steering of the vortex by geostrophic flow as a predictor, adjusting the prediction from errors detected in early forecasts. By contrast, NHC 73 (38) utilizes information from the 500 mb level of the US National Meteorological Center (NMC) Primitive Equation (PE) model, with some success.

Analysis of results with the various models discussed suggests that inclusion of synoptic data provides improved forecasts, especially after storm recurvature and as the forecast period is prolonged. On the other hand, empirical data over a short prediction time on non-recurved trajectories yield superior regression prediction. Zonal tracks are generally easier to predict than meridional ones. It must be said that, although much effort has been invested into this method of forecasting, substantially more research and testing must be done before systematically satisfactory forecasts can be produced.

Several limitations in the statistical process have been elucidated (37). Included among them are questions of statistical significance and predictor selection. It is not clear that the devices used to establish statistical significance are sensitive enough to sort out the most important and independent predictors, or their number. Indeed the method for predictor selection has become rather mechanical, based on "off the shelf" statistical tests. It should be evident that predictor selection must be related in some sense to the problem under study, in this case hurricane evolution. Other difficulties involve initial analyses, the problem with synoptic information already considered with regard to the dynamical method, i.e. availability and quality of initial data as well as its distribution and interpolation procedures. Overweighting of persistence may also be a flaw in statistical models, a use prompted by its success for short-term forecasts. Finally, statistical models may be burdened with too much attachment to "classical" techniques. New plausible hypotheses applied to statistical forecasting should be developed.

Results and potential

Having discussed the detailed construction of dynamical tropical prediction models suitable for numerical integration, we must consider what these models can accomplish and how they may best be

NUMERICAL HURRICANE PREDICTION 535

used. For a listing of specific models, the reader is directed to Table 11.2.

The quality of any model may be defined in terms of its output, i.e. how well the model simulates those processes for which it was designed. For example, the hurricane prediction model is constructed to reproduce those many features already discussed in previous chapters and which have been consolidated from observations. These factors include the different stages of the hurricane: developing, mature and dissipating. Within these stages the hurricane describes flow characteristics which can be assessed in models. Focus in modelling has been directed toward the developmental and mature stages; most models begin their integrations with an existing vortex. Features of the flow for which modellers strive, and against which they compare their forecasts include such predominant ones as strong inflow in the surface layer and outflow aloft, deep region of updraft and weak inflow/outflow (this represents the meridional circulation), frictional convergence in the boundary layer, sinking in the eye, intense tangential circulation developing and moving inward during intensification and spreading outward during the mature phase, substantial warming of the inner core environment as well as moistening during the developmental stage, spiral rain bands and many more.

Interestingly, many models show results which satisfy, to some extent, the observations denoted above. Yet it is evident from observations that the spectrum of hurricane behaviour is very broad. Maximum velocities exceed minimum limits by large amounts, as do storm sizes. Clearly, the size of a storm defines to some extent its energetics. If then a model simulates the gross properties of the atmosphere and, in particular, here the hurricane, that model may be used for sensitivity studies to isolate the various factors controlling the system and to establish the hurricane's sensitivity to them. This process should yield increased information about the physical and computational processes which drive the hurricane.

Additionally, we may ask if models exist which, when given adequate initial data taken from observations of the real atmosphere, are capable of giving satisfactory forecasts of the system both in its development and motion. From a social viewpoint, results of this inquiry determine the ultimate value of a model. Finally, should a model reproduce nature with sufficient fidelity, it might be used as an analogue for experimentation wherein physical changes could be tested for their response. An example of such a procedure is the simulation of cloud seeding in a hurricane model; forecasts

Table 11.2. List of hurricane models

Date	System Primitive Eq. or Bal.	Convective Parameterization	Moisture Prediction	Vertical Structure	Grid	Additional Features
SYMMETRIC MODELS						
1961 (18)	PE	No	No	50 mb levels	regular	
1964 (47)	PE	No	No	550 m levels	coarse	
1964 (9)	BAL	Yes	No			linear solution
1964 (39)	BAL	Yes	No	2-levels	fine	
1965 (22)	BAL	Yes	Yes	2-levels	variable	
1968 (59)	PE	Yes	No	13-layers	expanding	
1969 (41)	BAL	Yes	No	Incompressible layers	fine	
1970 (56, 57)	BAL	Yes	Yes	100 mb levels	regular	
1970 (48)	PE	Yes	Yes	7 levels	regular-fine	
1977 (60)	PE and nonhydrostatic	No	Yes	25 layers variable	ultra-fine	
1977 (2)	PE	Yes	Yes	4 layer σ-coords.	variable coarse	
1978 (50)	PE	No	Yes and Liquid water predict	12 layer σ	expanding	
3-DIMENSIONAL MODELS						
1972 (1)	PE	Yes	Yes	3-levels	staggered	
1972 (35)	PE	Yes	Yes	6-layer	coarse	real initial conditions
1973 (13)	PE	No	No	3-layer	nested	simulated forcing
1974 (25)	PE	Yes	Yes	11-layers	variable	
1974 (32)	PE	Yes	Yes	3-layer and PBL	nested	real initial conditions
1975 (29)	PE	Yes	No	3-layer	expanding	β-plane
1976 (26)	PE	No	No	3-levels	multi-nested	real initial conditions
1977 (16)	PE	Yes	Yes	3-levels	multi-nested moving	

incorporating such a process may provide insight into how nature might be altered in man's best interest.

Perhaps the single most important, as well as interesting feature which drives the hurricane is convection and its organization. The need for a model to organize convection on the hurricane scale became vividly apparent to the earliest modellers; their efforts at predicting convection led to uncontrolled amplification of the smallest scales, which were still substantially larger than those of a single convection element. Since instability increases with decreasing scale, scientists were led to repeat this experiment but with a larger scale model cut-off, thus reducing amplification of the smallest resolvable model scale (47). The rapid instability in the system was thus inhibited and some hurricane features surfaced, but no limit to system growth was observed.

This inability of models to span all scales and properly predict scale interactions led to the evolution of the CISK concept. Balanced symmetric models with cumulus parameterization (see Chapter 4) were produced which simulated the hurricane by organizing convection, and did so with very low vertical resolution (9, 40, 39). Indeed these models were able to achieve vortex velocities in excess of 40 mps within 60 hours of real time together with a sizeable meridional circulation. It took subsequent sophistication of the parameterization process to cut off the increased circulation as the environment became saturated, thus extending the integration process into the mature stage and beyond. Since small-scale instabilities due to convective gravitational activity were inhibited by the convective parameterization process, and since this parameterization seemed to provide adequate heating for the hurricane, all models utilized some form of it, most evolving from the developments by Kuo (22).

An experiment (60) of a non-hydrostatic model without convective parameterization suggests that the organization of convection can indeed be predicted by the model itself, i.e. CISK can occur by the model's internal structure. This result was achieved by integrating a very high resolution model (minimum scale of 400 m) in which convection was free to occur. Unfortunately, this resolution was constrained to a small part of the vortex and with it the resulting convective organization. Nevertheless, the implications of the study are significant; scale interaction can be forecast without parameterization given sufficient scale resolution.

Concurrently another study was reported (50) in which prediction of the evolution of a symmetric cyclone by a model without CISK parameterization, but with scale resolution of only 10 to 20 km was achieved. Since truncation errors occur with their maximum impact in the smallest scales, it is difficult to expect that this model can accurately describe the system to scales less than 50 km. Convection takes place on scales one order-of-magnitude less, however, and yet these results indicate appropriate convective organization and heating. The modeller, himself, has no immediate explanation for this success, but points out that more realistic moisture distribution initially reduces the potential gravitational instabilities of the smallest model scale.

One is left to conjecture at this point. After more than a decade of effort based on the conviction that convective parameterization was essential to hurricane modelling, the system was satisfactorily predicted without explicit parameterization. This does not rule out the possibility of implicit parameterization, a process which could well be transparently incorporated into almost any model. Such parameterization may come about through the computational scheme utilized in the forecasts. Most models utilize the upstream advective—forward time approach reviewed earlier. This scheme implies strong damping in the smallest scales, but it is these small scales which would lead to instability. Thus, by their removal or damping, the model is allowed to proceed in its time evolution on the larger scale. The larger scales pick up energy transports by non-linear advection and incorporate them into the physical processes and flow evolution. These processes may be sufficient to organize the system's energy into the proper scale for hurricane development, given that the other physical forces are properly expressed. The convective parameterization scheme does, after all, represent the small-scale processes in terms of the large-scale variables. The suggestion put forth here is that the model may do this job itself provided instability is removed. Although this interpretation is not informative of the physics, since it leans on computational processes and implicit truncation errors, it clearly calls for carefully considered alternatives.

A number of sensitivity studies testing variations in convective parameterization have been made in addition to those discussed above. Since these studies deal with variations in details substantially less profound than the difference between including and not including parameterization, their review here would not lead to additional insight.

Let us next consider the impact on hurricane prediction of variations in internal stress. Horizontal diffusion has generally been discredited in terms of its physical significance, yet it does have scale-dependent, dissipative properties which reduce the amplitude of the smallest scales, those which manifest the largest truncation errors. Rosenthal's experiment (50) also tested the effect of removing this type of diffusion entirely, and surprisingly found results not significantly different from comparable experiments which included diffusion. Vertical diffusion plays a more pronounced role in prediction systems, but its impact is predominantly in the boundary layer. Indeed some interesting experiments show dramatic changes in model prediction if the diffusion coefficient is introduced with altered values.

Since it is known that frictional inflow in the boundary layer produces the lifting of moist air required to organize convection, reduction or elimination of friction should inhibit the hurricane's development. This expectation has been substantiated by tests in which the drag coefficient has been reduced or set to zero. On the other hand, it has been shown (49) that too large a drag coefficient will cause an excess of energy dissipation and, thus, inhibit vortex intensification. Included in the argument of surface inflow is the requirement for moisture and heat addition from the surface. If this is not available, the convection process cannot proceed. Thus, if moisture flux from the oceans is diminished or eliminated (by specifying the coefficient to vanish), the vortex will not develop to hurricane strength (41). This sensitivity test has been performed by other modellers with comparable results. Indeed this is the device by which modellers bring the hurricane to its decaying state—by cutting off its supply of surface moisture, as would occur on landfall.

Observations indicate that hurricanes will not form if ocean temperatures are substantially lower than $26.5°C$. Experiments with models testing this hypothesis (56, 57) strongly support the observation. Moreover, they find that the primary inhibiting factor is reduced moisture flux rather than heat flux from the oceans at lower temperature. These results indicate that the representation utilized in models for describing fluxes from the surface have properties corresponding to reality.

Lateral boundary conditions depend to a great extent on the domain size. Models with closed boundary conditions show substantial differences from those with open conditions if the domain is small, i.e. approximately 500 km. This result is consistent with our expectations, since the domain is too small to confine the hurricane activity, especially in the outflow region aloft. As the boundary is moved outward, vortex development time decreases. If the domain is taken as very large, in excess of 3000 km, results of calculation with open or closed conditions are not significantly different, indicating that interactions outside that domain are negligible.

We have already considered some of the effects of truncation, a result of finite-difference operations. Most models are sufficiently damped on the small scale so that instability is not apparent. Truncation errors generally are not discussed. However, the effect of changing grid size from 20 to 10 km has been tested (48). A better definition of the eye region with its subsidence for the model with finer resolution was found, but the overall properties of hurricane development are not substantially altered. The technique of using an expanding grid length in symmetric models has become popular and appears to save computing time without ill effect.

The observation that hurricanes are not fully symmetric systems has led modellers, following their "successes" with symmetric models, to apply their efforts to asymmetric models. Those models, which initialize with symmetric conditions, do show an asymmetric development after some integration time. The vortex grows initially by symmetric processes similar to symmetric models and then develops asymmetries in the upper outflow region. Spiral rain bands are also predicted. Analysis of the asymmetric process suggests that it is caused by inertial instability so that waves grow at the expense of shearing motion (1, 16).

Symmetric hurricane models, despite the valuable information which they supply, are clearly not suitable for real prediction. Thus, experiments which alter the initial conditions of such models can merely tell us under which extreme circumstances the vortex will not evolve. No such conditions have been found, since given a suitable model, hurricanes develop even if there is no initial flow. The only variation in development is in evolution time. Since real hurricanes develop over widely varying periods, and only 1-10% of initial vortices do develop, this model information is not useful.

The desire to predict actual hurricane events led to the development of three-dimensional models. For computational economy, many of these models utilize the "nesting" scheme. Forecast results with these models are strongly dependent on initial conditions and, since these conditions are not accurately known, forecasts are difficult to assess. Because of insufficient tests with these models, we cannot determine whether forecasts are inadequate because of model deficiencies or initial conditions, or perhaps because of computational difficulties. Possibly information gleaned from symmetric models may be used to focus on specific limitations in three-dimensional models. Of particular interest here is hurricane movement and interaction with the environment. An interesting study involving a change in the Coriolis parameter (56, 57, 29) indicates that the hurricane can be steered northward by the β-effect as well as having its maximum intensity decreased more rapidly as it moves to larger f-values.

Is the hurricane well enough understood so that we can predict its behaviour under alteration? Assuming that this is the case, several models of the symmetric hurricane have incorporated external heating to simulate cloud seeding in an effort to predict the effects of this process. The results (49, 56, 57) are inconclusive at best, but do not suggest dramatic changes. There is, of course, no direct physical way to test these results, but since the models are quite limited (they are after all symmetric) we must await the future for definitive results on such prediction.

For tropical flow prediction on planetary scales, a model incorporating many of the facets of a hurricane model with the addition of radiative cooling and dry convective adjustment has been developed (20), but on a substantially larger minimum scale. Again forecast difficulties are focused on initial conditions, which are not adequate to meet the model's needs. In an effort to predict the movement of an easterly wave, the model established that most energy was derived from convection and not from shearing in the flow field. The model also suffers from lack of interaction with the middle latitudes.

The impact of this interaction (tropics with middle latitudes) was studied utilizing a primitive equation, general circulation model (51, 52). By pulsing the tropics in the model, the modeller was able to sense the consequence in middle latitudes and noted a definite response at all levels. The study clearly substantiates our expectation that the tropics do not operate on their own.

This fact is inherent in general circulation modelling, which must obviously include the tropics. Such models can only provide normal flow characteristics, since details of particular synoptic events are not a meaningful product of general circulation models. One such study discusses the tropics in detail (31). This model produces significant tropical Lows with energy characteristics comparable to observations. Details of such development on short time scales were assessed (36) and the results suggest that forecasting in the tropics suffers the same deterioration in quality as does middle latitude forecasting, with the added problem of inferior initial conditions.

Although we cannot say that numerical prediction of tropical flow and, more specifically, the hurricane is in its infancy, many issues are not yet clearly understood. A tentative attempt has been made at the challenging, but elusive, problem of hurricane formation (54). It is evident that many problems await the eager and talented student. The future for numerical prediction of the tropics and hurricanes appears bright, and because of the dominant impact of the tropics on the total atmosphere, we may expect the continued application of many resources toward further improved understanding.

References

(1) Anthes, R. A. (1972). Development of asymmetries in a three-dimensional model of the tropical cyclone. *Mon. Wea. Rev.* **100:** 461-476.
(2) Anthes, R. A. (1977). Hurricane model experiments with a new cumulus parameterization scheme. *Mon. Wea. Rev.* **105:** 287-300.
(3) Anthes, R. A., Rosenthal, S. L. and Trout, J. W. (1971). Preliminary results from an asymmetric model of the tropical cyclone. *Mon. Wea. Rev.* **99:** 744-758.
(4) Arakawa, A. (1966). Computational design for long-term numerical integration of the equations of fluid motion: two-dimensional incompressible flow. Part I. *J. Comput. Phys.* **1:** 119-143.
(5) Arakawa, A. and Schubert, W. H. Ref. 3, Chapter 4.
(6) Charney, J. G. (1948). On the scale of atmospheric motions. *Geof. Publik.* (Oslo) **17:** 17 pp.
(7) Charney, J. G., Fjortoft, R. and von Neumann, J. (1950). Numerical integration of the barotropic vorticity equation. *Tellus* **2:** 237-254.
(8) Charney, J. G. and Eliassen, A. Ref. 18, Chapter 4.
(9) Charney, J. G. and Eliassen, A. Ref. 19, Chapter 4.
(10) Staniforth, A. N. and Daley, R. W. (1977). A finite-element formulation for the vertical discretization of sigma-coordinate primitive equation models. *Mon. Wea. Rev.* **105:** 1108-1118.
(11) Deardorff, J. W. (1972). Parameterization of the planetary boundary layer for use in general circulation models. *Mon. Wea. Rev.* **100:** 93-106.
(12) Eliassen, A. (1949). The quasi-static equations of motion with pressure as independent variable. *Geof. Publik.* (Oslo) **17:** 44 pp.
(13) Harrison, Jr., E. J. (1973). Three-dimensional numerical simulations of tropical systems utilizing nested finite grids. *J. Atmos. Sci.* **30:** 1528-1543.
(14) Hope, J. R. and Neumann, C. J. (1970). An operational technique for relating the movement of existing tropical cyclones to past tracks. *Mon. Wea. Rev.* **98:** 925-933.
(15) Hope, J. R. and Neumann, C. J. (1977). A survey of world-wide tropical cyclone prediction models. pp. 367-374. 11th Technical Conference on Hurricanes and Tropical Meteorology, Dec. 13-16, 1977, Miami Beach, Amer. Meteor. Soc.
(16) Jones, R. W. (1977). A nested grid for a three-dimensional model of a tropical cyclone. *J. Atmos. Sci.* **34:** 1528-1553.
(17) Jordan, C. L. Ref. 16, Chapter 2.
(18) Kasahara, A. (1961). A numerical experiment on the development of a tropical cyclone. *J. Meteor.* **18:** 259-282.
(19) Kreiss, H. and Oliger, J. (1973). Methods for the Approximate Solution of Time Dependent Problems. GARP Publ. Series No. 10, World Meteor. Org., ICSU, Geneva. 107 pp.
(20) Krishnamurti, T. N., Kanamitsu, M., Ceselski, B. and Mathur, M. B. (1973). Florida State University's tropical prediction model. *Tellus* **25:** 523-535.

(21) Krishnamurti, T. N. and Kanamitsu, M. (1973). A study of a coasting easterly wave. *Tellus* **25:** 568-585.
(22) Kuo, H.-L. Ref. 38, Chapter 4.
(23) Kuo, H.-L. Ref. 39, Chapter 4.
(24) Kurihara, Y. and Holloway, J. L. (1967). Numerical integration of a nine-level global primitive equation model formulated by the box method. *Mon. Wea. Rev.* **95:** 509-530.
(25) Kurihara, Y. and Tuleya, R. E. (1974). Structure of a tropical cyclone developed in a three-dimensional numerical simulation model. *J. Atmos. Sci.* **31:** 893-919.
(26) Ley, G. W. and Elsberry, R. L. (1976). Forecasts of typhoon Irma using a nested grid model. *Mon. Wea. Rev.* **104:** 1154-1161.
(27) Lorenz, E. N. (1960). Energy and numerical weather prediction. *Tellus* **12:** 364-373.
(28) Machenhauer, B. (1974). On the Present State of Spectral Methods in Numerical Integrations of Global Atmospheric Models. GARP Prog. on Num. Exp., Report No. 7, pp. 1-21. Int. Symp. Copenhagen, 12-16 August 1974.
(29) Madala, R. V. and Paicsek, S. A. (1975). Numerical simulation of asymmetric hurricanes on a β-plane with vertical shear. *Tellus* **27:** 453-468.
(30) Manabe, S., Smagorinsky, J. and Holloway, Jr., J. L. (1965). Simulated climatology of a general circulation model with a hydrologic cycle. *Mon. Wea. Rev.* **93:** 769-798.
(31) Manabe, S., Holloway, Jr., J. L. and Stone, H. M. (1970). Tropical circulation in a time-integration of a global model of the atmosphere. *J. Atmos. Sci.* **27:** 580-613.
(32) Mathur, M. B. (1974). A multiple-grid primitive equation model to simulate the development of an asymmetric hurricane (Isbell, 1964). *J. Atmos. Sci.* **31:** 371-393.
(33) Mesinger, F. and Arakawa, A. (1976). Numerical methods used in Atmospheric Models. GARP Publ. Series No. 17, World Meteor Org., ICSU, Geneva. 64 pp.
(34) Miller, B. I., Hill, E. C. and Chase, P. P. (1968). Revised technique for forecasting hurricane motion by statistical methods. *Mon. Wea. Rev.* **96:** 540-548.
(35) Miller, B. I., Chase, P. P. and Jarvinen, B. R. (1972). Numerical prediction of tropical weather systems. *Mon. Wea. Rev.* **100:** 825-835.
(36) Miyakoda, J., Sadler, J. C. and Hembree, G. D. (1974). An experimental prediction of the tropical atmosphere for the case of March 1965. *Mon. Wea. Rev.* **102:** 571-591.
(37) Neumann, C. J. (1977). A Critical Look at Statistical Hurricane Prediction Models. pp. 375-380. 11th Technical Conference on Hurricanes and Tropical Meteorology, Dec. 13-16, 1977, Miami, Beach, Amer. Meteor. Soc.
(38) Neumann, C. J. and Lawrence, M. B. (1975). An operational experiment in the statistical-dynamical prediction of tropical cyclone motion. *Mon. Wea. Rev.* **103:** 665-673.
(39) Ogura, Y. (1964). Frictionally-controlled, thermally-driven circulations in a circular vortex with application to tropical cyclones. *J. Atmos. Sci.* **21:** 610-621.
(40) Ooyama, K. Ref. 63, Chapter 4.
(41) Ooyama, K. (1969). Numerical simulation of the life-cycle of tropical cyclones. *J. Atmos. Sci.* **26:** 3-40.
(42) Phillips, N. A. (1957). A coordinate system having some special advantages for numerical forecasting. *J. Meteor.* **14:** 184-185.

(43) Phillips, N. A. (1963). Geostrophic motion. *Rev. Geoph.* **1:** 123-176.
(44) Phillips, N. A. (1973). Principles of large-scale numerical weather prediction. *In* "Dynamic Meteorology." (P. Morel, Ed.) pp. 1-96. D. Reidel Publ. Co., Dordrecht-Holland. 622 pp.
(45) Renard, R. J., Colgan, S. G., Daley, M. J. and Rinard, S. K. (1973). Forecasting the motion of North Atlantic tropical cyclones by the objective MOHATT scheme. *Mon. Wea. Rev.* **101:** 206-214.
(46) Robert, A., Henderson, J. and Turnbull, C. (1972). An implicit time integration scheme for baroclinic models of the atmosphere. *Mon. Wea. Rev.* **100:** 329-335.
(47) Rosenthal, S. L. (1964). Some attempts to simulate the development of tropical cyclones by numerical methods. *Mon. Wea. Rev.* **92:** 1-21.
(48) Rosenthal, S. L. (1970). A circularly symmetric primitive equation model of tropical cyclone development containing an explicit water vapor cycle. *Mon. Wea. Rev.* **98:** 643-663.
(49) Rosenthal, S. L. (1971). A circularly symmetric primitive equation model of tropical cyclones and its response to artificial enhancement of the convective heating functions. *Mon. Wea. Rev.* **99:** 414-426.
(50) Rosenthal, S. L. Ref. 75, Chapter 4.
(51) Roundtree, P. R. (1976). Tropical forcing of atmospheric motions in a numerical model. *Quart. J. Roy. Meteor. Soc.* **102:** 583-606.
(52) Roundtree, P. R. (1976). Response of the atmosphere to a tropical Atlantic Ocean temperature anamaly. *Quart. J. Roy. Meteor. Soc.* **102:** 607-626.
(53) Sasamori, T. (1968). The radiative cooling calculation for application to general circulation experiments. *J. Appl. Meteor.* **7:** 721-729.
(54) Shapiro, L. J. (1977). Tropical storm formation from easterly waves: a criterion for development. *J. Atmos. Sci.* **34:** 1007-1021.
(55) Smagorinsky, T., Manabe, S. and Holloway, Jr., J. L. (1965). Numerical results from a nine-level general circulation model. *Mon. Wea. Rev.* **93:** 727-768.
(56) Sundqvist, H. (1970). Numerical simulation of the development of tropical cyclones with a ten-level model. Part I. *Tellus* **22:** 359-390.
(57) Sundqvist, H. (1970). Numerical simulation of the development of tropical cyclones with a ten-level model. Part II. *Tellus* **22:** 504-510.
(58) Syono, S. (1962). A Numerical Experiment of the Formation of Tropical Cyclones. Proc. Int. Symp. on Num. Wea. Pred. in Tokyo, 7-13 Nov. 1960, pp. 405-418. Meteor. Soc. Japan.
(59) Yamasaki, M. (1968). Detailed analysis of a tropical cyclone simulated with a 13-layer model. *Pap. Meteor. Geoph.* (Tokyo) **19:** 559-585.
(60) Yamasaki, M. (1977). A preliminary experiment of the tropical cyclone without parameterizing the effects of cumulus convection. *J. Meteor. Soc. Japan,* **55:** Ser II, 11-31.
(61) Riehl, H. Ref. 61, Chapter 9.
(62) Malkus, J. S. and Riehl, H. Ref. 43, Chapter 9.

12

The General Circulation

This book opened with the definition of absolute angular momentum about the earth's axis. In the absence of strong forces from space, the absolute angular momentum of the system earth-atmosphere is conserved on less than astronomical time scales. From its definition, the absolute angular momentum of the earth is greatest at the equator and becomes zero at the poles. The atmosphere need not conserve its momentum since it can be exchanged between earth and atmosphere back and forth at variable rates and in different areas. Since the mass of earth is 10^6 times that of the atmosphere, the momentum of the earth remains practically unchanged during such exchanges, while the momentum of the atmosphere can change by large fractions, especially the part in which our interest centres, the zonal wind components, easterly or westerly.

From Fig. 1.2, the surface global winds are definitely divided in two halves; easterlies between latitudes 30°N and S, westerlies from there poleward, at least to the margins of Arctic and Antarctic. The air has less angular momentum than the earth where the winds are easterly; its angular momentum is greater when the winds are westerly. From the actual location of the belts of easterlies and westerlies we concluded that the air in the tropics should have arrived there from higher latitudes; the air outside the tropics should have come from lower latitudes. Indeed, such is the case as demonstrated by Fig. 1.3. Considering the forces acting on air, only the pressure force can produce accelerations which will set the air in motion poleward or equatorward as observed. Therefore, high surface pressure should prevail in the subtropics, low pressure in the equatorial and subpolar zones. Such a pressure field is observed (Fig. 1.4). Invoking now the principle of conservation of mass, we deduced that ascending motion of air should take place in the equatorial and subpolar belts, descending motion in the subtropics. Since ascending motion leads to condensation and precipitation, descending motion to warming and drying, there should be two principal belts of precipitation on earth in the equatorial and subpolar regions with dryness interspersed in the subtropics. From Fig. 3.1 this supposition is largely verified. The variety of precipitation climates seen in Fig. 3.4 is laid to the distribution of continents and oceans.

In spite of this quick and successful presentation of the atmosphere's main climatic features, deduced from a small number of basic laws, we broke off the discussion at this point. Why not stop there and be satisfied? Undoubtedly, a share of the truth has been captured, but in order to become aware of some of the problems posed by the atmosphere, we shall inquire further into the mechanisms implicit in the classical general circulation hypothesis.

The classical model

All atmospheric motion starts from inequalities of heating and cooling, especially the net radiation difference between equator and pole. Since over the years temperature is nearly constant everywhere (climate changes apart), these differences must be balanced by heat exchange between high and low latitudes. Air currents set in motion poleward and equatorward largely effect the exchange; in modern times the role of ocean currents in transporting heat has been continually upgraded (57, 59).

Any general circulation machine must function so as to permit conversion of heat into mechanical energy, i.e. winds, else it would die away. The resulting wind system must satisfy fundamental laws of motion. By specifying that cold air sinks in the polar zones and that warm air rises in the equatorial belt, the classical theory attempted to satisfy the first requirement (Eq. (4.31)). Conservation of absolute angular momentum by air currents as leading dynamical principle is the means for satisfying the second.

If air ascends in the equatorial zone and moves poleward, it acquires an eastward component of motion under conservation of momentum. This component increases as the air penetrates to higher latitudes, and when sinking occurs there, air with westerly momentum is brought to the surface. Here friction becomes important in transferring the westerly momentum to the earth by slowing the winds

THE GENERAL CIRCULATION 547

down. As the air returns toward the tropics from the north with the meridional circulation, the westerlies decrease and eventually there will be a latitude where the zonal wind becomes zero, near latitudes 30°. Moving still farther equatorward, the air begins to lag the earth, so that easterly winds appear. The earth now acts to slow down the easterlies through friction, i.e. momentum in the tropics is transferred from earth to atmosphere. Eventually the meridional circuit is completed; it provides for heat exchange and for maintenance of the winds.

This is the grand design of a simple ageostrophic, two-dimensional and non-turbulent circulation, whose early conception can only be admired. Daring in parts from the days of Hadley (21), the model was expanded and perfected well into the twentieth century. Even Hadley was aware of some complications. The dryness of the subtropics was well known and he allowed for it by introducing a partial descent of the upper poleward current in the horse latitudes, a remarkable early insight. The single equator-to-pole circulation was replaced by three wheels, especially through the American meteorologist, Ferrell, of the nineteenth century, who noted that in the middle latitudes the surface winds did not blow from northwest all over, but predominantly from southwest in the Northern Hemisphere. Herewith, a reverse cell in middle latitudes was called for, which later became the object of much controversy. One principle was shared by all who built the classical theory: at first a "model" of the global mechanisms on the largest scale was evolved; smaller-size events, even the sometimes formidable middle latitude cyclones, were fitted into the world-wide framework.

Critique of the classical model: radiation

The classical theory tacitly assumes that solar radiation acts like the permanent flame of a laboratory experiment in heating the tropics. For the combined system earth-atmosphere this assumption is verified (Fig. 2.1). The reader may have noted that we carefully avoided a division into earth and atmosphere separately in Chapter 2; now, however, it must be our concern. As all radiation computations have shown conclusively, the atmosphere itself is a cold source and the strongest tropospheric cooling occurs in equatorial latitudes! The nature of radiation, and its transit through the atmosphere with its gaseous constituents and aerosols, is a large and complex subject treated in many books and articles; for the reader who desires a general survey, Chapters 8-10 of Hess (29) are recommended. The calculated radiation fluxes and the resulting cooling and warming have not remained constant over the years as improved information has become available. In Chapter 2 we noted the great impact of satellite measurements which, for instance, demonstrated that the albedo in the tropics previously had been overestimated. After a comprehensive calculation of the radiative energy budget of the atmosphere (43), seasonal vertical cross-sections of heating and cooling were published by Dopplick (17). Of these four charts, that for March to May has been picked at random for presentation (Fig. 12.1); in the tropics seasonal variation is very small, so that one season may represent all.

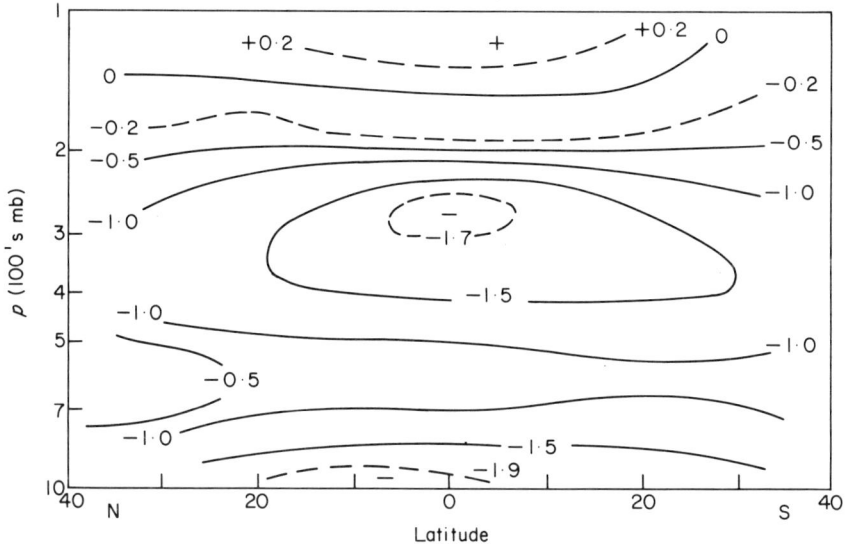

Fig. 12.1. Net daily heating or cooling (°C) in the troposphere, March to May (adapted from 17).

Besides the gross fact that net radiation cooling occurs everywhere and that it is greatest near the equator, one observes two maxima of cooling greater than 1·5°C per day: close to the surface and near 300 mb. In between, in the layer 700 to 500 mb, net cooling is slightly below 1°C per day. The upper maximum undoubtedly arises from the average height of tropical cloud covers; the low-level maximum represents a distinct departure from earlier estimates, where cooling was computed to increase upward from the ground to the middle troposphere. We have already come across this difference in seeking mechanisms for production of thermal instability for sustaining cumulus convection in Chapter 4 (Table 4.2).

The main point of overriding importance for the tropics is that solar radiation cannot be regarded as a constant flame under the tropical atmosphere. Rather there is strong net radiation cooling which would soon destroy the equator-pole temperature difference, were it not for the transfer of latent heat energy from the oceans to the tropical atmosphere, which sustains temperatures against radiation through release of latent heat and precipitation. Herewith, all that was said in Chapters 3, 4 and 8 about concentration of precipitation in narrow zones, the division of cumulus and cumulonimbus precipitation and the role of synoptic weather system in effecting the concentrated rain bands takes on new importance. These are not idle embroideries in a smoothly functioning general circulation. They are indeed, on the tropical side, the lifeline by which their aggregate, not just averages of basic parameters such as temperature, survives. They are therewith seen to have equality, if not precedence, in the order of wheels of the complex machinery that makes the atmosphere run.

Since the tropical rainfall events, as Chapter 8 has demonstrated, are in no way simple, obvious or reliable, the hypothesis of a steady energy input into the general circulation within the tropics cannot be accepted.

Critique of the classical model: motions

Whenever one can draw isotherms on a constant pressure surface, the potential energy of the atmosphere is elevated above the minimum state, when cold and warm (dense and less dense) air masses lie side by side. The atmosphere than has "available potential energy", a term coined by Lorentz (44). Available potential energy is somewhat similar to latent heat; the latter is carried around, often over long distances, before it is released in ascending currents. Similarly, one may draw isotherms on constant pressure charts for days without the available potential energy being released, in which event the isotherms would have to disappear. They remain, however, and this means that cold air remains in a raised state relative to warm air. Some mechanism must operate to uphold the cold air so that it will not simply sink down. Only some form of a rotational mechanism appears to offer a viable solution on the large scale of the atmosphere; magnetic fields, for instance, do not enter in tropospheric balances of forces.

Rossby number

A very clear and descriptive distinction between competing types of rotational forces can be made with introduction of the Rossby number (R_0) where

$$R_0 = \frac{U}{a\Omega}. \qquad (12.1)$$

Here, U is a characteristic velocity of the fluid analysed, a its radius and Ω is defined as some function of its rotation rate. Equation (12.1) permits comparison of circulations on bodies of such widely different size as sun, earth and laboratory experiment. Referring to the earth, U may be a typical zonal wind speed, perhaps 30 m/s outside the tropics, 10 m/s in the tropics; we choose $\Omega = \omega \sin \phi = f/2$. In a flat basin experiment, in contrast, Ω is the angular velocity of the basin, a constant. Equation (12.1) may now be rewritten as

$$R_0 = \frac{\frac{U^2}{a}}{\frac{Uf}{2}}, \qquad (12.2)$$

clearly a ratio of centrifugal (inertial) forces to Coriolis forces. We have already become acquainted with the hurricane as the prototype of a circulation which upholds its temperature field through the U^2/a [v_θ^2/r] term.

In a series of general circulation experiments in the hydrodynamics laboratory of D. Fultz at the University of Chicago (18), rotation and heating rates were varied to produce different general circulations. Interest here centres mainly on varying the rotation rate. Different rates in a flat basin may be compared to different latitude belts on the round earth since we have chosen $\Omega = \omega \sin \phi$. The experiments have demonstrated that a "high" Rossby number ($R_0 = 1$) corresponds to low rotation rates; a general circulation develops very similar to the classical tropical cell described above (76). As the rotation rate is increased the steady symmetrical circulation breaks down and goes over into a steady, westerly, long-wave pattern highly reminiscent of the wintertime subtropical jet stream (39); R_0 decreases to 0·3. With further acceleration of the rotation rate the steady wave pattern gives way to an unsteady regime with jet streams over variable orientation and strength, essentially passing through a "cyclus" of about 16 days, with surface low and high pressure systems on the inflection point of the upper long waves. Here a close analogue to the atmosphere in middle latitudes is attained; $R_0 = 0\cdot1$.

Based on the preceding hierarchy the tropics have been characterized as a high Rossby number, and the higher latitudes as a low Rossby-number regime. Suppose we take $R_0 = 0\cdot1$ at latitude 45° (sin $\phi = 0\cdot7$), U = 30 m/s, an acceptable value indicative of the strength of the upper westerlies. For $R_0 = 1$ and sin ϕ as little as 10^{-1} (latitude 6), $U = 47$ m/s, a speed typical only of hurricane interiors. For $U = 10$ m/s, an acceptable value for the tropics at large, $R_0 = 0\cdot2$ at latitude 6° and 0·85 at latitude 10°. $R_0 = 1$ is not obtained until latitudes 1° or 2°. Herewith the scaling of the tropics as a high Rossby rotation regime is negated, pointed out by Charney (10), who supplied a reason for this unexpected result. The tropical deep convection is concentrated in a very small fraction of the tropical area as stressed so widely in this text. Over the vast bulk of the tropics gravitational stability is high, i.e. air is constrained to move horizontally and is subject only to the very weak accelerations of the large-scale pressure fields. As emphasized, this argument is valid above the top of the cloud layer, roughly 700 mb. Vertical exchanges, adjusting the atmosphere to the underlying surface, dominate in subcloud and cloud layers and also in the deep adiabatic layer over dry land areas, especially deserts. There, wind variability is held largely to a minimum with wind steadiness (defined as $|\ \overline{V}\ |\ /\overline{V}$) of 80 to 90%. Above the cloud layer, say the trade inversion, steadiness decreases to 50% above 700 mb showing the influence of migratory upper large-scale troughs and ridges.

Instability of westerly rings of air

Several analyses have been offered why the expected high-rotation regime does not exist in the tropical upper troposphere. Given $U = 0$ at latitude 7°, it will be 80 m/s at latitude 25° and 120 m/s at latitude 30° as symmetrical rings of air contract poleward under conservation of absolute angular momentum (Fig. 12.2). A poleward pressure gradient must exist in sufficient strength to make such acceleration possible. Since the atmosphere is essentially hydrostatic, this means that the mean virtual temperature of the troposphere must decrease poleward sufficiently rapidly for the height of isobaric surface to slope at the appropriate rate. This rate must be at least slightly greater than that demanded for geostrophic balance (Fig. 12.3). Bjerknes et al. (7) suggested that it would be extremely difficult to maintain such pressure and temperature gradients. The thermal concentration

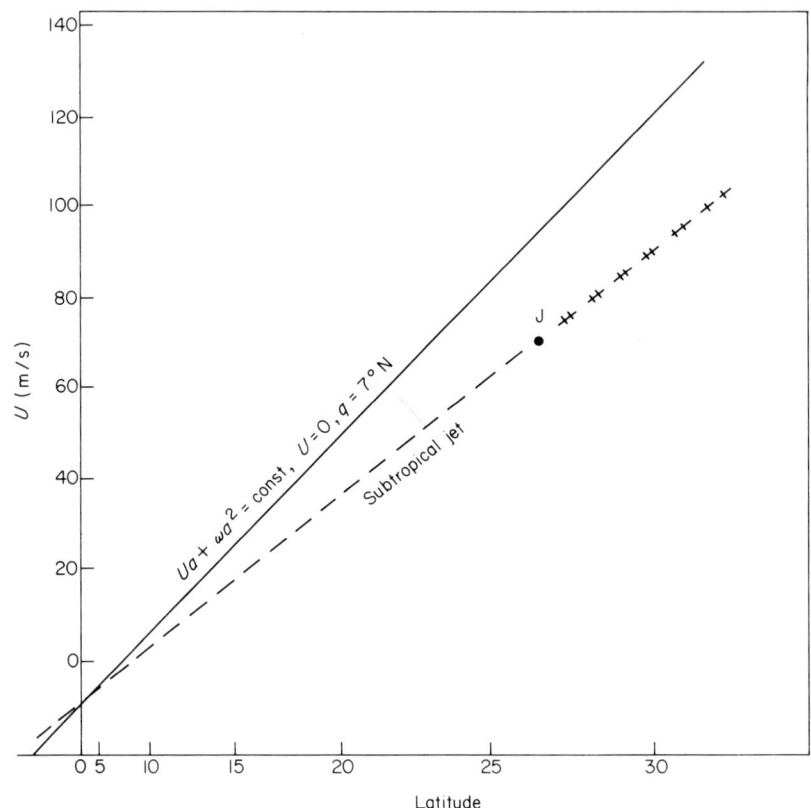

Fig. 12.2. Latitudinal profile of the zonal wind component U for constant absolute angular momentum, $U = 0$ at latitude 5°. Also profile for subtropical jet stream of winter in the Northern Hemisphere.

would fail, and therewith the simple general circulation wheel. If, however, the flow is arranged in several cells around the hemisphere, manifest at the surface in form of the subtropical high pressure centres, the air currents could flow long distances poleward and equatorward without the stringent requirement of conservation of absolute angular momentum. At the eastern cell ends, the pressure force accelerates the air eastward, i.e. adds to its angular momentum; at the western cells end the reverse occurs. Thereby, air can pass between tropics and high latitudes as a quasi-geostrophic or gradient current. In this analysis a dynamical reason for the existence of the subtropical Highs appears for the first time.

THE GENERAL CIRCULATION

At upper levels, vertical and horizontal shears of contracting zonal rings become very large; the question of the stability of such rings, i.e. their capability to remain rings, has often been subjected to theoretical analysis. One finds (cf. 7) that the rings will break down over certain ranges of wavelength and static gravitational stability and essentially yield waves corresponding to the long waves in the westerlies at latitude 40°. Herewith, not only the zonal rings are destroyed; both from the instability argument and that of Bjerknes *et al.*, active meridional heat exchange will be initiated through the north-south velocity components inherent in subtropical cells and upper waves, later termed "eddy heat" exchange. Through such meridional heat exchange the pole-to-equator temperature gradient is vastly reduced below that needed to support the meridional cells; therewith the possibility of existence of the meridional cell is reduced even further. Nevertheless, the classical picture is not altogether destroyed.

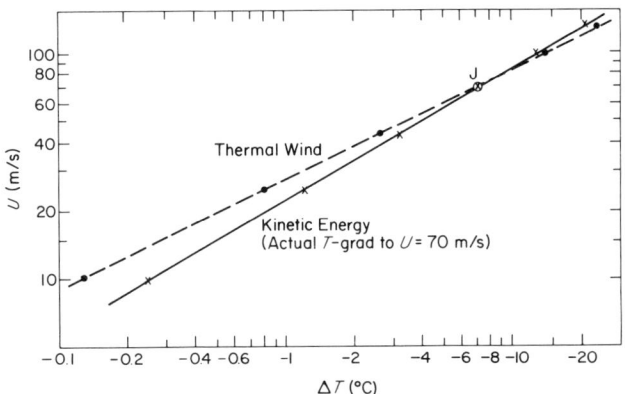

Fig. 12.3. Latitudinal profile of mean virtual temperature 850-200 mb computed from kinetic energy and geostrophic balance requirements for the subtropical jet stream of winter. Jet stream axis (J) is situated where the two profiles cross.

In Fig. 12.2 we see the observed velocity profile of the subtropical jet stream of winter and in Fig. 12.3 the associated temperature gradient above the low troposphere up to 850 mb, a demonstration of the degree of compromise which the actual atmosphere has achieved.

Thermal Rossby number

Further insight is provided by computing the "thermal Rossby number". If the thermal wind equation is integrated over the troposphere to, say, 200 mb (depth D), the characteristic value of U:

$$U_{200} = -\frac{g}{f\hat{T}} \frac{\partial \hat{T}}{\partial y} D,$$

where $\hat{T}$ is the vertically integrated, mean virtual temperature. We now introduce Eq. (12.1) using a as the same distance δy, over which the meridional temperature gradient is computed. Denoting the resulting expression by R_{0T},

$$R_{0T} = \frac{2gD}{\hat{T}a^2} \frac{\Delta \hat{T}}{f^2}. \qquad (12.3)$$

Given $D = 10$ km above the cloud layer and $\hat{T} = 250$ K, the factor in Eq. (12.3) is near unity if $a = 3000$ km and f is computed in units of 10^{-5}s^{-1}. The choice of $a = 3000$ km is convenient, since it permits us to compare the belts 0-30° and 30-60° directly. Only the ratio $\Delta \hat{T}/f^2$ need be considered.

Across the middle latitudes $\Delta \hat{T} = 30°$C in winter and $f = 10 \times 10^{-5}$s^{-1}. Then $\Delta \hat{T}/f^2 = 0.3$, the same magnitude as R_0. Across the tropics $\Delta \hat{T} \sim 6$-$8°$C, concentrated at the poleward margin (Fig. 2.25). Evaluating at latitude 15°, R_{0T} 0·35 to 0·40, again of the order of the middle latitude value. Under these circumstances the question of latitudinal variation of pressure force becomes definitely relevant. Using the above numbers, $fU = 3 \times 10^{-3}$m s^{-2} in middle latitudes and 2.5×10^{-4}m s^{-2} at latitude 10°. Here at last is the quantity which changes by an order of magnitude from high to low latitudes. For geostrophic flow as suggested by the R_0-numbers, the difference in pressure height will be 300 m at latitude 45° and 25 m at latitude 10°. If the wind is computed at 10 km and surface pressure gradients neglected, $\Delta \hat{T} = -8°$C over a 1000 km distance in middle latitude and $-0.6°$C at latitude 10°. Considering the accuracy of radiosonde instrumentation, the middle latitude temperature gradients should be readily measureable, whereas at low latitudes the gradient

THE GENERAL CIRCULATION

lies well within the instrumental error margin. This deduction is entirely in line with what the observations have shown and it furnishes independent substantiation of our chain of computations!

Conclusion

The analysis of R_0 and R_{0T} has brought out, surprisingly, that in the tropics as well as in high latitudes geostrophic balance of forces largely prevails. By this is meant, as stressed, that cold air is held up relative to warm air by the rotation of the earth rather than by inertial forces, as long as deep convection with high gravitational instability is confined in very small areas in the tropics, no doubt a mechanism of protection against entrainment. The only exception is the hurricane. Elsewhere, "available potential energy" can exist only in the measure that the Coriolis force is able to uphold cold domes. With the decrease of Coriolis force by an order of magnitude from high to low latitudes, only a very small amount of available potential energy can be maintained so that, in the first approximation, the ratio

$$\frac{g \frac{\partial h_p}{\partial y}}{fU} = \text{const}(y), \qquad (12.4)$$

where h_p is again the height of a constant pressure surface. The "feedback" is that, lacking available potential energy in the tropics, the actual release of kinetic energy is small, therewith wind speed itself, since the work to be done by pressure forces against friction is not latitude-dependent.

Equation (12.4) contains a singular point where U goes through zero between easterlies and westerlies. Here, the meridional pressure gradient must also be zero and p must go through a maximum for geostrophic motion. Thus, the subtropical high pressure belt appears as a necessary consequence of the separation of the world's wind systems into low latitude easterlies and high latitude westerlies, if these are to be geostrophically maintained. Earlier, the subtropical high pressure with sinking was inserted *ad hoc* to take account of subtropical dryness. In the present analysis it appears as a dynamical necessity for overall maintenance of the atmosphere's angular momentum budget at the ground; the subtropical dryness is an important but secondary consequence.

We must conclude that only a weak leftover of the simple meridional cell is found in Figs 12.2 and 12.3 and that the equations of motion which are valid must be the general quasi-geostrophic equations with small acceleration terms with special allowance for the small areas in which the tropical energy release occurs. We may also conclude that in a certain sense the observed general circulation of the atmosphere is accidental. If the rotation of the earth were faster or slower, or if the distance from the sun were greater or shorter, our weather may bear little or no resemblance to that which we know. The conflict between meridional cell and geostrophic control as defined here will pursue us further, especially in the energy budgets, since the Coriolis parameter does not enter directly into energy calculations except through "backdoor-feedbacks" as discussed.

Numerical modelling

In view of the great complexities of the general atmospheric circulation inside and outside the tropics, and failure of the noble attempt to order everything of importance into a simple, two-dimensional model, the reader will no doubt conclude that the obvious way forward points to numerical modelling, as presented in detail for the hurricane in Chapter 11. Such modelling indeed has been attempted and in the course of time has led to remarkable similarities with the observed general circulation. Along with computations to reproduce the present-day circulation, variations that would lead to insight on climate variations are being attempted in ever increasing numbers. It is unsatisfactory to present the numerical results and speculations based thereon as phenomena emanating from a "black box". As must be evident from Chapter 11, modelling depends heavily on many technological aspects of computer handling, damping factors, coordinate systems and other parameters. Detailed discussion of even the most relevant experiments for tropical meteorology would go far beyond the objectives of this book and will not be attempted.

While various articles carry extensive references, the paper by Washington and Kasahara (A-24) carries a comprehensive historical survey. The seasonal and interannual variation of the general circulation is treated in (51) and for the tropics in particular in (22, 47). Very detailed computations including hydrology are due to Manabe (46), where an ocean model is also included, and further studies of the tropical circulation are found in (2). An experiment with large-scale, seasonal forcing due to variable heating rates is described in (38); the simulation of the tropical climate in an ice age in (48). The list could be greatly expanded but will serve as starting point for those wishing to take up these subjects further. Most of the treatments contain evaluation of the classical theory, remnants of which are generally found, a treatment of the reverse, middle latitude Ferrel cell, which is beyond the immediate interest of this book, and calculations of precipitation and water balances. Many of these budgets bear close resemblance to those computed from the observations as shown in Chapter 5. It is of interest to read that the meridional cell plays an important role in budgets for the tropics, while it breaks down from an initial assumed state in the time integrations within two weeks (A-24). Here we shall utilize the empirical budget approach and demonstrate what has been achieved with the rather unspectacular method of fact finding. This road is also filled with some disappointment and pitfalls, notably due to lack of adequate observations, but quantity and quality of data over the world have improved, and so has the definite body of knowledge which has accumulated.

General circulation budgets

Absolute angular momentum

Perhaps the first to recognize that large-scale cyclones and anti-cyclones are integral parts of the general circulation machine and not just adornments was Defant (12). He proposed the definition of horizontal exchange (*Austausch*) coefficients on the scale of cyclones to effect the meridional heat exchange. Subsequently, Jeffreys (34, 35) considered the probable role of large-scale Reynolds stresses in maintaining the observed wind field against friction. By the late 1940s upper-air networks had increased sufficiently to permit computational tests of these ideas (4, 69). Initially, data were only marginally satisfactory for this purpose, especially since angular momentum transfer takes place mainly in the high troposphere where winds are strongest. Early and even present upper-air balloon, wind-measuring equipment often fails to measure peak winds, such as the subtropical jet stream maxima and the momentum transport by intense narrow ageostrophic wind belts. In this respect, the satellite and aircraft measured winds in the upper troposphere (Chapter 7) should provide considerable assistance, but the bias of the satellite winds in measuring only in cloudy areas must not be overlooked.

Latitudinal transfer

Given the earlier definition of absolute angular momentum, its transfer across a latitude circle over the troposphere is stated by

$$\Omega_\phi = \oint_0^{2\pi} \int_p^{p_0} (u\, a \cos\phi + \omega\, a^2 \cos^2\phi)\, v\, a \cos\phi\, \delta\lambda\, \frac{\delta p}{g}. \quad (12.5)$$

The usual procedure now is to divide the flow into stationary and "eddying" parts over, say, one month. The eddying part is seen as the synoptic weather systems, from cyclones and anticyclones near the ground to the great upper-tropospheric troughs and ridges. The stationary part may be subdivided by considering transport by the monthly average troughs and ridges, for instance, in addition to transport by the "mean meridional circulation". For the tropics this component is of relatively minor interest excepting no doubt the Andes Mountains; for simplicity it will be omitted. Then the definitions are $u = \bar{u} + u'$, $v = \bar{v} + v'$. Introducing these definitions into Eq. (12.5)

$$\Omega_\phi = \frac{2\pi\, a^2 \cos^2\phi}{g} \left[\int [\bar{u}\,\bar{v} + \overline{u'v'}]\, \delta p + a\omega \cos\phi \int \bar{v}\, \delta p \right]. \quad (12.6)$$

There should also be a term involving the mass field, i.e. the change in surface pressure which strictly vanishes only when integration is extended over the years, so that surface pressure may be taken as constant. However, the term is very small compared to the others and will be omitted.

The last term in Eq. (12.6) cannot contribute if surface pressure remains constant since then, by definition $\int \bar{v}\delta p = 0$; no net mass flow takes place through latitude circles. The momentum transport equation then reduces to

$$\Omega_\phi = \text{const}\, \cos^2\phi \int (\bar{u}\,\bar{v} + \overline{u'v'})\, \delta p. \quad (12.7)$$

The term $\bar{u}\,\bar{v}$ depends on a correlation between these wind components, each averaged around the globe, with pressure. Such a correlation does exist, for instance, in the Northern Hemisphere winter (Fig. 12.4). The term $\overline{u'v'}$ depends on a correlation between u' and v' around a latitude circle computed at any level at, say, longitude intervals of five degrees, and then integrated with respect to pressure. If a trough has northeast-southwest slant the flow ahead of the trough

THE GENERAL CIRCULATION

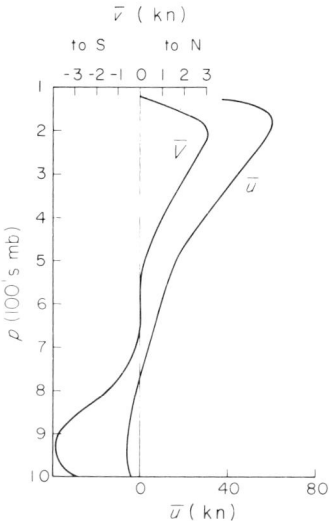

Fig. 12.4. Vertical profiles of zonal and meridional wind speed averaged around globe at latitude 20°N on 12 December, 1957 (14).

from the streamline orientation will carry higher relative momentum u than the northerly current and thus import momentum poleward. A trough with northwest-southeast orientation will do the opposite. Both orientations occur, but the northwest-southeast slant is relatively rare. Discussion about the relevance of the slope of troughs for the tropics

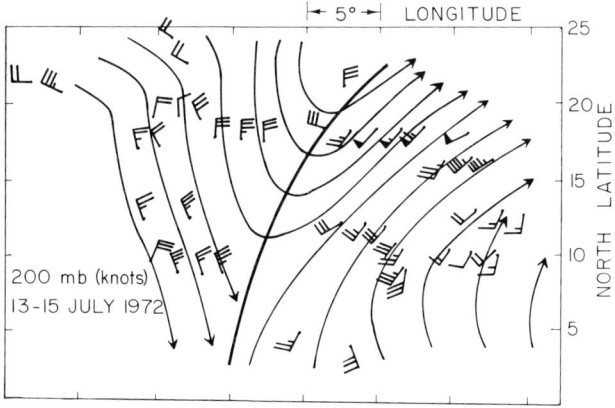

Fig. 12.5. 200 mb winds around trough moving westward across eastern and central Caribbean 13-15 July, 1972. Observations from five 12-hour periods have been composited (66).

was definitely resolved when data increased so that is became possible to draw charts like Fig. 12.5, obviously a model case for northward eddy momentum transport deep in the tropics.

The approximate angular momentum transfer in winter by the mean circulation and "eddies" (Fig. 12.6) is carried in equal amounts by these two mechanisms only in the lowest latitudes. Beyond latitude 15° the $\overline{u}\,\overline{v}$-term declines, demonstrating how the atmosphere avoids the problem of contracting rings of air (31, 59, 61). At latitude 30° the whole momentum exchange takes place via the alternating south and north currents. However, if one adopts coordinates following the subtropical jet stream rather than latitude-longitude coordinates, the picture changes (40). While the net momentum transport of necessity is the same in any coordinate system, the net mass transport across the jet stream axis accomplishes 75% of the flux in jet stream wave coordinates, in contrast to Fig. 12.6. A somewhat uneasy feeling arises when physical "mechanisms" appear to be subject to choice of coordinates. We shall inquire further into this problem, especially in connection with energy exchanges.

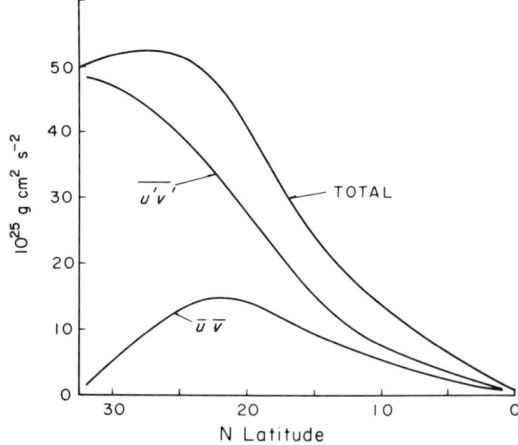

Fig. 12.6. Poleward transfer of absolute angular momentum in the tropics during winter by the mean meridional circulation and by eddies (after 49, 59, 71).

Vertical momentum transport

The latitudinal transport constitutes the bridge between the momentum source of the tropics and the sink in higher latitudes. For balance, source and sink must be equal and of the same value as the

THE GENERAL CIRCULATION   561

transport across the subtropics. The momentum exchange at the interface (Ω_z), positive for upward flow of westerly momentum from ground to air, is given by

$$\Omega_z = -\int a \cos\phi \, \tau_{x,0} \, \delta A \,, \tag{12.8}$$

where A may be the tropical or the extratropical region. If A is chosen as the area of the globe, the total transport must vanish. Introducing the "bulk formula" Eq. (4.9) at 10 m height

$$\Omega_z = -a\varrho_s C_D \int \cos\phi \, \mathbf{u}_{10} \, | \, V \, | \, \delta A \,, \tag{12.9}$$

Since concern here rests with the zonal stress the vector wind $\mathbf{u}$ must be retained explicitly to allow for changes of sign from west to east wind. For east wind (u negative), the angular momentum flow is upward; for west wind (u positive), it is directed from air to ground. Whenever the wind crosses the latitude circles at a small angle, the stress will be closely given by $\mathbf{u} \cdot | \, u \, |$ with magnitude u^2.

It is odd that the areas of east and west wind are equal on earth, since the stress depends on $u^2 \cos\phi$. For a given zonal wind speed, more momentum is exchanged at low than at high latitudes. For instance, assuming $U = -7$ m/s in the tropics where $\cos\phi = 1$, U must become $+10$ m/s (double the kinetic energy) if the reverse momentum transfer occurs at latitude 60° where $\cos\phi = $ one-half. The higher speeds offsets the closer positions with respect to the earth's axis. In view of the degree of freedom given by Eq. (12.8) there is really no obvious reason why the dividing line between easterlies and westerlies should be at latitudes 30°.

Equation (12.8) has proved difficult to evaluate. For possible variation of the drag coefficient see Eq. (9.11). Assuming that the magnitude of the stress is essentially proportional to u^2, one must use $\overline{u^2}$ and not $\bar{u}^2$ in evaluation. Climatically, only the latter value is readily available and, therefore, most published stress values are underestimated, though in the steady trade differences will be small. Both calculation (58) and measurement with aircraft gust probe† between the equator and 10°N have yielded stresses in the range 0·5 to 0·6 dyn/cm². In higher latitudes, especially the continental Northern Hemisphere, the transport across latitude 30°N has sometimes been used to estimate the surface stress.

†In 12 flights between Honolulu and Tahiti with aircraft of the National Oceanic and Atmospheric Administration, courtesy of B. R. Bean.

As may be expected from the vertical zonal wind profiles, nearly all momentum flux from tropics to high latitudes takes place in the high troposphere (Fig. 12.7). Therefore, a way must be found to transport momentum from ground to high troposphere in the tropics and downward again in high latitudes. Our concern is limited to the upward transport. Palmén (59) has found a very ingenious mechanism through the meridional transport of the earth's angular momentum,

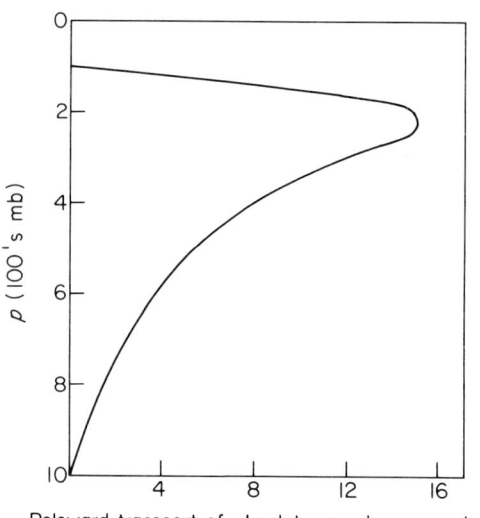

Fig. 12.7. Vertical distribution of absolute angular momentum transport across latitude 30°N in winter (10^{24} g cm^{2}s^{-2}/100 mb).

the last term in Eq. (12.5). Though this term cannot contribute to net momentum flow across latitude circles, there will be convergent and divergent fluxes between latitude circles (Fig. 12.8). In the mean tropical circulation mass rises much closer to the equator than it sinks in the subtropics. The difference is proportional to $\cos^2 \phi \sim 0.2$ between rising and sinking branches. When multiplied with the ascending and descending mass, the vertical transport values of Fig. 12.8 result.

Here, in contrast to the meridional transport, good use has been made of the circulation cell. Altogether, the net upward transport is 84 units. Of these, about 50 are exported poleward in winter (Fig. 12.6). The residual of 30 units must be drawn out of the poleward moving current by small-scale vertical stresses as given by

THE GENERAL CIRCULATION

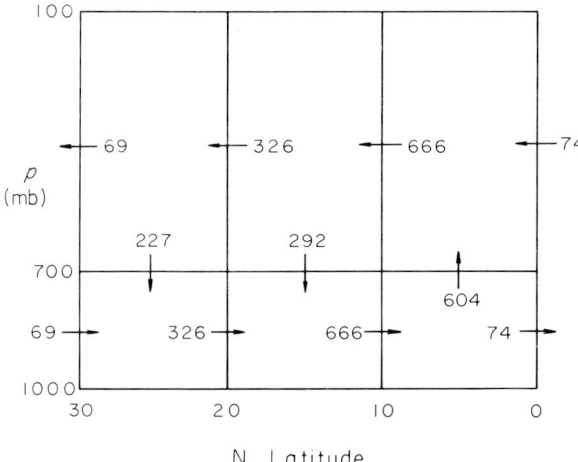

Fig. 12.8. Transfer of earth's angular momentum by the mean meridional circulation of the tropics in the northern winter and the balancing vertical transport. Units: 10^{25} g cm^2 sec^{-2} (59).

$$\Omega_z = - a \cos \phi \, \mu \, \frac{\partial \bar{u}}{\partial z} A.$$

Transfer may be upward into the stratosphere or downward back toward the ground. In the high troposphere the vertical shear in the subtropical jet stream is about 10^{-2}s^{-1}. If the stress is 0·3 dyn cm^{-2}—multiplying the surface stress of 0·5 dyn cm^{-2} with the ratio of 30 units transported down to 50 units exported—the coefficient μ = Transport/Gradient = 30 g cm^{-1}s^{-1}, a reasonable and small value for momentum flux in a gravitationally stable setting. A mechanism requiring cumulonimbus downdrafts need not be invoked. It is curious however, that most heavy precipitation events in the tropics occur with surface west winds (Chapters 2, 3, 8), i.e. they are efficient transporters of angular momentum upward given a "monsoon profile" of zonal wind (that is, westerlies decreasing upward and/or easterlies increasing upward). Especially in summer, ascending and descending branches of the tropical cell need not be arranged latitudinally; on account of the wide meanders of the equatorial trough zone, they may lie side by side at the same latitude (41). Then the Palmén effect would be weak or zero, and one can think of upward momentum transport by undilute towers. From an order-of-magnitude calculation (66, 25) they would be entirely able to take care of the whole transport.

Momentum transport in the Southern Hemisphere

The Southern Hemisphere may be a small momentum source for the Northern Hemisphere with flux of 4×10^{25} g cm^{2}s^{-2} northward determined from ten equatorial stations (28). In the Southern Hemisphere subtropics, angular momentum transport by eddies has been estimated as somewhat less than 30×10^{25} g cm^{2}s^{-2} with little seasonal variation (56). This value corresponds well to the annual transport across latitude 30°N. From the momentum transport, estimates of the mean meridional circulation of the Southern Hemisphere were derived (19, Chapter 1).

One large difference between the hemispheres is provided by the "mountain effect" on angular momentum sources and sinks. For the general circulation, this effect arises mainly due to the Rockies and Andes mountains, which block the path of the westerlies over large meridional distances. The winds cannot circumvent them like the Himalayas (Fig. 8.28). Damming up of the air on the windward side results in higher pressure on the western than on the eastern slope of the mountains—as happens when wind blows against a house. Air is decelerated on the upwind side, but the pressure difference across the obstacle is lost to the air. Thus, there is a "frictional" net loss of momentum to the ground; through the pressure difference across the mountains the atmosphere accelerates the earth and gives up part of its momentum in compensation.

The mountain torque may be deduced as follows from the momentum equation

$$\varrho \, a \cos \phi \, \frac{dU}{dt} = -a \cos \phi \frac{\partial p}{\partial x} + a \cos \phi \frac{\partial \tau_{x,z}}{\partial z}. \tag{12.10}$$

If, for present purposes, we set $dU/dt = 0$ and integrate around a latitude circle and over the height of the mountains (H)

$$\oint \frac{\partial p}{\partial x} \delta x \delta z = \oint \left(\tau_{x,H} - \tau_{x,o} \right) \delta x . \tag{12.11}$$

At the height H the mountain-produced stress will be zero. On the left side, the line integral normally vanishes as pressure comes back to itself. However, when a mountain range inttrudes, this will no longer be the case. If we denote by $\widehat{\Delta p}$ the mean differences of pressure across the mountains integrated from the surface to H,

$$\widehat{\Delta p}\frac{\delta z}{\delta x} = -\bar{\tau}_{xo} . \tag{12.12}$$

The mountain torque is thought to provide a considerable fraction of the transfer of absolute angular momentum from atmosphere to ground in middle latitudes of the Northern Hemisphere (82). It should not be overlooked, however, that in the Southern Hemisphere the high continuous chain of the Andes extends from 30° to 10°S. Therewith it provides a barrier to the westerlies equatorward of the latitude of maximum poleward transport of absolute angular momentum and acts against the accumulation of momentum in the equatorward flowing trades. The height (H) of the mountains may be taken as 4 km. The pressure difference is difficult to assess from the rather sketchy global chart but may be estimated as a drop of 4 mb from west to east side near the surface, decreasing linearly to zero at 4 km. Then, with $x = 40{,}000$ km, τ_{xo} in Eq. (12.12) is $+0\cdot2$ dyn/cm^2, i.e. acting in the sense of transporting westerly momentum into the ground, namely the mountains, as postulated initially. The magnitude is by no means small considering that the stress in the easterlies is about $-0\cdot5$ dyn/cm^2. If, in spite of this large brake, the poleward momentum export of Northern and Southern Hemispheres is about equal, this must be ascribed to the much larger oceanic area with trade winds in the Southern Hemisphere. However, the westerly circulation in the Southern Hemisphere is quite asymmetrical, with a marked 200 mb maximum over 50 m/s in the Australian sector especially in winter, and comparatively weaker winds over South America and South Africa of 35 to 40 m/s (78). The high-level westerlies extend close to the equator in all seasons; in the tropics a half-yearly oscillation with maxima above 500 mb in May and November is observed (77). A connection with the mountain abstraction of momentum may exist but has not been demonstrated. In any event, because of the small seasonal variation, the pressure gradient across the Andes persists throughout the year. Here is the reason for the lack of monsoonal seasonal changes over South America. Westward acceleration of air from the South Atlantic subtropical ridge toward the lee of the Andes occurs at all times. In winter, when there should be outflow even from a cool continent of small width, the lee-in-the-mountains low pressure is strongest and prevents any oceanward-directed monsoon from occurring.

Conclusion

Budgets of absolute angular momentum are popular. They are the easiest budgets to construct since momentum must always be accounted for by itself, in contrast to kinetic energy, heat and moisture. Thus, they provide a check on data quality as a basis for more complex budgets. Further, some insight has been gained: the inadequacy of the meridional cell for poleward flux and also its adequacy for vertical momentum transport, the question of the dividing line between easterlies and westerlies raised again by the additional degree of freedom furnished through Eq. (12.8) and the large impact of the mountain torque of the Andes against the general circulation in the Southern Hemisphere, which awaits further analysis.

Heat and moisture

At the beginning of Chapter 2 the seasonal course of net radiation energy gain or loss by the combined system earth-atmosphere was presented, followed by calculations of latitudinal energy transport to summer and winter poles for heat balance. As the discussion of momentum balance implied, a latitudinal temperature gradient much larger than observed would be required for operation of the simple meridional cell in winter. Therewith, seasonal temperature variations would be far in excess of those observed, especially at high latitudes. The transports of Figs 2.5 and 2.6 are observed only because the general circulation breaks down into large eddies capable of rather rapid meridional heat exchange. The general circulation so places a powerful constraint upon itself, but it is not the only constraint. We must consider radiational energy sources and sinks, their disposal within atmosphere and earth separately as demonstrated by Fig. 12.1 and the exchanges between earth and atmosphere. We shall examine the equatorial trough zone and, for contrast, the latitude belt 20-30°N in winter. In the latter belt, approximate radiation balance is observed, while the equatorial zone gains about $70W/m^2$ which, for balance, must be exported. Just why the energy storage has this particular value, remains uncertain, but we can say without hesitation, recalling Fig. 2.9, that if the latitude belt 20°N-S contained a solid continent, the picture of heat balance of earth plus atmosphere would be vastly different from that observed. On this subject more will be learned from observations on Mars and other planets. For the earth, the predominant oceanic coverage of the tropics is a major

THE GENERAL CIRCULATION

factor, as well as the presence of water vapour and clouds within the atmosphere. Therefore, heat and moisture budgets must be considered together.

Radiation fluxes

The disposal of incident solar radiation, valid for the year in the equatorial trough zone, is presented in Fig. 12.9 and that for the latitude belt 20-30°N in winter in Fig. 12.10 (for summer see (80)). The literature sources upon which these illustrations are based, have been given in Chapter 2 and need not be repeated explicitly here. It may be added, though, that the solar radiation incident on any latitude belt as a function of season and time of day, as well as other relevant information is contained in the Smithsonian Meteorological Tables, Smithsonian Institution, Washington, D.C., 1951, for any reader who desires to make calculations for his own location.

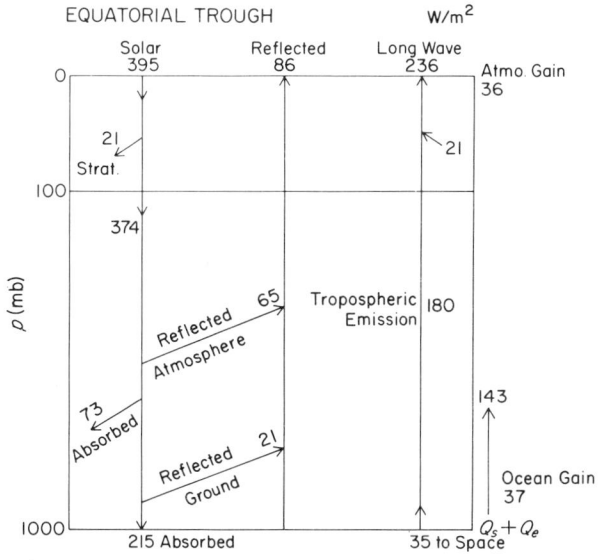

Fig. 12.9. Radiation balance of the equatorial trough zone (W/m²).

Of the incident radiation, a small fraction is absorbed and emitted by the stratosphere, a subject treated in texts on the physics of the upper atmosphere (cf. 11). Tropospheric absorption, especially due to aerosols in the lower atmosphere, has risen over the years in all estimates and is given as about 20% of the solar radiation penetrating into the troposphere, or 23-24% of the incident radiation after deduction of the tropospheric albedo. The actual aerosol content no doubt has risen in the twentieth century, but earlier estimates of the absorption probably were too low. At any rate one can no longer say flatly that the troposphere is "transparent" to solar radiation.

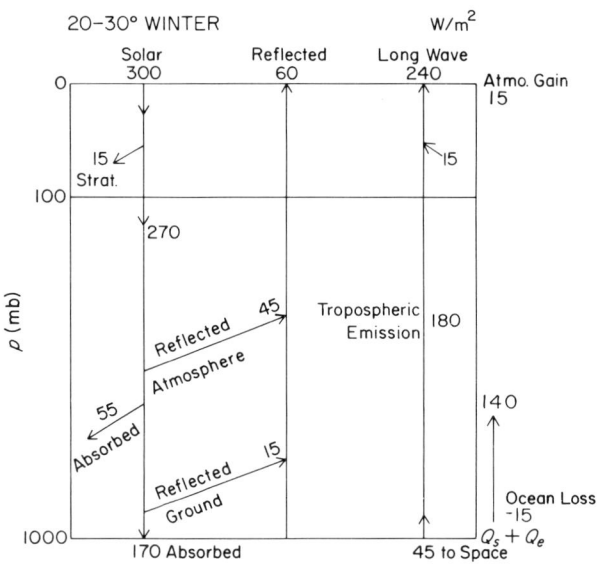

Fig. 12.10. Radiation balance of the latitude belt 20-30°N in winter (W/m²).

Tropospheric albedo, based on Fig. 2.2, is given as 17% for both latitude belts. One would have expected a higher albedo for the equatorial zone due to larger cloud cover, but Fig. 2.2 shows only a small, upward hump there. After allowance for surface albedo of 8-9% of radiation reaching the ground, solar radiation absorbed in the ground is 54-57% of that entering the outer atmosphere.

Turning to long-wave radiation, 16% of the absorbed radiation is returned to space in the equatorial zone and 26% in the belt 20-30°N in winter, reflecting the season and the higher continental fraction of total area in that belt. Now, if 70 W/m² is to remain as positive residual in the equatorial belt, while radiation balance prevails in the belt 20-30°N in winter, tropospheric emission will be 180 W/m² in both belts, which is equivalent to cooling of 1·73°C per day. Since the tropospheric absorption of solar energy is given as 30 to 40% of the long wave cooling, net cooling of the troposphere is 1·03°C for the equatorial zone and 1·2°C for the belt 20°-30°N. The latter computation agrees with Fig. 12.1, but the computed equatorial cooling is rather less and lies halfway between the values in (17) and (43). The residual uncertainty remains; here the computation in Fig. 12.9 will be accepted, especially as it lies in the middle of the range of computed cooling rates.

It remains to examine the energy exchange between earth and atmosphere, based on several estimates of Q_s and Q_e (8, 33). The equatorial net gain of 73 W/m² is split evenly between atmosphere and ocean, in accord with (79). In the belt 20-30°N the ocean suffers an estimated net loss of 15 W/m² equivalent to mean cooling of 0·2°C per month for a water column 50 m deep. In the atmosphere, seasonal cooling of 1°-2°C per month would equal about 5 W/m². The balance must be exported. Poleward transport of $(c_p T + gz)$ decreases from 4·86 to 2·84 × 10¹² kW between latitudes 20° and 30°N in winter (30). Moisture divergence from Fig. 3.2 and also from atmospheric divergence estimates (26) is +1 × 10¹² g/s in this belt, unfortunately given only on the annual basis. However, evaporation is largest and precipitation least in winter; hence, the annual divergence should underestimate the seasonal one. The latent heat equivalent of the annual divergence is 2·5 × 10¹² kW; therewith poleward energy export, including all energy forms, should rise at least from 2·94 to 3·42 × 10¹² kW between latitudes 20° and 30°. The difference of about 0·5 × 10¹² kW corresponds to 12 W/m², the right magnitude for the computed energy gain in the atmosphere to be exported.

Balance in the equatorial trough zone

The information just derived will now be used jointly with that on 'Energy Exchange by Cumulonimbus' (Chapter 4) and 'Trade Wind Energy Budgets' (Chapter 5) to obtain a complete overview of heat and moisture transports and balances in the equatorial trough zone, chosen for convenience as a belt of 10° latitude width adjoining the trough, which, from Chapter 2, is essentially a divide between Northern and Southern Hemisphere energy transports and balances. It may be well to re-emphasize, that the equatorial trough position is not identical with the equator. Many articles calculate energy transport across the equator, both seasonally and annually. Such transport is to be expected since the meteorological equator is at latitude 5°N for the year and since the seasonal position of the equatorial trough zone undergoes wide variations (Fig. 1.10). Reference may further be made to Fig. 1.11, which demonstrates the symmetry of surface variables with respect to the equatorial trough, not the equator.

The residual degree of uncertainty inherent in the budget computations, revolving around the tropospheric long wave emission, has just been stressed. Glancing at Fig. 12.11 on an overall basis, the total energy source from the ocean is estimated as 6·4 units as given in the diagram; the total radiation sink is 4·9 units, hence the net source to be exported is 1·5 units which corresponds to the 73 W/m² postulated from Fig. 2.1. The export is 23% of the source; most of the surface energy is used to uphold the tropical temperatures against radiation cooling. However, if the cooling should be 1·2°C/day (17), the residual for export would only be 12%; for radiation of 0·8°C/day (43) it would be as high as 42%, a large difference indeed, which, if realistic, would bring on or terminate ice ages very quickly. Clearly, knowledge about radiation needs to be tightened greatly before any realistic estimate of future climate changes can become meaningful! Here, we proceed on the basis of Fig. 12.9 and hope that our description is relevant to present climate!

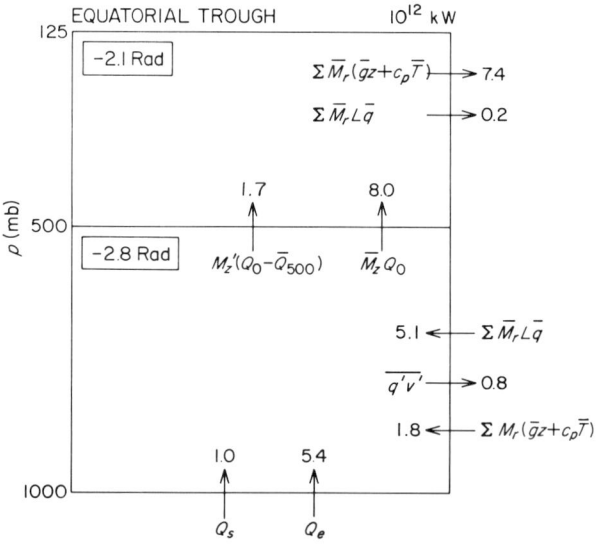

Fig. 12.11. Energy flow diagram, sources and sinks, for a circumpolar belt 10° latitude wide bordering on the equatorial trough.

The transport terms through the boundary of the area considered are defined by Eqs. (5.10) and (5.12) but, as for ATEX, cannot be treated as was done for a special simple case (the Northeast Pacific Trade) with Eqs. (5.13)-(5.16). The complete equations for steady-state energy balance within the equatorial trough zone are:

THE GENERAL CIRCULATION   571

$$Q_s + LP + R_a = 2\pi a \cos \phi \int \overline{(gz + c_p T) v} \frac{\delta p}{g}; \quad (12.13)$$

Also

$$L(E_w - P) = 2\pi a \cos \phi L \int \overline{qv} \frac{\delta p}{g}. \quad (12.14)$$

It is necessary to remember the definitions $\overline{qv} = \overline{q}\,\overline{v} + \overline{q'v'}$; $\overline{Tv} = \overline{T}\,\overline{v} + \overline{T'v'}$. The eddy term $\overline{gz'v'}$ is of small magnitude compared to $\overline{gz}\,\overline{v}$ for heat but not kinetic energy budgets. The eddy term $\overline{T'v'}$ appears small for the equatorial trough zone but will be shown to be very important in the next section.

The mean meridional circulation, a crucial quantity, was taken from (61) at latitude 5°N, which in winter lies 10° latitude north of the mean latitude of the equatorial trough zone (columns 2 and 3 of Table 12.1). When the lateral mass flow is multiplied with the climatic values of $\overline{gz} + \overline{c_p T}$ on the boundary (after subtraction of 300 J/g for reasons given in Ch. 4), one obtains the fluxes labelled $\Sigma\,\overline{M_r}(\overline{gz} + \overline{c_p T})$ in Fig. 12.11. An enormous poleward transport of 5·6 units appears, almost as large as the surface energy source. However, the result is deceiving; for a realistic assessment the latent heat flux must be included. As evident from Figs 3.1, the equatorial trough zone is a consumer of water vapour evaporated elsewhere; its role is to convert latent heat to other energy forms. Hence, $P - E > 0$; from Fig. 3.2 water vapour is imported into the trough zone; in columns 6 and 7 of Table 12.1 the energy transport is computed and appears in Fig. 12.11 as $\Sigma\,\overline{M_r}\,L\overline{q}$. Evidently, much vapour is imported and little vapour is exported in the high troposphere at very low values of q_s. Further, transport of cloud matter around the 10° latitude boundary may be neglected.

It follows that 6·9 units of total energy are imported and 7·6 units exported, difference 0·7 units or 11% of the surface energy source. If we were satisfied with the net radiation cooling of 1·2°C/day, we could stop here, but we would soon have problems at the subtropical ridge where the heat transfer required there could not be met by a factor of two. Since $\overline{T'v'}$ has been computed to be small, an eddy transport $\overline{q'v'}$ of 0·8 units northward must be assumed if the equatorial source is 1·5 units. A positive correlation $\overline{q'v'}$ is found at low latitudes. Our value is larger than that of earlier calculations (14, 72). One can readily

Table 12.1. Lateral energy export for a 10° wide belt adjoining the equatorial trough zone

1	2	3	4	5	6	7	8	9	10
p(100s mb)	v(m/s)	$M_r(10^{13}$ g/s)	$c_p\overline{T}+\overline{gz}$ (j/g, base 300)	$\overline{M_r(c_p\overline{T}+\overline{gz})}$ [10^{10} kW]	$L\overline{q}$(j/g)	$\overline{M_rLq}$ (10^{10} kW)	$\overline{M_r\overline{Q}}$ (10^{12} kW)	Eddy moisture export (10^{12} kW)	Net atmospheric export (10^{12} kW)
10-9	−2·2	−8·8	8·8	−77	37·6	−330	−4·07		
9-8	−1·4	−5·6	13·0	−73	27·5	−155	−2·28		
8-7	−0·4	−1·6	18·7	−30	18·4	−29	−0·59		
7-6	0	0	25·0	0	11·3	0	0		
6-5	0	0	30·5	0	7·1	0	0		
5-4	0·4	1·6	36·4	58	4·2	7	0·65		
4-3	1·2	4·8	42·8	204	2·1	10	2·14		
3-2	1·8	7·2	48·0	345	0·8	6	3·51		
2-1·25	1·2	2·4	55·3	132	—	—	1·32		
10-5				−180		−514	−6·9		
5-1·25				+739		23	7·6		
10-1·25				+559		−491	+0·7	+0·8	+1·5

Positive sign denotes outward flux.

THE GENERAL CIRCULATION 573

see the impact on world climate if variations of the magnitude here computed really took place.

The final step in Table 12.1 and Fig. 12.11, vertical transfer through the 500 mb surface, follows the model of Eq. (4.27), used also in (84) and other deep convection models, and need not be discussed again. Total precipitation heating from the moisture balance is ten units which reduces to 0·8 cm/day or twice the equatorial trough zone evaporation. It is seen that the latter does play a substantial (50%) role, thus variations in ocean temperature entering into Eq. (4.11) could become a potent factor in climate changes. Following the equatorial trough, annual precipitation is 290 cm and seasonal (3-month) precipitation 72 cm compared with 60 cm from Fig. 3.5. The latter number may well be an underestimate and should be upgraded by the satellite radiation estimates described in Chapters 3 and 7; a need for upgrading has become evident for the "maritime" continent of Indonesia. However, considering the range of uncertainty about precipitation, the calculations agree reasonably well with those of Hantel (23) who determined precipitation as residual from all source, sink and flux terms for each month of the year from equator to pole.

In summary, the computations based on new data with large changes in radiation storage, come out rather the same way as an earlier effort (65). The energy export by the mean circulation is very small and uncertain; one could picture it going to zero. The main function of the vertical overturning is transformation of latent heat into potential energy. Energy export is most reliably carried out by eddy moisture flux poleward; half of the export is carried by the ocean.

Energy balance of the tropical circulation cell

The upward energy transport in the equatorial trough zone by means of undilute towers takes place in a very small fraction of the tropical area emphasized. Yet, the impact of these transports must spread quickly, since tropical temperatures change very little from day to day. Due to leaning of the towers in the stronger wind shears usually encountered in the upper troposphere, erosion of heat from the towers has been estimated to provide direct compensation of 25% against radiation cooling (65). Another 25% may be derived from release of heat of fusion or sublimation in the high troposphere. For the balance of 50% of the radiation cooling (about $-0.5°C/day$) slow subsiding motion with magnitude given by Eq. (9.13) must be assumed against the upward mass transport in undilute towers. The descent would have the order of 0·15 cm/s at 500 mb. It is of interest that even from the radiation viewpoint statically stable and very nearly horizontal motion must exist over most of the tropical upper troposphere at all times, in agreement of the concept of a prevailing Rossby number R_0 = 0·1.

Beyond such immediate sinking within the deep tropics, the classical circulation theory envisions net ascending motion in the equatorial zone at warm and net descending motion in the subtropics at cooler temperatures. From this arrangement kinetic energy is generated in accord with Eq. (4.32). The circulation wheel, as stressed earlier in the chapter, rotates at a rate such that the generation of kinetic energy in the tropical cell equals dissipation by friction.

In computing descent on the equatorial side of the subtropical jet stream of winter, reasonable values of sinking motion, again using Eq. (9.13), were obtained giving a duration of one month for a full revolution of the circulation wheel (39). There has been a drastic change in estimates of vertical profiles of net cooling, however (Fig. 12.1); further, in clear areas radiometer sondes have not observed the secondary 300 mb cooling maximum, but rather a monotonous decrease of cooling rate with height, stabilizing the whole troposphere (Table 4.2). Where radiometer sondes pass through clouds and cloud layers (Figs. 4.19, 8.3), the former type of cooling increasing with height is seen but, with reduction of albedo and of the general estimate of amount of cloudiness in low latitudes, the dry type of radiation profile has become dominant. Certainly, Table 4.2 should be applicable to the region of sinking trade winds.

When a stepwise integration of Eq. (9.13) is performed with the net radiation data of Table 4.2, the descent time from 200 mb to the surface will still be nearly a month (Fig. 12.12). But not only is the radiation cooling cancelled; a temperature excess of up to 6°C over the mean tropical atmosphere of the trough zone develops! Equations (9.12)-(9.13) neglecting advection are unable to reproduce the observed temperature gradient. To the radiational cooling of about 35°C per month an additional cold source must be added; and this can only be supplied by the $\overline{T'v'}$ term acting across the subtropics by extracting warm air poleward through the insertion of long, cold troughs into the tropics (Figs 8.37 and 8.40). Only by aid of these troughs is the tropical circulation cell rescued from dying, as it was put earlier. Further, the introduction of available potential energy at high levels enables the intruding extratropical troughs to initiate a low-tropospheric weather system in the tropics (32). Here is thought to lie the mechansism for creation and maintenance of some, but not all, of the synoptic disturbances.

THE GENERAL CIRCULATION 575

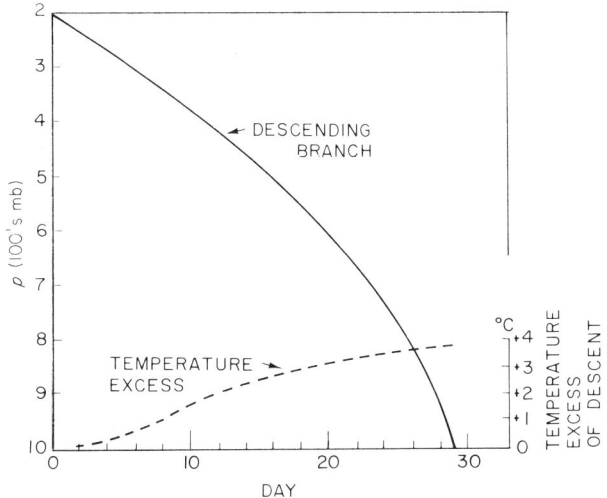

Fig. 12.12. Rate of descent in the subtropical descending branch of the tropical cell computed from radiation cooling only. Dashed: temperature excess of descending air over mean tropical atmosphere (66).

Kinetic energy

The horizontal kinetic energy equation per unit mass for the atmosphere is

$$\varrho \, dK/dt = - \mathbf{V} \cdot \nabla p + \mathbf{V} \cdot \varrho F. \tag{12.15}$$

This is a scalar equation in which the rotation of the earth does not appear; latitude plays no role and kinetic energy calculations are the same for the equator as for the pole. In the momentum equation the zonal pressure and frictional forces could act to augment and decrease the angular momentum of an air parcel. In their place production of kinetic energy by the whole pressure gradient, including the meridional component, and dissipation of kinetic energy by the whole frictional force appear in Eq. (12.15). Kinetic energy is generated from other energy forms and it again goes over into other energy forms—potential energy or heat.

The production term may be subdivided

$$- \mathbf{V} \cdot \nabla p = - \nabla \cdot (p\mathbf{V}) + p \nabla \cdot \mathbf{V}.$$

The first of these terms denotes pressure work done on any particular boundary; the second indicates a large-scale correlation between fields of pressure and horizontal divergence. In a global overview, the left side of Eq. (12.15) will vanish over, say, a year. Pressure boundary work is zero by definition when the whole atmosphere is considered. Thus, there is a balance between the production term $p \nabla \cdot \mathbf{V}$ and frictional dissipation on the boundary and within the atmosphere. In particular the role of the large subtropical anticyclones—yielding a positive correlation between pressure and divergence—as a prime kinetic energy source for the atmosphere has been stressed (70). Here, we shall limit discussion to the Northern Hemisphere winter and the exchange across the poleward boundary with higher latitudes.

Transport across the subtropics in winter

The term $\varrho \, dK/dt$ on the left side of Eq. (12.15) may be subdivided in customary notation

$$\varrho \frac{dK}{dt} = \varrho \left(\frac{\partial K}{\partial t} + \mathbf{V} \cdot \nabla K + w \frac{\partial K}{\partial z} \right).$$

After introduction of the mass continuity equation, the expression may be transformed to

$$\partial(\varrho K)/\partial t + \nabla \cdot (\varrho K \mathbf{V}) + \frac{\partial}{\partial z} (\varrho K w),$$

derived in theoretical texts. If kinetic energy in the tropical belt remains approximately constant, the left term vanishes. Further, one supposes that little kinetic energy is transferred directly to the stratosphere or the ocean, so that the third term is also zero. The middle term then is left; it gives the horizontal flux divergence which balances the right side of Eq. (12.15). If we integrate over the whole tropics and over the troposphere, say, from the surface pressure to 100 mb, and also neglect the small transport across the equatorial boundary, the transport across the subtropics (F_K) may be written as

$$F_K = \int \int K \, c_n \, \delta s \, \frac{\delta p}{g}. \tag{12.16}$$

Here c_n is the velocity component across a boundary s of arbitrary configuration. If one takes the boundary as a latitude circle, say 30°N, $\overline{Kc_n} = \overline{Kv} = \overline{K}\overline{v} + \overline{K'v'}$ after averaging around the circle. This is the formulation most often encountered in the literature (27, 30, 60). Since $\bar{v} \to 0$ at latitude 30°, the term $\overline{K'v'}$ dominates. Although variable from year to year and from month to month, a commonly quoted mean value is 20×10^{10} kW, remarkably high since it balances

THE GENERAL CIRCULATION

about half of the estimated frictional dissipation of the whole extratropical region (62).

The budget analysis may end with this extraordinary finding, but one should be interested to inquire a little more how the large transfer is brought about. We may suppose at once that the energy is drawn from the subtropical jet stream, a wavy current from Fig. 1.15 with velocity maximum of 70 m/s (Fig. 12.2) and vertical cross-section seen in Fig. 12.13 following the wavy axis. The extreme concentration of of kinetic energy in the layer near 200 mb is evident. Figure 12.14 shows spectacularly the build-up of the subtropical jet stream from rows of large, undilute cumulonimbi with sloping anvils above 400 mb along a polar trough intrusion into the tropics north of Puerto Rico (45). From the San Juan, Puerto Rico, rawinsonde observation, the root of the subtropical jet stream was situated at 400 mb and culminated with wind speeds of 45 m/s in the layer 200-150 mb. The distance of the observing aircraft from the clouds to the north was about 100 km; the length of the sloping anvils was also about 100 km, their top near 13 km and vertical motion in the anvil on the order of 10 m/s.

Since the position of the subtropical jet stream axis is relatively stable, we must suppose the steady state is achieved through large, poleward energy transfer extracted from the core. The s-axis in

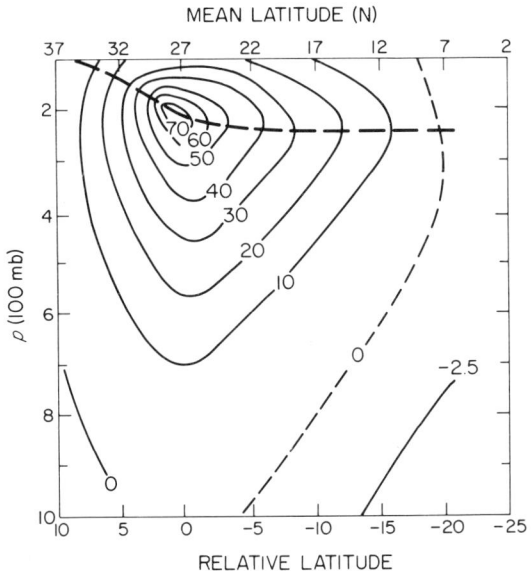

Fig. 12.13. Vertical cross-section of wind speed (m/s) following the axis of the subtropical jet stream (Fig. 1.15) (39).

Fig. 12.14. Shearing of large cumulonimbus anvils into subtropical jet stream north of Puerto Rico seen from aircraft looking toward north on 1 April, 1953 (45).

Eq. (12.16) may be defined as following the core and c_n as the wind component perpendicular to the axis. One observes a remarkable out-of-phase relation between the mean position of the three waves in the subtropical jet stream and the three waves found on the average in the polar westerlies (Fig. 12.15) (53). If the wave number one, clearly evident in the polar westerlies (1) is subtracted, two concentrated areas of interaction are found near the American and Asiatic east coasts, with a third weaker member near 45°E. Figure 12.16 illustrates the interaction there; where subtropical ridge and extratropical trough meet, and where subtropical jet stream speed is highest along the axis (Fig. 1.15), a channel for efflux from the tropics into the polar regions develops (52). Here the mass transport of Eq. (12.16) is concentrated carrying the maximum kinetic energy of the subtropical jet stream out of the tropics.

Due to the regional concentration one could readily think of a standing-eddy transport $[K'\ c_n']$ as accomplishing F_K without the need for recourse to net mass transfer through the jet stream axis. Such transport indeed makes a large contribution. Nevertheless,

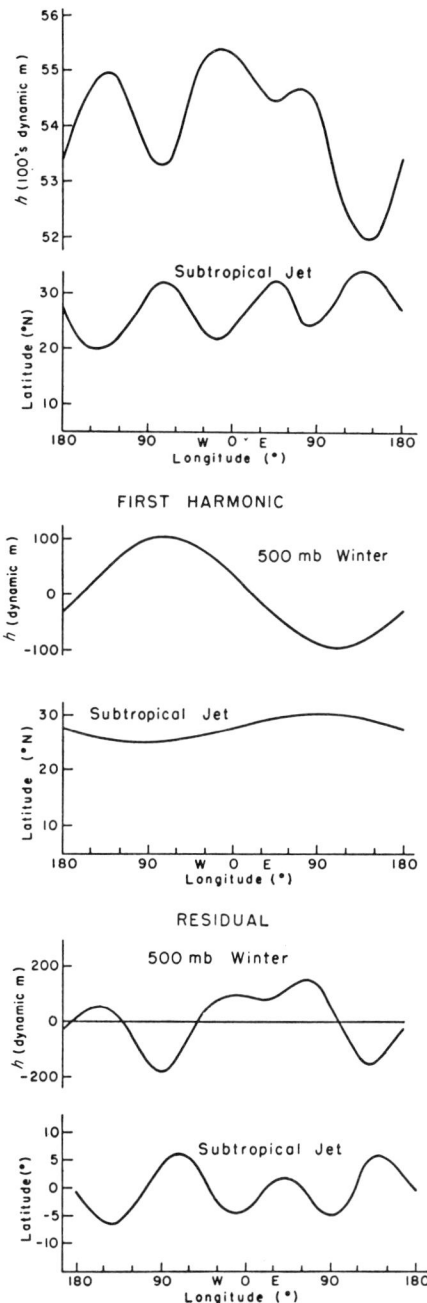

Fig. 12.15. The interaction between high and low latitudes in winter. Left: 500 mb height at 47·5°N and mean latitude of subtropical jet stream; centre: the first harmonic of the curves on the left; right: residual obtained by subtracting the first harmonic from the curves on the left.

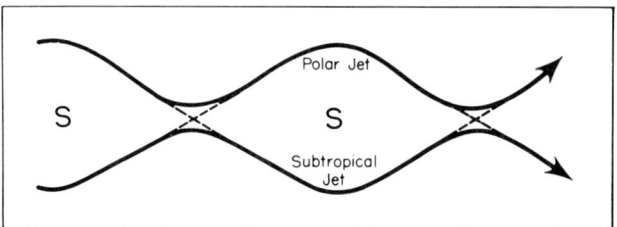

Fig. 12.16. Model of interaction between polar and subtropical jet streams at latitude 30°.

about half of the flux is accomplished by the net flux of mass $\bar{c}_n$ through the axis at these locations, which must be a net ageostrophic transport, thus $\bar{c}_n = \bar{c}_{na}$. In order to see how such transport can arise in a wavy, upper flow pattern, consider the gradient wind equation

$$K_t V^2 = -f(V - V_g) = -f V_a . \qquad (12.17)$$

Here K_t is the trajectory curvature, V_g geostrophic and V_a ageostrophic horizontal wind. If the pattern is stationary, $K_t = K_s$, the streamline curvature. If it is moving, a correction must be applied. K_t must be predominantly cyclonic west of a trough and anticyclonic to its east as in Fig. 12.5. Then $V < V_g$ west of the trough, $V_a = V - V_g$ is negative, i.e. directed backward along the flow (66). East of the trough, with anticyclonic flow, the opposite holds, and V_a is positive downstream. Thus V_a is directed away from the trough toward the ridges on both sides. There, net mass transfer across the axis will occur if the amplitude of the axis is less than that of the streamlines.

For an order-of-magnitude calculation using Eq. (12.16) consider $F_K = 10 \times 10^{10}$ kW or half of the transport, $\delta p/g = 2 \times 10^2$ g/cm², K from Fig. 12.13 $= (50 \text{ m/s})^2/2$ and $s = 40{,}000$ km. Then $\bar{c}_{na} = 1$ m/s, in good agreement with observed values (39). Higher values of transport, up to double that just used and therewith equal to total dissipation outside the tropics, have been obtained in some months (40). Variations of the general circulation on short and long time scales may well be discussed on the basis of Figs 12.15 and 12.16, as well as Eq. (12.16). A physical picture of the interaction between high and low latitudes is provided which the $\overline{v'K'}$ correlation does not yield. As seen in the section on momentum budget, another and possibly deeper view is obtained when Cartesian or spherical harmonic coordinates are replaced by coordinates expressive of the physical system considered.

Production and frictional dissipation in the tropics

Given adequate data, the production may be calculated directly by integrating the generation term in Eq. (12.15),

$$-\mathbf{V} \cdot \nabla p = -(u\frac{\partial p}{\partial x} + v\frac{\partial p}{\partial y})$$

over the volume of the tropics. Here again mean values around latitude circles, standing eddies, daily eddies etc. may be considered. It has proved to be an interesting exercise to make a simple compuation involving only the mean meridional circulation term

$$-\bar{v}\frac{\partial \bar{p}}{\partial y} \text{ or } -\varrho\,g\bar{v}\frac{\partial \bar{z}_p}{\partial y}$$

in isobaric coordinates. In the low levels, with the mean meridional flow from north and pressure decreasing toward the equator, this term give a positive contribution. In the high troposphere northward of latitude 10° $\bar{v}$ is positive, while the height of the isobaric surfaces, especially 200 mb, decreases northward. Again the contribution is positive excepting the interval 0-10°N (27). Following (60) the production, in units of 10^{10} kW, is seven units in the layer 1000-700 mb and 25 units in the layer 700-100 mb, total 31 units. Additionally, production is small in the layer 700-300 mb where the meridional circulation is weak (27). "Standing eddy" contribution in polar coordinates has been estimated as five units (26). In wave coordinates, production is larger, about 40 units compared to 30 in the latitude-longitude coordinate system. However, it must be remembered that the available computations cover different time periods and that difference between results are no larger than natural variations that may occur. All estimates agree that frictional dissipation is only about 10% of production, not surprising since the large, kinetic energies occur in the high troposphere where, with gravitaionally stable motion, internal friction may be assumed to be small.

What the computations really show, is the build-up of the subtropical jet stream toward the poleward margin of the tropics. For demonstration we may integrate Eq. (12.15) for steady and frictionless flow on a constant pressure surface, analogous to the calculation of hurricane outflow (Eq. (11.43)). We obtain the "horizontal Bernoulli" equation:

$$\frac{1}{2}(V_j^2 - V_o^2) = g(h_o - h_j)_p, \qquad (12.17)$$

where the subscripts j and o denote values at jet stream and starting point in the deep tropics. From Fig. 12.2 $V_0 = 0$ for a start at 7°N. Using $\overline{V}_j^2 = (70 \text{ m/s})^2$, $\overline{h}_o - \overline{h}_{j\ 200\ mb} = 250$ m, very nearly as observed; there is very little difference between $\overline{V}_j^2$ and $\overline{V}_{j'}^2$, somewhat surprisingly.

Introducing the hydrostatic equation differentiated between constant pressure surfaces (Eqs. (5.20), (11.46))

$$\tfrac{1}{2}\,\overline{V}_j^2 = g\,\frac{\hat{D}_m}{\hat{T}_m}\,(\hat{T}_o - \hat{T}_j), \qquad (12.18)$$

where the caret may indicate values between 850 and 200 mb, and $\hat{T}_m$ and $\hat{D}_m$ denote mean temperature and thickness of the layer 850-200 mb between latitude 7°N and the subtropical jet stream axis at 27°N average. From Fig. 12.3 $\hat{T}_o$-$\hat{T}_j = 7$°C using Eq. (12.18), just equal to the observed mean temperature difference over the latitude interval. The temperature curve for geostrophically balanced wind is also shown. The temperature gradient for thermal wind balance is less than that needed for kinetic energy production in the tropics. Near latitude 27°, however, the two curves cross and kinetic energy increase cannot continue with the requirement of geostrophic wind met at the same time. Here, then, is the limit which air columns from the tropics can attain except through the mechanism of Figs 12.15-12.16 involving interaction with the middle latitudes. Thus, Fig. 12.3 delineates the poleward boundary of the tropics and the physical mechanism imposing the boundary.

Potential energy

While the "eddy potential" energy transport $\overline{gz'v'}$ is negligible compared to $g\overline{z}\,\overline{v}$ in the heat budget as stated before, it may nevertheless have the magnitude of kinetic energy generation and transport. The latter, at latitude 10°, is about 10×10^{10} kW whereas heat transport is $7 \cdot 6 \times 10^{12}$ kW from Fig. 12.11. Therefore the kinetic energy transport is only a few per cent of the heat transport. In view of the problems with pressure-height data, especially in the upper troposphere, it is very difficult to make reliable estimates of $\overline{gz'v'}$. Nevertheless, this is what Holopainen attempted to do (32) by a roundabout method for daily values at 300 mb and more directly using monthly mean charts. The result at first sight is surprising; a substantial potential energy flux does exist across the subtropical ridge. It is directed into the tropics from higher latitudes and has the same magnitude as the reverse KE flow! Although the calculation rests on various assumptions, it is nevertheless very credible and quite in accord with the mechanisms of kinetic energy exchange advanced earlier: upper troughs carrying high potential energy elongate equatorward with sinking and higher h_p of the air west of the trough than on the east side where the air rises toward higher kinetic energy. This would be the means to provide a negative $z'v'$ correlation. Upon sinking of the cold air, kinetic energy is generated in an equal amount, so that a simple cycle $PE \rightarrow KE$ exists. As Holopainen puts it: "Maintenance of the large-scale eddies in the tropics [in winter] is mainly due to forcing by extratropical eddies [troughs]. This forcing occurs at 30°N as a southward eddy flux of potential energy." This statement completes the cycle of exchanges between high and low latitudes, here pictured beginning with the observations of low R_o and R_{oT} into the deep tropics. It also shows what is really implied when generation of kinetic energy by the mean meridional circulation is computed. For variations of the north-south exchanges, and the occasionally observed great upwelling of equatorial weather activity, the forcing from the extratropical side through the mechanism just pictured is a reasonable and feasible sequence. It is difficult to see how large-scale increases and decreases in tropical convection can be internally generated when there is so little available energy as we have seen, and when undilute cores do not just happen by themselves, but arise from forcing through convergence in a synoptic system which itself is generated through the latitudinal connections just analysed.

Variations of the general circulation

Competing circulation cell and hurricane modes

As noted in Chapter 9 the maximum radius observed in Pacific super-typhoons is about 1500 km. If ocean evaporation is 0·4 cm/day and an equivalent amount of water is condensed and precipitated by the typhoon, the area receiving one cm/day precipitation would have a 1000 km radius, or 2000 km diameter, of course with highly non-linear distribution. The amount of one cm/day is typical for precipitation in the equatorial trough zone; however, the latitudinal width of the zone would be 1000 rather than 2000 km. Thus, one huge hurricane could supplant the "normal" equatorial trough function over 30° longitude. Twelve such hurricanes operating continually with a 30° spacing could replace the whole equatorial trough zone mechanism of precipitation with its inefficient synoptic systems described in Chapter 8. Of course, such a grand arrangement is never observed. The energy release by one storm would be 700×10^{14} kj/day which is about twice the maximum computed for large hurricanes which, however, are not super-typhoons (Chapter 9). Yet, there is no obvious reason why a tropical general circulation of this type is not observed. It would be what a true "Hadley circulation" with $R_o = 1$ in the tropics demands. If it did exist, one may suggest that it would be accepted as the "natural" tropical circulation.

While this "ideal" state is never achieved, it is approached in the Atlantic and other oceans during the main hurricane season in some years, indicating a competitive mode between hurricanes and the usual synoptic systems. Further, from Fig. 9.2, long period trends on the order of half a century have occurred when hurricane frequency increased or decreased by a factor of two. Thus, the average role of the equatorial trough zone in the general circulation is not constant but subject to considerable variation. As yet no explanation can be furnished, but the fact is worth bringing out.

Some individual years in the Atlantic have brought no more than one or two hurricanes, others fifteen to twenty. The increased amount of upper-air observations in the tropics makes it possible at least to demonstrate the differences in upper-air structure even in the monthly average during hurricane-rich and hurricane-poor years. The hurricane season of 1969 was very active, that of 1972 almost suppressed. The difference of vertical zonal wind structure over northern Venezuela has been shown in Fig. 7.11—deep easterlies in 1969 and strong upper westerlies in 1972. We now enlarge on this picture to show the upper-air structure for the whole West Atlantic hurricane region.

A large, subtropical ridge overlay latitudes 20-25° in August 1969. Easterlies at 200 mb was located above the main hurricane spawning region (Fig. 12.17). Light cold-air advection from east took place through the layer 500-200 mb. Outflow from hurricanes could readily find a way into the main belts of westerlies; the circulations were connected with the main meridional temperature gradient of higher latitudes, thus the solenoidal circulation efficient in the sense described in Chapter 10.

In 1972, the 200 mb pattern was precisely inverse to that of 1969 (Fig. 12.18). A deep trough with east-west orientation was situated near latitude 20°, with cold air in the layer 500-200 mb. Sinking took place in this trough, and a large part of the tropical region

THE GENERAL CIRCULATION

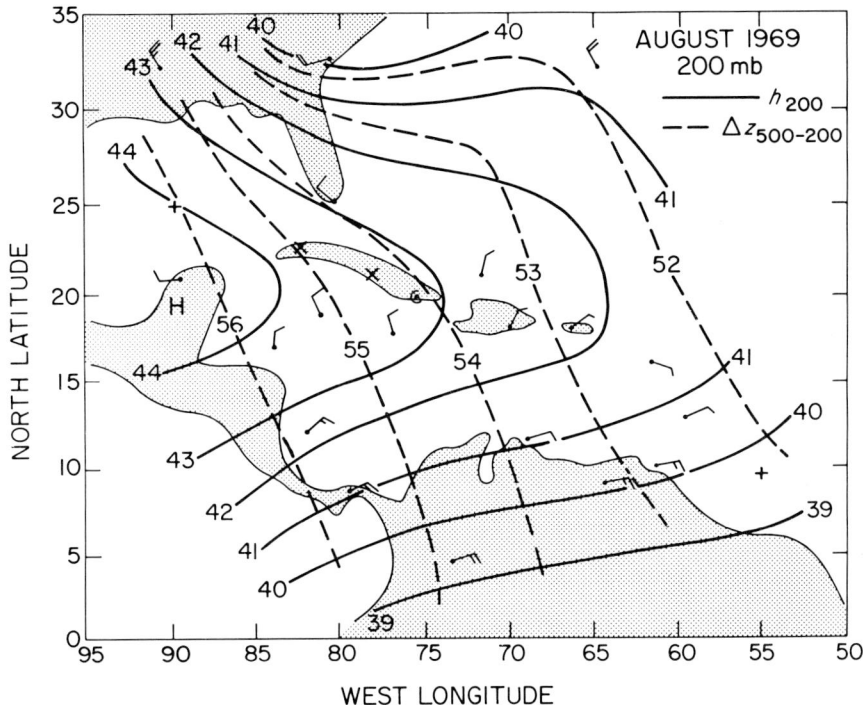

Fig. 12.17. 200 mb winds and pressure heights (10's m, 12 000+) for August 1969; thicknesses 500-200 mb (10's m, 6000 m+).

experienced drought. The trough has some resemblance to what may be expected at the poleward end of the tropical general circulation cell. With high-level convergence and radiational cooling, high cyclonic vorticity and sinking should be found. As noted throughout this chapter, the analogy contains some of the truth. The upper trough position is not far north of where the closest analogue to an equatorial trough could be found in 1969, at latitude 15° right through the middle of the Caribbean. In accordance, the equatorial rain belt remained pressed to the equator; in Venezuela, rainfall above average was measured only at latitude 5°N. Farther north, precipitation averaged 50% of average, a very large depression over so great an area. As noted variously in this book, the broad zone of maximum vertical instability is also the dry zone. Vertical stability is given by the 500-200 mb thickness lines, since the low troposphere remains constant.

Finally, great horizontal turbulence prevailed during August 1972 in the high troposphere, of which Fig. 12.5 was presented as one

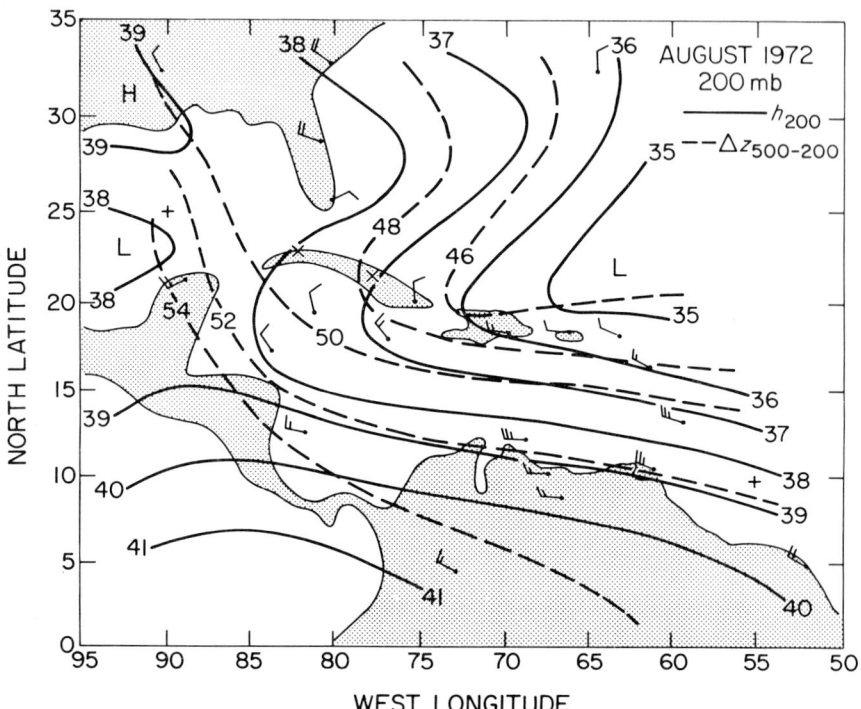

Fig. 12.18. Same for August 1972.

typical example. The upper-air trough cannot survive as a steady-state phenomenon (Fig. 12.12); a considerable number of cold air injections from north occurred during the month, importing potential energy, to keep the mean picture alive.

The pattern of Fig. 12.18, while a very good illustration of the general circulation mode, does not fully dominate all of the rainy season in dry years as it did in 1972, else there would be a very simple relation between seasonal upper flow pattern and precipitation. While the subject deserves more detailed investigation, periods of shorter or longer duration occur in many rainy seasons favouring the occurrence of one or more potent rain episodes not of the hurricane character, similar to the Manila episode of 1972. In the "continental" half of the equatorial trough zone where the alternation between seasons never is as extreme as in the Caribbean, more subtle changes must be found. An early postulate, namely that summer rains in West Africa are proportional to mass transport in the surface layer across the equator from the South Atlantic, does not hold up statistically (68). Rather,

THE GENERAL CIRCULATION 587

the upper air becomes drier and warmer during drought compared to wet years; the subtropical ridge line shifts several degrees farther southward and the strength of the upper easterlies diminishes.

There is no indication that energy exchange between tropics and higher latitudes changes depending on the tropical circulation "mode". However, an understanding of the two modes here illustrated and their competitive relation would undoubtedly contribute in large measure to comprehending short and long period climate variations in the tropics.

Competing meridional and zonal modes

As stressed in Chapter 2, the equatorial trough zone essentially separates energy transport to summer and winter poles. It is the general circulation divide between hemispheres rather than the geographical equator. Over the years, variations of the penetration of the zone into the one or the other summer hemisphere occur beyond the average position; in other years the zone does not attain its average latitude (20). Then one may fear drought in India and Africa; the secondary equatorial convergence zone of the South Indian Ocean (Fig. 8.30) may take on exceptional prominence at the expense of the main convergence zone over northern India. Conversely, during seasons when the southern winter circulation extends strongly into the Northern Hemisphere, one may find many examples of trajectories of Southern Hemisphere synoptic cloud masses across the equator, here shown for the period 18-22 June, 1969 (Fig. 12.19) which, on arrival in northern Venezuela, culminated in the heavy rain event shown in Fig. 3.21-23. Systematic studies of such transequator displacements can be made readily with world-wide satellite cloud observations.

An attractive idea (37) is that during a particularly cold winter in Antarctica the equatorial trough position would be forced farther north to increase the size of the tropical area, from which the heat source is funnelled southward. Opposition of temperature anomalies in Northern and Southern Hemispheres have been noted (54). On the other hand, in extreme drought seasons over India and Africa such as existed in 1972, the satellite photos show exceptionally intense typhoon activity in the Pacific. The Philippines had a very wet season, from which the example of the Manila rain episode was drawn in Chapter 8. The possibility of east-west compensation on a grand scale is suggested thereby, as a competitive mode for variations in the net meridional mass exchange.

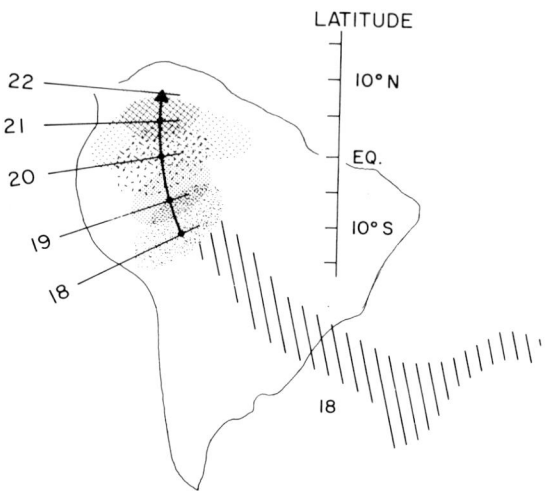

Fig. 12.19. Trajectory of large synoptic cloud mass tracked by ATS III from Brazil across the equator to Caribbean giving rise to rain event shown in Fig. 3.21. Open shading: cold front, from which cloud mass broke loose on 18 June, 1969.

The Southern Oscillation

The concept of large-scale, east-west negative correlations is an old one, which has been furthered considerably by concepts such as "teleconnections" over large distances (15). Since the rainfall of the Indian summer monsoon season is the most critical variable for the food production of one-sixth or more of the world's population, meteorologists have tried for many years to predict rainfall anomalies during the preceding spring. Most active in this endeavour have been G. C. T. Walker and his collaborators. Even though success in prediction has remained limited, Walker happened upon a remarkable circulation parameter for low latitudes which he termed the *Southern Oscillation* (81). He gave the following description: "In general terms, when pressure is high (relative to the mean) in the Pacific Ocean it tends to be low in the Indian Ocean from Africa to Australia; these conditions are associated with low temperature in both of these areas, and rainfall varies in the opposite direction to pressure. Conditions are related differently in winter and summer, and it is therefore necessary to examine separately the seasons December to February and June to August."

Further descriptions of the Southern Oscillation and its definition are given in (3, 75, 83); spatial and temporal variations (74); use in weather forecasting (63, 64); connections between the Southern Oscillation and the trade winds, especially of the North Pacific (4, 5, 6) and numerous other publications.

Ocean-atmosphere relations, El Niño

Closely interwoven, in fact inseparable from the foregoing, are the observations of an equatorial "dry zone" in Mid-Pacific, very cold ocean water along the equator but with substantial variations, and the occasional heavy rains that visit the deserts of the west coast of South America, notably Peru which, coming at the time when normally equatorial-trough rains occur in other southern climates, have been labelled "El Niño" (the child) for Christmas.

Interest in the Southern Oscillation (an imprecise east-west effect between the largest land mass and the largest ocean in the tropics) widened greatly when it became apparent that effects of the oscillation were felt across the whole Pacific Ocean as far as the Peruvian coast of South America and that it closely related to the cold ocean water and dry zone in the equatorial Pacific Ocean. Many of the ocean-atmosphere relations remain in need of clarification, but it may be worthwhile to present a brief summary of this oscillation, a large-scale manifestation of variation of the general circulation.

Walker's statement may be made more precise by noting that "high" and "low" pressure in the Pacific refer to size and intensity of the subtropical anticyclone in the southeastern Pacific, where Easter Island (29°S; 110°W) furnishes the most important regular pressure data. Essentially, then, one is concerned with the strength of the southeast trades which is distinctly variable and at irregular intervals becomes remarkably weak. The strength of these trades moving equatorward alongside the South American coast is related to the amount of upwelling of coastal water which gives the entire coastal strip from about 30°S equatorward its dry, cold climate with fog, low stratus and albino vegetation from the cloud droplets caught in coastal hills. A large tongue of cold water extends westward along the equator across more than half of the Pacific Ocean (Fig. 2.11). First thought to be merely a continuation of coastal advection of cold water, it soon turned out that closed, strong, negative temperature anomaly centres were found along the 10 000 km stretch of equator in the ocean as far as 180° longitude. Sverdrup and others postulated local upwelling along the equator due to divergent atmospheric wind stress upon the ocean as the source for the extensive cold water belt, especially since it is coupled with atmospheric dryness. Figure 3.24 has already shown the southern cut-off of precipitation in the eastern Atlantic where similar though less drastic upwelling takes place on the equator.

A series of cross-sections flown by B. R. Bean (NOAA) between Honolulu and Tahiti in late 1977-1978 has provided remarkable insight into the oceanic temperature structure and associated atmospheric events. The aircraft was equipped with gust probe measuring equipment for direct evaluation of Eqs. (4.1) and (4.2) plus "expendable bathythermograph instruments" released at frequent intervals. The aircraft then monitored the radio signals from the depth measurements down to 300 metres below the surface. One example from twelve rather similar sections (Fig. 12.20) shows the cold water on the equator with the warmest water near 7°N at the equatorial trough, the meeting point of southeast and northeast trades with maximum cloud and precipitation.

In the ocean, a weak, cold dome is indicated at the equator and a much larger one near latitude 10°N, rather similar to a famous cross-section constructed by Defant (13) and Sverdrup (73) from earlier oceanographic expedition data, notably the cruises of the *Carnegie*. The slope of the ocean surface in the north-south direction,

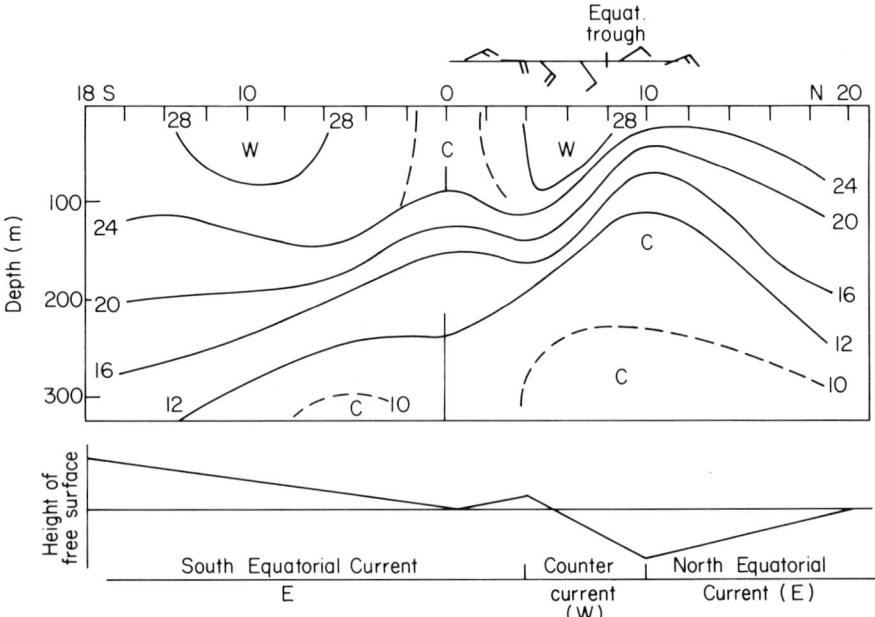

Fig. 12.20. North-south cross-section of temperature in Pacific Ocean at 158°W, 6 December, 1977. Bottom: Schematic indication of slope of free surface and location of principal ocean currents. Courtesy K. Wyrtki (Hawaii).

THE GENERAL CIRCULATION 591

comparable to an atmospheric pressure profile, is also sketched in Fig. 12.20. Where the upwelling occurs, surface heights must be lowest. Assuming that the slope indicates the zonal direction of water movement, an equatorial countercurrent from west appears between about 4-9°N. The equatorial convergence zone and the warmest water are located right atop the countercurrent which has been widely known for years and appears on ocean current charts (Fig. 2.12) including the Atlantic Ocean.

Considering the length and narrowness of the Pacific cold water zone, it is highly probable that a stage is set for large, interannual variations, and these are shown for two successive Octobers in Fig. 12.21. Where cold water anomaly dominated widely in October 1975,

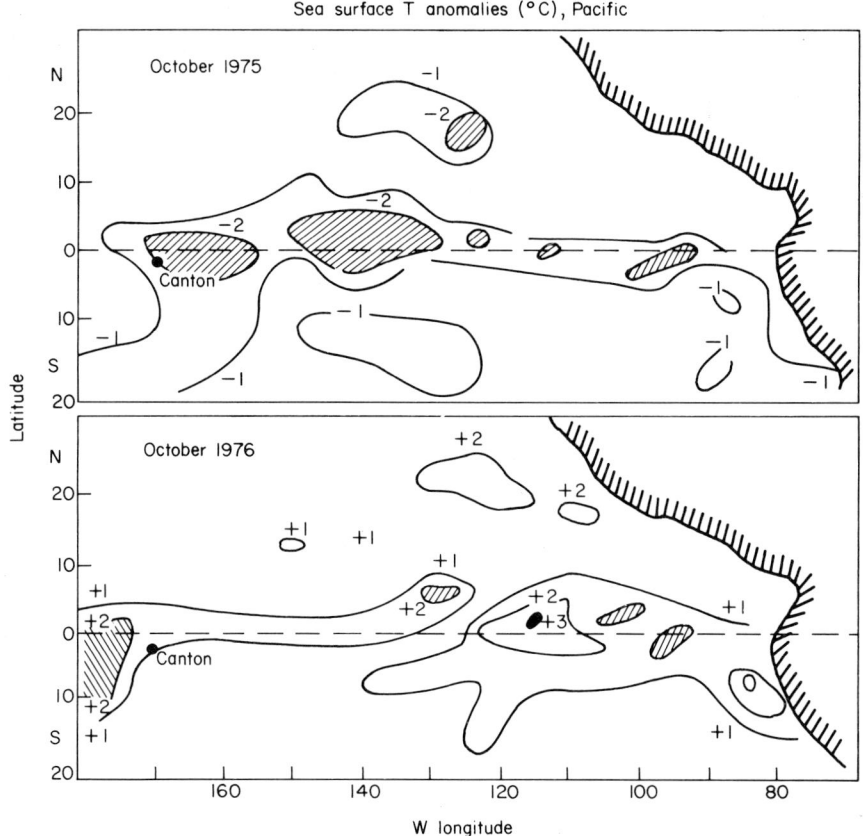

Fig. 12.21. Sea surface temperature anomalies (°C) over the tropical Pacific Ocean October 1975 and October 1976. Courtesy Environmental Research Labs., Nat. Ocean. Atmos. Admin., USA.

relatively warm water is in evidence in October 1976, associated with weakening of the southern trade winds, i.e. the phase of the Southern Oscillation when pressure is low in the Pacific and high in the Indian Ocean area. The large, oceanic temperature fluctuations and especially the cold water temperatures are related to oceanic equatorial upwelling over long distances, thought to be forced by atmospheric wind divergence, though this interpretation is not fully established. In any event, only a small latitudinal shift of the equatorial convergence zone will produce enormous rainfall variations in the mid-Pacific Islands. Indeed, these are so notorious for prolonged dryness, interrupted by occasional heavy rains, that we said in Chapter 3 that computation of climatic seasonal and annual means was an unrealistic statistical exercise. A mean curve was presented for Canton Island only (3°S, 172°W; Fig. 3.12), and this curve must now be examined further.

In Fig. 12.22 the course of ocean surface temperature and of surface air temperature is sketched for six years during the 1960s: it is in sharp contrast to that of Fig. 2.8. Usually air temperature exceeds that of the ocean, and dryness prevails, but at the end of 1963 and 1965 the ocean temperature rose to fully tropical values above 29°C, $T_w - T_{as}$ became positive by over 1°C and immense rains visited the island. From all that has been said in this book, no reader should imagine that the reversal of temperature gradient produced the rains all by itself.

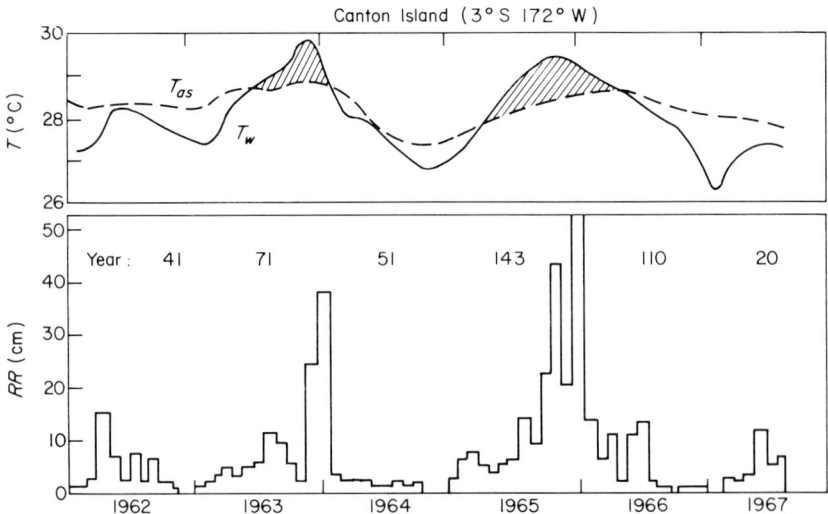

Fig. 12.22. Upper: Monthly ocean temperatures (T_w) and surface air temperatures (T_{as}) at Canton Island (3°S, 172°W). Lower: Monthly and annual precipitation.

All arguments of Chapter 3 related to precipitation occurring strictly within the confines of synoptic weather systems still hold. One must presume types of synoptic situations which introduce surface convergence, thereby stop the upwelling of cold water, remove the dry subsidence temperature inversion and produce the precipitation through the synoptic convergence. It has been observed that there is a positive correlation between heavy equatorial rains and the position of the subtropical jet stream (42, 55). The former paper suggests that dynamic instability in the westerlies of the subtropical jet stream plays an important role in regulating the intensity of the tropical circulation. One can think of the intrusion of troughs with high available potential energy discussed earlier in this chapter. There is no indication of a plausible mechanism derived within the equatorial zone itself.

Along the Peruvian coast the rise of temperature comparable or stronger than that of Fig. 12.21 occurs quite suddenly. How does the warm water get there so fast? Mere cessation of upwelling would not suffice, so interpretation tends in the direction of rapid transport of warm water from the west with the equatorial counter-current—as much as 10° longitude per month. If so, westward equatorial advance of warm water should foreshadow El Niño; continuous watch on the central Pacific Ocean is being kept, coupled with extensive research programmes in ocean and atmosphere.

Here, the phase of the Southern Oscillation becomes important. Normally, surface pressure decreases along the equator all the way from South America to Indonesia. In cold-water and dry years this pressure gradient is enhanced—strong South Pacific anticyclone, low pressure over Indonesia. In the years with opposite phase, the pressure difference is reduced, and the change of warm and wet conditions spreads eastward so rapidly along the equator that one can speak of "teleconnections" (16, 67). It may be added that the change in surface pressure difference is large, 7-8 mb in dry years, half that amount in wet years; an actual reversal of pressure gradient has not been observed as yet.

Secular rainfall changes

In this final section our outlook in time widens to the precipitation trends over a century. Precipitation obviously is the main weather element for which a long term trend is desired for the tropics and interest in its foreshadowing is intense. Perhaps it can be achieved, if ever, only through numerical models or statistical probability functions. Here our purpose will be limited to a small sample of observed trends. It may be noted that there is a close correlation between climatic variation and duration of record. This is perhaps inevitable. When the records extend to 200 or 500 years, longer "periods" than here presented appear.

A convenient tool for analysis is the method of cumulative residuals of precipitation, advocated by Kraus (36). Over the period of record, or over some chosen standard period, the mean precipitation or streamflow are determined, then deviations taken for individual months, seasons, or years, and these deviations are summed up cumulatively. In this way, yearly "turbulence" found in all records is suppressed pictorially and the long term trends are brought out clearly. At the same time, the data are not "massaged" in any way, for instance, by computing moving 10 year averages, so that no imputation of artificially introducing periods in random time series can be made. Especially when there is a network of observations, such allegations are rejected most readily. For streamflow, illustrated for the 150-year history of the Nile, or for hurricane frequency (Fig. 9.2) a network of course cannot be provided. The facts remain, however, and they show in both cases that sudden changes in the general circulation occurred. For a long time (approximately half a century) departures from the average are above or below the mean; then comes a sudden reversal to the opposite deviation which, both in case of the Nile and Atlantic hurricanes, means a one-third change in annual amount or frequency, which is a large change indeed, especially for water resources. The Nile record can be pursued another 100 years backward to 1700. There is every indication of another maximum like that of the year 1900 preceding the minimum near 1840; unfortunately, the record is broken for 25 years around 1800, so that a continuous quantitative backward calculation cannot be carried out. However, it is evident from this record, as well as that of other rivers, that large oscillations with duration of about a century are prevalent, and that computation of climatic averages on the basis of 30 or 40 years is generally highly misleading.

It is not possible to enter here into the large literature on climate changes, especially "desertification", and to what extent they are "natural" or "man-made". As a single example, consider Fig. 12.24, which shows precipitation profiles for the rainy season in northwestern Australia 1900-1920 and 1950-1970 based on a considerable number of long station records. Clearly, there is a large change. Early in the century summer rains ranged widely into the south, so that precipitation changed little over the latitude interval 14-18°S. In the later period, rainfall had increased in the north and

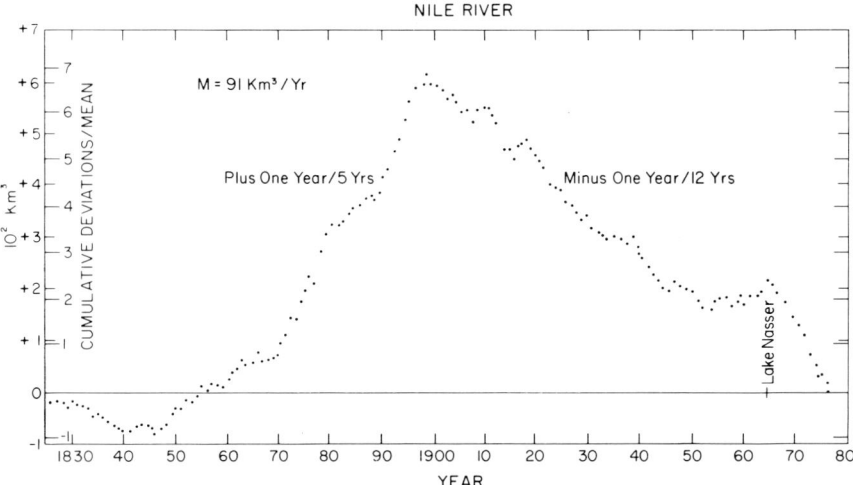

Fig. 12.23. Cumulative percent deviation of Nile River discharge from average 1825-1976. Year of completion of Lake Nasser indicated. Data after 1870 courtesy of High Dam Authority, Aswan, Egypt.

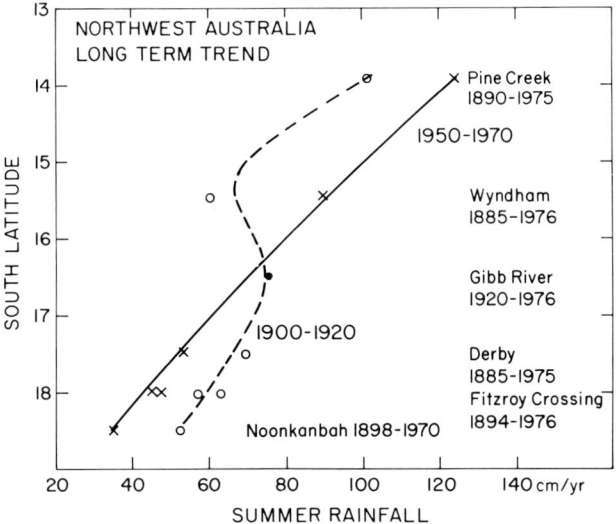

Fig. 12.24. Comparison of rainy season precipitation profile in northwestern Australia for 1900-1920 and 1950-1970. Stations used and length of record indicated. Data courtesy Bureau of Meteorology, Australia.

decreased in the south by considerable percentages; a one-third drop is observed in the south, certainly evidence of the southern desert creeping northward. Total precipitation over all of northwestern Australia has changed but little. There is a strong suggestion that a free path for the predominently westward travelling synoptic systems of the rainy season was open in the south early in the century, while storm tracks became crowded close to the north shore 50 years later. The change was rapid as in Figs 9.2 and 12.23 and occurred around 1930-1935; we find a nodal point at Gibb Riber where the record of yearly precipitation reveals only "white noise".

The suggestion of a shift in storm tracks indicates natural forcing, most likely by the subtropical jet stream which, as shown earlier, is strongest at present in the Australian sector of the Southern Hemisphere (78). Such an asymmetric distribution may not be part of the long-term, upper-air climatology of the Southern Hemisphere; alternatively, the asymmetry may change longitude with time. Blocking of the entrance of tropical rains into the latitude range 16-20°S by increasing strength of the subtropical jet stream of summer, situated over the southern part of the continent, appears to be the most plausible explanation that can be offered for what Fig. 12.24 demonstrates. Certainly, overgrazing by sheep in North Australia cannot be held responsible for the whole subtropical jet stream complex around the Southern Hemisphere.

References

(1) Barrett, E. W. (1961). Some applications of harmonic analysis to the study of the general circulation. *Beitr. Phys. Atmos.* **33:** 280-332.
(2) Baumhefner, D. P. (1968). Application of a diagnostic numerical model to the tropical atmosphere. *Mon. Wea. Rev.* **96:** 218-228.
(3) Berlage, H. P. and deBoer, H. J. (1959). On the Southern Oscillation, its way of operation and how it affects pressure patterns in the higher latitudes. *Geof. Pura Appl.* **46:** 329.
(4) Bjerknes, J. (1951). The maintenance of the zonal circulation of the atmosphere. Union Géod. et Géoph. Intern. Assoc. Météor., Adresse Presidentielle.
(5) Bjerknes, J. (1966). A possible response of the atmospheric Hadley circulation to equatorial anomalies of ocean temperature. *Tellus* **18:** 820-828.
(6) Bjerknes, J. (1969). Atmospheric teleconnections from the equatorial Pacific. *Mon. Wea. Rev.* **97:** 163-172.
(7) Bjerknes, V. *et al.* (1933). "Physikalische Hydrodynamik." Julius Springer Verlag, 797 pp.

(8) Budyko, M. I. Ref. 5, Chapter 2.
(9) Charney, J. G. (1947). The dynamics of long waves in a baroclinic westerly current. *J. Meteor.* **4:** 135-162.
(10) Charney, J. G. (1963). A note on large-scale motions in the tropics. *J. Atmos. Sci.* **20:** 607-609.
(11) Craig, R. A. (1965). "The Upper Atmosphere: Meteorology and Physics." Academic Press, New York and London. 509 pp.
(12) Defant, A. (1921). Die Zirkulation der Atmosphäre in den gemässigten Breiten der Erde. *Geographiska Ann.* **3:** 209-265.
(13) Defant, A. (1936). Schichtung und Zirkulation des Atlantischen Ozeans. Die Troposphaere. *Wiss. Ergebn. Deut. Atlant. Exped. "Meteor"* Berlin) **6:** 1.
(14) Defant, F. and van de Boogaard, H. E. M. (1963). The global circulation features of the troposphere between the equator and 40°N, based on a single day's data. *Tellus* **15:** 251-260.
(15) Doberitz, R., Flohn, H. and Schütte, K. (1967). Statistical Investigations of the Climatic Anomalies of the Equatorial Pacific. *Bonn. Meteor. Abh.* **7:** 76 pp.
(16) Doberitz, R. (1968). Cross Spectrum Analysis of Rainfall and Sea Temperature at the Equatorial Pacific Ocean. *Bonn. Meteor. Abh.* **8:** 53 pp.
(17) Dopplick, T. G. (1972). Radiative heating of the global atmosphere. *J. Atmos. Sci.* **29:** 1278-1294.
(18) Fultz, D. *et al.* (1959). Studies of Thermal Convection in a Rotating Cylinder with Some Implications for large-scale Atmospheric Motions. *Amer. Meteor. Soc. Monogr.* **4** (21): 104 pp.
(19) Gilman, P. A. (1965). The mean meridional circulation of the Southern Hemisphere inferred from momentum and mass balance. *Tellus* **17:** 281-284.
(20) Gruber, A. Ref. 15, Chapter 2.
(21) Hadley, G. (1735). Concerning the cause of the general trade-winds. *Phil. Trans.* **29:** 58-62.
(22) Hantel, M. (1974). A model of the tropical Hadley circulation. *Beitr. Phys. Atmos.* **47:** 90-118.
(23) Hantel, M. and Langholz, H. (1977). Precipitation flux climatology of the free atmosphere. *J. Atmos. Sci.* **34:** 713-719.
(24) Hantel, M. and Hacker, J. M. (1977). Vertical eddy flux of heat and momentum on a planetary scale. *Beitr. Phys. Atmos.* **50:** 134-142.
(25) Hantel, M. and Hacker, J. M. (1978). On the vertical eddy transports in the northern atmosphere. II. Vertical eddy momentum transport for summer and winter. *J. Geoph. Res.* **83:** 1305-1318.
(26) Hastenrath, S. L. (1958). A study of the atmospheric energy budget between equator and 60°N during the winter and summer seasons. I. The latitude-mean conditions. *Beitr. Phys. Atmos.* **41:** 157-183.
(27) Hastenrath, S. (1969). On the role of meridional circulations in the kinetic energy budget of the Northern Hemisphere. *Arch. Meteor. Geoph. Biokl.* (A) **18:** 1-16.
(28) Henning, D. (1968). Untersuchungen zur regionalen Verteilung von Transporten atmosphärischer Feldgrössen über den Äquator. *Beitr. Phys. Atmos.* **41:** 289-335.
(29) Hess, S. (1959). "Introduction to Theoretical Meterology." H. Holt & Co. 362 pp.
(30) Holopainen, E. O. (1965). On the role of mean meridional circulations in the energy balance of the atmosphere. *Tellus* **17:** 285-294.

(31) Holopainen, E. O. (1967). On the mean meridional circulation and the flux of angular momentum over the Northern Hemisphere. *Tellus* **19**: 1-13.
(32) Holopainen, E. O. (1969). On the maintenance of the atmosphere's kinetic energy over the Northern Hemisphere in winter. *Pure Appl. Geoph.* **77**: 104-121.
(33) Jacobs, W. C. (1951). "The Energy Exchange Between Sea and Atmosphere and Some of its Consequences." Univ. Calif. Press. 122 pp.
(34) Jeffreys, H. (1926). On the dynamics of geostrophic winds. *Quart. J. Roy. Meteor. Soc.* **52**: 251-261.
(35) Jeffreys, H. (1933). The Functions of cyclones in the general circulation. *Procés-verb. Ass. Intern. Météor. Phys. Atmos., Mem. Discuss.* (Lisbon) **2**.
(36) Kraus, E. B. (1955). Secular changes of tropical rainfall regimes. *Quart. J. Roy. Meteor. Soc.* **81**: 198-210.
(37) Kraus, E. B. Ref. 17, Chapter 2.
(38) Kraus, E. B. and Lorenz, E. N. (1966). Numerical experiments with large-scale seasonal forcing. *J. Atmos. Sci.* **23**: 3-12.
(39) Krishnamurti, T. N. Ref. 28, Chapter 1.
(40) Krishnamurti, T. N. (1961). On the role of the subtropical jet stream of winter in the atmospheric general circulation. *J. Meteor.* **18**: 657-670.
(41) Krishnamurti, T. N. (1971). Tropical east-west circulations during the northern summer. *J. Atmos. Sci.* **28**: 1342-1347.
(42) Krueger, A. F. and Winston, J. S. (1974). A comparison of flow over the tropics during two contrasting circulation regimes. *J. Atmos. Sci.* **31**: 358-369.
(43) London, J. and Sasamori, T. Ref. 20, Chapter 2.
(44) Lorenz, E. N. (1955). Available potential energy and the maintenance of the general circulation. *Tellus* **7**: 157-167.
(45) Malkus, J. S. and Ronne, C. (1954). On the structure of some cumulonimbus clouds which penetrated the high tropical troposphere. *Tellus* **6**: 351-366.
(46) Manabe, S. (1969). Climate and the ocean circulation. I. The atmospheric circulation and the hydrology of the earth's surface. *Mon. Wea. Rev.* **97**: 740-774.
(47) Manabe, S., Hahn, D. G. and Holloway, J. L. (1974). The seasonal variation of the tropical circulation as simulated by a global model of the atmosphere. *J. Atmos. Sci.* **31**: 43-83.
(48) Manabe, S. and Hahn, D. G. (1977). Simulation of the tropical climate of an Ice Age. *J. Geoph. Res.* **82**: 3889-3911.
(49) Mintz, Y. (1951). The geostrophic poleward flux of angular momentum in the month of January, 1949. *Tellus* **3**: 195-200.
(50) Mintz, Y. (1955). The Total Energy Budget of the Atmosphere. Dept. Meteorol. Univ. Calif. Los Angeles.
(51) Mintz, Y., Katayama, A. and Arakawa, A. (1972). Numerical Simulation of the Seasonally and Inter-Annually Varying Tropospheric Circulation. Climate Impact Assessment Program, Proc. Conf. February 15-16, 1972, pp. 194-216. Dept. Meteor., Univ. Calif.
(52) Mohri, K. (1953). On the fields of wind and temperature over Japan and adjacent waters during the winter of 1950. *Tellus* **5**: 340-358.
(53) Namias, J. and Clapp, P. (1944). Studies of the motion and development of long waves in the westerlies. *J. Meteor.* **1**: 57-77.
(54) Namias, J. (1963). Interactions of circulation and weather between Hemispheres. *Mon. Wea. Rev.* **91**: 482-486.
(55) Nicholls, N. Ref. 49, Chapter 8.

(56) Obasi, G. O. (1963). Poleward flux of atmospheric angular momentum in the Southern Hemisphere. *J. Atmos. Sci.* **20:** 516-528.
(57) Oort, A. H. and Vonder Haar, T. H. Ref. 23, Chapter 2.
(58) Palmén, E. and Alaka, M. A. (1952). On the budget of angular momentum in the zone between equator and 30°N. *Tellus* **4:** 324-331.
(59) Palmén, E. (1955). On the mean meridional circulation in low latitudes of the Northern Hemisphere and the associated meridional and vertical flux of angular momentum. *Soc. Scient. Fenn. Comm. Phys. Math.* **17:** 1-33.
(60) Palmén, E., Riehl, H. and Vuorela, L. A. (1958). On the meridional circulation and release of kinetic energy in the tropics. *J. Meteor.* **15:** 271-277.
(61) Palmén, E. and Vuorela, L. A. Ref. 32, Chapter 1.
(62) Pisharoty, P. R. (1955). The Kinetic Energy of the Atmosphere. Scient. Rep. No. 6. Dept. Meteor., Univ. Calif., Los Angeles. 61 pp.
(63) Quinn, W. H. and Burt, W. V. (1970). Prediction of abnormally heavy precipitation over the equatorial Pacific dry zone. *J. Appl. Meteor.* **9:** 20-28.
(64) Quinn, W. H. and Burt, W. V., (1972). Use of the Southern Oscillation in weather prediction. *J. Appl. Meteor.* **11:** 616-628.
(65) Riehl, H. and Malkus, J. S. Ref. 27, Chapter 2.
(66) Riehl, H. Ref. 28, Chapter 2.
(67) Schütte, K. (1958). Untersuchungen zur Meteorlogie und Klimatologie des El Niño-Phänomens in Ecuador und Nordperu. *Bonn. Meteor. Abh.* **9:** 152 pp.
(68) Schupelius, G. D. (1976). Monsoon rains over West Africa. *Tellus* **28:** 533-537.
(69) Starr, V. P. (1948). An essay on the general circulation of the earth's atmosphere. *J. Meteor.* **5:** 39-43.
(70) Starr, V. P. (1951). Applications of energy principles to the general circulation. In "Compendium of Meteorology." (T. F. Malone, Ed.) pp. 568-576. Amer. Meteor. Soc., Boston.
(71) Starr, V. P. and White, R. M. (1951). A hemispherical study of the atmospheric angular momentum balance. *Quart. J. Roy. Meteor. Soc.* **77:** 217-226.
(72) Starr, V. P. and Peixoto, J. P. (1964). The hemispheric eddy flux of water vapor and its implications for the mechanics of the general circulation. *Arch. Meteor. Geoph. Biokl.* (A) **14:** 111-130.
(73) Sverdrup, H. U. (1947). Wind-driven currents in a Baroclinic Ocean, with application to Equatorial Currents of the Eastern Pacific. *Proc. Acad. Sci.* **33**.
(74) Trenberth, K. E. (1976). Spatial and temporal variations of the Southern Oscillation. *Quart. J. Roy. Meteor. Soc.* **102:** 639-653.
(75) Troup, A. J. (1965). The Southern Oscillation. *Quart. J. Roy. Meteor.. Soc.* **91:** 490-506.
(76) Väisänen, A. (1961). A study of the symmetrical general circulation by the aid of a rotating water tank experiment. *Geophysica* (Helsinki) **8:** 39-61.
(77) Van Loon, H. and Jenne, R. L. (1969). The half-yearly oscillations in the tropics tropics of the Southern Hemisphere. *J. Atmos. Sci.* **26:** 218-232.
(78) Van Loon, H. (1973). A description of the geostrophic wind in the Southern Hemisphere. *Bonn. Meteor. Abh.* **17:** 223-237.
(79) Vonder Haar, T. H. and Oort, A. H. Ref. 38, Chapter 2.
(80) Vuorela, L. A. and Tuominen, I. (1964). On the mean zonal and meridional circulations and the flux of moisture in the Northern Hemisphere during the summer season. *Pure Appl. Geoph.* **57:** 167-180.
(81) Walker, G. C. T. (1924). World Weather II. *Mem. Indian Meteor. Dep.* **24:** 75.

(82) White, R. M. (1949). The role of mountains in the angular momentum balance of the atmosphere. *J. Meteor.* **6:** 353-355.
(83) Willett, H. C. and Bodurtha, F. T. (1952). An abbreviated Southern Oscillation. *Bull. Amer. Meteor. Soc.* **33:** 429-430.
(84) Yanai, M. et al. Ref. 95, Chapter 4.

Appendix A

Text books, monographs, and articles containing extensive data sources or comprehensive literature surveys

A-1   Angell, J. K. and Korshover, J. (1973). Quasi-Biennial and Long Term Fluctuations in Total Ozone. *Mon. Wea. Rev.* **101**: 426-443.
A-2   Brunt, K. (1944). "Physical and Dynamical Meteorology." 2nd ed. Cambridge Univ. Press (428 pp.)
A-3   Eagleson, P. S. (1970). Dynamic Hydrology. McGraw-Hill Book Co. (462 pp.)
A-4   Fletcher, N. H. (1962). "The Physics of Rainclouds." Cambridge Univ. Press (386 pp.)
A-5   Garratt, J. R. (1977). Review of Drag Coefficients over Oceans and Continents. *Mon. Wea. Rev.* **105**: 915-928.
A-6   Haurwitz, B. (1940). "Dynamic Meteorology." McGraw-Hill Book Co. (365 pp.)
A-7   Jaeger, L. (1976). Monatskarten des Niederschlags für die ganze Erde. Berichte des Deutschen Wetterdienstes No. 139. (38 pp.)
A-8   Jehn, K. H. (1973). A Sea Breeze Bibliography. Atmos. Science Group, Univ. Texas, Austin, Texas, Rep. No. 37. (50 pp.)
A-9   Kraus, E. B. (1967). Wind Stress Along the Sea Surface. *Advances in Geophysics* Vol. 12. pp. 213-225. Academic Press, New York and London.
A-10 Kraus, E. B. (1972). "Atmosphere-Ocean Interaction." Clarendon Press, Oxford. (275 pp.)
A-11 Landsberg, H. E. Ed. "World Survey of Climatology." A series of volumes. Elsevier Publishing Co.
A-12 Newell, R. E., Kidson, J. W., Vincent, D. G. and Boer, G. J. (1974). The General Circulation of the Tropical Atmosphere and its Interactions with Extratropical Latitudes. Mass. Inst. Tech. Press, Cambridge, Mass.
A-13 Palmén, E. and Newton, C. W. (1969). "Atmospheric Circulation Systems." Academic Press, London and New York. (603 pp.)
A-14 Priestley, C. H. B. (1959). "Turbulent Transfer in the Lower Atmosphere." Univ. Chicago Press. (130 pp.)
A-15 Ramage, C. S. (1971). "Monsoon Meteorology." Academic Press, London and New York. (296 pp.)
A-16 Reed, R. J. (1965). The Present Status of the 26-Month Oscillation. *Bull. Amer. Meteor. Soc.* **46**: 374-387.
A-17 Reiter, E. R. (1971). Atmospheric Transport Processes Part II. US Atomic Energy Commission. Div. Tech. Info., Washington, D.C.
A-18 Riehl, H. (1978). "Introduction to the Atmosphere." 3rd ed. McGraw-Hill Book Co. (410 pp.)
A-19 Riehl, H. (1977). On the Weather of Venezuela (English and Spanish). National Center for Atmospheric Research, Tech. Note NCAR/TN-126+STR. (41 pp.)
A-20 Roll, H. U. (1965). "Physics of the Marine Atmosphere." Academic Press, London and New York. (426 pp.)

A-21 Sadler, J. C. (1969). Average Cloudiness in the Tropics from Satellite Observations. East-West Center Press, Honolulu. (22 pp.)
A-22 Trewartha, G. T. (1961). "The Earth's Problem Climates." Univ. Wisconsin Press, Madison. (334 pp.)
A-23 van de Boogaard, H. (1977). The Mean Circulation of the Tropical and Subtropical Atmosphere—July. National Center for Atmospheric Research Technical Note TN-118+STR. (41 pp.)
A-24 Washington, W. M. and Kasahara, A. (1970). A January Simulation Experimentation with the two-layer version of the NCAR Global Circulation Model. *Mon. Wea. Rev.* **98:** 559-580.
A-25 Yanai, M. (1964). Formation of Tropical Cyclones. *Rev. Geophys.* **2:** 367-414.

Subject Index

A

Acceleration
  buoyancy, 139
    affected by weight of water, 154
  in hurricanes, 470
  in trade winds, 244
Air pollution, 284
Air-sea interaction
  formulae, 127-134, 253
  in general circulation, 567-571
  in hurricanes, 408, 417, 455
  in synoptic weather systems, 169, 326
  in trade winds, 234
Aircraft measurements
  in hurricanes,
    low levels, 401, 456
    upper levels, 413, 432
    temperature reduction, 293
    upper winds from, 298
Albedo
  in general circulation balance, 567
  global, in tropics, 41
Andes, 564
Angular momentum (*see* Momentum, angular)
Anticyclones
  above hurricanes, 419
  cellular, 552
  subtropical,
    surface, 11
    upper, 21
Atmosphere, mean tropical, 60

B

Bar of hurricane, 426
Basic currents, 24
Beta effect
  in hurricanes, 488, 541
  at trade inversion, 230

Bénard cells, 126
Boundary layer
  hurricane modelling of, 509, 513, 539
  subcloud, 134
Boundary layer modelling in hurricanes, 511-514
  bottom, 512, 522, 539
  closed, 512, 539
  hurricane vortex, 512
  inner, 525
  lateral, 511, 523, 524, 539
  open, 512, 539
  top, 524

C

Circulation, meridional (*see* Meridional circulation)
Climate, lakes and reservoirs
  precipitation over, 280
  stimulation, 282
Climate, local
  air pollution, 284
  cloudiness,
    diurnal variation, 257
    diurnal variation, over islands, 257
    seasonal, over islands, 260
  fog, 285
  land and sea breeze, 263
  moisture redistribution, 256
  mountain and valley breeze, 270
  radiation, surface, 254
  rainfall, diurnal,
    over lakes, 280
    over land, 272
    over sea, 271
  subcloud layer, 254
Climate, mountain
  mountain and valley breeze, 270
  precipitation, 276

Climate, mountain (cont.)
  variation with height, 3
Climate variation
  of general circulation, 583-596
  hurricanes, 396, 583
Clouds
  banded, 147
  cover across tropics, 41
  cumulonimbus (see Cumulonimbus clouds)
  diurnal change
    over coasts, 267
    over land, 257
  in hurricanes, 426
    ice cascades, 413, 429
  observations, 296
  trade wind, 224, 226
  types, 148-151
  whole sky code, 145
  wind shear, affected by, 162
Cold core synoptic systems
  extreme rainfall, with, 370
  upper cyclones, 353-358
    maintenance, 355
  waves, 322
Computers, electronic, 497
Condensation level
  of free convection, 134
  island heating, affected by, 259
  lifting, 141
Conditional instability, 468, 509, 537
Continuity equation of mass, 7, 500, 504
  between constant pressure surfaces, 245
  flux, 513
  transport, 532
Convection
  cells, 126
  cumulus, 162-166, 508-510
  cumulonimbus growth, 157
  deep, vertical mixing by, 167
  entrainment, 155
  general circulation, relation to, 160, 188
  life cycle, 162-166
  modelling, 184
  onset in subcloud layer, 125
  parcel ascent, 153
  and seasonal temperature lapse rates, 75

  shallow, heat transport by, 170
  slice method, 154
  undilute, 158
Coordinates of equations of motion
  cylindrical, 337, 441, 472
  prediction equations, 504
  natural, 244
  sigma, 502, 518
  vertical, 518
Cumulonimbus clouds
  downdrafts, 175
  energy exchange by, 179
  growth, 157
  outflow, 160
  in subtropical jet stream, 578
  in synoptic systems, 331
  tops, 159
  in tropical cyclones, 433
  undilute, 158
  wind shear, affected by, 163
Cyclones
  high tropospheric, 353
  monsoon depressions, 360
  subtropical, 358

D

Deformation of windfield, 507
Diffusion coefficients, 133, 506, 507
  K-theory, 506, 508
Divergence and convergence
  ATEX, determined in, 240
  and average precipitation, 89, 119
  equatorial, 18, 27, 589
  frictional, 509, 513, 539
  of surface meridional circulation, 7
  synoptic, 168
  in trade inversion, 207, 220
  in tropical cyclones,
    surface, 403, 406
    upper, 423
  in waves in easterlies, 349
Drag coefficient, 133, 446, 513, 514
Drainage of cold air, 258, 270
Drought, 94, 101, 119, 584
Dynamic instability, 472

E

Ekman pumping, 513
El Niño, 589

SUBJECT INDEX 605

Energy budgets,
 in ATEX, 240
 in equatorial trough zone, 566-573
 in hurricanes, 446-450
 in Pacific trade, 235
 in synoptic rain area, 188-197
 in tropical circulation cell, 573
Energy of motion
 in general circulation, 575
 production and dissipation, 581
 in synoptic systems, 351
 in tropical cyclones, 446-449, 486
  modelling, 526, 527
Energy, potential
 gain from kinetic energy, 141
 import into tropics, 582
 release of, 196
 in hurricanes, 473, 479
Energy, static
 defined, 80
  for ice processes, 375
 of mean tropical atmosphere, 62
 relative equatorial trough, 68
 resistance against overturning, 224
 synoptic mixing, 331
 in trade inversion, 208
 undilute convection, conserved in, 158, 430-434
Energy tubes, 190, 331
Entrainment
 in cloud layer, 155
 in subcloud layer, 135-139
Equator, meteorological, 9, 54
Equatorial low pressure trough
 energy balance through convection, 179, 569
 longitudinal variation of mean position, 17
 precipitation relative to, 86
 profiles, relative to, at surface, 18
 seasonal course, 16
 surface temperature at, 55
 temperature and humidity cross-sections, 62-66
 and tropical cyclone formation, 460
Equatorial westerlies
 in mean circulation, 12
 in stratosphere, 31
 synoptic, 346, 360
 and vertical momentum transport, 563

Evaporation
 compared to precipitation, 84
 determination
  over land, 297
  over sea, 129-133, 512, 514, 539
 in downdrafts, 175
 potential, 85
 profile, 84
 in trade winds, 235, 240
 in tropical storms, 408, 448, 451
Extended troughs, 386
Eye of tropical cyclones
 humidity, 436
 radar view, 309, 413
 rainbands, penetration into, 463
 static energy profile, 438
 surface structure, 412
 temperature, 430, 436
 vertical motion, 437
 wall structure, 428
  ice cascades, 429

F

Finite differences, for modelling, 515
 centred space, 516
 energy conserving, 518
 forward, 517
 Kuri-box, 518
 leapfrog, 517
 time, 520
 truncation, 540
 upstream advection, 517, 538
Fog, 285
Friction layer
 surface, 133
 trade winds, 244
 tropical storms, 445, 457

G

General circulation
 budgets,
  angular momentum, 557-566
  heat and moisture, 566-574
  kinetic energy, 575-583
 classical model, 2-8, 546
  critique of, 547-556
 cumulonimbus outflow into, 160, 187
 description,
  surface, 8-20
  troposphere, 21-29

General circulation *(cont.)*
  stratosphere, 29-36
  variations of, 583-596
Geostrophic wind in tropics, 549-553
  thermal, 554-556
  quasi-geostrophic assumption, 503
Gridding, for modelling, 518, 540
  fine, 520
  nesting, 519, 540
  three-dimensional, 519

H

Hail, 147
Heat transport
  in ATEX, 240
  by cumuli, 170
  by cumulonimbi, 158, 179, 188
  in downdrafts, 175
  horizontal, equations, 234, 514
  meridional, whole tropics, 43-48
  modelling, 184-188, 508-510
  from ocean, 127-134
  in Pacific trade, 235
  in trade inversion, 224
Himalayas, flow around, 363
Horizontal scaling, 518
Humidity
  diurnal change at surface, 255
  drop at trade wind inversion, 207
  in eye of hurricane, 436
  in mean tropical atmosphere, 61
  observations,
    surface, 293
    upper air, 301
  related to saturation, 63
  relative equatorial trough, 67
  in subcloud layer, 135
  synoptic systems cross-section, 325, 350
  trade winds, 235, 240
  transport by cumuli, 170
  in tropical cyclones, 408
Hurricanes (*see* Tropical cyclones)

I

Ice processes
  role in heavy rain events, 375
  in tropical cyclones, 431
Initial conditions, 520-521
  of hurricane vortex structure, 521

Islands
  diurnal heating over, 257
  precipitation affected by, 276
  windward and leeward sides, 279
  seasonal course of temperature, 260
Instability
  barotropic,
    in waves, 345
    in tropical cyclones, 472
  convective, in tropical cyclones, 468
Interaction high and low latitudes,
  across equator, 298-301
  across subtropics, 579
  extended troughs, 386
Intertropical convergence zone (*see* Equatorial trough)
Isothermal expansion, 408, 430, 455-458

J

Jet stream, subtropical,
  balance of forces, 555
  computation of highest speed, 581
  exchange across, 577
  profile maxima, 20, 553, 577
  seasonal change and monsoon onset, 363

K

Krakatao winds, 34

L

Lakes and reservoirs, climate (*see* Climate, lakes and reservoirs)
Land breeze, 264
Lapse rate of temperature
  in eye of tropical cyclones, 430, 438
  of mean tropical atmosphere, 61
  in subcloud layer,
    over sea, 135
    over land, 136-137
  superadiabatic, 136
  in trade inversion, 203, 208-211, 216-217
Layer, moist and dry, in trades, 231
Life cycle
  of cumuli, 166, 167
  of long-term climate trends, 583-596
  of tropical cyclones, 396
Linear momentum (*see* Momentum, linear)

SUBJECT INDEX

Local climate (*see* Climate, local)
Local precipitation (*see* Precipitation, local)
Local winds (*see* Winds, local)

M

Meridional circulation
 in classical general circulation theory, 546
 divergence of, 7
 heat transport by, 570
 momentum transport by, 563
 relative equatorial trough, 19, 180
 surface, 5
 troposphere, 27
 vertical motion of, 8
Mesoscale rain areas
 defined, 114
 ensembles, 117-119
 in extreme rain episode, 372
 radar echoes, related to, 115
 scattering, synoptic, 336-339
 in squall lines, 378
 in synoptic envelopes, 321
Meteorological equator (*see* Equator, meteorological)
Microanalysis, 307
Models
 boundary conditions, 511
 general circulation, 511, 541
 hurricane, 498, 521, 535, 540
 initial conditions, 520
 large-scale, 498, 540
 non-hydrostatic, 537
 sensitivity studies, 535, 538, 539
 steady-state, 521-532
 symmetric, 540, 504
 three-dimensional, 540
 for whole tropics, 541
Momentum, linear,
 surface exchange, 133
 transport by cumuli, 163
Momentum, angular,
 in general circulation, 2, 501, 522, 546, 551
  budget, 557
 in hurricanes, 421
  budget, 440, 523
 and hurricane formation, 465
 reduced by mountain ranges, 564
 rings with greater than earth momentum, 34
 in synoptic systems, 337
 in trade wind dynamic balance, 244
Monsoon
 depressions, 361
 onset, 361
 retreat, 369
 satellite view, 364
 seasonal temperatures, 58-59
 tropical cyclones, 459
 variations in season, 367
Mountain barriers
 Himalayas, flow around, 363
 Andes, braking westerly momentum, 564

N

Navier Stokes equations, 500
Nile River, secular change, 595
Numerical methods, 514-520
 consistency, 516
 convergence, 516
 finite differences, 515
 nesting, 540
 spectral, 515
 stability of, 516, 538

O

Ocean
 currents, 54
 diurnal radiation course, 251
 energy transport by, 567
 equatorial countercurrent, 567
 evaporation from, 129
 role in trade wind budgets, 235, 240
 sea surface isotherms, 54
 shore climate, 261
 temperature variations in Pacific, 591
 temperatures in tropical cyclones, 417
  and formation, 465
  and long period trends, 467
 upwelling,
  in Pacific, 589
  under trade inversion, 208

P

Parameterization, convective, 188-197, 508-510, 537

Parcel ascent, 153
Potential energy (*see* Energy, potential)
Potential temperature, equivalent, 80
  and sea breeze, 258
  at top of trade inversion, 214
  in tropical cyclones, 455
Precipitation
  annual distribution, 82-88
  annual and seasonal variability, 101-105
  as climatic element, 82
  composition, from daily rains, 106
    extremes, 100
  and mean divergence fields, 89, 119
  from mesoscale systems, 114, 321
  relative equatorial trough, 88
  secular changes, 593
  synoptic catastrophe, 370
  in tropical cyclones, 409
  along upper troughs, 351
  in waves in easterlies, 332
  in waves in westerlies, 383
Precipitation, local
  diurnal variation,
    over sea, 271
    over land and islands, 272
    on windward and leeward sides, 279
  variation with height, 278
Precipitation, seasonal march
  Africa, central, 92
  Africa, west, 90
  Atlantic Ocean, western shore, 92
  Indian Ocean, 99
  Maritime continent, 97
  Pacific Ocean, 96
Prediction methods
  analogue, 533
  dynamical, 498, 500
  dynamical-statistical, 499, 534
  general circulation, 498, 541
  hurricane, 497, 504, 540
  inductive, 499, 532-534
  regression equations, 533
  weather, numerical, 497, 540
Pressure
  changes, synoptic, 290, 328
  chart construction, 291
  cross-isobar flow and hurricane formation, 470
  diurnal variation, 286
  fast travelling waves, 291
  height, vertical profile, 326
    computations, 301
  in hurricanes, 399, 422
  observations, 290
  relative equatorial trough,
    surface, 19
    upper air, 67
  sea level profiles, 6
  wind relations,
    synoptic systems, 323
    in hurricanes, 407
  world distribution, seasonal, 13-16

Radar
  analysis, 308
  bands, 309, 428
  defining mesoscale rain areas, 115
  hurricane eye, seen by, 309, 413
  synoptic rain area, 324
  tropical cyclone cross-section, 435
Radiation
  balance,
    in equatorial trough, 567
    in subtropics, 568
  in cumulus skies, 173
  diurnal change, 251, 257
  effects in synoptic systems, 319
  in general circulation, 547
  satellite measurement of, 40-42
  in subsiding air, 174
  at trade inversion base, 209
  in tropical cyclones, 449
Radiosonde
  observations, 301
  network, 302
Rainstorms
  defined, 108
  examples, 109-114
  mesoscale concentrations, 114-120
Rainstorms, tropical
  frequencies, 340
  humidity structure, 168, 325
  life cycle, 331
  location in synoptic systems, 332-336
  Manila episode, 370
    ice processes in, 375
  propagation, 340
  scattering in synoptic envelopes, 336

SUBJECT INDEX 609

thermal structure, 158, 325
vertical cross-section, 322-325, 327
Rankine vortex, 439
Recurvature of tropical cyclones, 486
Rossby number, 550
  thermal, 554

S

Satellites
  hurricane views, 306, 395, 427, 460, 461
  radiation observations, 40-42
  rainstorm propagation, 341
  synoptic rain areas, 384, 387
  temperature profiles from, 302
  temperatures and monsoon advance, 364
  wind computations, 298
Sea breeze, 263
  factors in, 264
Sea surface
  rough and smooth, 134, 446
  state in tropical cyclones, 415
  temperatures, 53
  upwelling, 208, 589
Seasons in tropics, 39
Shearing stress
  at surface, 133
  components, 506, 507, 522
  diffusion by, 538, 539
  internal, 502, 505
  Reynolds, 505, 508
  in trade winds, 245
  in tropical cyclones, 445, 457
Slice method, 154
Southern Oscillation, 588
Squall lines, 378
Static energy (*see* Energy, static)
Stratosphere winds
  biennial oscillation, 31-34
  in hurricanes, 425
  tracers, 34
  unsteadiness, 29
  waves in, 352
Streamlines, stream function
  daily analysis, 303
  mean, surface, 14
Subcloud layer
  air-sea interaction, 127-134
  buoyancy, 125
  destroyed by moist convection, 144
  diurnal warming over land, 254
  dry convection cells, 126
  mixed, 134
  turbulent exchange in, 138
  vertical structure, 134-138
Sun, annual march, 16
Surge of trades, 386
Synoptic climatology, 310
  use in seasonal forecasting, 311

T

Temperature, dry
  and annual march of sun, 40
  cross-section relative equatorial trough, 62-66
  in GATE area, 66
  longitudinal profile in Pacific, 56
  observations, surface, 293
    satellite, 302
    upper air, 301
  mean vertical structure, 60-62
    regional contrasts of, 71
  ocean, whole tropics, 54
  seasonal changes at 500 mb, 73
    related to convection, 75-78
  surface, whole tropics, 48-54
  synoptic cross-section, 325
  through trade inversion, 207
  upper, in tropical cyclones, 430
Temperature, local changes
  diurnal, over islands, 257
    diurnal, land and sea, 255
  morning, over land, 254
  in mountain regions, 39
  seasonal, over islands, 260
Temperature, moist
  dewpoint observations, 293
  equivalent potential, 80
  virtual, 140
Temperature, seasonal march at surface
  Africa, 58
  Atlantic shore, 57
  monsoon, 58
  Pacific, 58
  whole tropics, 54
Tracers
  in troposphere, 28
  in stratosphere, 36

Thermals, 125
Thunderstorms, 147
  in tropical cyclones, 429
Tides, in tropical cyclones, 416
Time section analysis, 307
Tracks of tropical cyclones, 418, 487
Trade-wind inversion
  discovery of, 202
  dissolution mechanisms, 224
  diurnal variation, 267
  divergence in, 209, 218-222
  dynamic maintenance, 209, 244-248
  regional structure,
    Atlantic, 211-215, 240-243
    Gate area, 215
    Pacific, eastern, 221-233
  structure, thermal and moisture, 207
  winds, 205
Trade winds,
  basic current structure, 24
  conservation of vorticity in, 219
  deflected, 14
    momentum balance, 244
  sea breeze, influence on, 260
  seasonal changes, 9
  surge of, 386
Tropical atmosphere, mean, 61
  relative equatorial trough, 67
Tropical cyclones, structure
  budget of angular momentum, 440
    kinetic energy, 446, 457
    thermal and moisture, 447
  cloud systems, 426
  dynamic relations, 439
  efficiency, 446
  eye,
    surface structure, 312
    upper structure, 436
  humidity, low levels, 408, 455
  life cycle, 396
  long period trends, 396
  model cross-section, 434
  outer limits, 450
  precipitation, 409
  pressure, surface, 399
  state of sea,
    waves, 415
    ocean temperatures, 417
  surface stress, 445
  temperature,
    low levels, 408, 455
    upper levels, 429
  winds,
    low levels, 401, 530, 531
    upper levels, 418-426
Tropical cyclones, formation, late stages
  areas of development, 459
  barotropic instability, 472
  frontal hypothesis, 471
  general requirements, 464
  intensity and mass flow, 463
  maximum intensity, 482, 525, 535
  mode of general circulation, 583
  non-development, 475
  potential energy release, 479
  precursors of, 462
  revival, late, 485
  weakening over land, 483
Tropical cyclones, motion
  forecasting, 491
  internal forces, 488
  recurvature latitude, 484
  steering, 488
  tracks, 418-487
  vortex interaction, 489
Tropical rainstorms (*see* Rainstorms, tropical)
Tropics
  defined, 1
  hierarchy of scales, 317
  seasonal cycle
    of temperature, 39
    of humidity, 40
Turbulence
  in convection,
    deep, 167, 175
    shallow, 170
  scales of, 316
  shearing stresses in trade inversion, 244
  in squall lines, 379
  in subcloud layer, 130, 142
  in trade inversion, 231
  transport measurements, 129-133
  in tropical cyclones, 433
Typhoon (*see* Tropical cyclones)

V

Vertical motion
  computed from mass continuity, 8

SUBJECT INDEX 611

in cumulonimbus, 158, 377
effect of weight of water, 154
in general circulation, 6, 573
over islands, 259, 264
from mean divergence fields, 8, 119
related to cloud belts, 8
in squall lines, 379
subsiding, in trades, 218
synoptic concentration, 329
in tropical cyclones,
  low levels, 410
  modelling, 434
  in eye, 437
in waves in easterlies, 350
Volcanic eruptions, 34
Vortex trains, upper, 354
Vorticity, horizontal
conservation of, 165
and cumulus shearing, 162
role in tropical cyclone formation, 469
Vorticity, vertical
conservation in trade winds, 218
in synoptic systems, 327
in tropical cyclones,
  surface, 403, 407
  upper, 421
in upper waves, 344, 354

W

Water balance
vapour transport, 84
whole tropics, 81
Waves
in easterlies, 322-327, 332, 342, 353
  origin, 345
in stratosphere, 352
in westerlies, 382
Weather analysis
large-scale, 303
local, 307
spectral method, 515
Weather prediction
local, 307
using synoptic climatology, 310

Winds
basic currents, 24
cross-sections, mean wind components, 25
mean zonal surface, 3
  geostrophic, 10
meridional, surface, 4
  relative equatorial trough, 19
observations,
  surface, 295
  upper, 297
profiles, across GATE area, 66
sea breeze, large-scale, 271
in stratosphere, 29-36
subtropical jet stream, 20, 552, 553
from tracer trajectories, 28, 34
in trade inversion, 205
in tropical cyclones,
  surface, 401-407
  extreme, 401
world distribution,
  surface, 11-16
  upper, 21-24
Winds, local
dynamic effects, 267
land and sea breeze, 263
mountain and valley breeze, 270
Winds, upper, synoptic
charts, data coverage, 297
extreme rain episode, 374
squall lines, 379
stratosphere, 425
surge of trades, 386
thermal, 327
tropical cyclones, 414, 416-426
  vertical profiles, 420, 422
  vertical shear, 421
  vorticity, 421
upper Lows, 353
upper outflow charts, 424, 432
variability, 317
waves,
  in easterlies, 322, 327, 333, 343-349
  in westerlies, 343-349